BASELINE WATER QUALITY DATA

INVENTORY AND ANALYSIS

Badlands National Park

WATER RESOURCES DIVISION AND SERVICEWIDE INVENTORY AND MONITORING PROGRAM

WATER RESOURCES DIVISION

National Park Service - Department of the Interior
Fort Collins - Denver - Washington

The National Park Service Water Resources Division is responsible for providing water resources management policy and guidelines, planning, technical assistance, training, and operational support to units of the National Park System. Program areas include water rights, water resources planning, regulatory guidance and review, hydrology, water quality, watershed management, watershed studies, and aquatic ecology.

Technical Reports

The National Park Service disseminates the results of biological, physical, and social research through the Natural Resources Technical Report Series. Natural resources inventories and monitoring activities, scientific literature reviews, bibliographies, and proceedings of technical workshops and conferences are also disseminated through this series.

Mention of trade names or commercial products does not constitute endorsement or recommendation for use by the National Park Service.

Copies of this report are available from the following:

Technical Information Center   (303) 969-2130
Denver Service Center
P.O. Box 25287
Denver, CO 80225-0287

U. S. Department of Commerce   (703) 487-4650
National Technical Information Service
5285 Port Royal Road
Springfield, VA 22161

BASELINE WATER QUALITY DATA

INVENTORY AND ANALYSIS

BADLANDS NATIONAL PARK

National Park Service
Water Resources Division
Fort Collins, CO 80525

Technical Report NPS/NRWRD/NRTR-98/161

OCTOBER 1998

United States Department of the Interior
National Park Service
Washington, D.C.

EXECUTIVE SUMMARY

This document presents the results of surface-water-quality data retrievals for Badlands National Park (BADL) from six of the United States Environmental Protection Agency's (EPA) national databases: (1) Storage and Retrieval (STORET) water quality database management system; (2) River Reach File (RF3); (3) Industrial Facilities Discharge (IFD); (4) Drinking Water Supplies (DRINKS); (5) Water Gages (GAGES); and (6) Water Impoundments (DAMS). This document is one product resulting from a cooperative contractual endeavor between the National Park Service's (NPS) Servicewide Inventory and Monitoring Program, the National Park Service's Water Resources Division (WRD), and Horizon Systems Corporation to retrieve, format, and analyze surface water quality data for all units of the National Park System containing significant water resources. The primary goal of the project is to provide descriptive water quality information in a manner and format that is both consistent with the goals of the Servicewide Inventory and Monitoring Program and useable by park resource managers. The document provides: (1) a complete inventory of all retrieved water quality parameter data, water quality stations, and the entities responsible for the data collection; (2) descriptive statistics and appropriate graphical plots of water quality data characterizing period of record, annual, and seasonal central tendencies and trends; (3) a comparison of the park's water quality data to relevant EPA and WRD water quality screening criteria; and (4) an Inventory Data Evaluation and Analysis (IDEA) to determine what Servicewide Inventory and Monitoring Program "Level I" water quality parameters have been measured within the study area. Accompanying the report are disks containing digital copies of all data used in the report, as well as all components of the report (tables, figures, etc.).

The results of the retrievals for the study area from the IFD, DRINKS, GAGES, and DAMS databases located two industrial/municipal dischargers; no drinking water intakes; nine active or inactive U. S. Geological Survey (USGS) stream gages; and 24 water impoundments. The results of the STORET retrieval for the study area yielded 6,896 observations for 197 separate parameters collected by the NPS, USGS, EPA, South Dakota Department of Water and Natural Resources, and South Dakota Department of Environment and Natural Resources at 31 monitoring stations from 1962 through 1996. Approximately 94 percent of the 6,896 observations collected within the study area were reported by the USGS from 1962 through 1996. Of the 31 monitoring stations, eight stations were located within the park boundary (see Station Period of Record Tabulation). Of the eight stations within the park boundary, seven were located in the Stronghold Unit and one was located in the North Unit.

Many of the monitoring stations represent either one-time or intensive single-year sampling efforts by the collecting agencies. Ten stations within the study area (four within the park boundary) yielded longer-term records consisting of multiple observations for several important water quality parameters (see Station Period of Record Tabulation). The stations yielding the longer-term records within the park boundary are: (1) White River near Rockyford, South Dakota (BADL 0011); (2) 41N46W 9BBCA (BADL 0021); (3) 41N45W24ACCA (BADL 0014); and (4) 41N46W12DACD (BADL 0018). The stations yielding the longer-term records within the study area, but outside of the park boundary, are: (1) 42N42W 2DACA (BADL 0008); (2) 41N44W31CBCC (BADL 0013); (3) 41N46W24DCBA (BADL 0019); (4) 42N48W25ABCC (BADL 0031); (5) 41N45W23BBCB (BADL 0015); and (6) 41N45W27BBAA (BADL 0016)[†].

Screening criteria consisting of published EPA water-quality criteria and instantaneous concentration values selected by the WRD were used to identify potential water quality problems within the study area. While the criteria represent important threshold concentrations of pollutants, it is important to remember that criteria may have been exceeded due to any number of natural or anthropogenic factors, including errors in field, laboratory, and/or recording procedures. The reader is advised to read the Introduction for additional caveats in interpreting the exceeded criteria in this report. The results of the BADL water quality criteria screen found eight groups of parameters that exceeded screening criteria at least once within the study area. The pH and copper exceeded their respective EPA criteria for the protection of freshwater aquatic life. Sulfate, nitrite, and lead exceeded their

---

[†]Water quality station location descriptions are verbatim from STORET. Any misspellings and abbreviations in STORET are replicated in this document.

respective EPA drinking water criteria. Fecal-indicator bacteria concentrations (total coliform and fecal coliform) and turbidity exceeded the WRD screening limits for freshwater bathing and aquatic life, respectively.

The pH was measured 235 times at 30 monitoring stations from 1962 through 1995. Of the 170 observations used in the criteria analysis (see Composite Type Screen in the Methodology for explanation), three observations at three lake stations in eastern Pennington County (BADL 0002, BADL 0003, BADL 0004) were outside the pH range of 6.5 to 9.0 standard units (SU) (EPA chronic criteria for freshwater aquatic life). All three observations were greater than or equal to a pH of 9.0 SU. The highest reported pH was 9.3 SU in Pike Pond (BADL 0002) in June 1978.

Turbidity was measured six times at three monitoring stations (BADL 0023, BADL 0026, BADL 0028) from 1971 through 1994. Four observations, ranging from 51 Jackson Candle Turbidity Units (JTU) to 1,055 JTU, at two stations, in Battle Creek (BADL 0023) and French Creek (BADL 0028), exceeded the WRD screening criterion of 50 Jackson Candle/Formazin/Nephelometric Turbidity Units. Three of the four observations exceeding the criterion were recorded in Battle Creek above the confluence with the Cheyenne River (BADL 0023) during 1971 and 1972, including the highest concentration of 1,055 JTU in June 1971.

Total coliform concentrations were measured 26 times at nine monitoring stations during 1994 and 1995. Of the 22 observations used in the criteria analysis (see Remark Code Screen in the Methodology for explanation), six observations, ranging from 1,200 Colony Forming Units per 100 milliliters (CFU/100 ml) to 11,800 CFU/100 ml, at five spring stations in the southern half of the study area (BADL 0008, BADL 0013, BADL 0014, BADL 0021, BADL 0031), exceeded the WRD bathing water screening criterion of 1,000 Colony Forming Units/Most Probable Number per 100 milliliters (CFU/MPN/100 ml) during 1994 and 1995. Three of the observations exceeding the criterion were reported at two spring stations within the Stronghold Unit (BADL 0014, BADL 0021), including the highest value of 11,800 CFU/100 ml in the southwestern section of the Stronghold Unit (BADL 0021) in July 1995. Fecal coliform concentrations were determined 34 times at 13 monitoring stations from 1978 through 1995. Of the 29 observations used in the criteria analysis (see Remark Code Screen in the Methodology for explanation), 11 observations, ranging from 207 CFU/100 ml to 8,800 CFU/100 ml, at seven stations in the southern half of the study area (BADL 0008, BADL 0014, BADL 0015, BADL 0019, BADL 0021, BADL 0026, BADL 0031) exceeded the WRD bathing water screening criterion of 200 CFU/MPN/100 ml during 1994 and 1995. Three of the observations exceeding the criterion were reported at two spring stations within the Stronghold Unit (BADL 0014, BADL 0021), including the highest concentration of 8,800 CFU/100 ml in the southwestern section of the Stronghold Unit (BADL 0021) in July 1995.

Sulfate concentrations (including dissolved and total) were measured 169 times at 28 monitoring stations from 1962 through 1995. Of the 104 observations used in the criteria analysis (see Composite Type Screen in the Methodology for explanation), 13 total observations, at four stations in the southern half of the study area (BADL 0011, BADL 0024, BADL 0027, BADL 0030), exceeded the secondary drinking water criterion of 250 milligrams per liter (mg/L) from 1965 through 1994. Six of the observations exceeding the criterion were reported at two stations within the Stronghold Unit (BADL 0011, BADL 0024) from 1965 through 1992. The highest reported concentration was 1,300 mg/L in the Cheyenne River near Fairburn (BADL 0027) in August 1988.

Dissolved nitrite concentrations (as N and NO_2) were measured 42 times at 14 monitoring stations from 1964 through 1995. Five observations of nitrite as NO_2, ranging from 4.0 mg/L to 9.0 mg/L, in the White River within the Stronghold Unit (BADL 0011), exceeded the drinking water criterion of 3.3 mg/L for nitrite as NO_2 during 1964 and 1965. The highest reported concentration of 9.0 mg/L was reported in September 1965.

Copper concentrations (including dissolved and total) were measured 53 times at 19 monitoring stations from 1967 through 1995. One dissolved observation, 80 micrograms per liter (µg/L), in the White River within the Stronghold Unit (BADL 0011), exceeded the acute freshwater criterion of 18 µg/L in January 1967.

Dissolved lead concentrations were measured 50 times at 16 monitoring stations from 1988 through 1995. Four observations, ranging from 20 µg/L to 30 µg/L, at four spring stations in the southern half of the study area (BADL 0014, BADL 0015, BADL 0019, BADL 0031), exceeded the drinking water criterion of 15 µg/L during 1995.

The highest value of 30 μg/L was reported twice at two stations south of the Stronghold Unit (BADL 0015, BADL 0019) in May 1995; and once in the southeastern section of the Stronghold Unit (BADL 0014) in April 1995.

The IDEA conducted for BADL indicates that STORET data exist for 12 of the 13 Level I parameter groups in the study area. No STORET data exist for the parameter group Chlorophyll. For the groups Alkalinity and Sulfates/Total Dissolved Solids/Hardness, less than 25 percent of the observations were recorded since 1985. Relative to other parameter groups, data were limited for Dissolved Oxygen, Clarity/Turbidity, Phosphate/Phosphorus, and Bacteria. Data for the two groups Clarity/Turbidity and Bacteria were recorded at less than half of the 31 monitoring stations. For the groups, Bacteria and Toxic Elements, more than 90 percent of the observations were recorded since 1985. Results for 13 of the 126 EPA priority toxic pollutants (consisting of inorganic parameters and metals) were retrieved from STORET. Approximately 51 percent of the 6,896 total observations reported in the study area were collected at one station in the White River within the Stronghold Unit (BADL 0011) from 1964 through 1967.

Surface water resources in the BADL study area include the Cheyenne and White Rivers; Battle, Cedar, Palmer, Sage, and several other smaller and intermittent creeks; many springs; and numerous small reservoirs and wildlife support dams. Based on the data inventories and analyses contained in this report, surface water quality within the study area, especially at several of the spring stations monitored by the USGS in the southern half of the study area, appears to have been impacted by human activities. Potential anthropogenic sources of contaminants include municipal wastewater discharges; stormwater runoff; agricultural activities; recreational use; and atmospheric deposition.

TABLE OF CONTENTS

INTRODUCTION

The National Park Service's (NPS) Organic Act of 1916 states that the mission of the NPS is to promote and regulate the use of national parks, monuments, and other units "... to conserve the scenery and the natural and historic objects and wildlife therein and to provide for the enjoyment of the same in such a manner and by such means as will leave them unimpaired for the enjoyment of future generations." One task embodied by this mission is preserving and protecting water resources and water dependent environments in parks. Ensuring the integrity of park water quality, due to its importance in sustaining natural, aquatic park ecosystems and supporting human consumptive and recreational use, is fundamental to successfully addressing this task. The first step in ensuring the integrity of park water quality is defining historic and extant water quality.

This document represents one product of an ongoing effort by the NPS Water Resources Division (WRD) and the Servicewide Inventory and Monitoring Program to characterize baseline water quality using existing data at park units containing significant natural resources. This effort was initiated in 1993 by the award of a contract to Horizon Systems Corporation to retrieve, format, and analyze surface water quality data from the Environmental Protection Agency's (EPA) Storage and Retrieval (STORET) database system. The scope of work identified in the Request For Proposals outlined several sequential, interrelated project phases, including, but not limited to: (1) determining the water quality retrieval/query area around each park; (2) downloading and assessing the quality of the data from STORET; (3) generating basic water quality summary statistics and graphic plots; (4) reformatting water quality data for compatibility with the park-based Water Quality Data Management System presently under-development; and (5) providing recommendations concerning possible hardware, software, and personnel options for storing combined park databases in a centralized NPS water quality database. This report documents the results of phases one through four of this effort for this park unit.

Goal

The goal of this document is to provide descriptive water quality information in a format usable for park planning purposes (eg. Water Resources Management Plans, Resource Management Plans, and General Management Plans). The report is designed to characterize baseline water quality rather than assess specific water quality problems at a park. This is consistent with the Servicewide Inventory and Monitoring Program's goal of obtaining basic, "Level I", water quality parameters for key waterbodies at each park (National Park Service 1993). Consequently, this report is best used as a reference document to help design new goal-driven water quality monitoring programs rather than as conclusive evidence of previous or existing water quality problems.

Purpose

The purpose of this report is to inventory existing park water quality data; establish baseline water quality at the park; identify potential water quality problems; and establish a park water quality database. This report is intended to enable park resource managers to compare and contrast water quality data collected as part of ongoing inventory and monitoring programs with historical water quality trends. Additionally, this report is intended to foster better designed park-based water quality inventory and monitoring programs in the future. The water quality databases which accompany this report will also lay the groundwork for establishing a NPS water quality database that will allow Regions and Washington Offices to generate regional and national assessments of park water quality.

Objectives

Specific objectives of the study documented in this report are to:

1.   Retrieve water quality and related data from the EPA's STORET and other database systems;

2.   Develop a complete inventory of all retrieved data;

3. Produce descriptive statistics and appropriate time series and box-and-whiskers plots of water quality data to characterize period of record, annual, and seasonal central tendencies and trends;

4. Compare water quality data with relevant national EPA water quality criteria on a station-by-station and study area basis;

5. Determine the presence and/or absence of the Servicewide Inventory and Monitoring Program's "Level I" water quality parameters within the study area; and

6. Reformat water quality and other related data for use in the park-based Water Quality Data Management System, presently under-development, and other appropriate analytical tools.

Document Overview

This report is comprised of five chapters. The first chapter, this Introduction, provides a brief statement of the study's background; goal, purpose, and objectives; and the key personnel who helped produce the document. This chapter also contains this brief overview of the document's contents and important interpretive caveats to consider when referring to and using this document. The second chapter focuses on the methods, procedures, and databases that were employed to retrieve and analyze water quality data for the park. The third chapter is the user's interpretive guide to chapter four. Chapter three explains how to interpret all the tables and figures presented in chapter four. Chapter four, which likely comprises the majority of the document (unless there isn't much water quality data for the park), contains detailed inventories, descriptive statistics, graphics, and national EPA water quality criteria comparisons characterizing the park unit's water quality data on a station-by-station basis and over the entire study area. This chapter also contains a comparison of park water quality data with the Servicewide Inventory and Monitoring Program's "Level I" water quality inventory parameters and a listing of water quality observations that were outside the STORET edit criteria range. Chapter five, the Appendices, contains more specialized materials such as the file names and database structures included on floppy disk(s) with this report; STORET edit criteria; national EPA water quality criteria; Servicewide Inventory and Monitoring Program's "Level I" water quality inventory parameters; selected water quality references; and other materials which provide background on the methods, procedures, and databases used or produced by this study.

The water quality and other related data referenced in this report accompany the document on floppy disk. The water quality parameter data file is in DBASE III+[1] format and will be useable in the park-based Water Quality Data Management System presently under-development. The water quality stations, industrial facilities discharges, drinking water intakes, water gages, water impoundments, and River Reach databases are also in DBASE III+ and/or ASCII format for ready-use in Geographic Information Systems (GIS), Computer-Aided Design Systems, or Desktop Mapping Systems.

Caveats

While intended primarily as a reference document, it is important that users peruse the first three chapters and Appendices of this report to better understand and interpret the results presented in chapter four. As a means for identifying potential areas for more intensive study, comparisons of the park's water quality data with relevant national EPA water quality criteria for appropriate designated uses[2] and with the Servicewide Inventory and

---

[1]The use and/or mention of specific proprietary hardware or software packages is for informational purposes only and is not intended to connote or denote an endorsement.

[2]The Environmental Protection Agency's Quality Criteria for Water 1995 Final Draft (Silver Book) was the primary source of water quality criteria. In the spirit of the other caveats offered in this section, it is important to recognize that water quality criteria are often revised when new or better information become available.

Monitoring Program's "Level I" water quality inventory parameters have been made. Extreme caution must be exercised in interpreting the results of these comparisons. Observations that exceed water quality criteria may have occurred due to any number of natural or anthropogenic factors, as well as other reasons. For example, STORET is a "user-beware" water quality database system. While there is some rudimentary edit (bounds) checking of any data entered in STORET (See Appendix C), users are basically free to enter their own data. Beyond data entry errors, the possibility of inaccurate data entering the system due to inappropriate measurement techniques, sample mistreatment, and other reasons is a serious concern. Consequently, if observations for a particular parameter frequently exceed the EPA water quality criterion over a prolonged time period, the best approach is to examine in detail the data exceeding the criterion. Questions which should be asked regarding the data include: What water source(s) are manifesting the problem? Does the data make sense? Was it collected by a reputable organization following a sound study plan and employing accepted techniques? If the answers to these questions still cause concern, a specific cause and effect water quality investigation focusing on the parameters of concern may be warranted. Similarly, the absence of particular Servicewide Inventory and Monitoring Program "Level I" water quality parameters from the park only means that no entity or organization has collected and entered this data into the EPA's STORET database. Too frequently, data that are collected in and around NPS units never make it into the EPA's national water quality database. These data may exist in published or unpublished reports, file cabinets, or other databases. Before definitively concluding that no baseline data exist for a particular parameter, these alternative resting grounds for data should be investigated. Such a detailed exploration, however, was beyond the scope of this study.

Key Personnel

Many individuals contributed to the design and implementation of this project. The primary contributors and their roles in the project are briefly mentioned below.

National Park Service, Water Resources Division:

Dean Tucker was the Contracting Officer's Technical Representative responsible for designing, coordinating, and implementing all aspects of this effort.

Mike Matz coordinated and managed the team which prepared all components of the report.

Gary Rosenlieb provided administrative oversight and was involved in quality control for all tasks related to this project.

Barry Long and Roy Irwin reviewed technical tasks and provided water quality expertise related to data analysis.

Donnie Dustin, Greg Harp, and Clint Bassett helped prepare reports and write the Executive Summaries.

Elizabeth Eisenhauer, Robert Flynn, Dawn Grandbois, Bill Folsom, Dana Griffin, Jonathan Duran, and Aymn Elhaddad provided digital cartographic support, both in determining retrieval/query areas and producing maps and graphics.

Kelli O'Connor, Mary Beth Talty, Curtis Cooper, Paul McElvery, J. Chris Echohawk, Kristie Maczko, Adam Henson, Shawndra Mawhorter, Lisa Smith, Eric Janney, Ryan Shy, Lisa Dummer, Eric Lord, Adriane Petersen, and Margaret Matter uploaded water quality data to STORET prior to report preparation.

Jacquie Nolan designed the cover.

Horizon Systems:

Cindy McKay served as Project Manager for Horizon Systems, performed the initial requirements analysis, and was involved in all quality control tasks related to the project.

Alan Cahoon was responsible for automating the procedures which produced the water quality databases and Water Quality Results chapter.

Sue Hanson, P.E., provided technical advice for writing this document.

Dr. Jim Loftis was the data quality analyst for the project.

Armando F. Ballofet, P.E., served as the local technical liaison between Horizon Systems and the NPS.

Other National Park Service:

Several other individuals provided invaluable technical review, comments, administrative support, and/or other assistance, including: Dan Kimball, Bill Jackson, Mark Flora, Gary Williams, John Karish, Brendhan Zubricki, Richard Hammerschlag, Randy Ferrin, Gary Vequist, Mike Martin, Kevin Berghoff, and Dyra Monroe.

METHODOLOGY

This section provides an overview of the procedures and criteria used to retrieve and analyze water quality data for each park unit. Generating baseline water quality data inventories and analyses for all NPS units is a monumental task. To accomplish this undertaking given a very limited budget, the procedures employed to produce each report had to be as generic and automated as possible. Consequently, customization of reports to individual park needs and issues was not feasible. Moreover, such customization was beyond the scope of this effort which was simply intended to produce baseline water quality data inventories for all parks rather than customized issue-driven reports. During the procedure-development stages of the project, specifications for the final product evolved, within the context of the aforementioned resource constraints, to focus on comprehensive water quality baseline data inventories and concise, descriptive statistical examinations of the available water quality data for each park unit. Detailed below are the data sources and final methods and procedures that were used to create the baseline water quality inventories, analyses, databases, and other products for each park unit. A thorough understanding of the limitations of the data sources and procedures described in this chapter and the next (Interpretive Guide to Water Quality Results) is a prerequisite to intelligent use of the results presented in this document.

Delineation of Park Study Area

The first step in retrieving water resources-related data for each park was deciding on a procedure to determine the study area boundary. Since water flows through parks, utilizing the park boundary as a simple query/study area was deemed inadequate. On the other end of the continuum, using the entire watershed as the study area was considered superfluous given: (1) the areal extent of certain park watersheds (eg. the entire Mississippi River); (2) the sheer volume of potentially irrelevant data such a large study area could generate; and (3) the resources required to specify the watershed for each park unit. The approach which was ultimately adopted - a modified hydrologic boundary - reflects a compromise between the park boundary and the entire watershed. Thus the study area employed for each park is an area extending at least three miles upstream and one mile downstream from the park boundary. Although these distances are somewhat arbitrary, this approach is easy to automate and was felt to limit the data retrieved, in most instances, to that of most importance to the park. Extending the query area one mile downstream of the park was intended to capture any data immediately downstream of the park which may reflect the quality of the water in the park. A current (as possible) copy of each park's boundary was obtained in digital format directly from the park or digitized from Regional land status maps, U.S. Geological Survey (USGS) quadrangles, or other sources. Using GIS techniques, the boundary was used to create the three miles upstream, one mile downstream buffer. For a few parks with which WRD water quality specialists were very familiar with potential water quality threats and/or valuable sources of data that may lie just outside the study area, the study area may have been tweaked (enlarged) to cover these areas of concern or interest. Unfortunately, a customized study area was not feasible for all park units. Hence, the three miles upstream, one mile downstream buffer was the primary study area employed for most parks. This study area was transferred to the EPA mainframe computer and used as the basis for all water resources-related data retrievals from the data sources described below.

Data Sources

The EPA maintains many mainframe data systems related to national water resources (U.S. Environmental Protection Agency 1992). Six of these data systems were used for this project:

- STOrage and RETrieval System (STORET) - water quality parameter data, locations of sampling stations, descriptive elements about stations and parameters;

- Industrial Facilities Discharge (IFD) - locations of industrial and municipal point source discharge facilities;

- Drinking Water Supplies (DRINKS) - locations of intake pipes for drinking water supplies;

- Water Gages (GAGES) - locations of USGS and other water gages;

- Water Impoundments (DAMS) - locations of most large water impoundments (greater than 10,000 acre feet at normal pool volume) and many smaller impoundments; and

- River Reach File, Version 3 (RF3) - 1:100,000 scale geographical representation of surface waters (rivers, lakes, etc.) with a unique identifier assigned to each surface water segment and connectivity information useful for routing and navigation.

STORET is the national water quality data repository (U.S. Environmental Protection Agency 1989). Water quality data is entered in STORET by public agencies (federal, state, or local) that collect water samples and/or perform laboratory analysis. As such, STORET is a "user-beware" data system. Although the EPA manages the STORET data system and, since November 1983, has imposed some minimum quality control criteria on the data (See Appendix C), data are generated and input to STORET by the "owner" agencies. Consequently, the EPA does not certify any data within STORET. Currently, there are over 800,000 active and inactive sampling stations and more than 225 million observations covering in excess of 13,000 water quality parameters entered in STORET. The earliest data dates back to the turn of the century. Using the bi-monthly update cycle, user agencies may store results of recent monitoring activities in STORET. Included in STORET is USGS WATSTORE water quality data, which is updated on a monthly basis. Although STORET contains a phenomenal amount of data, it is important to note that data exist in STORET only if the collectors decide to upload their data to the system. Since many agencies and researchers do not upload their data to STORET, the absence of water quality data in the system for a particular area doesn't mean that there has never been any water quality data collected for the area. The data may exist in published or unpublished reports, file cabinets, or in agency-specific databases. Identifying and retrieving these other sources of data were beyond the scope of the present effort. All parameter data and water quality station location data downloaded from STORET within the park's study area are included in DBASE III+ format files on disk(s) accompanying this report (See Appendices A and B).

The data within the IFD database are extracted from the EPA's Permit Compliance System (PCS). IFD contains the facility locations of all industrial and municipal dischargers which require a National Pollutant Discharge Elimination System (NPDES) permit to operate. Over 7,100 municipal, federal, and industrial facilities discharging into the waters of the United States are tracked by PCS and IFD. If any industrial facilities discharges exist within the study area, a file in DBASE III+ format documenting a variety of information about each discharge accompanies this report on disk (See Appendices A and B).

The EPA DRINKS database identifies locations of drinking water supply intakes. This file contains data for 850 supplies which serve more than 25,000 people, and 6,800 supplies which serve between 1,000 and 25,000 people. If any drinking water intakes exist within the study area, a file in DBASE III+ format documenting a variety of information about each intake accompanies this report on disk (See Appendices A and B).

The GAGES data originates primarily with the USGS and copies are maintained on the EPA mainframe computer for ease of integration with other EPA national data systems. Although other agency's water gages, as well as some artificial gages, may appear in GAGES, the vast majority of gages are stream gages belonging to the USGS. The GAGES database contains approximately 36,000 records for both active and inactive gaging stations. If any USGS or other agency stream gages occur within the study area, a file in DBASE III+ format documenting several fields of information about each gage accompanies this report on disk (See Appendices A and B).

The Water Impoundment database was originally compiled by the U.S. Army Corps of Engineers in response to a Congressional inquiry on dam safety hazards (GKY and Associates 1990). The EPA subsequently modified the database for use in water quality investigations. Of the 68,155 dams in the database, 2,125 are considered large (impounding 10,000 acre feet or more at normal pool volume). It is important to note that while the database includes entries for 66,030 smaller dams, estimates place the actual number of dams in the U.S. at several million (including small farm ponds). If any water impoundments occur within the study area, a file in DBASE III+ format documenting several fields of information about each impoundment accompanies this report on disk (See Appendices A and B).

The RF3 data system is a hydrologic database of surface water features across the U.S. (excluding, at present, Idaho, Oregon and Washington, which currently operate a different system - although this data is expected to be converted to RF3 soon, Alaska and Hawaii). RF3 was created primarily from 1:100,000 scale USGS Digital Line Graph data. RF3 is made up of over 3,000,000 individual "reaches". A reach is generally defined as a portion of surface water between two confluences (U.S. Environmental Protection Agency 1993). The linework underlying RF3 contains over 95,000,000 coordinate points. RF3 is designed to facilitate hydrologic routing, identifying upstream and downstream elements, and specifying the exact location of any point on a stream network. RF3 data exists as a series of traces with associated attributes. The EPA project which is producing RF3 is being conducted in three phases: Compilation, Assessment, and Revision. The Compilation phase is complete except for Idaho, Washington, Oregon, and Alaska. The Assessment phase was completed during the first half of 1994; while the Revision phase was begun in March 1994. One important outcome of the Revision phase is that the reach codes which uniquely identify each surface water feature will change. Consequently, these codes should not be used, at this time, as keys for relating other data to RF3. The RF3 data provided with this document is provisional and should be used only to provide a geographic backdrop for the park's water quality data. RF3 data covering each USGS catalog unit (a geographic area representing a single or multiple drainage basin(s), or some other distinct hydrologic feature (U.S. Geological Survey 1982)) touched by the park's study area is included in ASCII export and DBASE III+ formats on the disk(s) accompanying this report (See Appendices A and B).

For additional information on any of these data systems, contact the EPA Office of Water at (202) 260-7028.

Data Retrieval and Analysis Procedures

The six EPA data systems discussed above reside on the EPA mainframe computer located in Research Triangle Park, N.C. Horizon Systems used a dedicated, leased telephone line with a data transfer rate of 9600 bits per second to download data occurring within the park's study area from all the databases. The bisynchronous communication software and hardware provided error checking during all data transfer procedures.

As described above, the park study/query area boundary was used to select the water quality stations, industrial facilities discharges, drinking water intakes, water gages, water impoundments, and river reaches associated with the park unit. For various reasons, screening criteria (described later in this section) were employed to select appropriate water quality stations, parameters, and observations. Horizon Systems wrote several mainframe programs to automate, to the greatest extent feasible, the STORET data retrieval and storage procedures. Once the data were extracted from the EPA data systems, they were downloaded to a microcomputer for statistical analyses and reformatted into DBASE III+ compatible format.

Specifically, once on the PC, the data were processed to:

(1)   Reformat the data into DBASE III+ format and other database structures;
(2)   Eliminate questionable data outside the STORET edit criteria ranges (See Appendix C);
(3)   Display on a map the location of water quality monitoring stations and other water resources themes;
(4)   Determine the frequency of water quality observations by station, parameter, and station/parameter;
(5)   Generate descriptive period-of-record water quality statistics in a tabular format;
(6)   Generate appropriate descriptive annual and seasonal analyses of the water quality data in a tabular format;
(7)   Plot appropriate period of record time series and annual and seasonal box-and-whisker graphs;
(8)   Compare the water quality data against relevant EPA national criteria; and

(9)   Compare the water quality data against the NPS Servicewide Inventory and Monitoring Program's "Level I" water quality parameters.

Special customized microcomputer programs (primarily written in Clipper and Microsoft Professional BASIC) and procedures were created to address each of these tasks. All reformatted database files are included on disk(s) accompanying this document. The contents of these databases are described briefly below. Complete database structures are included in Appendices A and B. The descriptive water quality tabular statistics (see "Statistical Analyses" below) were computed based upon NPS specifications. Command or batch files were generated to drive STATGRAPHICS 7.0 in order to produce all the time series and box-and-whiskers plots.

Park Unit Databases

Up to seven digital databases in DBASE III+ and other formats have been created for the park by querying the water resources-related data sources described above. The disk(s) containing these databases accompany the report. The contents of each of these databases are discussed briefly below. More detailed documentation of these databases is included in Appendices A and B.

(A)   Water Quality Parameter Data: This database includes all the water quality parameter data downloaded from STORET that passed the STORET Edit Criteria, Date, Station Type, and Phase 0 Parameter screens (described below) and is summarized tabularly and graphically in this document. This constitutes the park's baseline water quality data. Since it is already in digital format, more sophisticated analysis of the data is possible than the descriptive statistics and graphics presented here.

(B)   Water Quality Station Locations: This database consists of the STORET header information describing each station where water quality data was collected. As the latitude and longitude of the station are included in the database, this file is easily imported into the park's GIS.

(C)   Industrial Facility Discharge Locations: This database includes any industrial or municipal point source discharges located within the park's study area. As the latitude and longitude of each discharge facility are included in the database, this file is easily imported into the park's GIS.

(D)   Drinking Water Intake Locations: This database includes any drinking water intakes located within the park's study area. As the latitude and longitude of each intake are included in the database, this file is easily imported into the park's GIS.

(E)   Water Gage Locations: This database includes water (stream, lake, estuary, well, spring, climate, or other) gages located within the park's study area. Most of the gages will likely be stream gages belonging to the USGS. As the latitude and longitude of each gage are included in the database, this file is easily imported into the park's GIS.

(F)   Water Impoundment Locations: This database includes any water impoundments (dams) located within the park's study area. As the latitude and longitude of each impoundment are included in the database, this file is easily imported into the park's GIS.

(G)   River Reach Data: This database includes all stream traces (1:100,000 scale) and attributes for reaches falling within any USGS catalog unit that touches the park's study area. The traces are geo-referenced in ASCII format. The attributes are in both ASCII export and DBASE III+ formats. This information is also readily incorporated into the park's GIS.

The absence of any of these seven files from the disk(s) accompanying the report indicates that there was either no data of this type within the park's study area or the data was unavailable. Several other files are included on the disk(s) accompanying this report, including digital copies of all the figures and tables contained in the document and some other items. Refer to Appendices A and B for detailed documentation of these files. Not included on disk is an Encyclopedia File (for WRD reference) that documents the minimum and maximum values for each water quality parameter and the parks in which those values were recorded. When Baseline Water Quality Data Inventory and Analysis reports have been completed for all parks, this Encyclopedia File will be available upon request from the NPS WRD.

Screening Methodologies and Procedures

Developing automated or semi-automated procedures to produce baseline water quality inventories and analyses for all national park units required constant testing and debugging of procedures. Three parks, Rock Creek Park, Yellowstone National Park, and Indiana Dunes National Lakeshore, were used to pilot test and refine the automated procedures. It became evident, after a preliminary analysis of all the downloaded STORET data, especially for Indiana Dunes National Lakeshore, that the specifications for the graphical analyses could generate hundreds (possibly thousands) of plots, many of which would not necessarily be useful. Also, there were many stations; parameters; and/or observations downloaded that were not part of the study's objectives; not overly useful; or of dubious quality. In order to reduce the number of graphical plots (time series, annual and seasonal box-and-whiskers) to fit within project resources, various screening criteria were investigated. Ultimately, a comprehensive set of screening criteria were developed to reduce the number of graphical plots. After initial counts of the total number of possible time series and annual and seasonal box-and-whiskers plots were generated, these counts were used to decide which screening criteria would be applied to limit the number of these plots produced for the park unit. Additional screening criteria were employed to restrict the tabular descriptive statistics results to only those deemed useful to the park. Table A provides the categories of screening criteria and to which analyses the screens were applied. A "yes" entry in the table means that the screening category eliminated or prevented data from appearing in certain tables and plots contained in the document. Consequently, in understanding how data from STORET was used in this report, it may be helpful to keep in mind the three general types of screening criteria: (1) screens that apply to stations; (2) screens that apply to certain parameters at stations; and/or (3) screens that apply only to particular observations of parameters at stations. A detailed description of each of the screening criteria categories follows this table. *It is important to note that statistics in "Inventory" reports may not be consistent with statistics in "Overview" reports since different categories of screening criteria were applied.* Also, if attempting to replicate the results of the statistical and graphical analyses presented in this document, be sure to follow the same screening methodologies.

STORET Edit Criteria

As mentioned previously, STORET is a "user-beware" data system. As the EPA doesn't certify any data in STORET, public agencies enter and are responsible for the quality of their own data. Only data entered since November 1983 have been subjected to any rudimentary edit/bounds checking. Agencies entering data since this date can elect to override the edit/bounds checking for individual observations. USGS WATSTORE water quality data is entered into STORET without any EPA edit/bounds checking to ensure data integrity between WATSTORE and STORET. Unfortunately, during the course of our pilot tests, erroneous USGS and EPA water quality data values were discovered. In order to eliminate as much "bad" data as possible, all water quality data downloaded from STORET was subjected to automatic edit/bounds checking (STORET Edit Criteria contained in Appendix C) for the 190 most common parameters. Observations falling outside the STORET Edit Criteria were documented (See the Water Quality Observations Outside STORET Edit Criteria for Park section in the Water Quality Results chapter) and then retained or discarded from the database and all tables and plots based on whether the value was judged as being in the realm of possibility. Although the STORET Edit Criteria screen likely removed some "bad" data for these common parameters, the probability of other erroneous data in the database is high. Be sure to consult the Caveat section in the Introduction.

Table A   Categories of Screening Criteria and to Which Output Products They Apply (A "yes" Entry Means the Screening Category Eliminated or Prevented Data From Being Used in the Product):

Screening Category	Data Download	Overview Tables	Inventory Tables	Annual Tables	Seasonal Tables	Standards Tables	Plots (All)
STORET Edit Criteria	yes	yes	yes	yes	yes	yes	yes
Date	yes	yes	yes	yes	yes	yes	yes
Station Type	yes	yes	yes	yes	yes	yes	yes
Phase 0 Parameter	yes	yes	yes	yes	yes	yes	yes
Phase 1 Parameter	no	no	yes	yes	yes	yes	yes
Media Type	no	no	yes	yes	yes	yes	yes
Remark Codes	no	no	yes	yes	yes	yes	yes
Composite Type	no	no	yes	yes	yes	yes	yes
Phase 2 Parameter	no	no	no	no	no	no	yes
Observations/Period of Record	no	no	no	yes	yes	no	yes

Date Screen

Every water quality observation in STORET typically has a sampling date associated with it. Unfortunately, STORET does not prevent users from entering incorrect dates. Consequently, any water quality observation with an incorrect and/or suspect date (eg. a month greater than 12; a day greater than 31; or a sample date later than the STORET retrieval date) were discarded.

Station Type Screen

STORET contains data from a wide variety of stations classified by the type of waterbody in which samples were collected. As this project's purpose was to inventory and analyze surface-water quality, the following surface-water station types were retrieved (clarification provided in parentheses):

Station Types Included In Retrieval
(a) STREAM
(b) CANAL
(c) LAKE
(d) RESERV (Reservoir)
(e) SPRING
(f) FWTLND (Fresh Water Wetland)
(g) SWTLND (Salt Water Wetland)
(h) ESTURY (Estuary)
(i) OCEAN

Ground water and/or other station type data may have been retrieved if the entering agency classified the station type incorrectly. Rectifying this error was beyond the scope and resources of this project.

Phase 0 Parameter Screen

Nearly all water quality parameters associated with each station type listed above were retrieved. The only exception to this was the exclusion of most of the STORET administrative parameters. A complete list of STORET administrative parameters is included in Appendix D. The few administrative parameters that were included in the retrievals are as follows:

Code	STORET Administrative Parameter Description
00027	Code No. for Agency Collecting Sample
00028	Code No. for Agency Analyzing Sample
00063	Sampling Points, Number of In a Cross Section
00111	Ratio of Fecal Coliform to Fecal Streptococci
00115	Sample Treatment Code (1=Raw, 2=Treated)
34772	NPDES Number, Cross Reference
45580	Method of Analysis
74065	Stream Flow Class
74066	Annual Runoff
74067	Soil Classification
74068	Water Quality Designated Use Classification

Phase 1 Parameter Screen

Some of the data retrieved from STORET was not suitable for statistical or graphical analysis. Consequently, this screening criterion eliminated all parameters which were not suitable for statistical or graphical analysis within the context of this project. The full list of these parameters is presented in Appendix E. Examples of parameters excluded from statistical and graphical analysis include the administrative parameters mentioned above, land use acreage, encoded values, dates, latitude/longitude, etc. Excluded parameters do, however, appear in the Parameter Period of Record and Station/Parameter Period of Record (two of the "Overview" Tables), as well as in the water quality parameter file included on disk(s) accompanying this report.

Media Type Screen

Water quality samples can be taken in a variety of aqueous media. Water quality data were retrieved from STORET only if the media were WATER or VERT (vertically integrated). WATER and VERT samples comprise the overwhelming majority of samples in STORET. The media screen eliminated the following water quality sampling media:

Media Screen	Description
BOTTOM	Sampled At the Bottom
DREDGE	Sampled By Dredge
PORE	Pore Sample
CORE	Core Sample

Remark Code Screen

STORET enables the agency collecting water quality samples to provide a qualifying remark for each parameter observation. These remarks provide additional information about the measured or observed value entered into STORET (See Appendix B - Parameter Data File for a complete listing and description of all remark codes). Based on the STORET remark codes, two potential screens were applied to water quality observations based on whether the measured value was used in subsequent analyses: (1) Elimination or (2) Modification/Inclusion.

Elimination:

Non-composite water quality parameters with the remark codes presented in Table B were eliminated from the period of record, annual, and seasonal descriptive statistics and graphics. Not including observations with these remarks was justified by the fact that most of the remarks: (A) indicate either less confidence in the measured value; (B) are remarks for nominal or categorical data that doesn't lend itself to statistical analysis; or, (C) complicate the statistical analysis beyond the scope of this effort. Observations containing these remark codes comprise a very small fraction of the data. Although statistical analyses weren't undertaken on this data, all water quality observations, regardless of remark code, are included on disk(s) accompanying this report. If you re-analyze this data in order to replicate the results presented here, be sure to eliminate all non-composite observations with the remark codes presented in Table B.

Table B.	Non-composite Parameters With the Following Remark Codes Were Eliminated From Statistical and Graphical Analysis:	
Remark Code	**Description of STORET Remark Code**
F	Female Species.
J	Estimated, Not the Result of Analytic Measurement.
M	Presence Verified, But Not Quantified, Below Quantification Limit. For Species, Male. For Oxygen Reduction Potential, Indicates Negative Value.
N	Presumptive Evidence of Presence.
O	Analysis Lost.
V	Analyte Was Detected In Sample and Method Blank.
W	Less Than Lowest Value Reportable Under Remark "T".
Z	Too Many Colonies Were Present to Count (TNTC), Value Represents Filtration Value.

Modification/Inclusion:

Water quality parameter observations with the remark codes presented in Table C were halved prior to inclusion in period of record, annual, and seasonal descriptive statistics and graphics. These remark codes deal with observations that were below the detection limit for the parameter. The common water quality data analysis convention for these remark codes is to use half of the detection limit in statistical analyses (Ward, Loftis, and McBride 1990; Gilbert 1987). Although this is a somewhat defensible treatment of observations below the detection limit, the statistics that may be computed using these halved values may not be defensible. Consequently, any computed statistics in inventory, annual, or seasonal tables that are comprised of 50% or more K, T, and U remark codes are footnoted "Computed with 50% or more of the total observations as values that were half the detection limit." This will provide the user with some caution in using and interpreting these results. Water quality data included on disk(s) accompanying this report that may have these remark codes are stored as the original entry (detection limit). If you re-analyze this data in order to replicate the results presented here, be sure to substitute half the detection limit value in the database whenever these remark codes are encountered.

Table C.	The Value of Water Quality Parameters With the Following Remark Codes Were Halved (Half of the Detection Limit Entered In STORET) Prior to Inclusion In Descriptive Statistics and Graphics:
Remark Code	**Description of STORET Remark Code**
K	Off-scale Low. Actual Value Not Known, But Known to Be Less Than Value Shown.
T	Less Than Detection Criteria.
U	Analyzed For But Not Detected. Value is Detection Limit For Process Used. If Species, Undetermined.

Composite Type Screen

Sometimes data entered in STORET represent something other than a single measurement at one location at one point in time. These samples are typically referred to as composite samples due to the fact that they vary temporally and spatially. Consequently, the observation entered into STORET for composite data is typically a computed value that summarizes the data over time and/or space. Such data complicate statistical and graphical analyses and must be handled separately. Such treatment was beyond the scope of this study; although composite values typically represent only a fraction of STORET observations. The composite type screen eliminates all composite observations from statistical and graphical analyses, except those with a composite type code of "A" that have a one day or less sampling period and those with a composite type code "D". All water quality observations, regardless of composite type code, are included on disk(s) accompanying this report. If you re-analyze this data in order to replicate the results presented here, be sure to exclude all composite observations except those with a code of "A" that have a one day or less sampling period and those with a code of "D". Table D presents a list of possible STORET composite type codes.

Table D. Possible STORET Composite Type Codes	
Composite Type Code	**STORET Composite Type Description**
A	Average
H	Maximum
L	Minimum
N	Number of Observations
#	Number of Observations
S	Standard Deviation
U	Sum of Squares
V	Variance
C	Coefficient of Error
X	Coefficient of Variance
E	Skewness
F	Kurtosis
Z	Number of Obs. That Exceed An Established Limit
%	Precision
$	Accuracy
B	N/A
D	Indicates Replicate Sample

Phase 2 Parameter Screen

Due to budgetary limitations, the number of graphical plots (time series, annual and seasonal box-and-whiskers) produced had to be manageable - typically no more than 100 total plots. After scrutinizing the results of the pilot tests and the Baseline Water Quality Data Inventory and Analysis Reports produced for the first group of parks, the 19 parameters which, typically, were the most frequently measured at nearly all stations were water temperature, stage, discharge, and various meteorological measurements (See Table E). Consequently, most of the graphical plots produced would be of water temperature, stage, discharge, and meteorological conditions. Although these are important parameters, particularly in conjunction with other water quality parameters, it was felt that plotting resources would be better allocated to other water quality parameters. Consequently the STORET parameter codes listed in Table E never generated graphical plots. It is important to note, however, that these parameters are included in all other aspects of the project, including all applicable period of record, annual, and seasonal descriptive statistics tables.

Table E.	Frequently Measured STORET Codes That Were Prevented From Generating Plots
STORET Parameter Code	**STORET Parameter Description**
00003	Sampling Station Location, Vertical (Feet)
00010	Water Temperature (Degrees Centigrade)
00020	Temperature, Air (Degrees Centigrade)
00021	Temperature, Air (Degrees Fahrenheit)
00025	Barometric Pressure (MM of HG)
00032	Cloud Cover (Percent)
00035	Wind Velocity (Miles Per Hour)
00036	Wind Direction in Degrees from Trun N (Clockwise)
00040	Wind Direction (Azimuth)
00045	Precipitation, Total (Inches Per Day)
00046	Precipitation, Total (Inches Per Week)
00052	Humidity, Relative (Percent)
00061	Stream Flow, Instantaneous (CFS)
00065	Stream Stage (Feet)
81903	Depth of Bottom of Water @ Sample Site (Feet)
82553	Rainfall In 1 Day Inclusive Prior to Sample (Inches)
82554	Rainfall In 7 Days Inclusive Prior to Sample (Inches)
82371	Rainfall In 3 Days Inclusive Prior to Sample (Inches)
82372	Rainfall In 14 Days Inclusive Prior to Sample (Inches)
85599	Precipitation, Total/Period-Rain Equivalent (Cm/Sample)

Observations/Period of Record Screen

Despite never plotting water temperature, stage, discharge, and meteorological measurements, the number of plots generated by some parks still exceeded the 100 plot limit. Also, some rationale was needed to plot only those parameters with sufficient data density to make a meaningful statistical graphic. For example, time series plots comprised of only a few observations or annual or seasonal box-and-whiskers plots with limited observations and/or data in only one or two years or seasons are not very informative. Consequently, a number of plotting criteria were developed to limit the number of time series and box-and-whiskers plots to, at most, 100 informative graphics by using each parameter's number of observations and period of record. Similar, albeit less stringent criteria, were used for including results of annual and seasonal analyses in descriptive statistics tables. Consequently, there are more summaries of annual and seasonal results in tables than in graphics. Whenever an entry in an annual or seasonal table generated a plot, this entry was footnoted to notify the reader of the presence of the graphic. Due to differing quantities of data at parks, different screening criteria were employed. The same criteria for appearance in seasonal and annual tables were used for all parks. Table F presents the least stringent plot screens.

Table F. Least Stringent Plot Screening Criteria Used to Limit the Number of Plots Generated

Time Series:

To generate a time series plot, a station/parameter combination must have a period of record of at least 2 years and a total of at least 8 observations.

Annual Analysis:

To generate an annual box-and-whiskers plot, a station/parameter combination must have at least 9 observations in each of at least 4 years. The years do not have to be consecutive.

Seasonal Analysis:

To generate a seasonal box-and-whiskers plot, a station/parameter combination must have at least 9 observations in each of 2 seasons and a period of record of at least 6 years and observations in at least 3 of the 6 years. The years do not have to be consecutive.

The exact three plot screens used varied by park unit and are documented in the Overview section of the Water Quality Results chapter. If your park's plotting criteria deviated from these least stringent criteria, it is because too many plots would have been generated using these criteria.

The criteria used for appearance of station/parameter combinations in annual and seasonal analysis tables are presented in Table G. These tabular criteria, which are actually the least stringent plotting criteria, were constant from park to park.

Table G. Criteria Used for Generating Entries in Annual and Seasonal Analysis Tables

Annual Analysis:

For an entry to appear in an annual table, a station/parameter combination must have at least 9 observations in each of at least 4 years. The years do not have to be consecutive.

Seasonal Analysis:

For an entry to appear in a seasonal table, a station/parameter combination must have at least 9 observations in each of 2 seasons and a period of record of at least 6 years and observations in at least 3 of the 6 years. The years do not have to be consecutive.

Statistical Definitions

Since this report is intended only to characterize historical and/or existing water quality at the park rather than address specific water quality problems, only simple descriptive statistics are presented. Inferential and non-parametric statistical analysis to examine relationships and trends were beyond the scope of the study. The complete water quality dataset is provided on disk accompanying this report to afford the opportunity for more detailed exploratory data analysis. The descriptive statistics are included in the inventory, annual, and seasonal tables. Table H provides a brief definition of each descriptive statistic provided for each parameter at a station.

Table H. Definition of Descriptive Statistics Contained in Inventory, Annual, and Seasonal Tables

Observations:	The number of samples collected.
Median:	The median is the 50th percentile or the value in a dataset sorted in ascending order that exceeds 50% of all observations, yet is also exceeded by the remaining 50% of all observations.
Mean:	The sum of all observations collected divided by the number of observations.
Maximum:	The maximum value observed.
Minimum:	The minimum value observed.
Variance:	This is a measure of variability or dispersion of the observations; or, in other words, describes how many observations are close (or far), from the mean. It is calculated as the weighted average of the squared deviations from the mean.
Standard Deviation:	The positive square root of the variance.
10th Percentile:	The value in a dataset sorted in ascending order that exceeds 10% of all observations, yet is itself exceeded by the remaining 90% of all observations.
25th Percentile:	The value in a dataset sorted in ascending order that exceeds 25% of all observations, yet is itself exceeded by the remaining 75% of all observations. The 25th percentile is also known as the first quartile.
75th Percentile:	The value in a dataset sorted in ascending order that exceeds 75% of all observations, yet is itself exceeded by the remaining 25% of all observations. The 75th percentile is also known as the third quartile.
90th Percentile:	The value in a dataset sorted in ascending order that exceeds 90% of all observations, yet is itself exceeded by the remaining 10% of all observations.

As with the tabular descriptive statistics, the scope of the project limited the generation of exploratory graphics to time series plots and annual and seasonal box-and-whiskers plots. Plots were only generated, however, provided the parameter met or exceeded the relevant plotting criteria specified in the previous section.

Time series plots display the parameter concentration on the Y-axis and the date on the X-axis. This provides the user with a visual feeling for not only the parameter's concentration and variability over time, but also the density of data in different time periods. The time series plots provide a visual representation of the data in the basic station inventory. Due to software limitations, a line connects each measured value in sequence regardless of the time period between samples. Readers are cautioned not to assume that the concentration of the parameter between any two data points can be represented by a straight line. It is likely that the concentration varied between any two observations, particularly if the observations are separated by a significant time period.

The annual and seasonal box-and-whisker plots provide a graphical overview of the measured data and give the user a better understanding of the data's distribution and possible outliers. In essence, the box-and-whisker plots provide a visual representation of the data contained in the annual and/or seasonal tables. The interpretation of the boxes is provided in the figure to the right. Each box encompasses the middle 50 percent of measured values (from the 75th to 25th percentiles). The difference between the 75th and 25th percentiles is also known as the interquartile range. The horizontal line inside each box is the median or 50th percentile. The lines which extend out from each end of the box are the whiskers. The whiskers extend out from first quartile (25th percentile) and third quartile (75th percentile) to the smallest data point within 1.5 interquartile ranges from the first and third quartiles. Observations that extend beyond the whiskers are known as outliers. Far outliers are observations whose values lie more than three interquartile ranges below the first quartile or above the third quartile. These are designated with plus signs.

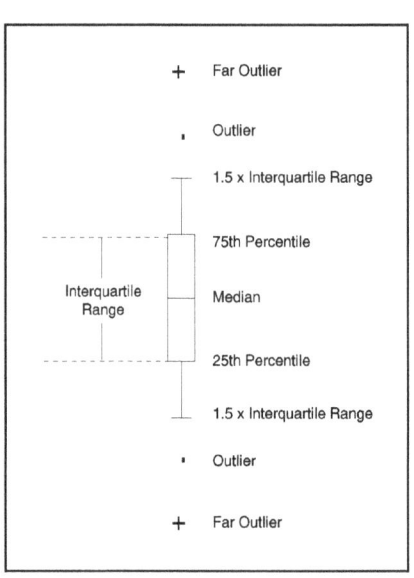

INTERPRETIVE GUIDE

TO WATER QUALITY RESULTS

This interpretive guide discusses each of the products presented in the next chapter - Water Quality Results. This chapter highlights how each of the tables and figures were prepared and how they can be used. Each subheading in this chapter corresponds to a particular product in the subsequent Water Quality Results chapter.

Overview

The Overview provides a brief one-page summary of the results of the various database retrievals for both the study area and the park. The study area results include the park results since the study area encompasses the park and all lands and waters within at least 3 miles upstream and 1 mile downstream of the park. Thus, the GIS estimated acreage of the study area should always be greater than the park acreage. The park acreage was computed from the digital boundary that was obtained for the park. More than likely this acreage will differ, perhaps significantly, from the "official" published acreage for the park due to the spatial and temporal accuracy of the digital boundary, treatment of inholdings, and other concerns. The number of STORET stations is the number of locations within the study area and park where an agency monitored (or intended to monitor) water quality. The number of stations with no data reveals the number of stations created in STORET for which water quality data were never entered. The number of stations with no statistical analysis reports the number of stations in the study area and park that contain data not amenable to normal parametric statistics. The number of longer term stations indicates the number of stations in the study area and park with at least 6 parameters having periods-of-record extending 2 years with an average of at least 1 observation per year over the period-of-record. The date of STORET retrieval is the calendar date when Horizon Systems downloaded all the data from STORET. Thus, the report documents all data entered in STORET prior to the retrieval date. Keep in mind that an agency can upload archival data at any time. Consequently, a retrieval date only guarantees that as of that date, this report contains all the data that had been entered into STORET. The period of record is the earliest date for which water quality data exist in STORET for the study area and park up to the date when the most recent data were entered prior to the retrieval date. The number of parameters measured is the number of unique water quality parameters measured within the study area and park and entered in STORET. The number of water quality observations is the sum of the total number of observations across all parameters within the study area and park. The number of industrial/municipal facilities discharges, drinking water intakes, water gages, and water impoundments are the number of each of these entities found within the study area and park. The number of time series, annual, and seasonal plots are the number of these different types of graphics produced by station/parameter combinations within the study area and park using the plotting criteria described in the previous chapter. The hydrologic seasons, described below, are the seasons used for the seasonal water quality data analysis. The time series, annual, and seasonal criteria are the plot and tabular screening criteria described in the previous chapter.

Regional Location Map

The Regional Location Map provides a small scale, general representation of the park and study area location within the United States. Digital, reproducible copies of this graphic are included on the disk(s) accompanying this report.

Water Quality Monitoring Locations Map(s)

The Water Quality Monitoring Locations Map(s) usually provides a larger scale representation of the park and study area than the Regional Location Map. This map indicates the locations within the study area where water quality has been monitored and the data entered into STORET. The water quality monitoring stations are labelled sequentially with the rightmost significant digits. The station names were assigned in numerically ascending order by latitude (for parks with a greater north-south extent than east-west) or longitude (for parks with a greater east-

19

west extent than north-south). Thus, this map serves as a visual index to the water quality data contained in the report. Since the 1:100,000 scale hydrography (from the River Reach File Ver. 3.0 or other sources) is displayed on the map, users can refer to the map to locate the station number on the reach in which they are interested and then find the appropriate section in the report that documents the water quality at that station. If the scale allows, USGS catalog units are also displayed on the map to provide an approximation of drainage basins. More than one Water Quality Monitoring Location map may be presented if the scale requires breaking the area into multiple maps for legibility. If multiple maps are necessary, an index map showing the geographic extent of each sub-map or panel will be present. Digital, reproducible copies of this graphic are included on the disk(s) accompanying this report. The digital, geo-referenced data files documented in Appendices A and B will allow the park to create water quality monitoring stations as a coverage in their GIS.

Dischargers, Drinking Intakes, Gages, and Impoundments Map(s)

The Dischargers, Drinking Intakes, Gages, and Impoundments Map(s) displays the same information as the Water Quality Monitoring Location Map(s) except the water quality stations are replaced by industrial/municipal facilities discharges, drinking water intakes, active and inactive gage locations, and water impoundments. This map also serves as a visual index allowing the user to determine the identification code of each discharger, drinking intake, gage, or impoundment. This number can then be used to obtain additional information about the entity on the following page of the report or to refer to the more detailed database files accompanying the report on disk. These more detailed database files are geo-referenced (See Appendices A and B), thus allowing the park to create these coverages in their GIS. More than one Dischargers, Drinking Intakes, Gages, and Impoundments map may be presented if the scale requires breaking the area into multiple maps for legibility. If multiple maps are necessary, an index map showing the geographic extent of each sub-map or panel will be present. Digital, reproducible copies of this graphic are also included on the disk(s) accompanying this report.

Industrial Facilities Discharges, Drinking Water Intakes, Water Gages, and Water Impoundments Table

This table provides some additional information about each of the discharges, drinking intakes, water gages, and water impoundments displayed on the previous map(s). This information generally includes the site identification number; the station or facility name; an address or some other indication of location; and some other pertinent information. More detailed information about each of these entities is contained in the database files on disk accompanying the report (See Appendices A and B).

Representative Mean Annual Hydrograph for Seasonal Analysis

One component of the water quality data analysis contained in the document is a seasonal analysis of the data (where adequate data exist). In order to undertake this analysis, some representation of the park's seasons was required. Seasons can be based on many factors (eg. hydrologic, climatic, recreational use, etc.). Since project resources did not allow us to contact every park and discuss with resource management staff what appropriate seasons may be for the park, WRD staff elected to adopt primarily a hydrologic/climatic definition of the seasons which uses a process of hydrograph separation to glean seasons from stream discharge patterns. The procedure employed to make these determinations was as follows:

(1) Find the nearest USGS Hydro-Climatic Data Network (HCDN) station (U.S. Geological Survey 1992) to the park that is most representative of streamflow conditions at the park. The HCDN is basically a subset of USGS streamflow stations, including only those stations that are unaffected by artificial diversions, storage, or other disruptions of the natural channel. All HCDN stations generally have at least a 20 year period of record. Consequently, discharge patterns at these stations should reflect only hydrologic and climatic influences. For the most part, selected HCDN sites were typically within 15-20 miles of the park. In some parks where WRD staff were aware of the existence of a stream gage located within the park that would be more representative of park waters even though it wasn't an HCDN site, this gage was selected.

(2) Retrieve the daily discharge values for the selected station from the USGS Daily Values File and generate a mean annual hydrograph and a box-and-whiskers plot of daily flows by month.

(3) Interpret the plots based on our knowledge of the hydrologic regime at these parks and assign seasons.

This approach, used for the majority of parks, assumes that most water quality data at the park will be found in streams and that the discharge pattern of the selected stream is representative of the seasons for all park waterbodies. Although this assumption may be weak for certain parks, project resources did not allow a more thorough investigation. For parks where there wasn't any stream gage (HCDN or otherwise) deemed representative of park waters, precipitation records from a nearby meteorological station were obtained from the National Climatic Data Center. Plotting daily average precipitation and box-and-whiskers of monthly precipitation sums allowed WRD hydrologists to make a rough approximation of climatic seasons for use in analyzing the water quality data.

Again, it is important to note the many ways of defining "seasons" and thus the limitations of the seasonal analysis contained in this document. For certain parks it may be more useful to perform a seasonal analysis with seasons defined by recreational use patterns or some other natural or anthropogenic factor. This option is available to the park since all the water quality data analyzed in this document is contained on disk(s) accompanying this report. Digital, reproducible copies of this seasonal analysis graphic are also included on the disk(s) accompanying this report.

Contacts for Agency Codes Retrieved

This table provides a list of the organizations who have entered data into STORET. A contact name at the organization and a phone number are also supplied. The agency code in the first column is the key for identifying which stations belong to that agency. This code will appear in the first line of each station's inventory. Although the agencies listed in this table are potential partners for future water quality monitoring or management endeavors, don't be surprised if the name of the contact and/or the telephone number is out of date. This information is entered when an agency first creates a station. The agency may not update this information when the initial contact moves on or the telephone number changes. Nonetheless, it is likely that the contact or someone else at the agency may be able to provide you with project reports or other information relative to the agency's data. A digital copy of this table accompanies this report on disk (See Appendices A and B).

Quantity of Data Retrieved by Agency Code

This table displays the period-of-record; numbers of water quality stations, longer-term stations, and stations without data; total number of water quality observations; and the number of unique water quality parameters measured by each agency within the study area and park boundary. Using this table, a park can quickly determine which agencies collect the most data in and around the park and whether they have monitored recently. A digital copy of this table accompanies this report on disk (See Appendices A and B).

Station Period of Record Tabulation

The Station Period of Record Tabulation provides a quick overview of the names of all the stations within the study area where water quality has been monitored and data entered into STORET. It also furnishes the total number of observations taken at each station and the frequency of observations between certain dates: (1) 01/01/85 until the most recent date data were measured; (2) 01/01/75 - 12/31/84; and (3) prior to 01/01/75. The station identification number, the four character park abbreviation code followed by a four digit number, provides the means to jump from a particular station in the table to the statistical and graphical analyses for this station contained in the Station-By-Station Results section. The Station Period of Record Tabulation reveals which water quality stations were situated within the park as defined by the park's GIS boundary. The Station Period of Record Tabulation also footnotes longer-term water quality stations. Longer-term stations are those that have at least 6 parameters with an average of one or more observations per year for those parameters during a period of record extending at least two years. Note that although a station may not be flagged as longer-term, it can still harbor much important data (albeit for only a few parameters or over a very long term with just a few observations). A digital copy of this table accompanies this report on disk (See Appendices A and B).

Parameter Period of Record Tabulation

The Parameter Period of Record Tabulation provides a complete listing of every water quality parameter ever measured in the study area and entered into STORET. This table is a summation of all the water quality observations for each parameter across all stations in the study area. Like the Station Period of Record Tabulation, the total number of observations for each parameter and the frequency of observations between: (1) 01/01/85 until the most recent date data were measured; (2) 01/01/75 - 12/31/84; and (3) prior to 01/01/75 are provided. This table is handy for quickly assessing whether particular parameters have been measured in the study area. The Parameter Period of Record Tabulation also shows how many in-park (and total) water quality stations contained data for each parameter. Some administrative parameters and parameters not suitable for statistical analysis within the context of this project (as discussed in the Screening Methodologies and Procedures section of the Methodology chapter) are listed in the Parameter Period of Record Tabulation, but not in the Station-By-Station Results section. A digital copy of this table accompanies this report on disk (See Appendices A and B).

Station/Parameter Period of Record Tabulation

The Station/Parameter Period of Record Tabulation combines the information found in the Station Period of Record Tabulation and the Parameter Period of Record Tabulation. This table provides a listing of all the stations where a particular water quality parameter was measured in the study area and the data entered into STORET. The table provides the start and end dates of the period of record of each parameter at each station; the number of years of measurement (computed from the start and end dates); whether the station/parameter combination occurred within the park boundary; the total number of observations for each parameter at each station, and whether a time series (T), annual (A), and/or seasonal (S) plot was generated for the station/parameter combination in the Station-By-Station Results section. This table is very useful when you need to determine at which locations within the study area (or park) particular parameters were monitored and how much data was collected there. Some administrative parameters and parameters not suitable for statistical analysis within the context of this project (as discussed in the Screening Methodologies and Procedures section of the Methodology chapter) are listed in the Station/Parameter Period of Record Tabulation, but not in the Station-By-Station Results section. A digital copy of this table accompanies this report on disk (See Appendices A and B).

Station-By-Station Results

Probably the most voluminous portion of the document is the Station-By-Station Results. Here the results of the water quality analyses for each station are presented in sequence. The results include the station inventory; parameter inventory; EPA water quality criteria analysis; and, as applicable, time series graphics and annual and seasonal tables and box-and-whiskers graphics. Each of these products are discussed below.

Station Inventory for Station

Each station's data commences with its Station Inventory. The Station Inventory provides the descriptive attributes about each water quality monitoring station contained in STORET. This includes a variety of locational information such as a verbal description, the Federal Information Processing codes for county and state, latitude and longitude, and other items; the station type (stream, spring, estuary, etc.); monitoring agency; creation date; indices to the River Reach File; whether the station lies within the park boundary; and several other attributes. This water quality station location data is also contained on disk(s) accompanying the report (See Appendices A and B).

Parameter Inventory for Station

Following the descriptive attributes about a station is the Parameter Inventory for the station. The Parameter Inventory provides a complete inventory and descriptive summary of all the water quality parameter data for the station. This table furnishes the parameter STORET code and name; the period of record for this parameter at this station; and the descriptive statistics defined in the Statistical Definitions in the previous chapter. Three different footnotes can appear on a parameter's descriptive statistics. Two asterisks (**) in the 10th, 25th, 75th, or 90th percentile columns indicates that there was insufficient data to compute these statistics for this parameter. Percentiles were not computed unless the parameter had at least 9 observations. Two number signs (##) next to the number of observations indicates that more than 50 percent of the observations entered into the computations as values that were taken to be half the detection limit. Caution should be employed in interpreting and using statistical results when more than half the values are set to half the detection limit. The letter "p" following a numeric STORET parameter code in the Parameter Inventory indicates that a time series plot was produced for this parameter at this station. Digital, reproducible copies of the Parameter Inventory tables are contained on the disk(s) accompanying this report.

Two downloaded parameter groups, pH and bacteriological, received special treatment whenever descriptive statistics were computed in the Parameter Inventory (as well as subsequent annual and seasonal tables). Whenever pH appears in a descriptive statistics table, the entry is increased to 3 entries: (1) the original pH entry; (2) pH computed from conversion to and from $\mu eq/l\ H^+$; and (3) $\mu eq/l\ H^+$. The reason for these conversions is that pH is actually the negative logarithm of the hydrogen ion concentration. To be technically correct in computing descriptive statistics, pH values must be converted to $\mu eq/l\ H^+$ (Kunkle and Wilson 1984). Once the descriptive statistics are computed using the pH values expressed as $\mu eq/l\ H^+$, the results can be converted back to pH. The three pH entries in the descriptive statistics table will all have the same STORET code.

Whenever a bacteriological parameter appears in a descriptive statistics table, the entry is increased to 3 entries: (1) the original bacteriological entry; (2) an entry computed using the log of each measured value; and (3) an entry that simply reports the geometric mean. The reason for converting to logs and displaying the geometric mean is convention. Bacteriological water quality standards typically reference the geometric mean rather than the arithmetic. The three bacteriological entries in the descriptive statistics tables will all have the same STORET code.

EPA Water Quality Criteria Analysis for Station

The EPA Water Quality Criteria Analysis table follows the Parameter Inventory. This table presents a comparison between the station's STORET water quality data and applicable national water quality criteria for freshwater and marine aquatic organisms; drinking water; and other concerns. Comparison against applicable State water quality criteria was not feasible given project resources. Appendix F provides the relevant national EPA water quality criteria values. In most cases, the EPA water quality criteria values are single sample concentrations that can be directly compared to single sample STORET entries. There are, however, two notable exceptions to this single sample/single value comparison: ammonia and fecal-indicator bacteria. For these two parameters, criteria are either derived from or depend on the results of other chemical characteristics of the water or require a time series statistical treatment of multiple samples to determine whether the criterion has been exceeded. The EPA ammonia criterion is pH and temperature dependent. To calculate the criterion for each ammonia sample value was beyond

23

the scope of this project. Consequently, ammonia criteria were not included in Appendix F or the EPA Water Quality Criteria Analyses. Un-ionized ammonia criteria can be determined from formula table values included in the EPA Silver Book (Environmental Protection Agency 1995).

For the purposes of this project, fecal-indicator bacteria data were flagged as exceeding criteria when their concentrations exceeded 200, 1000, 126, and 33 (fresh)/35 (salt) colony forming units or most probable number for single samples of fecal coliform, total coliform, E. coli, and enterococci, respectively. These values represent only approximations of the criteria for primary contact recreation waters where criteria are typically expressed in terms of a geometric mean computed with no less than 5 samples during a given month. When a fecal-indicator bacterial observation exceeds a criterion in the EPA Water Quality Criteria Analysis section, the reader should refer to the corresponding geometric mean calculations in the preceding Parameter Inventory. Long-term geometric means that exceed the respective water quality criteria for multiple samples are more indicative of chronic bacteriological problems than single sample values.

Water quality observations carrying non-detection or below-detection limit remark codes (K, T, and U) required special treatment in the EPA Water Quality Criteria Analysis. As with the statistics in the Parameter Inventory, half the detection limit was the value used in the EPA Water Quality Criteria Analysis. For certain observations, however, half the detection limit may exceed a water quality criterion. For those observations it would be inappropriate to classify them as exceeding a criterion since the actual value wasn't known. Thus, it was decided that any below detection limit or non-detect observations that exceed a water quality criterion using half the detection value would be excluded from the EPA Water Quality Criteria Analysis. If non-detect or below detection limit values are excluded from the EPA Water Quality Criteria Analysis for a particular parameter, the total observations for that parameter will be footnoted with an ampersand (&). This will also explain the difference between the total observations in the Parameter Inventory and the EPA Water Quality Criteria Analysis. Non-detect or below detection limit values are included in the EPA Water Quality Criteria Analysis, however, if half the detection limit doesn't exceed the parameter's criterion.

The EPA Water Quality Criteria Analysis for each station lists the parameter; the standard type and value; the total number of observations for the parameter at this station; the number of observations that exceeded the standard value; and the proportion of observations that exceeded the standard value. Water quality observations are considered as having exceeded a criterion regardless of whether the criterion represents a maximum acceptable value or a minimum acceptable value. The table also breaks down the water quality criteria analysis on a seasonal basis to allow the reader to discern whether parameter observations tend to exceed criteria during only certain seasons or year round. Although the EPA Water Quality Criteria Analysis table is a good starting point for assessing potential water quality problems at the station, the reader is strongly encouraged to read the caveat section in the Introduction concerning drawing conclusions about water quality problems from this table. Digital, reproducible copies of these tables accompany the report on disk (See Appendices A and B).

Time Series Plots for Station

Following the EPA Water Quality Criteria analysis will be any Time Series Plots for each parameter that met the time series plot screening criterion selected for the park unit. If a time series plot is generated for a particular parameter at a station, a "p" will appear next to the STORET parameter code in the Parameter Inventory. If no time series plots are present for the particular station, the data did not meet the time series screening criterion listed in the Overview section of the Water Quality Results chapter. The x-axis on these plots is the period of record, listing only the 2-digit calendar year for clarity (i.e. 1983 is presented as 83). The y-axis is the concentration of the selected parameter in its measurement units. In general, the units for a given parameter are given either on the y-axis or in the parameter description in the subtitle of the graph. Subtitle and/or y-axis parameter descriptions may be truncated on the plots so as to not exceed the maximum number of plotting characters. Y-axis values less than zero are sometimes shown for better representation of the entire plot. The station identification code, parameter description, and parameter STORET code are presented in the main title. The footnote provides a descriptive location name. Observations on the plot are represented as squares. Lines are drawn connecting each successive observation. As mentioned previously in the Statistical Definitions section of the Methodology chapter, the interconnecting line is drawn only for ease of reading and provides no indication of what the actual parameter values were between the two observed measurements. Digital, reproducible copies of all time series plots accompany the report on disk (See Appendices A and B).

For time series plots of pH, the original pH values are plotted. For time series plots of bacteriological data, the log of the measured value is plotted. Hence, the y-axis of a time series plot for bacteriological parameters is log-linear.

Annual Analysis for Station

If more than 9 observations exist in each of at least 4 years for a particular parameter at a station, an Annual Analysis table will be generated. Entries will be made in the table for each parameter having more than 9 observations in each of at least 4 years. The Annual Analysis presents the same descriptive statistics as the Parameter Inventory table, except that it provides the statistics by year, rather than the entire period of record. Although some of the years may not contain 9 observations, these years still have an entry in the table. A parameter needs only to have 9 observations in any 4 years of its period of record to qualify for the Annual Analysis table. Like the Parameter Inventory, percentiles with fewer than 9 observations are not computed and entries computed with greater than 50 percent of the data values set to half the detection limit are flagged. Entries in the Annual Analysis table that also meet the annual analysis box-and-whisker plot screening criterion will be flagged with a "p" next to the STORET code. Digital, reproducible copies of these tables accompany the report on disk (See Appendices A and B).

Annual Box-and-Whiskers Plots for Station

Entries in the Annual Analysis table that meet the annual box-and-whisker plot screening criterion will generate Annual Box-and-Whiskers Plots. The interpretation of box-and-whiskers plots is explained in the Statistical Definitions section of the Methodology chapter. A box is generated for each year of the period of record, even if less than 9 observations were recorded in the year. The axis labeling and plot titling is the same as for the time series plots. Digital, reproducible copies of these graphics accompany the report on disk (See Appendices A and B).

For annual box-and-whiskers plots of pH, μeq/l H^+ are plotted. For annual box-and-whiskers plots of bacteriological data, the log of the measured value is plotted. Hence, the y-axis of an annual box-and-whiskers plot for bacteriological parameters is log-linear.

Seasonal Analysis for Station

As explained above, a park's hydrologic seasons for seasonal water quality analysis were determined using a process of hydrograph separation and other techniques. If a parameter has more than 9 observations in each of 2 seasons with a period of record of at least 6 years and observations in at least 3 of the 6 years, a Seasonal Analysis table will be generated for the station. The Seasonal Analysis presents the same descriptive statistics as the Parameter Inventory table, except that it provides the statistics by season, rather than the entire period of record. Although certain parameters for a season at a station may not contain 9 observations, these parameters can still have an entry in the table. A parameter needs only to have 9 observations in each of 2 seasons with a period of record of at least 6 years and observations in at least 3 of the 6 years to qualify for the Seasonal Analysis table. Consequently, some of the parameters could have fewer than 9 observations in a particular season but still generate a table entry. Like the Parameter Inventory and Annual Analysis, percentiles with fewer than 9 observations are not computed and entries computed with greater than 50 percent of the data values set to half the detection limit are flagged. Entries in the Seasonal Analysis table that also meet the seasonal analysis box-and-whisker plot screening criterion will be flagged with a "p" next to the STORET code. Digital, reproducible copies of these tables accompany the report on disk (See Appendices A and B).

Seasonal Box-and-Whiskers Plots for Station

Entries in the Seasonal Analysis table that meet the seasonal box-and-whisker plot screening criterion will generate Seasonal Box-and-Whiskers Plots. The interpretation of box-and-whiskers plots is explained in the Statistical Definitions section of the Methodology chapter. A box is generated for each season of the period of record, even if less than 9 observations were recorded in the season. On the x-axis, the seasons are labeled 1 through the number of seasons defined for the park through hydrograph separation. The actual calendar dates that correspond to these numerically labeled seasons exist in the Overview section and the Seasonal Analysis tables in the Water Quality Results chapter. The axis labeling and plot titling are the same as for the time series and annual box-and-whiskers plots. Digital, reproducible copies of these graphics accompany the report on disk (See Appendices A and B).

For seasonal box-and-whiskers plots of pH, μeq/l H^+ are plotted. For seasonal box-and-whiskers plots of bacteriological data, the log of the measured value is plotted. Hence, the y-axis of a seasonal box-and-whiskers plot for bacteriological parameters is log-linear.

EPA Water Quality Criteria Analysis for Entire Park Study Area

This table essentially summarizes all the individual station-by-station EPA water quality criteria analyses in the study area. (Refer to the EPA Water Quality Criteria Analysis for Station section above for more detailed information on the treatment of special cases in the EPA Water Quality Criteria Analysis for Entire Park Study Area.) This table presents a comparison between the study area's STORET water quality data and applicable national water quality criteria for freshwater and marine aquatic organisms; drinking water; and other concerns. Comparison against applicable State water quality criteria was not feasible given project resources. Appendix F provides the relevant national EPA water quality criteria values. The EPA Water Quality Criteria Analysis for the Entire Park Study Area lists the parameter; the standard type and value; the total number of observations for the parameter at this station; the number of observations that exceeded the standard value; and the proportion of observations that exceeded the standard value. Water quality observations are considered as having exceeded a criterion regardless of whether the criterion represents a maximum acceptable value or a minimum acceptable value. The table also breaks down the water quality criteria analysis on a seasonal basis to allow the reader to discern whether parameter observations tend to exceed criteria during only certain seasons or year round. Although the EPA Water Quality Criteria Analysis for the Entire Park Study Area is a good starting point for assessing potential water quality problems at the park, the reader is strongly encouraged to read the caveat section in the Introduction before drawing conclusions about water quality problems from this table. A digital, reproducible copy of this table accompanies the report on disk (See Appendices A and B).

NPS Servicewide Inventory and Monitoring Program
Level I Water Quality Inventory Data Evaluation and Analysis (IDEA)

One of the objectives of this Baseline Water Quality Data Inventory and Analysis project is to perform an IDEA - an Inventory Data Evaluation and Analysis - to determine the presence and/or absence of Servicewide Inventory and Monitoring Program "Level I" water quality parameter groups in the park's study area. The Strategic Plan for Conducting Baseline Natural Resource Inventories in the National Park Service (National Park Service 1993) identified the basic water quality parameters displayed in Table I as the parameters that all parks must have for "key" waterbodies (determined on the basis of size, uniqueness, threats, etc.) within park boundaries. Since these parameters can be measured in different ways and with different units, there are multiple STORET codes associated with each parameter; hence the concept of parameter groups. The Strategic Plan distinguishes between those parameter groups required for all parks and parameter groups required only on a case-by-case basis.

The IDEA basically compares the parameters listed in the Parameter Period of Record Tabulation and Station/Parameter Period of Record Tabulation with the "Level I" Servicewide Inventory and Monitoring water quality parameter groups, listed in Table I and in Appendix G, and notes, not only the presence or absence of each parameter group, but the total number of observations for each parameter present in the group; the number of observations between certain time periods; and the total number of stations within the study area at which the parameter was measured. The total number of different (unique) stations measuring parameters for the group is in parentheses on each parameter group's summary line.

The first page of the IDEA lists the missing Servicewide Inventory and Monitoring Program "Level I" groups. If a parameter group appears on this list, no data for any of the parameters defining the group (See Appendix G) was retrieved for it within the study area. So-called non-priority parameter groups may appear in the missing list. Non-priority parameters are park-specific parameters (case-by-case) which may not be applicable to your park. Consequently, if you believe a particular parameter, not included in IDEA (See Appendix G), to be important for your park, you will have to consult the Parameter and Station/Parameter Period of Record Tabulations to determine the presence or absence of this parameter for the park. Although considered a "Level I" parameter, biological data, obtained through rapid bioassessment or other means, is not considered in this report which deals specifically with surface water chemistry. Following the Missing Level I Group list is the Present Level I Group list which displays the summary results for each Servicewide Inventory and Monitoring "Level I" water quality parameter group that was found.

Table I.   Basic "Level I" Water Quality Parameters Identified as Required and Optional By the Servicewide Inventory and Monitoring Program for "Key" Park Waterbodies

Required Parameter Groups:
(1) Alkalinity
(2) pH
(3) Conductivity
(4) Dissolved Oxygen
(5) Rapid Bioassessment Baseline (EPA/State protocols, involving fish and macroinvertebrates)
(6) Temperature
(7) Flow

Case-By-Case Parameters Groups:
(8) Toxic Elements
(9) Clarity/Turbidity
(10) Nitrate/Nitrogen
(11) Phosphate/Phosphorus
(12) Chlorophyll
(13) Sulfates
(14) Bacteria

The last page of the IDEA summarizes the information from the Missing and Present Level I Group lists. This page provides information on the temporal and spatial distributions of the data. Included in this table are the total number of observations for each parameter group; the number of observations since January 1, 1985; the percent of the total observations since January 1, 1985; the number of stations measuring each parameter group; the percent of the total number of stations with data measuring the parameter group; the number of observations per station with data; the period-of-record for this parameter group; and the average number of observations per year of the period-of-record.

In interpreting the results of the IDEA, the reader should first consult the Missing Level I Group list. For the parameter groups listed, there was no baseline water quality data within the study area entered in STORET. Consequently, these parameter groups could be a higher priority for data collection. It is important, however, to realize that data within these parameter groups may have been already collected but not entered into STORET. The resources for this project did not enable us to pursue thorough literature and file cabinet reviews to dredge up every last iota of data. If data exists for certain Servicewide Inventory and Monitoring Program "Level I" water quality parameter groups in a park's file cabinet, it is the park's responsibility to factor that data into their IDEA. Consequently, the listing of a parameter group on the Missing "Level I" Group list is not a WRD endorsement to launch a study to collect these data. The IDEA is intended to simply note that no data exist for these parameter groups in STORET for the park. It is the park's responsibility to ascertain whether such data has already been collected by the park or other entities before embarking on a new study. In fact, in the future the WRD will require that any park study plan proposing to collect baseline water quality data show that they have consulted their Baseline Water Quality Data Inventory and Analysis report and searched in other locations (file cabinets, published literature, etc.) for the data they propose to collect. A similar interpretation springs from the Present "Level I" Group list. Insufficient data density in certain time periods for particular parameter groups is not necessarily cause for launching a new inventory and/or monitoring program. The park should still consult with other potential sources of data. Again, the IDEA is designed to provide only a quick check on data in STORET for the Servicewide Inventory and Monitoring Program "Level I" water quality parameter groups.

Water Quality Observations Outside STORET Edit Criteria for Park

STORET data entered after November 1983 were subjected to rudimentary edit/bounds checking for 190 common parameters (See the STORET Edit Criteria in Appendix C). None of the data entered into STORET prior to that time has been subjected to edit/bounds checking. Moreover, to maintain exact comparability with USGS WATSTORE data, WATSTORE data entered into STORET has never been subjected to the EPA edit/bounds checking. During the pilot test phase of this project, obviously incorrect data was identified from both USGS and other agency data in STORET. As a consequence, all data downloaded from STORET was filtered through the STORET edit criteria to identify parameter observation values that fall outside any edit criterion ranges. This section documents the station name, parameter, date, time, parameter value, agency, and STORET station name of every observation that fell outside the range of an edit criterion. Not all data falling outside an edit criterion are necessarily incorrect. Such data may represent unique or special conditions. Consequently, every observation falling outside a STORET edit criterion was scrutinized to determine, in our best professional judgement, whether the value was in the realm of possibility or obviously incorrect. Water quality observations that appeared to be obviously incorrect are marked with an "X" in the Disposition column of this table. These values were not retrieved or included in any of the inventory tables or graphs. Water quality values outside a STORET edit criterion but within the realm of possibility were retained and included in inventory tables and graphs. The Water Quality Observations Outside STORET Edit Criteria for Park table documents all values that were outside an edit criterion range. This documentation is also necessitated by the fact that agencies can override the STORET edit criteria for individual observations. Although the edit criteria eliminate some potentially "bad" data from the report, the probability of other incorrect data, for both the 190 parameters that are edit/bound checked and all the other STORET parameters that aren't error checked, is high. Readers should consult the Caveat section in the Introduction for guidelines on the use and interpretation of STORET data. The responsibility for correcting these observations rests with the collecting agency.

WATER QUALITY RESULTS

OVERVIEW FOR BADL

Study Area Boundary Description

The study area includes the park and all areas within at least 3 miles upstream of the park unit boundary and at least 1 mile downstream.

	Study Area	Park
GIS Estimated Acreage:	904666	241453
# STORET Stations:	31	8
# Stations With No Data:	0	0
# Stations With No Stat. Analysis:	0	0
# Longer Term Stations:	10	4
Date of STORET Retrieval:	02/02/98	02/02/98
Period of Record:	11/29/62-09/10/96	08/13/64-08/28/95
# Parameters Measured:	197	128
# Water Quality Observations:	6896	4269
# Industrial/Municipal Facilities:	2	0
# Drinking Water Intakes:	0	0
# Water Gages:	9	3
# Water Impoundments:	24	2
# Total Plots:	36	33
# Time Series:	34	33
# Annual:	1	0
# Seasonal:	1	0

Hydrologic Definition of Seasons:

1. October 1 - January 31
2. February 1 - April 14
3. April 15 - June 30
4. July 1 - September 30

Time Series Plot Criteria:

To be included in the time series plots, a station/parameter combination must have at least 2 years and at least 8 observations.

Annual Analysis Criteria:

To be included in the annual box-and-whisker plots, a station/parameter combination must have at least 9 observations in each of at least 4 years.

To be included in the annual analysis tables, a station/parameter combination must have at least 9 observations in each of at least 4 years.

Seasonal Analysis Criteria:

To be included in the seasonal box-and-whisker plots, a station/parameter combination must have at least 9 observations in each of 2 seasons and a period of record of at least 6 years and observations in at least 3 of the 6 years.

To be included in the seasonal analysis tables, a station/parameter combination must have at least 9 observations in each of 2 seasons and a period of record of at least 6 years and observations in at least 3 of the 6 years.

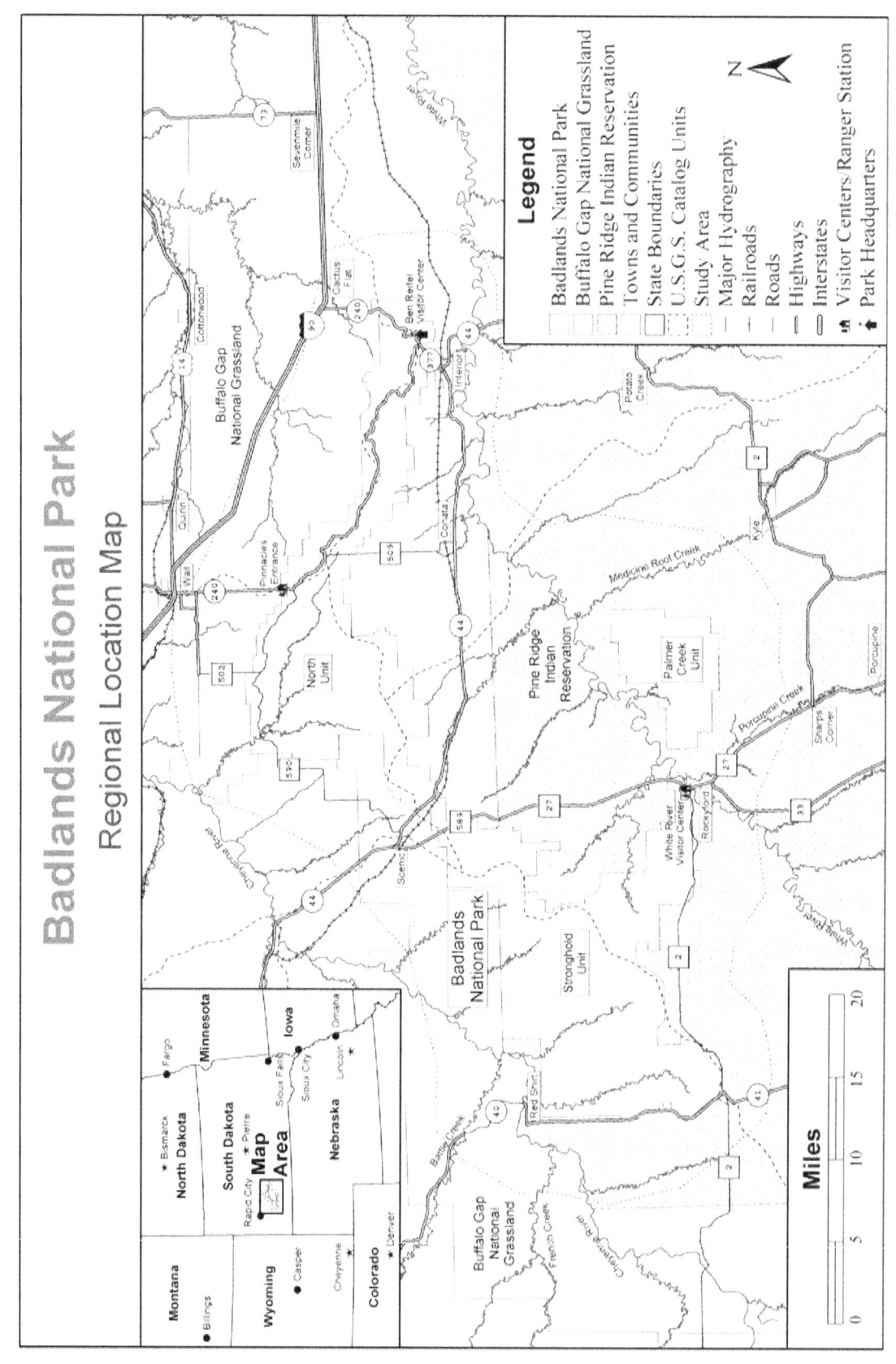

Badlands National Park

Regional Location Map

32

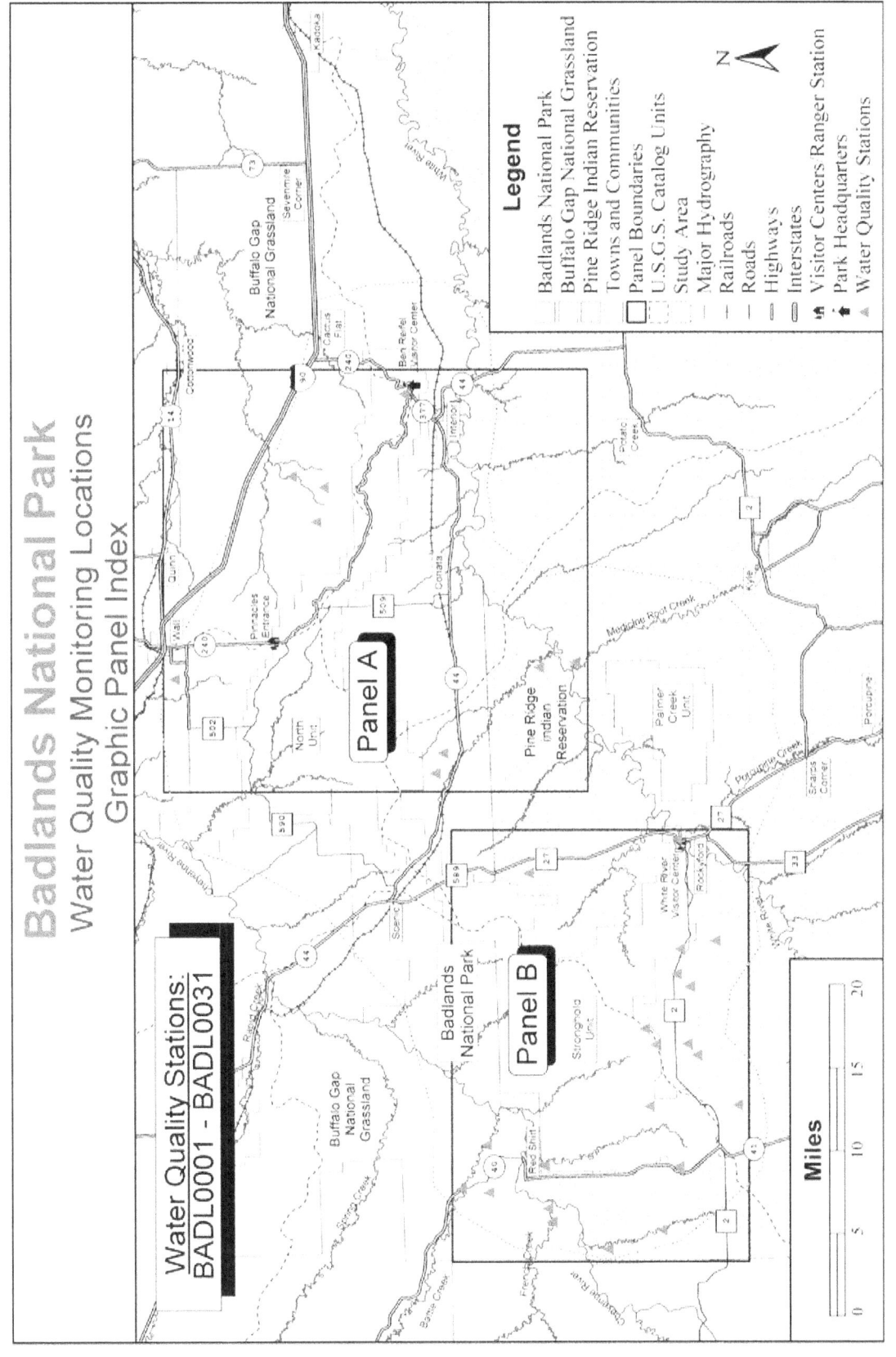

Badlands National Park
Water Quality Monitoring Locations
Graphic Panel Index

Water Quality Stations:
BADL0001 - BADL0031

Panel A

Panel B

Legend

Badlands National Park
Buffalo Gap National Grassland
Pine Ridge Indian Reservation
Towns and Communities
Panel Boundaries
U.S.G.S. Catalog Units
Study Area
Major Hydrography
Railroads
Roads
Highways
Interstates
Visitor Centers Ranger Station
Park Headquarters
Water Quality Stations

Miles

0   5   10   15   20

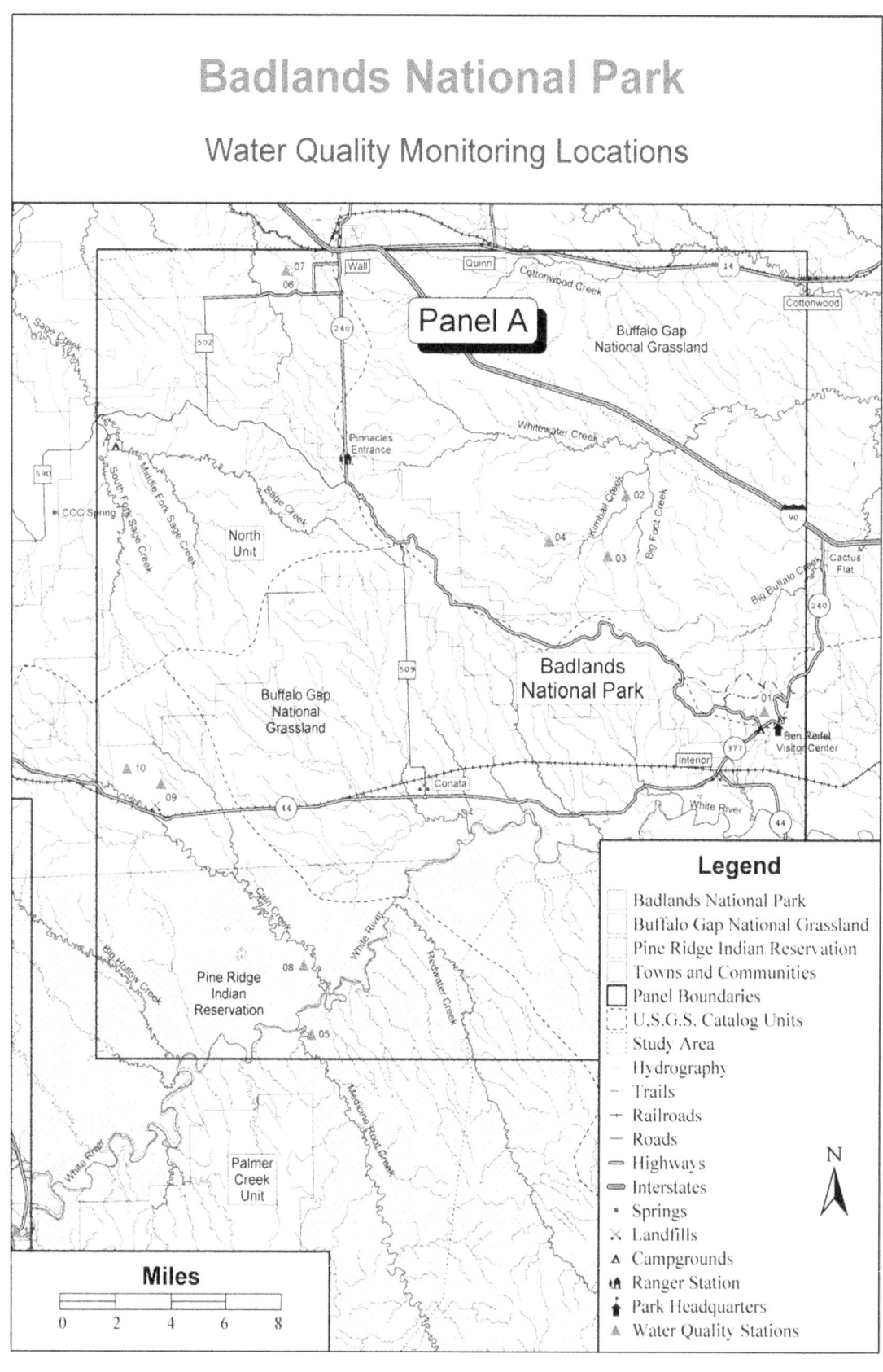

Badlands National Park

Water Quality Monitoring Locations

Panel A

Legend

Badlands National Park
Buffalo Gap National Grassland
Pine Ridge Indian Reservation
Towns and Communities
Panel Boundaries
U.S.G.S. Catalog Units
Study Area
Hydrography
Trails
Railroads
Roads
Highways
Interstates
Springs
Landfills
Campgrounds
Ranger Station
Park Headquarters
Water Quality Stations

Miles

0   2   4   6   8

N

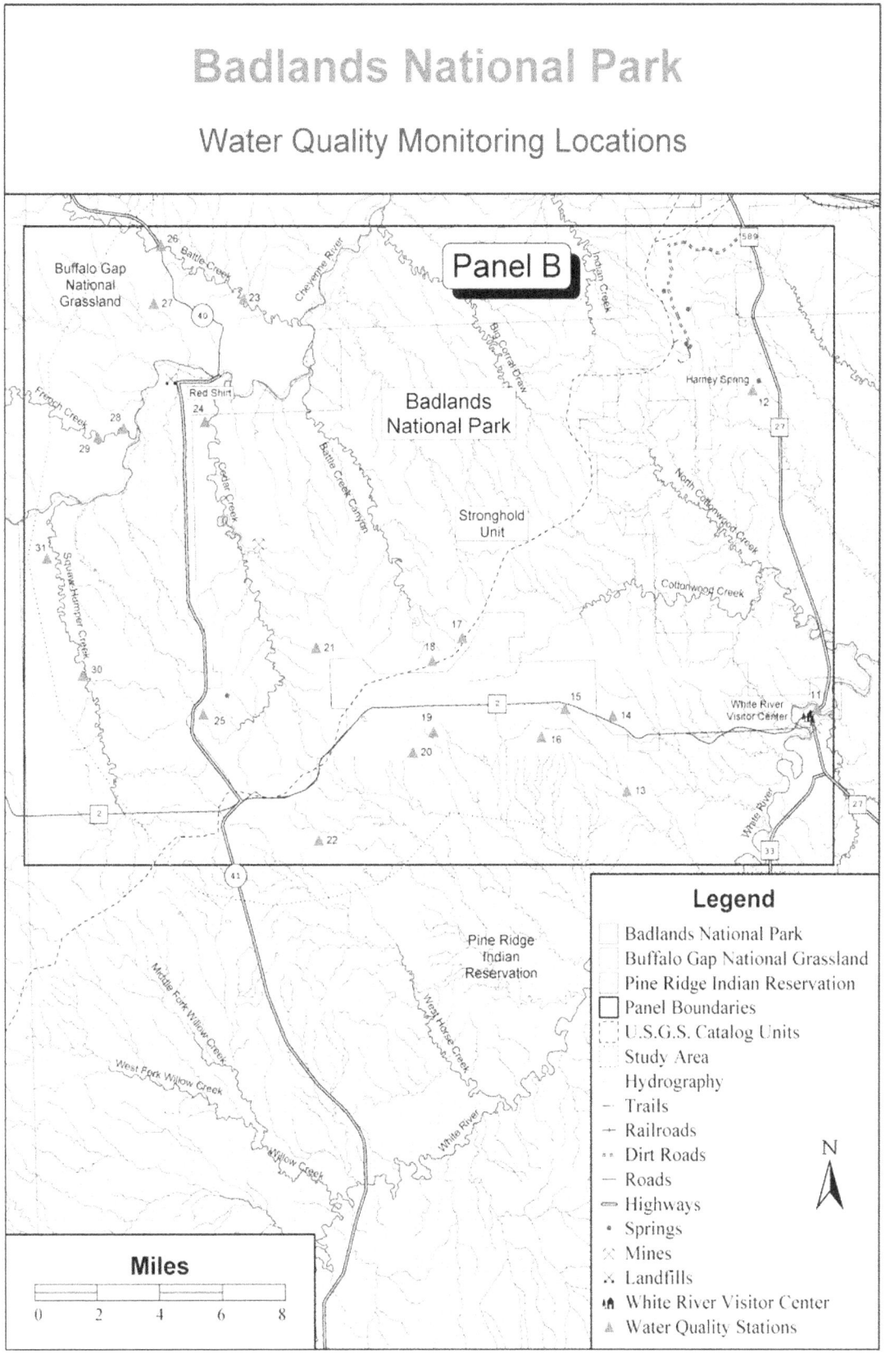

Badlands National Park

Water Quality Monitoring Locations

Panel B

Buffalo Gap National Grassland

Badlands National Park

Stronghold Unit

Pine Ridge Indian Reservation

Red Shirt

Harney Spring

Legend

Badlands National Park
Buffalo Gap National Grassland
Pine Ridge Indian Reservation
Panel Boundaries
U.S.G.S. Catalog Units
Study Area
Hydrography
- Trails
+ Railroads
·· Dirt Roads
— Roads
⊸ Highways
· Springs
✕ Mines
⋏ Landfills
ᴧ White River Visitor Center
▲ Water Quality Stations

N

Miles

0   2   4   6   8

Badlands National Park

Dischargers, Drinking Intakes, Water Gages, & Water Impoundments
Graphic Panel Index

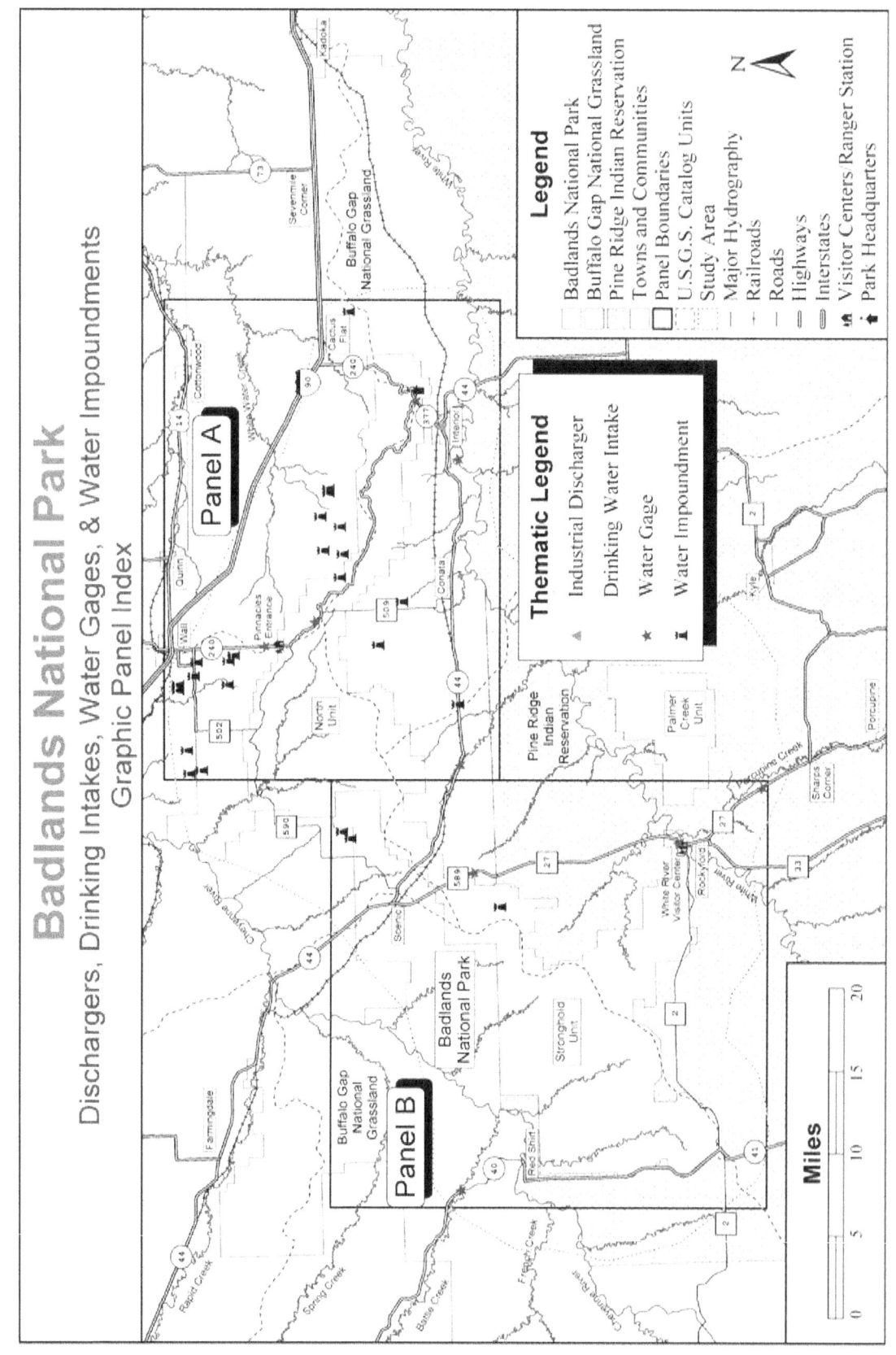

Badlands National Park

Dischargers, Drinking Intakes, Water Gages, & Water Impoundments

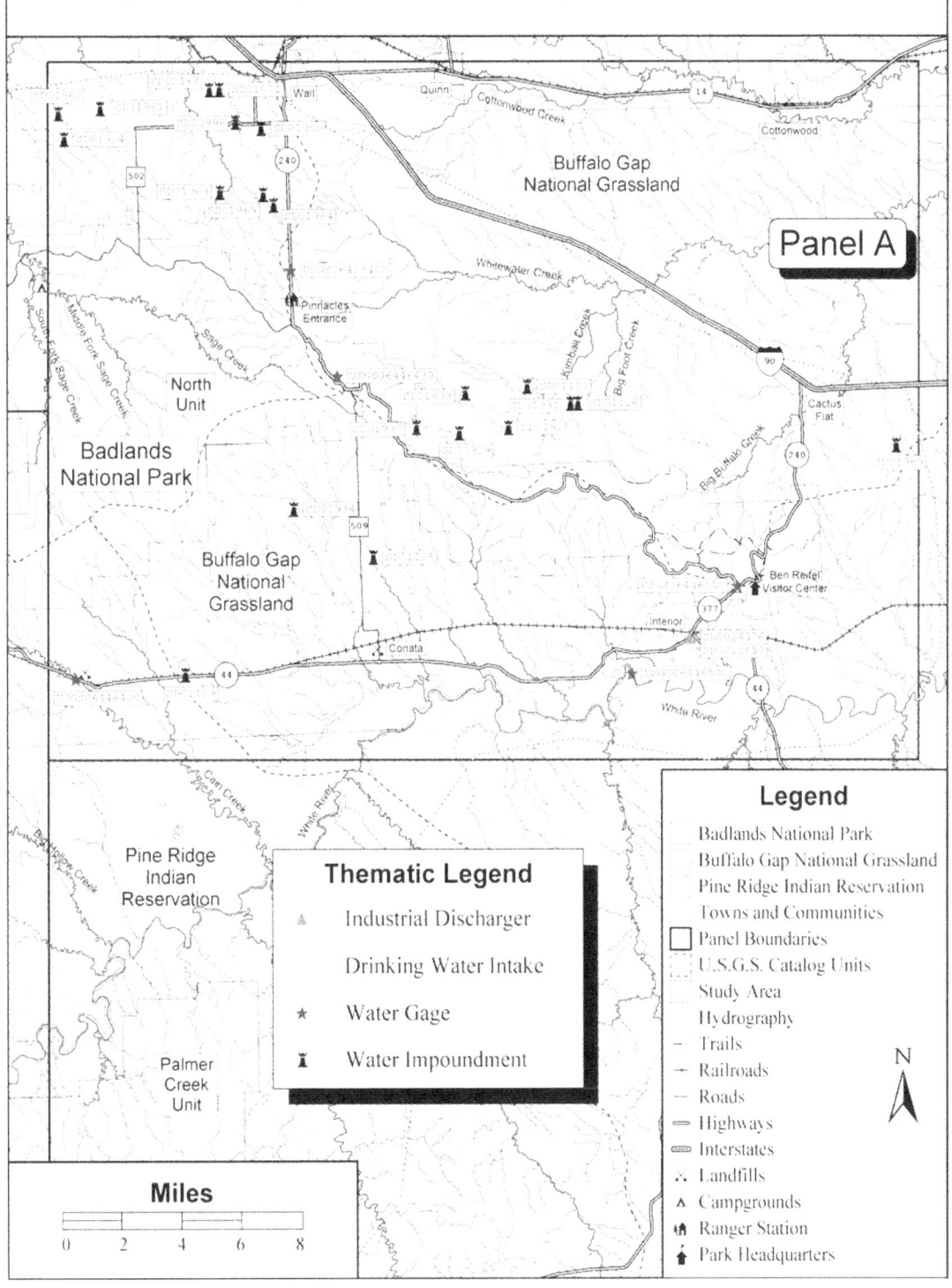

Panel A

Thematic Legend

△ Industrial Discharger

Drinking Water Intake

★ Water Gage

⛭ Water Impoundment

Legend

Badlands National Park
Buffalo Gap National Grassland
Pine Ridge Indian Reservation
Towns and Communities
☐ Panel Boundaries
U.S.G.S. Catalog Units
Study Area
Hydrography
- Trails
+ Railroads
- Roads
═ Highways
═ Interstates
∴ Landfills
∧ Campgrounds
⚑ Ranger Station
⛿ Park Headquarters

N

Miles

0 2 4 6 8

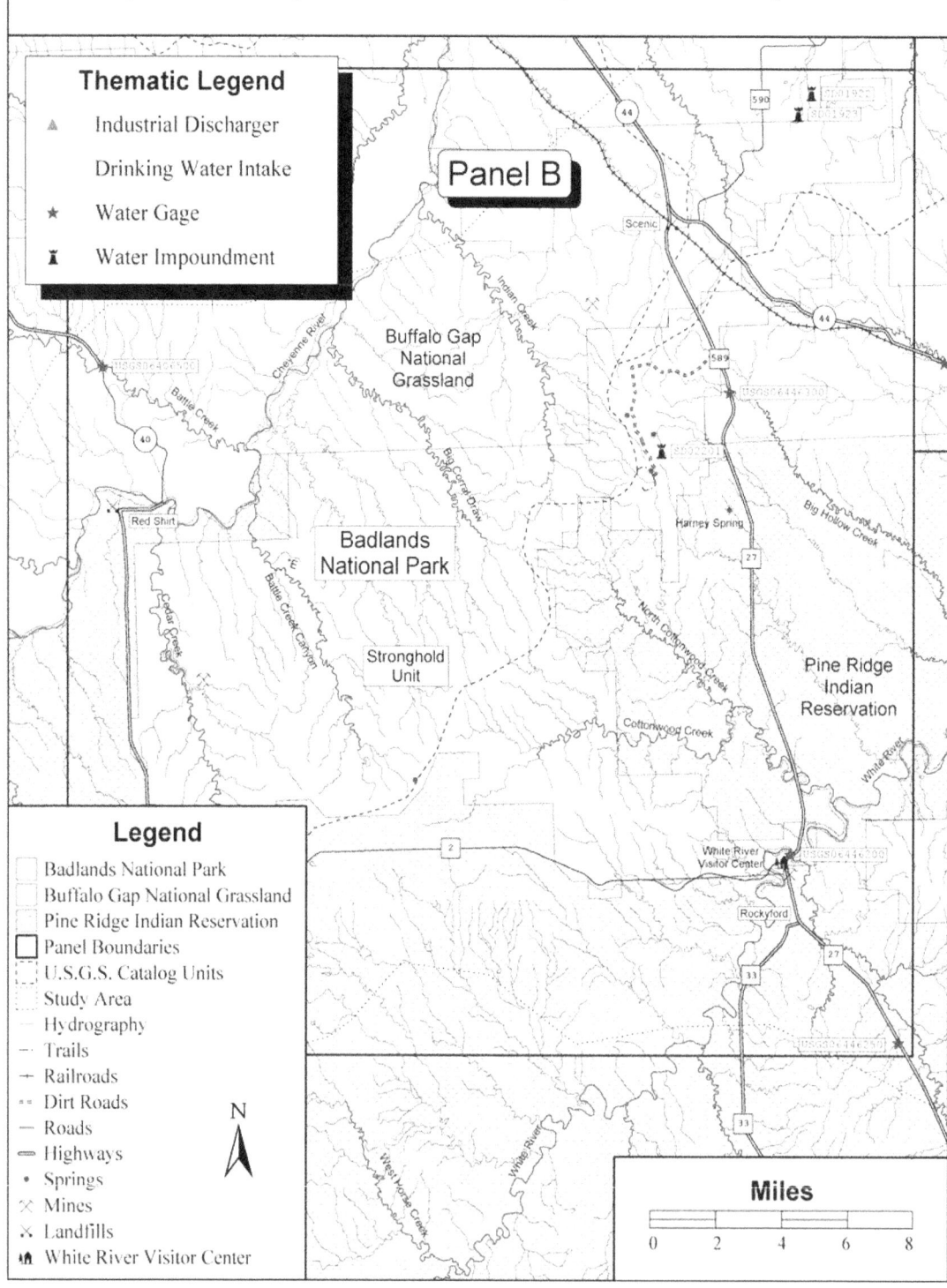

Badlands National Park

Dischargers, Drinking Intakes, Water Gages, & Water Impoundments

Thematic Legend

▲ Industrial Discharger

Drinking Water Intake

★ Water Gage

♟ Water Impoundment

Panel B

Buffalo Gap National Grassland

Badlands National Park

Stronghold Unit

Pine Ridge Indian Reservation

Scenic

Red Shirt

Harney Spring

White River Visitor Center

Rockyford

Legend

☐ Badlands National Park
☐ Buffalo Gap National Grassland
☐ Pine Ridge Indian Reservation
☐ Panel Boundaries
⬚ U.S.G.S. Catalog Units
Study Area
— Hydrography
-- Trails
+ Railroads
** Dirt Roads
— Roads
— Highways
• Springs
✕ Mines
✕ Landfills
⛹ White River Visitor Center

N

Miles

0 2 4 6 8

Industrial Facility Discharges, Drinking Water Intakes,
Water Gages, and Water Impoundments Within the BADL Study Area

Industrial Facility Discharges

Site ID	Station/Facility Name	Address	City	Facility Receiving Water Name
SD0021857	INTERIOR TOWN OF	....	INTERIOR	....
SD0024376	USNPS BADLANDS NM	P O BOX 6	INTERIOR	

Drinking Water Intakes

Site ID	Station/Facility Name	City	Population Served	Avg. Daily Production (Gal./Day)

No drinking water intakes available for this study area.

Water Gages

Site ID	Station Name	Site Type	Drainage Area (Square Miles)	Begin Year	End Year
USGS06406500	BATTLE CR BELOW HERMOSA SD	Stream	285.00	1951	1997
USGS06423400	BULL CR TRIB NEAR WALL SD	Stream	.39	1970	1978
USGS06446200	WHITE R NEAR ROCKYFORD SD		3000.00	1964	1973
USGS06446250	PORCUPINE CR TRIB NEAR ROCKYFORD SD	Stream	1.65	1968	1979
USGS06446300	BIG HOLLOW CR TRIB NEAR SCENIC SD	Stream	2.71	1968	1976
USGS06446400	CAIN CR TRIB AT IMLA	Stream	15.80	1956	1980
USGS06446430	WHITE RIVER TRIB NE		.17	1956	1973
USGS06446500	WHITE RIVER NEAR IN		4120.00	1904	1942
USGS06446550	WHITE R TRIB NEAR IN	Stream	.32	1956	1980

Water Impoundments

Site ID	Impoundment Name	Owner	Primary Purpose	Type of Dam	Downstream Hazard	Year Completed
SD00996	BRUCE DAM	ROBERT E HAYS	Rec.	Earth	Low	1937
SD01000	CROWNDAM	MERLECROWN	Supply	Earth	Low	1950
SD01071	NEWWALLLAKE	STATE OF SOUTH DAKOTA	Rec.	Earth	Low	1947
SD01126	SOUTH WHITEWATER 251	USDA FS	Other	Earth	Low	1936
SD01127	MISSILE 249	USDA FS	Other	Earth	Low	1936
SD01128	CONATA EAST 302	USDA FS	Other	Earth	Low	1938
SD01129	SAGE CREEK 344	USDA FS	Other	Earth	Low	1937
SD01922	LARSEN LYMAN 1	WALTER E WITCHER	Supply	Earth	Low	1958
SD01923	LARSEN LYMAN 2	WALTER E WITCHER	Supply	Earth	Low	1959
SD01924	PERCY DAM	MARY PERCY	Supply	Earth	Low	1956
SD01925	SDONAME 173	FRED WOLF	Supply	Earth	Low	1957
SD01929	VIRGIL HORTON 1	VIRGIL HORTON	Supply	Earth	Low	1916
SD01930	VIRGIL HORTON 2	VIRGIL HORTON	Supply	Earth	Low	1937
SD01931	US GOVT DAM	USDA FS	Supply	Earth	Low	1936
SD01933	VIRGIL HORTON 4	VIRGIL HORTON	Supply	Earth	Low	1952
SD01934	WILLIAM HUETHER	USDA FS	Supply	Earth	Low	1958
SD01961	ROY SHULL NO 1	ROY SHULL	Supply	Earth	Low	1957
SD01962	GLEN LAKNER	GLEN LAKNER	Supply	Earth	Low	1947
SD01963	GEORGE KNAPP	H AND K RANCH	Irrig.	Earth	Low	1955

Industrial Facility Discharges, Drinking Water Intakes,
Water Gages, and Water Impoundments Within the BADL Study Area

Water Impoundments

Site ID	Impoundment Name	Owner	Primary Purpose	Type of Dam	Downstream Hazard	Year Completed
SD01964	ED FRIEN DAM	DONALD-ROBERT KELLY	Supply	Earth	Low	1958
SD01965	ROY SHULL NO 2	H AND K RANCH DAM	Supply	Earth	Low	1943
SD01989	MARTIN DAM	BUCKLES	Supply	Earth	Low	1940
SD02167	OLD TOWN DAM	HARVEY STONE	Supply	Earth	Low	1941
SD02201	BADLANDS NAT.PK.NONAME DAMS 1 ------		Farm	Earth	Low	1950

REPRESENTATIVE MEAN ANNUAL HYDROGRAPH FOR SEASONAL ANALYSIS

BADLANDS NATIONAL PARK
White River near Oglala, SD
06446000, 42 year record

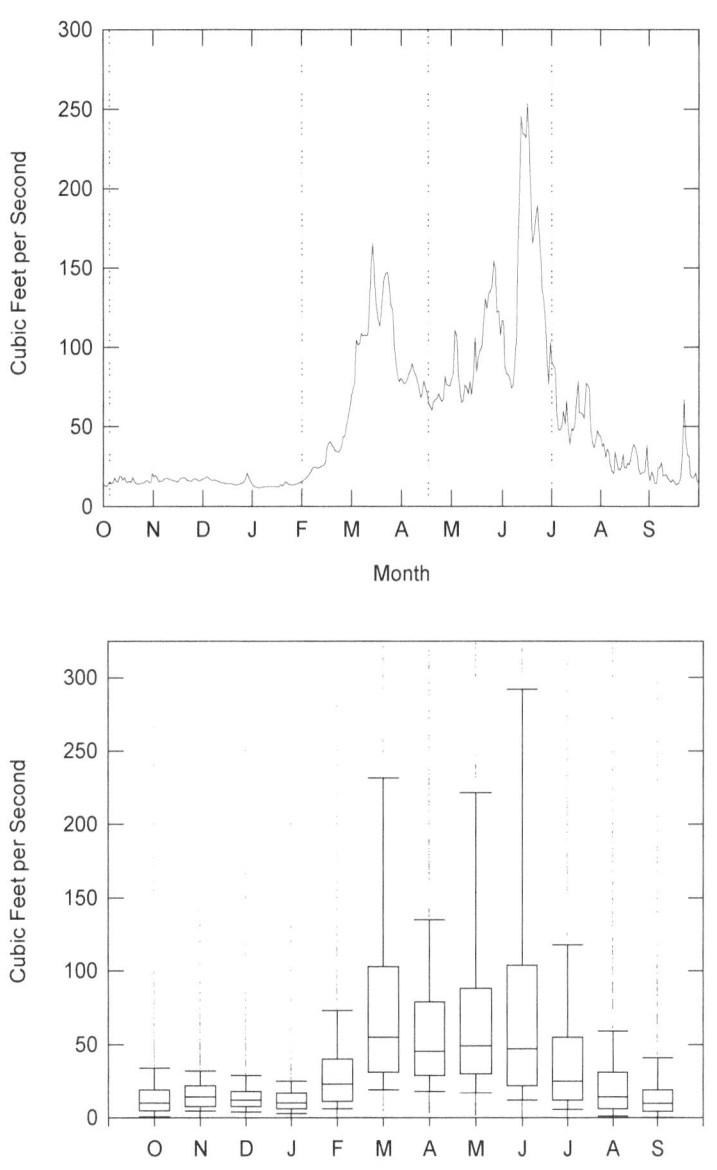

Representative mean annual hydrograph (top) and distr bution of daily flows by month (bottom) for hydrologic season determination. Box and whiskers represent a five number summary; bottom whisker cap is 10th percentile, bottom of box is 25th percentile, internal line is median, top of box is 75th percentile, and top whisker is 90th percentile. Hydrologic seasons for Badlands National Park are: Oct. 1 to Jan. 31, Feb. 1 to Apr. 14, Apr. 15 to Jun. 30, and Jul. 1 to Sep. 30.

CONTACTS FOR AGENCY CODES RETRIEVED FOR BADL

AGENCY	PRIMARY CONTACT NAME	ORGANIZATION	PHONE NUMBER(S)	
112WRD	BRIGGS, JOHN	US GEOLOGICAL SURVEY	(703)648-5624	
31BLHICD	STORET USER ASSISTANCE	USEPA HQ	(202)260-7050	(800)424-9067
* DATA FOR 31BLHICD HAS BEEN 'RETIRED' AT THE REQUEST OF STORET USER ASSISTANCE (703)883-8861 ON 03/14/86				
11NPSWRD	TUCKER, DEAN	NATIONAL PARK SERVICE	(970)225-3516	(970)225-3518
21SDAK01	BARON, LEE	SD DEPT WATER & NAT RES	(605)773-3812	
21SDLASS	REPSYS, ANDREW	SD DENR	(605)773-3696	

QUANTITY OF DATA RETRIEVED FOR BADL BY AGENCY CODE

WITHIN THE ENTIRE STUDY AREA (S.A.) AND JUST WITHIN THE PARK

Agency	Organization	Period of Record		Water Quality Stations		Longer Term Stations[1]		No Data Stations		Water Quality Observations		Water Quality Parameters	
		Study Area	Park Only	S.A.	Park	S.A.	Park	S.A.	Park	S.A.	Park	S.A.	Park
112WRD	US GEOLOGICAL SURVEY	11/29/62-09/10/96	08/13/64-08/28/95	21	7	10	4	0	0	6464	4251	149	119
31BLHICD	USEPA HQ	06/01/71-08/21/72	No Data in Park	2	0	0	0	0	0	119	0	30	0
11NPSWRD	NATIONAL PARK SERVICE	06/29/78-04/23/79	06/29/78-06/29/78	3	1	0	0	0	0	88	18	52	18
21SDAK01	SD DEPT WATER & NAT RES	12/19/77-09/19/78	No Data in Park	3	0	0	0	0	0	161	0	24	0
21SDLASS	SD DENR	07/12/89-08/16/89	No Data in Park	2	0	0	0	0	0	64	0	17	0
Totals		11/29/62-09/10/96	08/13/64-08/28/95	31	8	10	4	0	0	6896	4269	197	128

[1]Station With At Least 6 Parameters Having An Average of 1 Or More Observations Per Year During a Period of Record Extending At Least 2 Years

Station Period of Record Tabulation
From 11/29/62 To 09/10/96

Station Ident	Location Description	In Park	Total Obs	01/01/85 to 09/10/96	01/01/75 to 12/31/84	Before 01/01/75
BADL0001	SPRING NW OF BADLANDS NP HEADQUARTERS	Yes	18	0	18	0
BADL0002	PIKE IN PENNINGTON COUNTY	No	40	0	40	0
BADL0003	N WHITE WATER IN PENN COUNTY	No	81	0	81	0
BADL0004	MISSLE ALLOTMENT IN PENN COUNTY	No	40	0	40	0
BADL0005	42N42W 2BB JOHN POURIER	No	30	0	0	30
BADL0006	NEW WALL-E PENNINGTON CO	No	33	33	0	0
BADL0007	NEW WALL LAKE-E PENNINGTON CO	No	31	31	0	0
BADL0008	42N42W 2DACA	No	207	207	0	0
BADL0009	402462	No	35	0	35	0
BADL0010	401905	No	35	0	35	0
BADL0011	WHITE R NEAR ROCKYFORD SD	Yes	3493	0	0	3493
BADL0012	43N44W35BAD	No	30	0	0	30
BADL0013	41N44W31CBCC	No	206	206	0	0
BADL0014	41N45W24ACCA	Yes	199	199	0	0
BADL0015	41N45W23BBCB	No	204	204	0	0
BADL0016	41N45W27BBAA	No	198	198	0	0
BADL0017	41N45W 6DDCC	Yes	38	38	0	0
BADL0018	41N46W12DACD	Yes	195	195	0	0
BADL0019	41N46W24DCBA	No	205	205	0	0
BADL0020	41N46W25BCA W DAGMAN	No	30	0	0	30
BADL0021	41N46W 9BBCA	Yes	251	251	0	0
BADL0022	40N46W18BBAD	No	38	38	0	0
BADL0023	BATTLE CR ABV CONF WITH CHEYENNE	No	89	0	0	89
BADL0024	42N47W 2ACAA	Yes	37	37	0	0
BADL0025	41N47W23BAAC	Yes	38	38	0	0
BADL0026	BATTLE CR BELOW HERMOSA SD	No	553	553	0	0
BADL0027	CHEYENNE RIVER NR FAIRBURN, SD	No	249	249	0	0
BADL0028	FRENCH CREEK ABV CONF W CHEYENNE	No	30	0	0	30
BADL0029	FRENCH CREEK NEAR RED SHIRT, SD	No	27	27	0	0
BADL0030	41N47W 7DCBA	No	34	34	0	0
BADL0031	42N48W25ABCC	No	202	202	0	0

[1] Longer Term Station With At Least 6 Parameters Having An Average of 1 Or More Observations Per Year During a Period of Record Extending At Least 2 Years

Parameter Period of Record Tabulation
From 11/29/62 To 09/10/96

Parameter Code	Name	Total Obs	01/01/85 to 09/10/96	01/01/75 to 12/31/84	Before 01/01/75	Stations Total	Park
00003	SAMPLING STATION LOCATION, VERTICAL (FEET)	3	0	0	3	2	0
00010	TEMPERATURE, WATER (DEGREES CENTIGRADE)	247	123	7	117	26	7
00011	TEMPERATURE, WATER (DEGREES FAHRENHEIT)	8	0	8	0	3	0
00020	TEMPERATURE, AIR (DEGREES CENTIGRADE)	124	115	5	4	21	6
00021	TEMPERATURE, AIR (DEGREES FAHRENHEIT)	12	4	8	0	5	0
00025	BAROMETRIC PRESSURE (MM OF HG)	21	21	0	0	16	6
00027	CODE NO FOR AGENCY COLLECTING SAMPLE-SEE APPEND	318	126	0	192	21	7
00028	CODE NO FOR AGENCY ANALYZING SAMPLE (SEE APPEND)	318	126	0	192	21	7
00060	FLOW, STREAM, MEAN DAILY CFS	193	0	0	193	3	1
00061	FLOW, STREAM, INSTANTANEOUS CFS	121	121	0	0	13	4
00063	SAMPLING POINTS, NUMBER OF IN A CROSS SECTION	3	0	0	3	1	1
00065	STAGE, STREAM (FEET)	20	20	0	0	6	2
00070	TURBIDITY, (JACKSON CANDLE UNITS)	4	0	0	4	2	0
00076	TURBIDITY,HACH TURBIDIMETER (FORMAZIN TURB UNIT)	2	2	0	0	1	0
00078	TRANSPARENCY, SECCHI DISC (METERS)	2	2	0	0	1	0
00080	COLOR (PLATINUM-COBALT UNITS)	36	0	0	36	1	1
00095	SPECIFIC CONDUCTANCE (UMHOS/CM @ 25C)	238	122	8	108	28	7
00300	OXYGEN, DISSOLVED MG/L	43	28	8	7	19	4
00400	PH (STANDARD UNITS)	172	54	10	108	29	7
00403	PH, LAB, STANDARD UNITS SU	63	54	9	0	22	7
00405	CARBON DIOXIDE (MG/L AS CO2)	3	0	0	3	3	0
00410	ALKALINITY, TOTAL (MG/L AS CACO3)	113	12	12	89	19	1
00411	ALKALINITY,METHYLORANGE MG/L	1	0	1	0	1	1
00415	ALKALINITY, PHENOLPHTHALEIN (MG/L)	4	4	0	0	2	0
00419	ALKALINITY,CARBONATE,INCREMENTAL TITR FIELD MG/L	4	4	0	0	4	1
00430	ALKALINITY, CARBONATE (MG/L AS CACO3)	4	0	0	4	2	0
00440	BICARBONATE ION (MG/L AS HCO3)	110	6	0	104	10	1
00445	CARBONATE ION (MG/L AS CO3)	105	1	0	104	5	1
00447	CARBONATE,INCREMENTAL TITRATION,(CO3) FIELD MG/L	3	3	0	0	3	1
00450	BICARBONATE,INCREMENTAL TITRATION,(HCO3) FIELDMG/L	3	3	0	0	3	1
00452	CARBONATE,WATER,DISS,INCR TIT, FIELD, AS CO3, MG/L	4	4	0	0	3	1
00453	BICARBONATE,WATER,DISS,INCR TIT,FIELD,AS HCO3,MG/L	8	8	0	0	7	2
00500	RESIDUE, TOTAL (MG/L)	16	4	8	4	7	0
00505	RESIDUE, TOTAL VOLATILE (MG/L)	4	0	0	4	2	0
00515	RESIDUE, TOTAL FILTRABLE (DRIED AT 105C),MG/L	12	0	8	4	5	0
00520	RESIDUE, VOLATILE FILTRABLE (MG/L)	5	0	1	4	3	1
00530	RESIDUE, TOTAL NONFILTRABLE (MG/L)	14	6	8	0	6	0
00535	RESIDUE, VOLATILE NONFILTRABLE (MG/L)	1	0	1	0	1	1
00608	NITROGEN, AMMONIA, DISSOLVED (MG/L AS N)	32	30	2	0	11	3
00610	NITROGEN, AMMONIA, TOTAL (MG/L AS N)	10	4	6	0	5	0
00613	NITRITE NITROGEN, DISSOLVED (MG/L AS N)	28	20	8	0	13	3
00618	NITRATE NITROGEN, DISSOLVED (MG/L AS N)	4	0	1	3	4	1
00623	NITROGEN, KJELDAHL, DISSOLVED (MG/L AS N)	30	30	0	0	10	3
00625	NITROGEN, KJELDAHL, TOTAL, (MG/L AS N)	12	4	8	0	5	0
00630	NITRITE PLUS NITRATE, TOTAL 1 DET (MG/L AS N)	16	4	8	4	7	0
00631	NITRITE PLUS NITRATE, DISS 1 DET (MG/L AS N)	34	34	0	0	11	3
00650	PHOSPHATE, TOTAL (MG/L AS PO4)	16	0	1	15	4	2
00660	PHOSPHATE, ORTHO (MG/L AS PO4)	5	0	1	4	3	1
00665	PHOSPHORUS, TOTAL (MG/L AS P)	6	4	2	0	4	0
00666	PHOSPHORUS, DISSOLVED (MG/L AS P)	30	30	0	0	10	3
00671	PHOSPHORUS, DISSOLVED ORTHOPHOSPHATE (MG/L AS P)	29	24	5	0	15	3
00723	CYANIDE, DISSOLVED STD METHOD (UG/L)	1	1	0	0	1	0
00900	HARDNESS, TOTAL (MG/L AS CACO3)	109	0	1	108	7	2
00901	HARDNESS, CARBONATE (MG/L AS CACO3)	5	0	1	4	3	1
00902	HARDNESS, NON-CARBONATE (MG/L AS CACO3)	108	0	0	108	6	1
00915	CALCIUM, DISSOLVED (MG/L AS CA)	161	50	1	110	23	8
00916	CALCIUM, TOTAL (MG/L AS CA)	2	0	2	0	2	0
00925	MAGNESIUM, DISSOLVED (MG/L AS MG)	161	50	1	110	23	8
00929	SODIUM, TOTAL (MG/L AS NA)	2	0	2	0	2	0
00930	SODIUM, DISSOLVED (MG/L AS NA)	161	50	1	110	23	8
00931	SODIUM ADSORPTION RATIO	104	0	0	104	4	1
00932	SODIUM, PERCENT	85	0	0	85	4	1
00935	POTASSIUM, DISSOLVED (MG/L AS K)	158	49	1	108	23	8
00937	POTASSIUM, TOTAL MG/L AS K)	2	0	2	0	2	0
00940	CHLORIDE,TOTAL IN WATER MG/L	166	50	8	108	25	7
00941	CHLORIDE, DISSOLVED IN WATER MG/L	1	0	1	0	1	1
00945	SULFATE, TOTAL (MG/L AS SO4)	166	50	8	108	27	7
00946	SULFATE, DISSOLVED (MG/L AS SO4)	3	0	3	0	2	1
00950	FLUORIDE, DISSOLVED (MG/L AS F)	147	43	0	104	18	7
00955	SILICA, DISSOLVED (MG/L AS SI02)	154	44	0	110	21	7
01000	ARSENIC, DISSOLVED (UG/L AS AS)	50	50	0	0	16	6

Parameter Period of Record Tabulation
From 11/29/62 To 09/10/96

Parameter Code	Name	Total Obs	01/01/85 to 09/10/96	01/01/75 to 12/31/84	Before 01/01/75	Stations Total	Park
01002	ARSENIC, TOTAL (UG/L AS AS)	4	2	2	0	3	0
01005	BARIUM, DISSOLVED (UG/L AS BA)	44	44	0	0	15	6
01007	BARIUM, TOTAL (UG/L AS BA)	2	0	2	0	2	0
01010	BERYLLIUM, DISSOLVED (UG/L AS BE)	44	44	0	0	15	6
01012	BERYLLIUM, TOTAL (UG/L AS BE)	2	0	2	0	2	0
01020	BORON, DISSOLVED (UG/L AS B)	153	49	0	104	19	6
01022	BORON, TOTAL (UG/L AS B)	2	0	2	0	2	0
01025	CADMIUM, DISSOLVED (UG/L AS CD)	50	50	0	0	16	6
01027	CADMIUM, TOTAL (UG/L AS CD)	35	35	0	0	14	6
01030	CHROMIUM, DISSOLVED (UG/L AS CR)	50	50	0	0	16	6
01034	CHROMIUM, TOTAL (UG/L AS CR)	3	0	2	1	3	1
01035	COBALT, DISSOLVED (UG/L AS CO)	43	43	0	0	15	6
01037	COBALT, TOTAL (UG/L AS CO)	2	0	2	0	2	0
01040	COPPER, DISSOLVED (UG/L AS CU)	51	50	0	1	17	7
01042	COPPER, TOTAL (UG/L AS CU)	2	0	2	0	2	0
01045	IRON, TOTAL (UG/L AS FE)	8	2	2	4	5	0
01046	IRON, DISSOLVED (UG/L AS FE)	45	44	1	0	16	7
01049	LEAD, DISSOLVED (UG/L AS PB)	50	50	0	0	16	6
01055	MANGANESE, TOTAL (UG/L AS MN)	40	0	2	38	5	1
01056	MANGANESE, DISSOLVED (UG/L AS MN)	44	43	1	0	15	7
01057	THALLIUM, DISSOLVED (UG/L AS TL)	18	18	0	0	9	3
01060	MOLYBDENUM, DISSOLVED (UG/L AS MO)	49	49	0	0	16	6
01062	MOLYBDENUM, TOTAL (UG/L AS MO)	2	0	2	0	2	0
01065	NICKEL, DISSOLVED (UG/L AS NI)	43	43	0	0	15	6
01067	NICKEL, TOTAL (UG/L AS NI)	2	0	2	0	2	0
01075	SILVER, DISSOLVED (UG/L AS AG)	43	43	0	0	15	6
01077	SILVER, TOTAL (UG/L AS AG)	2	0	2	0	2	0
01080	STRONTIUM, DISSOLVED (UG/L AS SR)	43	43	0	0	15	6
01082	STRONTIUM, TOTAL (UG/L AS SR)	2	0	2	0	2	0
01085	VANADIUM, DISSOLVED (UG/L AS V)	49	49	0	0	16	6
01087	VANADIUM, TOTAL (UG/L AS V)	2	0	2	0	2	0
01090	ZINC, DISSOLVED (UG/L AS ZN)	51	50	0	1	17	7
01092	ZINC, TOTAL (UG/L AS ZN)	4	2	2	0	3	0
01095	ANTIMONY, DISSOLVED (UG/L AS SB)	2	2	0	0	1	0
01105	ALUMINUM, TOTAL (UG/L AS AL)	2	0	2	0	2	0
01106	ALUMINUM, DISSOLVED (UG/L AS AL)	2	2	0	0	1	0
01112	CERIUM, TOTAL (UG/L AS CE)	2	0	2	0	2	0
01130	LITHIUM, DISSOLVED (UG/L AS LI)	44	44	0	0	15	6
01132	LITHIUM, TOTAL (UG/L AS LI)	2	0	2	0	2	0
01142	SILICON, TOTAL (UG/L AS SI)	2	0	2	0	2	0
01145	SELENIUM, DISSOLVED (UG/L AS SE)	47	43	0	4	16	4
01147	SELENIUM, TOTAL (UG/L AS SE)	4	2	2	0	3	0
01152	TITANIUM, TOTAL (UG/L AS TI)	2	0	2	0	2	0
01189	SCANDIUM, TOTAL (UG/L AS SC)	2	0	2	0	2	0
01203	YTTRIUM, TOTAL (UG/L AS Y)	2	0	2	0	2	0
01239	NIOBIUM, TOTAL UG/L	2	0	2	0	2	0
03515	BETA, DISSOLVED GROSS, AS CS-137, PC/L	12	12	0	0	10	3
04126	ALPHA, DISSOLVED, WATER (AS TH-230) PCI/L	12	12	0	0	10	3
09510	RADIUM 226, DISSOLVED, PLANCHET COUNT	10	10	0	0	9	3
22703	URANIUM, NATURAL, DISSOLVED	18	18	0	0	11	3
31501	COLIFORM,TOT,MEMBRANE FILTER,IMMED M-ENDO MED,35C	17	17	0	0	9	3
31503	COLIFORM,TOT,MEMBR FILTER,DELAYED,M-ENDO MED,35 C	6	6	0	0	6	2
31504	COLIFORM,TOT,MEMBR FILTER,IMMED,LES ENDO AGAR,35C	3	3	0	0	3	1
31616	FECAL COLIFORM,MEMBR FILTER,M-FC BROTH,44 5 C	7	0	7	0	3	0
31625	FECAL COLIFORM, MF,M-FC, 0 7 UM	27	27	0	0	10	3
31633	E COLI,THERMOTOL,MF,M-TEC,IN SITU UREASE #/100ML	25	25	0	0	9	3
31673	FECAL STREPTOCOCCI, MBR FILT,KF AGAR,35C,48HR	27	27	0	0	10	3
32730	PHENOLICS, TOTAL, RECOVERABLE (UG/L)	47	2	0	45	2	1
39024	PROPAZINE,COULSON CONDUCTIVITY,WATER SAMPL(UG/L)	4	4	0	0	1	0
39030	TREFLAN, MICROCOULOMETRIC,WATER SAMPLE (UG/L)	4	4	0	0	1	0
39051	METHOMYL IN WHOLE WATER (UG/L)	4	4	0	0	1	0
39052	PROPHAM IN WHOLE WATER (UG/L)	4	4	0	0	1	0
39054	SIMETRYNE IN WHOLE WATER (UG/L)	4	4	0	0	1	0
39055	SIMAZINE IN WHOLE WATER (UG/L)	4	4	0	0	1	0
39056	PROMETONE IN WHOLE WATER (UG/L)	4	4	0	0	1	0
39057	PROMETRYNE IN WHOLE WATER (UG/L)	4	4	0	0	1	0
39086	ALKALINITY,WATER,DISS,INCR TIT,FIELD,AS CACO3,MG/L	8	8	0	0	7	2
39630	ATRAZINE(AATREX) IN WHOLE WATER SAMPLE (UG/L)	4	4	0	0	1	0
39750	SEVIN IN WHOLE WATER SAMPLE (UG/L)	4	4	0	0	1	0
70300	RESIDUE,TOTAL FILTRABLE (DRIED AT 180C),MG/L	158	54	0	104	22	7
70301	SOLIDS, DISSOLVED-SUM OF CONSTITUENTS (MG/L)	3	0	0	3	3	0

Parameter Period of Record Tabulation
From 11/29/62 To 09/10/96

Parameter Code	Name	Total Obs	01/01/85 to 09/10/96	01/01/75 to 12/31/84	Before 01/01/75	Stations Total	Park
70302	SOLIDS, DISSOLVED-TONS PER DAY	101	0	0	101	1	1
70303	SOLIDS, DISSOLVED-TONS PER ACRE-FT	101	0	0	101	1	1
70331	SUSPENDED SED SIEVE DIAMETER,% FINER THAN 062MM	21	0	0	21	1	1
70337	SUS SED FALL DIA(DISTLD WATER)%FINER THAN 002MM	31	0	0	31	1	1
70338	SUS SED FALL DIA(DISTLD WATER)%FINER THAN 004MM	31	0	0	31	1	1
70339	SUS SED FALL DIA(DISTLD WATER)%FINER THAN 008MM	16	0	0	16	1	1
70340	SUS SED FALL DIA(DISTLD WATER)%FINER THAN 016MM	30	0	0	30	1	1
70341	SUS SED FALL DIA(DISTLD WATER)%FINER THAN 031MM	15	0	0	15	1	1
70342	SUS SED FALL DIA(DISTLD WATER)%FINER THAN 062MM	17	0	0	17	1	1
70343	SUS SED FALL DIA(DISTLD WATER)%FINER THAN 125MM	10	0	0	10	1	1
70344	SUS SED FALL DIA(DISTLD WATER)%FINER THAN 250MM	7	0	0	7	1	1
70345	SUS SED FALL DIA(DISTLD WATER)%FINER THAN 500MM	5	0	0	5	1	1
70346	SUS SED FALL DIA(DISTLD WATER)%FINER THAN 1 00MM	1	0	0	1	1	1
70347	SUS SED FALL DIA(DISTLD WATER)%FINER THAN 2 00MM	1	0	0	1	1	1
70505	PHOSPHATE,TOTAL,COLORIMETRIC METHOD (MG/L AS P)	8	0	8	0	3	0
70507	PHOSPHORUS,IN TOTAL ORTHOPHOSPHATE (MG/L AS P)	3	0	3	0	3	0
71832	HYDROXIDE,INCREMENTAL TITRATION,(OH) FIELD MG/L	1	1	0	0	1	0
71846	NITROGEN, AMMONIA, DISSOLVED (MG/L AS NH4)	18	0	0	18	1	1
71851	NITRATE NITROGEN, DISSOLVED (MG/L AS NO3)	104	0	0	104	4	1
71856	NITRITE NITROGEN, DISSOLVED (MG/L AS NO2)	14	0	0	14	1	1
71870	BROMIDE (MG/L AS BR)	19	0	0	19	1	1
71883	MANGANESE, TOTAL ELEMENTAL (UG/L AS MN)	3	0	0	3	3	0
71885	IRON (UG/L AS FE)	59	0	0	59	4	1
71890	MERCURY, DISSOLVED (UG/L AS HG)	50	50	0	0	16	6
72000	ELEVATION OF LAND SURFACE DATUM (FT ABOVE MSL)	3	0	0	3	3	0
75986	ALPHA GROSS,1 SIGMA PRC EST AS NAT U,DISS,WTR UG/L	12	12	0	0	10	3
75987	ALPHA GROSS,DISS,1 SIGMA PRC EST AS TH230,WTR PC/L	12	12	0	0	10	3
75988	BETA GROSS,DISS,1 SIGMA PRC EST AS SR90/Y90 PC/L	12	12	0	0	10	3
75989	BETA GROSS,1 SIGMA PRC EST AS CS-137,DISS,WTR PC/L	12	12	0	0	10	3
75990	URANIUM,NATURAL,1 SIGMA PRC EST,DISS,WATER UG/L	14	14	0	0	11	3
76001	RADIUM 226,1 SIGMA PRC EST,DISSOLVED,WATER PC/L	10	10	0	0	9	3
77825	ALACHLOR WHOLE WATER,UG/L	4	4	0	0	1	0
80030	ALPHA,DISSOLVED GROSS,AS URANIUM-NATURAL,UG/L	12	12	0	0	10	3
80050	BETA,DISSOLVED GROSS,AS SR-Y-90, PC/L	12	12	0	0	10	3
80154	SUSP SEDIMENT CONCENTRATION-EVAP AT 110C (MG/L)	87	1	0	86	2	1
80155	SUSPENDED SEDIMENT DISCHARGE (TONS/DAY)	86	0	0	86	1	1
80158	BED MATERIAL FALL DIAMETER, % FINER THAN 062MM	1	0	0	1	1	1
80159	BED MATERIAL FALL DIAMETER, % FINER THAN 125MM	1	0	0	1	1	1
80160	BED MATERIAL FALL DIAMETER, % FINER THAN 250MM	2	0	0	2	1	1
80161	BED MATERIAL FALL DIAMETER, % FINER THAN 500MM	2	0	0	2	1	1
80162	BED MATERIAL FALL DIAMETER, % FINER THAN 1 00MM	2	0	0	2	1	1
80163	BED MATERIAL FALL DIAMETER, % FINER THAN 2 00MM	2	0	0	2	1	1
80170	BED MATERIAL SIEVE DIAMETER,% FINER THAN 4 00MM	3	0	0	3	1	1
80171	BED MATERIAL SIEVE DIAMETER,% FINER THAN 8 00MM	3	0	0	3	1	1
80172	BED MATERIAL SIEVE DIAMETER,% FINER THAN 16 0MM	3	0	0	3	1	1
80173	BED MATERIAL SIEVE DIAMETER,% FINER THAN 32 0MM	3	0	0	3	1	1
81757	CYANAZINE IN THE WHOLE WATER SAMPLE UG/L	4	4	0	0	1	0
82033	MAGNESIUM - TOTAL UG/L(AS MG)	2	0	2	0	2	0
82184	AMETRYNE (GESAPAX OR EVIK) TOTAL UG/L	4	4	0	0	1	0
82233	SILICON (SI) TOTAL IN WATER MG/L AS (SIO2)	1	0	1	0	1	1
82364	THORIUM, TOTAL IN WATER UG/L	2	0	2	0	2	0
82611	METRIBUZIN, WHOLE WATER, TOTAL RECOVERABLE UG/L	4	4	0	0	1	0
82612	METOLACHLOR, WHOLE WATER, TOTAL RECOVERABLE UG/L	4	4	0	0	1	0
84000	GEOLOGIC AGE CODE (SEE USGS CATALOG)	16	13	0	3	10	2
84001	AQUIFER NAME CODE (SEE USGS CATALOG)	16	13	0	3	10	2

Station	In Park	Code	Name	Start - End	Years	Obs	Plots[1]
BADL0023	No	00003	SAMPLING STATION LOCATION, VERTICAL (FEET)	06/01/71-07/29/71	0	2	
BADL0028	No	00003	SAMPLING STATION LOCATION, VERTICAL (FEET)	06/03/71-06/03/71	0	1	
BADL0002	No	00010	TEMPERATURE, WATER (DEGREES CENTIGRADE)	06/08/78-06/08/78	0	1	
BADL0003	No	00010	TEMPERATURE, WATER (DEGREES CENTIGRADE)	12/19/77-06/08/78	0	3	
BADL0004	No	00010	TEMPERATURE, WATER (DEGREES CENTIGRADE)	06/08/78-06/08/78	0	1	
BADL0006	No	00010	TEMPERATURE, WATER (DEGREES CENTIGRADE)	07/12/89-08/16/89	0	2	
BADL0007	No	00010	TEMPERATURE, WATER (DEGREES CENTIGRADE)	07/12/89-08/16/89	0	2	
BADL0008	No	00010	TEMPERATURE, WATER (DEGREES CENTIGRADE)	11/12/92-07/11/95	2	4	
BADL0009	No	00010	TEMPERATURE, WATER (DEGREES CENTIGRADE)	04/23/79-04/23/79	0	1	
BADL0010	No	00010	TEMPERATURE, WATER (DEGREES CENTIGRADE)	04/18/79-04/18/79	0	1	
BADL0011	Yes	00010	TEMPERATURE, WATER (DEGREES CENTIGRADE)	08/13/64-09/19/67	3	113	
BADL0013	No	00010	TEMPERATURE, WATER (DEGREES CENTIGRADE)	09/23/92-06/21/95	2	4	
BADL0014	Yes	00010	TEMPERATURE, WATER (DEGREES CENTIGRADE)	09/22/92-08/28/95	2	4	
BADL0015	No	00010	TEMPERATURE, WATER (DEGREES CENTIGRADE)	09/25/92-08/28/95	2	4	
BADL0016	No	00010	TEMPERATURE, WATER (DEGREES CENTIGRADE)	09/30/92-07/10/95	2	4	
BADL0017	Yes	00010	TEMPERATURE, WATER (DEGREES CENTIGRADE)	09/21/92-09/21/92	0	1	
BADL0018	Yes	00010	TEMPERATURE, WATER (DEGREES CENTIGRADE)	09/21/92-06/21/95	2	4	
BADL0019	No	00010	TEMPERATURE, WATER (DEGREES CENTIGRADE)	10/08/92-08/29/95	2	3	
BADL0021	Yes	00010	TEMPERATURE, WATER (DEGREES CENTIGRADE)	10/01/92-07/13/95	2	4	
BADL0022	No	00010	TEMPERATURE, WATER (DEGREES CENTIGRADE)	09/29/92-09/29/92	0	1	
BADL0023	No	00010	TEMPERATURE, WATER (DEGREES CENTIGRADE)	06/01/71-08/21/72	1	3	
BADL0024	Yes	00010	TEMPERATURE, WATER (DEGREES CENTIGRADE)	09/30/92-09/30/92	0	1	
BADL0025	Yes	00010	TEMPERATURE, WATER (DEGREES CENTIGRADE)	09/18/92-09/18/92	0	1	
BADL0026	No	00010	TEMPERATURE, WATER (DEGREES CENTIGRADE)	03/09/89-09/10/96	7	71	
BADL0027	No	00010	TEMPERATURE, WATER (DEGREES CENTIGRADE)	05/06/88-09/08/94	6	6	
BADL0028	No	00010	TEMPERATURE, WATER (DEGREES CENTIGRADE)	06/03/71-06/03/71	0	1	
BADL0029	No	00010	TEMPERATURE, WATER (DEGREES CENTIGRADE)	06/30/95-11/21/95	0	4	
BADL0031	No	00010	TEMPERATURE, WATER (DEGREES CENTIGRADE)	11/05/92-07/12/95	2	3	
BADL0002	No	00011	TEMPERATURE, WATER (DEGREES FAHRENHEIT)	06/08/78-09/19/78	0	2	
BADL0003	No	00011	TEMPERATURE, WATER (DEGREES FAHRENHEIT)	12/19/77-09/19/78	0	4	
BADL0004	No	00011	TEMPERATURE, WATER (DEGREES FAHRENHEIT)	06/08/78-09/19/78	0	2	
BADL0002	No	00020	TEMPERATURE, AIR (DEGREES CENTIGRADE)	06/08/78-06/08/78	0	1	
BADL0003	No	00020	TEMPERATURE, AIR (DEGREES CENTIGRADE)	12/19/77-06/08/78	0	3	
BADL0004	No	00020	TEMPERATURE, AIR (DEGREES CENTIGRADE)	06/08/78-06/08/78	0	1	
BADL0008	No	00020	TEMPERATURE, AIR (DEGREES CENTIGRADE)	11/12/92-07/11/95	2	4	
BADL0013	No	00020	TEMPERATURE, AIR (DEGREES CENTIGRADE)	09/23/92-06/21/95	2	4	
BADL0014	Yes	00020	TEMPERATURE, AIR (DEGREES CENTIGRADE)	09/22/92-08/28/95	2	4	
BADL0015	No	00020	TEMPERATURE, AIR (DEGREES CENTIGRADE)	09/25/92-08/28/95	2	4	
BADL0016	No	00020	TEMPERATURE, AIR (DEGREES CENTIGRADE)	09/30/92-07/10/95	2	4	
BADL0017	Yes	00020	TEMPERATURE, AIR (DEGREES CENTIGRADE)	09/21/92-09/21/92	0	1	
BADL0018	Yes	00020	TEMPERATURE, AIR (DEGREES CENTIGRADE)	09/21/92-06/21/95	2	4	
BADL0019	No	00020	TEMPERATURE, AIR (DEGREES CENTIGRADE)	10/08/92-08/29/95	2	3	
BADL0021	Yes	00020	TEMPERATURE, AIR (DEGREES CENTIGRADE)	10/01/92-07/13/95	2	3	
BADL0022	No	00020	TEMPERATURE, AIR (DEGREES CENTIGRADE)	09/29/92-09/29/92	0	1	
BADL0023	No	00020	TEMPERATURE, AIR (DEGREES CENTIGRADE)	06/01/71-08/21/72	1	3	
BADL0024	Yes	00020	TEMPERATURE, AIR (DEGREES CENTIGRADE)	09/30/92-09/30/92	0	1	
BADL0025	Yes	00020	TEMPERATURE, AIR (DEGREES CENTIGRADE)	09/18/92-09/18/92	0	1	
BADL0026	No	00020	TEMPERATURE, AIR (DEGREES CENTIGRADE)	03/09/89-09/10/96	7	68	
BADL0027	No	00020	TEMPERATURE, AIR (DEGREES CENTIGRADE)	05/06/88-09/08/94	6	6	
BADL0028	No	00020	TEMPERATURE, AIR (DEGREES CENTIGRADE)	06/03/71-06/03/71	0	1	
BADL0029	No	00020	TEMPERATURE, AIR (DEGREES CENTIGRADE)	06/30/95-07/23/96	1	4	
BADL0031	No	00020	TEMPERATURE, AIR (DEGREES CENTIGRADE)	11/05/92-07/12/95	2	3	
BADL0002	No	00021	TEMPERATURE, AIR (DEGREES FAHRENHEIT)	06/08/78-09/19/78	0	2	
BADL0003	No	00021	TEMPERATURE, AIR (DEGREES FAHRENHEIT)	12/19/77-09/19/78	0	4	
BADL0004	No	00021	TEMPERATURE, AIR (DEGREES FAHRENHEIT)	06/08/78-09/19/78	0	2	
BADL0006	No	00021	TEMPERATURE, AIR (DEGREES FAHRENHEIT)	07/12/89-08/16/89	0	2	
BADL0007	No	00021	TEMPERATURE, AIR (DEGREES FAHRENHEIT)	07/12/89-08/16/89	0	2	
BADL0008	No	00025	BAROMETRIC PRESSURE (MM OF HG)	11/12/92-08/02/94	1	2	
BADL0013	No	00025	BAROMETRIC PRESSURE (MM OF HG)	09/23/92-09/23/92	0	1	
BADL0014	Yes	00025	BAROMETRIC PRESSURE (MM OF HG)	09/22/92-09/22/92	0	1	
BADL0015	No	00025	BAROMETRIC PRESSURE (MM OF HG)	09/25/92-09/25/92	0	1	
BADL0016	No	00025	BAROMETRIC PRESSURE (MM OF HG)	09/30/92-09/30/92	0	1	
BADL0017	Yes	00025	BAROMETRIC PRESSURE (MM OF HG)	09/21/92-09/21/92	0	1	
BADL0018	Yes	00025	BAROMETRIC PRESSURE (MM OF HG)	09/21/92-07/14/94	1	2	
BADL0019	No	00025	BAROMETRIC PRESSURE (MM OF HG)	10/08/92-10/08/92	0	1	
BADL0021	Yes	00025	BAROMETRIC PRESSURE (MM OF HG)	10/01/92-10/01/92	0	1	
BADL0022	No	00025	BAROMETRIC PRESSURE (MM OF HG)	09/29/92-09/29/92	0	1	
BADL0024	Yes	00025	BAROMETRIC PRESSURE (MM OF HG)	09/30/92-09/30/92	0	1	
BADL0025	Yes	00025	BAROMETRIC PRESSURE (MM OF HG)	09/18/92-09/18/92	0	1	
BADL0026	No	00025	BAROMETRIC PRESSURE (MM OF HG)	09/07/93-08/30/94	0	2	
BADL0027	No	00025	BAROMETRIC PRESSURE (MM OF HG)	04/20/94-09/08/94	0	2	
BADL0030	No	00025	BAROMETRIC PRESSURE (MM OF HG)	11/04/92-11/04/92	0	1	

[1]T=Times Series Plot, A=Annual Plot, and S=Seasonal Plot

Station	In Park	Code	Name	Start - End	Years	Obs	Plots[1]
BADL0031	No	00025	BAROMETRIC PRESSURE (MM OF HG)	11/05/92-08/03/94	1	2	
BADL0005	No	00027	CODE NO FOR AGENCY COLLECTING SAMPLE-SEE APPEND	11/30/62-11/30/62	0	1	
BADL0008	No	00027	CODE NO FOR AGENCY COLLECTING SAMPLE-SEE APPEND	11/12/92-07/11/95	2	4	
BADL0011	Yes	00027	CODE NO FOR AGENCY COLLECTING SAMPLE-SEE APPEND	08/13/64-09/19/67	3	189	
BADL0012	No	00027	CODE NO FOR AGENCY COLLECTING SAMPLE-SEE APPEND	11/30/62-11/30/62	0	1	
BADL0013	No	00027	CODE NO FOR AGENCY COLLECTING SAMPLE-SEE APPEND	09/23/92-06/21/95	2	4	
BADL0014	Yes	00027	CODE NO FOR AGENCY COLLECTING SAMPLE-SEE APPEND	09/22/92-08/28/95	2	4	
BADL0015	No	00027	CODE NO FOR AGENCY COLLECTING SAMPLE-SEE APPEND	09/25/92-08/28/95	2	4	
BADL0016	No	00027	CODE NO FOR AGENCY COLLECTING SAMPLE-SEE APPEND	09/30/92-07/10/95	2	4	
BADL0017	Yes	00027	CODE NO FOR AGENCY COLLECTING SAMPLE-SEE APPEND	09/21/92-09/21/92	0	1	
BADL0018	Yes	00027	CODE NO FOR AGENCY COLLECTING SAMPLE-SEE APPEND	09/21/92-06/21/95	2	4	
BADL0019	No	00027	CODE NO FOR AGENCY COLLECTING SAMPLE-SEE APPEND	10/08/92-08/29/95	2	4	
BADL0020	No	00027	CODE NO FOR AGENCY COLLECTING SAMPLE-SEE APPEND	11/29/62-11/29/62	0	1	
BADL0021	Yes	00027	CODE NO FOR AGENCY COLLECTING SAMPLE-SEE APPEND	10/01/92-07/13/95	2	5	
BADL0022	No	00027	CODE NO FOR AGENCY COLLECTING SAMPLE-SEE APPEND	09/29/92-09/29/92	0	1	
BADL0024	Yes	00027	CODE NO FOR AGENCY COLLECTING SAMPLE-SEE APPEND	09/30/92-09/30/92	0	1	
BADL0025	Yes	00027	CODE NO FOR AGENCY COLLECTING SAMPLE-SEE APPEND	09/18/92-09/18/92	0	1	
BADL0026	No	00027	CODE NO FOR AGENCY COLLECTING SAMPLE-SEE APPEND	03/09/89-09/10/96	7	73	
BADL0027	No	00027	CODE NO FOR AGENCY COLLECTING SAMPLE-SEE APPEND	05/06/88-09/08/94	6	6	
BADL0029	No	00027	CODE NO FOR AGENCY COLLECTING SAMPLE-SEE APPEND	06/30/95-07/23/96	1	5	
BADL0030	No	00027	CODE NO FOR AGENCY COLLECTING SAMPLE-SEE APPEND	11/04/92-11/04/92	0	1	
BADL0031	No	00027	CODE NO FOR AGENCY COLLECTING SAMPLE-SEE APPEND	11/05/92-07/12/95	2	4	
BADL0005	No	00028	CODE NO FOR AGENCY ANALYZING SAMPLE (SEE APPEND)	11/30/62-11/30/62	0	1	
BADL0008	No	00028	CODE NO FOR AGENCY ANALYZING SAMPLE (SEE APPEND)	11/12/92-07/11/95	2	4	
BADL0011	Yes	00028	CODE NO FOR AGENCY ANALYZING SAMPLE (SEE APPEND)	08/13/64-09/19/67	3	189	
BADL0012	No	00028	CODE NO FOR AGENCY ANALYZING SAMPLE (SEE APPEND)	11/30/62-11/30/62	0	1	
BADL0013	No	00028	CODE NO FOR AGENCY ANALYZING SAMPLE (SEE APPEND)	09/23/92-06/21/95	2	4	
BADL0014	Yes	00028	CODE NO FOR AGENCY ANALYZING SAMPLE (SEE APPEND)	09/22/92-08/28/95	2	4	
BADL0015	No	00028	CODE NO FOR AGENCY ANALYZING SAMPLE (SEE APPEND)	09/25/92-08/28/95	2	4	
BADL0016	No	00028	CODE NO FOR AGENCY ANALYZING SAMPLE (SEE APPEND)	09/30/92-07/10/95	2	4	
BADL0017	Yes	00028	CODE NO FOR AGENCY ANALYZING SAMPLE (SEE APPEND)	09/21/92-09/21/92	0	1	
BADL0018	Yes	00028	CODE NO FOR AGENCY ANALYZING SAMPLE (SEE APPEND)	09/21/92-06/21/95	2	4	
BADL0019	No	00028	CODE NO FOR AGENCY ANALYZING SAMPLE (SEE APPEND)	10/08/92-08/29/95	2	4	
BADL0020	No	00028	CODE NO FOR AGENCY ANALYZING SAMPLE (SEE APPEND)	11/29/62-11/29/62	0	1	
BADL0021	Yes	00028	CODE NO FOR AGENCY ANALYZING SAMPLE (SEE APPEND)	10/01/92-07/13/95	2	5	
BADL0022	No	00028	CODE NO FOR AGENCY ANALYZING SAMPLE (SEE APPEND)	09/29/92-09/29/92	0	1	
BADL0024	Yes	00028	CODE NO FOR AGENCY ANALYZING SAMPLE (SEE APPEND)	09/30/92-09/30/92	0	1	
BADL0025	Yes	00028	CODE NO FOR AGENCY ANALYZING SAMPLE (SEE APPEND)	09/18/92-09/18/92	0	1	
BADL0026	No	00028	CODE NO FOR AGENCY ANALYZING SAMPLE (SEE APPEND)	03/09/89-09/10/96	7	73	
BADL0027	No	00028	CODE NO FOR AGENCY ANALYZING SAMPLE (SEE APPEND)	05/06/88-09/08/94	6	6	
BADL0029	No	00028	CODE NO FOR AGENCY ANALYZING SAMPLE (SEE APPEND)	06/30/95-07/23/96	1	5	
BADL0030	No	00028	CODE NO FOR AGENCY ANALYZING SAMPLE (SEE APPEND)	11/04/92-11/04/92	0	1	
BADL0031	No	00028	CODE NO FOR AGENCY ANALYZING SAMPLE (SEE APPEND)	11/05/92-07/12/95	2	4	
BADL0011	Yes	00060	FLOW, STREAM, MEAN DAILY CFS	08/13/64-09/19/67	3	189	T
BADL0023	No	00060	FLOW, STREAM, MEAN DAILY CFS	06/01/71-08/21/72	1	3	
BADL0028	No	00060	FLOW, STREAM, MEAN DAILY CFS	06/03/71-06/03/71	0	1	
BADL0008	No	00061	FLOW, STREAM, INSTANTANEOUS CFS	11/12/92-07/11/95	2	4	
BADL0013	No	00061	FLOW, STREAM, INSTANTANEOUS CFS	09/23/92-06/21/95	2	4	
BADL0014	Yes	00061	FLOW, STREAM, INSTANTANEOUS CFS	09/22/92-08/28/95	2	4	
BADL0015	No	00061	FLOW, STREAM, INSTANTANEOUS CFS	09/25/92-08/28/95	2	4	
BADL0016	No	00061	FLOW, STREAM, INSTANTANEOUS CFS	09/30/92-07/10/95	2	4	
BADL0017	Yes	00061	FLOW, STREAM, INSTANTANEOUS CFS	09/21/92-09/21/92	0	1	
BADL0018	Yes	00061	FLOW, STREAM, INSTANTANEOUS CFS	09/21/92-06/21/95	2	4	
BADL0019	No	00061	FLOW, STREAM, INSTANTANEOUS CFS	10/08/92-08/29/95	2	4	
BADL0021	Yes	00061	FLOW, STREAM, INSTANTANEOUS CFS	10/01/92-07/13/95	2	4	
BADL0026	No	00061	FLOW, STREAM, INSTANTANEOUS CFS	03/09/89-09/10/96	7	73	
BADL0027	No	00061	FLOW, STREAM, INSTANTANEOUS CFS	05/06/88-09/08/94	6	6	
BADL0029	No	00061	FLOW, STREAM, INSTANTANEOUS CFS	06/30/95-07/23/96	1	5	
BADL0031	No	00061	FLOW, STREAM, INSTANTANEOUS CFS	11/05/92-07/12/95	2	4	
BADL0011	Yes	00063	SAMPLING POINTS, NUMBER OF IN A CROSS SECTION	07/27/65-09/08/66	1	3	
BADL0008	No	00065	STAGE, STREAM (FEET)	08/02/94-07/11/95	0	2	
BADL0014	Yes	00065	STAGE, STREAM (FEET)	07/12/94-08/28/95	1	2	
BADL0015	No	00065	STAGE, STREAM (FEET)	07/13/94-08/28/95	1	2	
BADL0016	No	00065	STAGE, STREAM (FEET)	07/27/94-07/27/94	0	1	
BADL0021	Yes	00065	STAGE, STREAM (FEET)	07/13/95-07/13/95	0	1	
BADL0026	No	00065	STAGE, STREAM (FEET)	10/01/92-08/30/94	1	12	
BADL0023	No	00070	TURBIDITY, (JACKSON CANDLE UNITS)	06/01/71-08/21/72	1	3	
BADL0028	No	00070	TURBIDITY, (JACKSON CANDLE UNITS)	06/03/71-06/03/71	0	1	
BADL0026	No	00076	TURBIDITY,HACH TURBIDIMETER (FORMAZIN TURB UNIT)	09/07/93-08/30/94	0	2	
BADL0006	No	00078	TRANSPARENCY, SECCHI DISC (METERS)	07/12/89-08/16/89	0	2	
BADL0011	Yes	00080	COLOR (PLATINUM-COBALT UNITS)	08/13/64-09/18/67	3	36	T
BADL0002	No	00095	SPECIFIC CONDUCTANCE (UMHOS/CM 25C)	06/08/78-09/19/78	0	2	

[1]T=Times Series Plot, A=Annual Plot, and S=Seasonal Plot

Station/Parameter Period of Record Tabulation
From 11/29/62 To 09/10/96

Station	In Park	Code	Name	Start - End	Years	Obs	Plots[1]
BADL0003	No	00095	SPECIFIC CONDUCTANCE (UMHOS/CM @ 25C)	12/19/77-09/19/78	0	4	
BADL0004	No	00095	SPECIFIC CONDUCTANCE (UMHOS/CM @ 25C)	06/08/78-09/19/78	0	2	
BADL0005	No	00095	SPECIFIC CONDUCTANCE (UMHOS/CM @ 25C)	11/30/62-11/30/62	0	1	
BADL0006	No	00095	SPECIFIC CONDUCTANCE (UMHOS/CM @ 25C)	08/16/89-08/16/89	0	1	
BADL0007	No	00095	SPECIFIC CONDUCTANCE (UMHOS/CM @ 25C)	08/16/89-08/16/89	0	1	
BADL0008	No	00095	SPECIFIC CONDUCTANCE (UMHOS/CM @ 25C)	11/12/92-07/11/95	2	4	
BADL0011	Yes	00095	SPECIFIC CONDUCTANCE (UMHOS/CM @ 25C)	08/13/64-09/18/67	3	101	T
BADL0012	No	00095	SPECIFIC CONDUCTANCE (UMHOS/CM @ 25C)	11/30/62-11/30/62	0	1	
BADL0013	No	00095	SPECIFIC CONDUCTANCE (UMHOS/CM @ 25C)	09/23/92-06/21/95	2	4	
BADL0014	Yes	00095	SPECIFIC CONDUCTANCE (UMHOS/CM @ 25C)	09/22/92-08/28/95	2	4	
BADL0015	No	00095	SPECIFIC CONDUCTANCE (UMHOS/CM @ 25C)	07/13/94-08/28/95	1	3	
BADL0016	No	00095	SPECIFIC CONDUCTANCE (UMHOS/CM @ 25C)	09/30/92-07/10/95	2	4	
BADL0017	Yes	00095	SPECIFIC CONDUCTANCE (UMHOS/CM @ 25C)	09/21/92-09/21/92	0	1	
BADL0018	Yes	00095	SPECIFIC CONDUCTANCE (UMHOS/CM @ 25C)	09/21/92-06/21/95	2	4	
BADL0019	No	00095	SPECIFIC CONDUCTANCE (UMHOS/CM @ 25C)	10/08/92-08/29/95	2	4	
BADL0020	No	00095	SPECIFIC CONDUCTANCE (UMHOS/CM @ 25C)	11/29/62-11/29/62	0	1	
BADL0021	Yes	00095	SPECIFIC CONDUCTANCE (UMHOS/CM @ 25C)	10/01/92-07/13/95	2	5	
BADL0022	No	00095	SPECIFIC CONDUCTANCE (UMHOS/CM @ 25C)	09/29/92-09/29/92	0	1	
BADL0023	No	00095	SPECIFIC CONDUCTANCE (UMHOS/CM @ 25C)	06/01/71-08/21/72	1	3	
BADL0024	Yes	00095	SPECIFIC CONDUCTANCE (UMHOS/CM @ 25C)	09/30/92-09/30/92	0	1	
BADL0025	Yes	00095	SPECIFIC CONDUCTANCE (UMHOS/CM @ 25C)	09/18/92-09/18/92	0	1	
BADL0026	No	00095	SPECIFIC CONDUCTANCE (UMHOS/CM @ 25C)	03/09/89-09/10/96	7	69	A,S
BADL0027	No	00095	SPECIFIC CONDUCTANCE (UMHOS/CM @ 25C)	05/06/88-09/08/94	6	6	
BADL0028	No	00095	SPECIFIC CONDUCTANCE (UMHOS/CM @ 25C)	06/03/71-06/03/71	0	1	
BADL0029	No	00095	SPECIFIC CONDUCTANCE (UMHOS/CM @ 25C)	06/30/95-11/21/95	0	4	
BADL0030	No	00095	SPECIFIC CONDUCTANCE (UMHOS/CM @ 25C)	11/04/92-11/04/92	0	1	
BADL0031	No	00095	SPECIFIC CONDUCTANCE (UMHOS/CM @ 25C)	11/05/92-07/12/95	2	4	
BADL0002	No	00300	OXYGEN, DISSOLVED MG/L	06/08/78-09/19/78	0	2	
BADL0003	No	00300	OXYGEN, DISSOLVED MG/L	12/19/77-09/19/78	0	4	
BADL0004	No	00300	OXYGEN, DISSOLVED MG/L	06/08/78-09/19/78	0	2	
BADL0006	No	00300	OXYGEN, DISSOLVED MG/L	07/12/89-08/16/89	0	2	
BADL0007	No	00300	OXYGEN, DISSOLVED MG/L	07/12/89-08/16/89	0	2	
BADL0008	No	00300	OXYGEN, DISSOLVED MG/L	11/12/92-07/11/95	2	3	
BADL0011	Yes	00300	OXYGEN, DISSOLVED MG/L	01/31/67-07/22/67	0	3	
BADL0013	No	00300	OXYGEN, DISSOLVED MG/L	07/26/94-06/21/95	0	2	
BADL0014	Yes	00300	OXYGEN, DISSOLVED MG/L	07/12/94-08/28/95	1	2	
BADL0015	No	00300	OXYGEN, DISSOLVED MG/L	08/28/95-08/28/95	0	1	
BADL0016	No	00300	OXYGEN, DISSOLVED MG/L	07/27/94-07/10/95	0	2	
BADL0018	Yes	00300	OXYGEN, DISSOLVED MG/L	06/21/95-06/21/95	0	1	
BADL0019	No	00300	OXYGEN, DISSOLVED MG/L	07/27/94-08/29/95	1	2	
BADL0021	Yes	00300	OXYGEN, DISSOLVED MG/L	07/13/95-07/13/95	0	1	
BADL0023	No	00300	OXYGEN, DISSOLVED MG/L	06/01/71-08/21/72	1	3	
BADL0026	No	00300	OXYGEN, DISSOLVED MG/L	09/07/93-08/30/94	0	2	
BADL0027	No	00300	OXYGEN, DISSOLVED MG/L	05/06/88-09/08/94	6	6	
BADL0028	No	00300	OXYGEN, DISSOLVED MG/L	06/03/71-06/03/71	0	1	
BADL0031	No	00300	OXYGEN, DISSOLVED MG/L	08/03/94-07/12/95	0	2	
BADL0002	No	00400	PH (STANDARD UNITS)	06/08/78-09/19/78	0	2	
BADL0003	No	00400	PH (STANDARD UNITS)	12/19/77-09/19/78	0	4	
BADL0004	No	00400	PH (STANDARD UNITS)	06/08/78-09/19/78	0	2	
BADL0005	No	00400	PH (STANDARD UNITS)	11/30/62-11/30/62	0	1	
BADL0006	No	00400	PH (STANDARD UNITS)	07/12/89-08/16/89	0	2	
BADL0007	No	00400	PH (STANDARD UNITS)	07/12/89-08/16/89	0	2	
BADL0008	No	00400	PH (STANDARD UNITS)	11/12/92-07/11/95	2	4	
BADL0009	No	00400	PH (STANDARD UNITS)	04/23/79-04/23/79	0	1	
BADL0010	No	00400	PH (STANDARD UNITS)	04/18/79-04/18/79	0	1	
BADL0011	Yes	00400	PH (STANDARD UNITS)	08/13/64-09/18/67	3	101	T
BADL0012	No	00400	PH (STANDARD UNITS)	11/30/62-11/30/62	0	1	
BADL0013	No	00400	PH (STANDARD UNITS)	09/23/92-06/21/95	2	4	
BADL0014	Yes	00400	PH (STANDARD UNITS)	09/22/92-08/28/95	2	4	
BADL0015	No	00400	PH (STANDARD UNITS)	09/25/92-08/28/95	2	4	
BADL0016	No	00400	PH (STANDARD UNITS)	09/30/92-07/10/95	2	4	
BADL0017	Yes	00400	PH (STANDARD UNITS)	09/21/92-09/21/92	0	1	
BADL0018	Yes	00400	PH (STANDARD UNITS)	09/21/92-06/21/95	2	4	
BADL0019	No	00400	PH (STANDARD UNITS)	10/08/92-08/29/95	2	4	
BADL0020	No	00400	PH (STANDARD UNITS)	11/29/62-11/29/62	0	1	
BADL0021	Yes	00400	PH (STANDARD UNITS)	10/01/92-07/13/95	2	5	
BADL0022	No	00400	PH (STANDARD UNITS)	09/29/92-09/29/92	0	1	
BADL0023	No	00400	PH (STANDARD UNITS)	06/01/71-08/21/72	1	3	
BADL0024	Yes	00400	PH (STANDARD UNITS)	09/30/92-09/30/92	0	1	
BADL0025	Yes	00400	PH (STANDARD UNITS)	09/18/92-09/18/92	0	1	
BADL0026	No	00400	PH (STANDARD UNITS)	09/07/93-08/30/94	0	2	
BADL0027	No	00400	PH (STANDARD UNITS)	05/06/88-09/08/94	6	6	

[1] T=Times Series Plot, A=Annual Plot, and S=Seasonal Plot

Station	In Park	Code	Name	Start - End	Years	Obs	Plots[1]
BADL0028	No	00400	PH (STANDARD UNITS)	06/03/71-06/03/71	0	1	
BADL0030	No	00400	PH (STANDARD UNITS)	11/04/92-11/04/92	0	1	
BADL0031	No	00400	PH (STANDARD UNITS)	11/05/92-07/12/95	2	4	
BADL0001	Yes	00403	PH, LAB, STANDARD UNITS SU	06/29/78-06/29/78	0	1	
BADL0002	No	00403	PH, LAB, STANDARD UNITS SU	06/08/78-09/19/78	0	2	
BADL0003	No	00403	PH, LAB, STANDARD UNITS SU	12/19/77-09/19/78	0	4	
BADL0004	No	00403	PH, LAB, STANDARD UNITS SU	06/08/78-09/19/78	0	2	
BADL0006	No	00403	PH, LAB, STANDARD UNITS SU	07/12/89-08/16/89	0	2	
BADL0007	No	00403	PH, LAB, STANDARD UNITS SU	07/12/89-08/16/89	0	2	
BADL0008	No	00403	PH, LAB, STANDARD UNITS SU	11/12/92-07/11/95	2	4	
BADL0013	No	00403	PH, LAB, STANDARD UNITS SU	09/23/92-06/21/95	2	4	
BADL0014	Yes	00403	PH, LAB, STANDARD UNITS SU	09/22/92-08/28/95	2	4	
BADL0015	No	00403	PH, LAB, STANDARD UNITS SU	09/25/92-08/28/95	2	4	
BADL0016	No	00403	PH, LAB, STANDARD UNITS SU	09/30/92-07/10/95	2	4	
BADL0017	Yes	00403	PH, LAB, STANDARD UNITS SU	09/21/92-09/21/92	0	1	
BADL0018	Yes	00403	PH, LAB, STANDARD UNITS SU	09/21/92-06/21/95	2	4	
BADL0019	No	00403	PH, LAB, STANDARD UNITS SU	10/08/92-08/29/95	2	4	
BADL0021	Yes	00403	PH, LAB, STANDARD UNITS SU	10/01/92-07/13/95	2	5	
BADL0022	No	00403	PH, LAB, STANDARD UNITS SU	09/29/92-09/29/92	0	1	
BADL0024	Yes	00403	PH, LAB, STANDARD UNITS SU	09/30/92-09/30/92	0	1	
BADL0025	Yes	00403	PH, LAB, STANDARD UNITS SU	09/18/92-09/18/92	0	1	
BADL0026	No	00403	PH, LAB, STANDARD UNITS SU	09/07/93-08/30/94	0	2	
BADL0027	No	00403	PH, LAB, STANDARD UNITS SU	05/06/88-09/08/94	6	6	
BADL0030	No	00403	PH, LAB, STANDARD UNITS SU	11/04/92-11/04/92	0	1	
BADL0031	No	00403	PH, LAB, STANDARD UNITS SU	11/05/92-07/12/95	2	4	
BADL0005	No	00405	CARBON DIOXIDE (MG/L AS CO2)	11/30/62-11/30/62	0	1	
BADL0012	No	00405	CARBON DIOXIDE (MG/L AS CO2)	11/30/62-11/30/62	0	1	
BADL0020	No	00405	CARBON DIOXIDE (MG/L AS CO2)	11/29/62-11/29/62	0	1	
BADL0002	No	00410	ALKALINITY, TOTAL (MG/L AS CACO3)	06/08/78-09/19/78	0	2	
BADL0003	No	00410	ALKALINITY, TOTAL (MG/L AS CACO3)	12/19/77-09/19/78	0	4	
BADL0004	No	00410	ALKALINITY, TOTAL (MG/L AS CACO3)	06/08/78-09/19/78	0	2	
BADL0005	No	00410	ALKALINITY, TOTAL (MG/L AS CACO3)	11/30/62-11/30/62	0	1	
BADL0006	No	00410	ALKALINITY, TOTAL (MG/L AS CACO3)	07/12/89-08/16/89	0	2	
BADL0007	No	00410	ALKALINITY, TOTAL (MG/L AS CACO3)	07/12/89-08/16/89	0	2	
BADL0008	No	00410	ALKALINITY, TOTAL (MG/L AS CACO3)	11/12/92-08/02/94	1	2	
BADL0009	No	00410	ALKALINITY, TOTAL (MG/L AS CACO3)	04/23/79-04/23/79	0	2	
BADL0010	No	00410	ALKALINITY, TOTAL (MG/L AS CACO3)	04/18/79-04/18/79	0	2	
BADL0011	Yes	00410	ALKALINITY, TOTAL (MG/L AS CACO3)	10/01/65-09/18/67	1	82	
BADL0012	No	00410	ALKALINITY, TOTAL (MG/L AS CACO3)	11/30/62-11/30/62	0	1	
BADL0013	No	00410	ALKALINITY, TOTAL (MG/L AS CACO3)	07/26/94-07/26/94	0	1	
BADL0016	No	00410	ALKALINITY, TOTAL (MG/L AS CACO3)	07/27/94-07/27/94	0	1	
BADL0019	No	00410	ALKALINITY, TOTAL (MG/L AS CACO3)	07/27/94-07/27/94	0	1	
BADL0020	No	00410	ALKALINITY, TOTAL (MG/L AS CACO3)	11/29/62-11/29/62	0	1	
BADL0023	No	00410	ALKALINITY, TOTAL (MG/L AS CACO3)	06/01/71-08/21/72	1	3	
BADL0027	No	00410	ALKALINITY, TOTAL (MG/L AS CACO3)	06/20/88-10/31/88	0	2	
BADL0028	No	00410	ALKALINITY, TOTAL (MG/L AS CACO3)	06/03/71-06/03/71	0	1	
BADL0031	No	00410	ALKALINITY, TOTAL (MG/L AS CACO3)	08/03/94-08/03/94	0	1	
BADL0001	Yes	00411	ALKALINITY,METHYLORANGE MG/L	06/29/78-06/29/78	0	1	
BADL0006	No	00415	ALKALINITY, PHENOLPHTHALEIN (MG/L)	07/12/89-08/16/89	0	2	
BADL0007	No	00415	ALKALINITY, PHENOLPHTHALEIN (MG/L)	07/12/89-08/16/89	0	2	
BADL0014	Yes	00419	ALKALINITY,CARBONATE,INCREMENTAL TITR FIELD MG/L	08/28/95-08/28/95	0	1	
BADL0015	No	00419	ALKALINITY,CARBONATE,INCREMENTAL TITR FIELD MG/L	08/28/95-08/28/95	0	1	
BADL0019	No	00419	ALKALINITY,CARBONATE,INCREMENTAL TITR FIELD MG/L	08/29/95-08/29/95	0	1	
BADL0027	No	00419	ALKALINITY,CARBONATE,INCREMENTAL TITR FIELD MG/L	06/20/88-06/20/88	0	1	
BADL0023	No	00430	ALKALINITY, CARBONATE (MG/L AS CACO3)	06/01/71-08/21/72	1	3	
BADL0028	No	00430	ALKALINITY, CARBONATE (MG/L AS CACO3)	06/03/71-06/03/71	0	1	
BADL0005	No	00440	BICARBONATE ION (MG/L AS HCO3)	11/30/62-11/30/62	0	1	
BADL0008	No	00440	BICARBONATE ION (MG/L AS HCO3)	08/02/94-08/02/94	0	1	
BADL0011	Yes	00440	BICARBONATE ION (MG/L AS HCO3)	08/13/64-09/18/67	3	101	T
BADL0012	No	00440	BICARBONATE ION (MG/L AS HCO3)	11/30/62-11/30/62	0	1	
BADL0013	No	00440	BICARBONATE ION (MG/L AS HCO3)	07/26/94-07/26/94	0	1	
BADL0016	No	00440	BICARBONATE ION (MG/L AS HCO3)	07/27/94-07/27/94	0	1	
BADL0019	No	00440	BICARBONATE ION (MG/L AS HCO3)	07/27/94-07/27/94	0	1	
BADL0020	No	00440	BICARBONATE ION (MG/L AS HCO3)	11/29/62-11/29/62	0	1	
BADL0027	No	00440	BICARBONATE ION (MG/L AS HCO3)	06/20/88-06/20/88	0	1	
BADL0031	No	00440	BICARBONATE ION (MG/L AS HCO3)	08/03/94-08/03/94	0	1	
BADL0005	No	00445	CARBONATE ION (MG/L AS CO3)	11/30/62-11/30/62	0	1	
BADL0011	Yes	00445	CARBONATE ION (MG/L AS CO3)	08/13/64-09/18/67	3	101	T
BADL0012	No	00445	CARBONATE ION (MG/L AS CO3)	11/30/62-11/30/62	0	1	
BADL0020	No	00445	CARBONATE ION (MG/L AS CO3)	11/29/62-11/29/62	0	1	
BADL0027	No	00445	CARBONATE ION (MG/L AS CO3)	06/20/88-06/20/88	0	1	
BADL0014	Yes	00447	CARBONATE,INCREMENTAL TITRATION,(CO3) FIELD MG/L	08/28/95-08/28/95	0	1	

[1]T=Times Series Plot, A=Annual Plot, and S=Seasonal Plot

Station	In Park	Code	Name	Start - End	Years	Obs	Plots[1]
BADL0015	No	00447	CARBONATE,INCREMENTAL TITRATION,(CO3) FIELD MG/L	08/28/95-08/28/95	0	1	
BADL0019	No	00447	CARBONATE,INCREMENTAL TITRATION,(CO3) FIELD MG/L	08/29/95-08/29/95	0	1	
BADL0014	Yes	00450	BICARBONATE,INCREMENTAL TITRATION,(HCO3) FIELDMG/L	08/28/95-08/28/95	0	1	
BADL0015	No	00450	BICARBONATE,INCREMENTAL TITRATION,(HCO3) FIELDMG/L	08/28/95-08/28/95	0	1	
BADL0019	No	00450	BICARBONATE,INCREMENTAL TITRATION,(HCO3) FIELDMG/L	08/29/95-08/29/95	0	1	
BADL0008	No	00452	CARBONATE,WATER,DISS,INCR TIT, FIELD, AS CO3, MG/L	07/11/95-07/11/95	0	1	
BADL0018	Yes	00452	CARBONATE,WATER,DISS,INCR TIT, FIELD, AS CO3, MG/L	06/21/95-06/21/95	0	1	
BADL0027	No	00452	CARBONATE,WATER,DISS,INCR TIT, FIELD, AS CO3, MG/L	10/31/88-04/20/94	5	2	
BADL0008	No	00453	BICARBONATE,WATER,DISS,INCR TIT,FIELD,AS HCO3,MG/L	07/11/95-07/11/95	0	1	
BADL0013	No	00453	BICARBONATE,WATER,DISS,INCR TIT,FIELD,AS HCO3,MG/L	06/21/95-06/21/95	0	1	
BADL0016	No	00453	BICARBONATE,WATER,DISS,INCR TIT,FIELD,AS HCO3,MG/L	07/10/95-07/10/95	0	1	
BADL0018	Yes	00453	BICARBONATE,WATER,DISS,INCR TIT,FIELD,AS HCO3,MG/L	06/21/95-06/21/95	0	1	
BADL0021	Yes	00453	BICARBONATE,WATER,DISS,INCR TIT,FIELD,AS HCO3,MG/L	07/13/95-07/13/95	0	1	
BADL0027	No	00453	BICARBONATE,WATER,DISS,INCR TIT,FIELD,AS HCO3,MG/L	10/31/88-04/20/94	5	2	
BADL0031	No	00453	BICARBONATE,WATER,DISS,INCR TIT,FIELD,AS HCO3,MG/L	07/12/95-07/12/95	0	1	
BADL0002	No	00500	RESIDUE, TOTAL (MG/L)	06/08/78-09/19/78	0	2	
BADL0003	No	00500	RESIDUE, TOTAL (MG/L)	12/19/77-09/19/78	0	4	
BADL0004	No	00500	RESIDUE, TOTAL (MG/L)	06/08/78-09/19/78	0	2	
BADL0006	No	00500	RESIDUE, TOTAL (MG/L)	07/12/89-08/16/89	0	2	
BADL0007	No	00500	RESIDUE, TOTAL (MG/L)	07/12/89-08/16/89	0	2	
BADL0023	No	00500	RESIDUE, TOTAL (MG/L)	06/01/71-08/21/72	1	3	
BADL0028	No	00500	RESIDUE, TOTAL (MG/L)	06/03/71-06/03/71	0	1	
BADL0023	No	00505	RESIDUE, TOTAL VOLATILE (MG/L)	06/01/71-08/21/72	1	3	
BADL0028	No	00505	RESIDUE, TOTAL VOLATILE (MG/L)	06/03/71-06/03/71	0	1	
BADL0002	No	00515	RESIDUE, TOTAL FILTRABLE (DRIED AT 105C),MG/L	06/08/78-09/19/78	0	2	
BADL0003	No	00515	RESIDUE, TOTAL FILTRABLE (DRIED AT 105C),MG/L	12/19/77-09/19/78	0	4	
BADL0004	No	00515	RESIDUE, TOTAL FILTRABLE (DRIED AT 105C),MG/L	06/08/78-09/19/78	0	2	
BADL0023	No	00515	RESIDUE, TOTAL FILTRABLE (DRIED AT 105C),MG/L	06/01/71-08/21/72	1	3	
BADL0028	No	00515	RESIDUE, TOTAL FILTRABLE (DRIED AT 105C),MG/L	06/03/71-06/03/71	0	1	
BADL0001	Yes	00520	RESIDUE, VOLATILE FILTRABLE (MG/L)	06/29/78-06/29/78	0	1	
BADL0023	No	00520	RESIDUE, VOLATILE FILTRABLE (MG/L)	06/01/71-08/21/72	1	3	
BADL0028	No	00520	RESIDUE, VOLATILE FILTRABLE (MG/L)	06/03/71-06/03/71	0	1	
BADL0002	No	00530	RESIDUE, TOTAL NONFILTRABLE (MG/L)	06/08/78-09/19/78	0	2	
BADL0003	No	00530	RESIDUE, TOTAL NONFILTRABLE (MG/L)	12/19/77-09/19/78	0	4	
BADL0004	No	00530	RESIDUE, TOTAL NONFILTRABLE (MG/L)	06/08/78-09/19/78	0	2	
BADL0006	No	00530	RESIDUE, TOTAL NONFILTRABLE (MG/L)	07/12/89-08/16/89	0	2	
BADL0007	No	00530	RESIDUE, TOTAL NONFILTRABLE (MG/L)	07/12/89-08/16/89	0	2	
BADL0026	No	00530	RESIDUE, TOTAL NONFILTRABLE (MG/L)	09/07/93-08/30/94	0	2	
BADL0001	Yes	00535	RESIDUE, VOLATILE NONFILTRABLE (MG/L)	06/29/78-06/29/78	0	1	
BADL0003	No	00608	NITROGEN, AMMONIA, DISSOLVED (MG/L AS N)	12/19/77-06/08/78	0	2	
BADL0008	No	00608	NITROGEN, AMMONIA, DISSOLVED (MG/L AS N)	08/02/94-07/11/95	0	3	
BADL0013	No	00608	NITROGEN, AMMONIA, DISSOLVED (MG/L AS N)	07/26/94-06/21/95	0	3	
BADL0014	Yes	00608	NITROGEN, AMMONIA, DISSOLVED (MG/L AS N)	07/12/94-08/28/95	1	3	
BADL0015	No	00608	NITROGEN, AMMONIA, DISSOLVED (MG/L AS N)	07/13/94-08/28/95	1	3	
BADL0016	No	00608	NITROGEN, AMMONIA, DISSOLVED (MG/L AS N)	07/27/94-07/10/95	0	3	
BADL0018	Yes	00608	NITROGEN, AMMONIA, DISSOLVED (MG/L AS N)	07/14/94-06/21/95	0	3	
BADL0019	No	00608	NITROGEN, AMMONIA, DISSOLVED (MG/L AS N)	07/27/94-08/29/95	1	3	
BADL0021	Yes	00608	NITROGEN, AMMONIA, DISSOLVED (MG/L AS N)	07/28/94-07/13/95	0	4	
BADL0026	No	00608	NITROGEN, AMMONIA, DISSOLVED (MG/L AS N)	09/07/93-08/30/94	0	2	
BADL0031	No	00608	NITROGEN, AMMONIA, DISSOLVED (MG/L AS N)	08/03/94-07/12/95	0	3	
BADL0002	No	00610	NITROGEN, AMMONIA, TOTAL (MG/L AS N)	06/08/78-09/19/78	0	2	
BADL0003	No	00610	NITROGEN, AMMONIA, TOTAL (MG/L AS N)	03/15/78-09/19/78	0	2	
BADL0004	No	00610	NITROGEN, AMMONIA, TOTAL (MG/L AS N)	06/08/78-09/19/78	0	2	
BADL0006	No	00610	NITROGEN, AMMONIA, TOTAL (MG/L AS N)	07/12/89-08/16/89	0	2	
BADL0007	No	00610	NITROGEN, AMMONIA, TOTAL (MG/L AS N)	07/12/89-08/16/89	0	2	
BADL0002	No	00613	NITRITE NITROGEN, DISSOLVED (MG/L AS N)	06/08/78-09/19/78	0	2	
BADL0003	No	00613	NITRITE NITROGEN, DISSOLVED (MG/L AS N)	12/19/77-09/19/78	0	4	
BADL0004	No	00613	NITRITE NITROGEN, DISSOLVED (MG/L AS N)	06/08/78-09/19/78	0	2	
BADL0008	No	00613	NITRITE NITROGEN, DISSOLVED (MG/L AS N)	05/17/95-07/11/95	0	2	
BADL0013	No	00613	NITRITE NITROGEN, DISSOLVED (MG/L AS N)	04/27/95-06/21/95	0	2	
BADL0014	Yes	00613	NITRITE NITROGEN, DISSOLVED (MG/L AS N)	04/24/95-08/28/95	0	2	
BADL0015	No	00613	NITRITE NITROGEN, DISSOLVED (MG/L AS N)	05/11/95-08/28/95	0	2	
BADL0016	No	00613	NITRITE NITROGEN, DISSOLVED (MG/L AS N)	04/14/95-07/10/95	0	2	
BADL0018	Yes	00613	NITRITE NITROGEN, DISSOLVED (MG/L AS N)	04/27/95-06/21/95	0	2	
BADL0019	No	00613	NITRITE NITROGEN, DISSOLVED (MG/L AS N)	05/12/95-08/29/95	0	2	
BADL0021	Yes	00613	NITRITE NITROGEN, DISSOLVED (MG/L AS N)	05/16/95-07/13/95	0	2	
BADL0026	No	00613	NITRITE NITROGEN, DISSOLVED (MG/L AS N)	09/07/93-08/30/94	0	2	
BADL0031	No	00613	NITRITE NITROGEN, DISSOLVED (MG/L AS N)	05/16/95-07/12/95	0	2	
BADL0001	Yes	00618	NITRATE NITROGEN, DISSOLVED (MG/L AS N)	06/29/78-06/29/78	0	1	
BADL0005	No	00618	NITRATE NITROGEN, DISSOLVED (MG/L AS N)	11/30/62-11/30/62	0	1	
BADL0012	No	00618	NITRATE NITROGEN, DISSOLVED (MG/L AS N)	11/30/62-11/30/62	0	1	
BADL0020	No	00618	NITRATE NITROGEN, DISSOLVED (MG/L AS N)	11/29/62-11/29/62	0	1	

[1]T=Times Series Plot, A=Annual Plot, and S=Seasonal Plot

Station/Parameter Period of Record Tabulation
From 11/29/62 To 09/10/96

Station	In Park	Code	Name	Start - End	Years	Obs	Plots[1]
BADL0008	No	00623	NITROGEN, KJELDAHL, DISSOLVED (MG/L AS N)	08/02/94-07/11/95	0	3	
BADL0013	No	00623	NITROGEN, KJELDAHL, DISSOLVED (MG/L AS N)	07/26/94-06/21/95	0	3	
BADL0014	Yes	00623	NITROGEN, KJELDAHL, DISSOLVED (MG/L AS N)	07/12/94-08/28/95	1	3	
BADL0015	No	00623	NITROGEN, KJELDAHL, DISSOLVED (MG/L AS N)	07/13/94-08/28/95	1	3	
BADL0016	No	00623	NITROGEN, KJELDAHL, DISSOLVED (MG/L AS N)	07/27/94-07/10/95	0	3	
BADL0018	Yes	00623	NITROGEN, KJELDAHL, DISSOLVED (MG/L AS N)	07/14/94-06/21/95	0	3	
BADL0019	No	00623	NITROGEN, KJELDAHL, DISSOLVED (MG/L AS N)	07/27/94-08/29/95	1	3	
BADL0021	Yes	00623	NITROGEN, KJELDAHL, DISSOLVED (MG/L AS N)	07/28/94-07/13/95	0	4	
BADL0026	No	00623	NITROGEN, KJELDAHL, DISSOLVED (MG/L AS N)	09/07/93-08/30/94	0	2	
BADL0031	No	00623	NITROGEN, KJELDAHL, DISSOLVED (MG/L AS N)	08/03/94-07/12/95	0	3	
BADL0002	No	00625	NITROGEN, KJELDAHL, TOTAL, (MG/L AS N)	06/08/78-09/19/78	0	2	
BADL0003	No	00625	NITROGEN, KJELDAHL, TOTAL, (MG/L AS N)	12/19/77-09/19/78	0	4	
BADL0004	No	00625	NITROGEN, KJELDAHL, TOTAL, (MG/L AS N)	06/08/78-09/19/78	0	2	
BADL0006	No	00625	NITROGEN, KJELDAHL, TOTAL, (MG/L AS N)	07/12/89-08/16/89	0	2	
BADL0007	No	00625	NITROGEN, KJELDAHL, TOTAL, (MG/L AS N)	07/12/89-08/16/89	0	2	
BADL0002	No	00630	NITRITE PLUS NITRATE, TOTAL 1 DET (MG/L AS N)	06/08/78-09/19/78	0	2	
BADL0003	No	00630	NITRITE PLUS NITRATE, TOTAL 1 DET (MG/L AS N)	12/19/77-09/19/78	0	4	
BADL0004	No	00630	NITRITE PLUS NITRATE, TOTAL 1 DET (MG/L AS N)	06/08/78-09/19/78	0	2	
BADL0006	No	00630	NITRITE PLUS NITRATE, TOTAL 1 DET (MG/L AS N)	07/12/89-08/16/89	0	2	
BADL0007	No	00630	NITRITE PLUS NITRATE, TOTAL 1 DET (MG/L AS N)	07/12/89-08/16/89	0	2	
BADL0023	No	00630	NITRITE PLUS NITRATE, TOTAL 1 DET (MG/L AS N)	06/01/71-08/21/72	1	3	
BADL0028	No	00630	NITRITE PLUS NITRATE, TOTAL 1 DET (MG/L AS N)	06/03/71-06/03/71	0	1	
BADL0008	No	00631	NITRITE PLUS NITRATE, DISS 1 DET (MG/L AS N)	08/02/94-07/11/95	0	3	
BADL0013	No	00631	NITRITE PLUS NITRATE, DISS 1 DET (MG/L AS N)	07/26/94-06/21/95	0	3	
BADL0014	Yes	00631	NITRITE PLUS NITRATE, DISS 1 DET (MG/L AS N)	07/12/94-08/28/95	1	3	
BADL0015	No	00631	NITRITE PLUS NITRATE, DISS 1 DET (MG/L AS N)	07/13/94-08/28/95	1	3	
BADL0016	No	00631	NITRITE PLUS NITRATE, DISS 1 DET (MG/L AS N)	07/27/94-07/10/95	0	3	
BADL0018	Yes	00631	NITRITE PLUS NITRATE, DISS 1 DET (MG/L AS N)	07/14/94-06/21/95	0	3	
BADL0019	No	00631	NITRITE PLUS NITRATE, DISS 1 DET (MG/L AS N)	07/27/94-08/29/95	1	3	
BADL0021	Yes	00631	NITRITE PLUS NITRATE, DISS 1 DET (MG/L AS N)	07/28/94-07/13/95	0	4	
BADL0026	No	00631	NITRITE PLUS NITRATE, DISS 1 DET (MG/L AS N)	09/07/93-08/30/94	0	2	
BADL0027	No	00631	NITRITE PLUS NITRATE, DISS 1 DET (MG/L AS N)	05/06/88-10/31/88	0	4	
BADL0031	No	00631	NITRITE PLUS NITRATE, DISS 1 DET (MG/L AS N)	08/03/94-07/12/95	0	3	
BADL0001	Yes	00650	PHOSPHATE, TOTAL (MG/L AS PO4)	06/29/78-06/29/78	0	1	
BADL0011	Yes	00650	PHOSPHATE, TOTAL (MG/L AS PO4)	11/04/64-07/22/67	2	11	T
BADL0023	No	00650	PHOSPHATE, TOTAL (MG/L AS PO4)	06/01/71-08/21/72	1	3	
BADL0028	No	00650	PHOSPHATE, TOTAL (MG/L AS PO4)	06/03/71-06/03/71	0	1	
BADL0001	Yes	00660	PHOSPHATE, ORTHO (MG/L AS PO4)	06/29/78-06/29/78	0	1	
BADL0023	No	00660	PHOSPHATE, ORTHO (MG/L AS PO4)	06/01/71-08/21/72	1	3	
BADL0028	No	00660	PHOSPHATE, ORTHO (MG/L AS PO4)	06/03/71-06/03/71	0	1	
BADL0006	No	00665	PHOSPHORUS, TOTAL (MG/L AS P)	07/12/89-08/16/89	0	2	
BADL0007	No	00665	PHOSPHORUS, TOTAL (MG/L AS P)	07/12/89-08/16/89	0	2	
BADL0009	No	00665	PHOSPHORUS, TOTAL (MG/L AS P)	04/23/79-04/23/79	0	1	
BADL0010	No	00665	PHOSPHORUS, TOTAL (MG/L AS P)	04/18/79-04/18/79	0	1	
BADL0008	No	00666	PHOSPHORUS, DISSOLVED (MG/L AS P)	08/02/94-07/11/95	0	3	
BADL0013	No	00666	PHOSPHORUS, DISSOLVED (MG/L AS P)	07/26/94-06/21/95	0	3	
BADL0014	Yes	00666	PHOSPHORUS, DISSOLVED (MG/L AS P)	07/12/94-08/28/95	1	3	
BADL0015	No	00666	PHOSPHORUS, DISSOLVED (MG/L AS P)	07/13/94-08/28/95	1	3	
BADL0016	No	00666	PHOSPHORUS, DISSOLVED (MG/L AS P)	07/27/94-07/10/95	0	3	
BADL0018	Yes	00666	PHOSPHORUS, DISSOLVED (MG/L AS P)	07/14/94-06/21/95	0	3	
BADL0019	No	00666	PHOSPHORUS, DISSOLVED (MG/L AS P)	07/27/94-08/29/95	1	3	
BADL0021	Yes	00666	PHOSPHORUS, DISSOLVED (MG/L AS P)	07/28/94-07/13/95	0	4	
BADL0026	No	00666	PHOSPHORUS, DISSOLVED (MG/L AS P)	09/07/93-08/30/94	0	2	
BADL0031	No	00666	PHOSPHORUS, DISSOLVED (MG/L AS P)	08/03/94-07/12/95	0	3	
BADL0002	No	00671	PHOSPHORUS, DISSOLVED ORTHOPHOSPHATE (MG/L AS P)	06/08/78-06/08/78	0	1	
BADL0003	No	00671	PHOSPHORUS, DISSOLVED ORTHOPHOSPHATE (MG/L AS P)	12/19/77-06/08/78	0	3	
BADL0004	No	00671	PHOSPHORUS, DISSOLVED ORTHOPHOSPHATE (MG/L AS P)	06/08/78-06/08/78	0	1	
BADL0006	No	00671	PHOSPHORUS, DISSOLVED ORTHOPHOSPHATE (MG/L AS P)	07/12/89-08/16/89	0	2	
BADL0007	No	00671	PHOSPHORUS, DISSOLVED ORTHOPHOSPHATE (MG/L AS P)	07/12/89-08/16/89	0	2	
BADL0008	No	00671	PHOSPHORUS, DISSOLVED ORTHOPHOSPHATE (MG/L AS P)	05/17/95-07/11/95	0	2	
BADL0013	No	00671	PHOSPHORUS, DISSOLVED ORTHOPHOSPHATE (MG/L AS P)	04/27/95-06/21/95	0	2	
BADL0014	Yes	00671	PHOSPHORUS, DISSOLVED ORTHOPHOSPHATE (MG/L AS P)	04/24/95-08/28/95	0	2	
BADL0015	No	00671	PHOSPHORUS, DISSOLVED ORTHOPHOSPHATE (MG/L AS P)	05/11/95-08/28/95	0	2	
BADL0016	No	00671	PHOSPHORUS, DISSOLVED ORTHOPHOSPHATE (MG/L AS P)	04/14/95-07/10/95	0	2	
BADL0018	Yes	00671	PHOSPHORUS, DISSOLVED ORTHOPHOSPHATE (MG/L AS P)	04/27/95-06/21/95	0	2	
BADL0019	No	00671	PHOSPHORUS, DISSOLVED ORTHOPHOSPHATE (MG/L AS P)	05/12/95-08/29/95	0	2	
BADL0021	Yes	00671	PHOSPHORUS, DISSOLVED ORTHOPHOSPHATE (MG/L AS P)	05/16/95-07/13/95	0	2	
BADL0026	No	00671	PHOSPHORUS, DISSOLVED ORTHOPHOSPHATE (MG/L AS P)	09/07/93-08/30/94	0	2	
BADL0031	No	00671	PHOSPHORUS, DISSOLVED ORTHOPHOSPHATE (MG/L AS P)	05/16/95-07/12/95	0	2	
BADL0026	No	00723	CYANIDE, DISSOLVED STD METHOD (UG/L)	09/07/93-09/07/93	0	1	
BADL0001	Yes	00900	HARDNESS, TOTAL (MG/L AS CACO3)	06/29/78-06/29/78	0	1	
BADL0005	No	00900	HARDNESS, TOTAL (MG/L AS CACO3)	11/30/62-11/30/62	0	1	

[1]T=Times Series Plot, A=Annual Plot, and S=Seasonal Plot

Station/Parameter Period of Record Tabulation
From 11/29/62 To 09/10/96

Station	In Park	Code	Name	Start - End	Years	Obs	Plots[1]
BADL0011	Yes	00900	HARDNESS, TOTAL (MG/L AS CACO3)	08/13/64-09/18/67	3	101	T
BADL0012	No	00900	HARDNESS, TOTAL (MG/L AS CACO3)	11/30/62-11/30/62	0	1	
BADL0020	No	00900	HARDNESS, TOTAL (MG/L AS CACO3)	11/29/62-11/29/62	0	1	
BADL0023	No	00900	HARDNESS, TOTAL (MG/L AS CACO3)	06/01/71-08/21/72	1	3	
BADL0028	No	00900	HARDNESS, TOTAL (MG/L AS CACO3)	06/03/71-06/03/71	0	1	
BADL0001	Yes	00901	HARDNESS, CARBONATE (MG/L AS CACO3)	06/29/78-06/29/78	0	1	
BADL0023	No	00901	HARDNESS, CARBONATE (MG/L AS CACO3)	06/01/71-08/21/72	1	3	
BADL0028	No	00901	HARDNESS, CARBONATE (MG/L AS CACO3)	06/03/71-06/03/71	0	1	
BADL0005	No	00902	HARDNESS, NON-CARBONATE (MG/L AS CACO3)	11/30/62-11/30/62	0	1	
BADL0011	Yes	00902	HARDNESS, NON-CARBONATE (MG/L AS CACO3)	08/13/64-09/18/67	3	101	T
BADL0012	No	00902	HARDNESS, NON-CARBONATE (MG/L AS CACO3)	11/30/62-11/30/62	0	1	
BADL0020	No	00902	HARDNESS, NON-CARBONATE (MG/L AS CACO3)	11/29/62-11/29/62	0	1	
BADL0023	No	00902	HARDNESS, NON-CARBONATE (MG/L AS CACO3)	06/01/71-08/21/72	1	3	
BADL0028	No	00902	HARDNESS, NON-CARBONATE (MG/L AS CACO3)	06/03/71-06/03/71	0	1	
BADL0001	Yes	00915	CALCIUM, DISSOLVED (MG/L AS CA)	06/29/78-06/29/78	0	1	
BADL0005	No	00915	CALCIUM, DISSOLVED (MG/L AS CA)	11/30/62-11/30/62	0	1	
BADL0008	No	00915	CALCIUM, DISSOLVED (MG/L AS CA)	11/12/92-07/11/95	2	4	
BADL0011	Yes	00915	CALCIUM, DISSOLVED (MG/L AS CA)	08/13/64-09/18/67	3	103	T
BADL0012	No	00915	CALCIUM, DISSOLVED (MG/L AS CA)	11/30/62-11/30/62	0	1	
BADL0013	No	00915	CALCIUM, DISSOLVED (MG/L AS CA)	09/23/92-06/21/95	2	4	
BADL0014	Yes	00915	CALCIUM, DISSOLVED (MG/L AS CA)	09/22/92-08/28/95	2	4	
BADL0015	No	00915	CALCIUM, DISSOLVED (MG/L AS CA)	09/25/92-08/28/95	2	4	
BADL0016	No	00915	CALCIUM, DISSOLVED (MG/L AS CA)	09/30/92-07/10/95	2	4	
BADL0017	Yes	00915	CALCIUM, DISSOLVED (MG/L AS CA)	09/21/92-09/21/92	0	1	
BADL0018	Yes	00915	CALCIUM, DISSOLVED (MG/L AS CA)	09/21/92-06/21/95	2	4	
BADL0019	No	00915	CALCIUM, DISSOLVED (MG/L AS CA)	10/08/92-08/29/95	2	4	
BADL0020	No	00915	CALCIUM, DISSOLVED (MG/L AS CA)	11/29/62-11/29/62	0	1	
BADL0021	Yes	00915	CALCIUM, DISSOLVED (MG/L AS CA)	10/01/92-07/13/95	2	5	
BADL0022	No	00915	CALCIUM, DISSOLVED (MG/L AS CA)	09/29/92-09/29/92	0	1	
BADL0023	No	00915	CALCIUM, DISSOLVED (MG/L AS CA)	06/01/71-08/21/72	1	3	
BADL0024	Yes	00915	CALCIUM, DISSOLVED (MG/L AS CA)	09/30/92-09/30/92	0	1	
BADL0025	Yes	00915	CALCIUM, DISSOLVED (MG/L AS CA)	09/18/92-09/18/92	0	1	
BADL0026	No	00915	CALCIUM, DISSOLVED (MG/L AS CA)	09/07/93-08/30/94	0	2	
BADL0027	No	00915	CALCIUM, DISSOLVED (MG/L AS CA)	05/06/88-09/08/94	6	6	
BADL0028	No	00915	CALCIUM, DISSOLVED (MG/L AS CA)	06/03/71-06/03/71	0	1	
BADL0030	No	00915	CALCIUM, DISSOLVED (MG/L AS CA)	11/04/92-11/04/92	0	1	
BADL0031	No	00915	CALCIUM, DISSOLVED (MG/L AS CA)	11/05/92-07/12/95	2	4	
BADL0009	No	00916	CALCIUM, TOTAL (MG/L AS CA)	04/23/79-04/23/79	0	1	
BADL0010	No	00916	CALCIUM, TOTAL (MG/L AS CA)	04/18/79-04/18/79	0	1	
BADL0001	Yes	00925	MAGNESIUM, DISSOLVED (MG/L AS MG)	06/29/78-06/29/78	0	1	
BADL0005	No	00925	MAGNESIUM, DISSOLVED (MG/L AS MG)	11/30/62-11/30/62	0	1	
BADL0008	No	00925	MAGNESIUM, DISSOLVED (MG/L AS MG)	11/12/92-07/11/95	2	4	
BADL0011	Yes	00925	MAGNESIUM, DISSOLVED (MG/L AS MG)	08/13/64-09/18/67	3	103	T
BADL0012	No	00925	MAGNESIUM, DISSOLVED (MG/L AS MG)	11/30/62-11/30/62	0	1	
BADL0013	No	00925	MAGNESIUM, DISSOLVED (MG/L AS MG)	09/23/92-06/21/95	2	4	
BADL0014	Yes	00925	MAGNESIUM, DISSOLVED (MG/L AS MG)	09/22/92-08/28/95	2	4	
BADL0015	No	00925	MAGNESIUM, DISSOLVED (MG/L AS MG)	09/25/92-08/28/95	2	4	
BADL0016	No	00925	MAGNESIUM, DISSOLVED (MG/L AS MG)	09/30/92-07/10/95	2	4	
BADL0017	Yes	00925	MAGNESIUM, DISSOLVED (MG/L AS MG)	09/21/92-09/21/92	0	1	
BADL0018	Yes	00925	MAGNESIUM, DISSOLVED (MG/L AS MG)	09/21/92-06/21/95	2	4	
BADL0019	No	00925	MAGNESIUM, DISSOLVED (MG/L AS MG)	10/08/92-08/29/95	2	4	
BADL0020	No	00925	MAGNESIUM, DISSOLVED (MG/L AS MG)	11/29/62-11/29/62	0	1	
BADL0021	Yes	00925	MAGNESIUM, DISSOLVED (MG/L AS MG)	10/01/92-07/13/95	2	5	
BADL0022	No	00925	MAGNESIUM, DISSOLVED (MG/L AS MG)	09/29/92-09/29/92	0	1	
BADL0023	No	00925	MAGNESIUM, DISSOLVED (MG/L AS MG)	06/01/71-08/21/72	1	3	
BADL0024	Yes	00925	MAGNESIUM, DISSOLVED (MG/L AS MG)	09/30/92-09/30/92	0	1	
BADL0025	Yes	00925	MAGNESIUM, DISSOLVED (MG/L AS MG)	09/18/92-09/18/92	0	1	
BADL0026	No	00925	MAGNESIUM, DISSOLVED (MG/L AS MG)	09/07/93-08/30/94	0	2	
BADL0027	No	00925	MAGNESIUM, DISSOLVED (MG/L AS MG)	05/06/88-09/08/94	6	6	
BADL0028	No	00925	MAGNESIUM, DISSOLVED (MG/L AS MG)	06/03/71-06/03/71	0	1	
BADL0030	No	00925	MAGNESIUM, DISSOLVED (MG/L AS MG)	11/04/92-11/04/92	0	1	
BADL0031	No	00925	MAGNESIUM, DISSOLVED (MG/L AS MG)	11/05/92-07/12/95	2	4	
BADL0009	No	00929	SODIUM, TOTAL (MG/L AS NA)	04/23/79-04/23/79	0	1	
BADL0010	No	00929	SODIUM, TOTAL (MG/L AS NA)	04/18/79-04/18/79	0	1	
BADL0001	Yes	00930	SODIUM, DISSOLVED (MG/L AS NA)	06/29/78-06/29/78	0	1	
BADL0005	No	00930	SODIUM, DISSOLVED (MG/L AS NA)	11/30/62-11/30/62	0	1	
BADL0008	No	00930	SODIUM, DISSOLVED (MG/L AS NA)	11/12/92-07/11/95	2	4	
BADL0011	Yes	00930	SODIUM, DISSOLVED (MG/L AS NA)	08/13/64-09/18/67	3	103	T
BADL0012	No	00930	SODIUM, DISSOLVED (MG/L AS NA)	11/30/62-11/30/62	0	1	
BADL0013	No	00930	SODIUM, DISSOLVED (MG/L AS NA)	09/23/92-06/21/95	2	4	
BADL0014	Yes	00930	SODIUM, DISSOLVED (MG/L AS NA)	09/22/92-08/28/95	2	4	
BADL0015	No	00930	SODIUM, DISSOLVED (MG/L AS NA)	09/25/92-08/28/95	2	4	

[1]T=Times Series Plot, A=Annual Plot, and S=Seasonal Plot

Station/Parameter Period of Record Tabulation
From 11/29/62 To 09/10/96

Station	In Park	Code	Name	Start - End	Years	Obs	Plots[1]
BADL0016	No	00930	SODIUM, DISSOLVED (MG/L AS NA)	09/30/92-07/10/95	2	4	
BADL0017	Yes	00930	SODIUM, DISSOLVED (MG/L AS NA)	09/21/92-09/21/92	0	1	
BADL0018	Yes	00930	SODIUM, DISSOLVED (MG/L AS NA)	09/21/92-06/21/95	2	4	
BADL0019	No	00930	SODIUM, DISSOLVED (MG/L AS NA)	10/08/92-08/29/95	2	4	
BADL0020	No	00930	SODIUM, DISSOLVED (MG/L AS NA)	11/29/62-11/29/62	0	1	
BADL0021	Yes	00930	SODIUM, DISSOLVED (MG/L AS NA)	10/01/92-07/13/95	2	5	
BADL0022	No	00930	SODIUM, DISSOLVED (MG/L AS NA)	09/29/92-09/29/92	0	1	
BADL0023	No	00930	SODIUM, DISSOLVED (MG/L AS NA)	06/01/71-08/21/72	1	3	
BADL0024	Yes	00930	SODIUM, DISSOLVED (MG/L AS NA)	09/30/92-09/30/92	0	1	
BADL0025	Yes	00930	SODIUM, DISSOLVED (MG/L AS NA)	09/18/92-09/18/92	0	1	
BADL0026	No	00930	SODIUM, DISSOLVED (MG/L AS NA)	09/07/93-08/30/94	0	2	
BADL0027	No	00930	SODIUM, DISSOLVED (MG/L AS NA)	05/06/88-09/08/94	6	6	
BADL0028	No	00930	SODIUM, DISSOLVED (MG/L AS NA)	06/03/71-06/03/71	0	1	
BADL0030	No	00930	SODIUM, DISSOLVED (MG/L AS NA)	11/04/92-11/04/92	0	1	
BADL0031	No	00930	SODIUM, DISSOLVED (MG/L AS NA)	11/05/92-07/12/95	2	4	
BADL0005	No	00931	SODIUM ADSORPTION RATIO	11/30/62-11/30/62	0	1	
BADL0011	Yes	00931	SODIUM ADSORPTION RATIO	08/13/64-09/18/67	3	101	T
BADL0012	No	00931	SODIUM ADSORPTION RATIO	11/30/62-11/30/62	0	1	
BADL0020	No	00931	SODIUM ADSORPTION RATIO	11/29/62-11/29/62	0	1	
BADL0005	No	00932	SODIUM, PERCENT	11/30/62-11/30/62	0	1	
BADL0011	Yes	00932	SODIUM, PERCENT	10/01/65-09/18/67	1	82	
BADL0012	No	00932	SODIUM, PERCENT	11/30/62-11/30/62	0	1	
BADL0020	No	00932	SODIUM, PERCENT	11/29/62-11/29/62	0	1	
BADL0001	Yes	00935	POTASSIUM, DISSOLVED (MG/L AS K)	06/29/78-06/29/78	0	1	
BADL0005	No	00935	POTASSIUM, DISSOLVED (MG/L AS K)	11/30/62-11/30/62	0	1	
BADL0008	No	00935	POTASSIUM, DISSOLVED (MG/L AS K)	11/12/92-07/11/95	2	4	
BADL0011	Yes	00935	POTASSIUM, DISSOLVED (MG/L AS K)	08/13/64-09/18/67	3	101	T
BADL0012	No	00935	POTASSIUM, DISSOLVED (MG/L AS K)	11/30/62-11/30/62	0	1	
BADL0013	No	00935	POTASSIUM, DISSOLVED (MG/L AS K)	09/23/92-06/21/95	2	4	
BADL0014	Yes	00935	POTASSIUM, DISSOLVED (MG/L AS K)	09/22/92-08/28/95	2	4	
BADL0015	No	00935	POTASSIUM, DISSOLVED (MG/L AS K)	09/25/92-08/28/95	2	4	
BADL0016	No	00935	POTASSIUM, DISSOLVED (MG/L AS K)	09/30/92-07/10/95	2	4	
BADL0017	Yes	00935	POTASSIUM, DISSOLVED (MG/L AS K)	09/21/92-09/21/92	0	1	
BADL0018	Yes	00935	POTASSIUM, DISSOLVED (MG/L AS K)	09/21/92-06/21/95	2	4	
BADL0019	No	00935	POTASSIUM, DISSOLVED (MG/L AS K)	10/08/92-08/29/95	2	4	
BADL0020	No	00935	POTASSIUM, DISSOLVED (MG/L AS K)	11/29/62-11/29/62	0	1	
BADL0021	Yes	00935	POTASSIUM, DISSOLVED (MG/L AS K)	10/01/92-07/13/95	2	4	
BADL0022	No	00935	POTASSIUM, DISSOLVED (MG/L AS K)	09/29/92-09/29/92	0	1	
BADL0023	No	00935	POTASSIUM, DISSOLVED (MG/L AS K)	06/01/71-08/21/72	1	3	
BADL0024	Yes	00935	POTASSIUM, DISSOLVED (MG/L AS K)	09/30/92-09/30/92	0	1	
BADL0025	Yes	00935	POTASSIUM, DISSOLVED (MG/L AS K)	09/18/92-09/18/92	0	1	
BADL0026	No	00935	POTASSIUM, DISSOLVED (MG/L AS K)	09/07/93-08/30/94	0	2	
BADL0027	No	00935	POTASSIUM, DISSOLVED (MG/L AS K)	05/06/88-09/08/94	6	6	
BADL0028	No	00935	POTASSIUM, DISSOLVED (MG/L AS K)	06/03/71-06/03/71	0	1	
BADL0030	No	00935	POTASSIUM, DISSOLVED (MG/L AS K)	11/04/92-11/04/92	0	1	
BADL0031	No	00935	POTASSIUM, DISSOLVED (MG/L AS K)	11/05/92-07/12/95	2	4	
BADL0009	No	00937	POTASSIUM, TOTAL MG/L AS K)	04/23/79-04/23/79	0	1	
BADL0010	No	00937	POTASSIUM, TOTAL MG/L AS K)	04/18/79-04/18/79	0	1	
BADL0002	No	00940	CHLORIDE,TOTAL IN WATER MG/L	06/08/78-09/19/78	0	2	
BADL0003	No	00940	CHLORIDE,TOTAL IN WATER MG/L	12/19/77-09/19/78	0	4	
BADL0004	No	00940	CHLORIDE,TOTAL IN WATER MG/L	06/08/78-09/19/78	0	2	
BADL0005	No	00940	CHLORIDE,TOTAL IN WATER MG/L	11/30/62-11/30/62	0	1	
BADL0008	No	00940	CHLORIDE,TOTAL IN WATER MG/L	11/12/92-07/11/95	2	4	
BADL0011	Yes	00940	CHLORIDE,TOTAL IN WATER MG/L	08/13/64-09/18/67	3	101	T
BADL0012	No	00940	CHLORIDE,TOTAL IN WATER MG/L	11/30/62-11/30/62	0	1	
BADL0013	No	00940	CHLORIDE,TOTAL IN WATER MG/L	09/23/92-06/21/95	2	4	
BADL0014	Yes	00940	CHLORIDE,TOTAL IN WATER MG/L	09/22/92-08/28/95	2	4	
BADL0015	No	00940	CHLORIDE,TOTAL IN WATER MG/L	09/25/92-08/28/95	2	4	
BADL0016	No	00940	CHLORIDE,TOTAL IN WATER MG/L	09/30/92-07/10/95	2	4	
BADL0017	Yes	00940	CHLORIDE,TOTAL IN WATER MG/L	09/21/92-09/21/92	0	1	
BADL0018	Yes	00940	CHLORIDE,TOTAL IN WATER MG/L	09/21/92-06/21/95	2	4	
BADL0019	No	00940	CHLORIDE,TOTAL IN WATER MG/L	10/08/92-08/29/95	2	4	
BADL0020	No	00940	CHLORIDE,TOTAL IN WATER MG/L	11/29/62-11/29/62	0	1	
BADL0021	Yes	00940	CHLORIDE,TOTAL IN WATER MG/L	10/01/92-07/13/95	2	5	
BADL0022	No	00940	CHLORIDE,TOTAL IN WATER MG/L	09/29/92-09/29/92	0	1	
BADL0023	No	00940	CHLORIDE,TOTAL IN WATER MG/L	06/01/71-08/21/72	1	3	
BADL0024	Yes	00940	CHLORIDE,TOTAL IN WATER MG/L	09/30/92-09/30/92	0	1	
BADL0025	Yes	00940	CHLORIDE,TOTAL IN WATER MG/L	09/18/92-09/18/92	0	1	
BADL0026	No	00940	CHLORIDE,TOTAL IN WATER MG/L	09/07/93-08/30/94	0	2	
BADL0027	No	00940	CHLORIDE,TOTAL IN WATER MG/L	05/06/88-09/08/94	6	6	
BADL0028	No	00940	CHLORIDE,TOTAL IN WATER MG/L	06/03/71-06/03/71	0	1	
BADL0030	No	00940	CHLORIDE,TOTAL IN WATER MG/L	11/04/92-11/04/92	0	1	

[1]T=Times Series Plot, A=Annual Plot, and S=Seasonal Plot

Station/Parameter Period of Record Tabulation
From 11/29/62 To 09/10/96

Station	In Park	Code	Name	Start - End	Years	Obs	Plots[1]
BADL0031	No	00940	CHLORIDE,TOTAL IN WATER MG/L	11/05/92-07/12/95	2	4	
BADL0001	Yes	00941	CHLORIDE, DISSOLVED IN WATER MG/L	06/29/78-06/29/78	0	1	
BADL0002	No	00945	SULFATE, TOTAL (MG/L AS SO4)	06/08/78-09/19/78	0	2	
BADL0003	No	00945	SULFATE, TOTAL (MG/L AS SO4)	03/15/78-09/19/78	0	2	
BADL0004	No	00945	SULFATE, TOTAL (MG/L AS SO4)	06/08/78-09/19/78	0	2	
BADL0005	No	00945	SULFATE, TOTAL (MG/L AS SO4)	11/30/62-11/30/62	0	1	
BADL0008	No	00945	SULFATE, TOTAL (MG/L AS SO4)	11/12/92-07/11/95	2	4	
BADL0009	No	00945	SULFATE, TOTAL (MG/L AS SO4)	04/23/79-04/23/79	0	1	
BADL0010	No	00945	SULFATE, TOTAL (MG/L AS SO4)	04/18/79-04/18/79	0	1	
BADL0011	Yes	00945	SULFATE, TOTAL (MG/L AS SO4)	08/13/64-09/18/67	3	101	T
BADL0012	No	00945	SULFATE, TOTAL (MG/L AS SO4)	11/30/62-11/30/62	0	1	
BADL0013	No	00945	SULFATE, TOTAL (MG/L AS SO4)	09/23/92-06/21/95	2	4	
BADL0014	Yes	00945	SULFATE, TOTAL (MG/L AS SO4)	09/22/92-08/28/95	2	4	
BADL0015	No	00945	SULFATE, TOTAL (MG/L AS SO4)	09/25/92-08/28/95	2	4	
BADL0016	No	00945	SULFATE, TOTAL (MG/L AS SO4)	09/30/92-07/10/95	2	4	
BADL0017	Yes	00945	SULFATE, TOTAL (MG/L AS SO4)	09/21/92-09/21/92	0	1	
BADL0018	Yes	00945	SULFATE, TOTAL (MG/L AS SO4)	09/21/92-06/21/95	2	4	
BADL0019	No	00945	SULFATE, TOTAL (MG/L AS SO4)	10/08/92-08/29/95	2	4	
BADL0020	No	00945	SULFATE, TOTAL (MG/L AS SO4)	11/29/62-11/29/62	0	1	
BADL0021	Yes	00945	SULFATE, TOTAL (MG/L AS SO4)	10/01/92-07/13/95	2	5	
BADL0022	No	00945	SULFATE, TOTAL (MG/L AS SO4)	09/29/92-09/29/92	0	1	
BADL0023	No	00945	SULFATE, TOTAL (MG/L AS SO4)	06/01/71-08/21/72	1	3	
BADL0024	Yes	00945	SULFATE, TOTAL (MG/L AS SO4)	09/30/92-09/30/92	0	1	
BADL0025	Yes	00945	SULFATE, TOTAL (MG/L AS SO4)	09/18/92-09/18/92	0	1	
BADL0026	No	00945	SULFATE, TOTAL (MG/L AS SO4)	09/07/93-08/30/94	0	2	
BADL0027	No	00945	SULFATE, TOTAL (MG/L AS SO4)	05/06/88-09/08/94	6	6	
BADL0028	No	00945	SULFATE, TOTAL (MG/L AS SO4)	06/03/71-06/03/71	0	1	
BADL0030	No	00945	SULFATE, TOTAL (MG/L AS SO4)	11/04/92-11/04/92	0	1	
BADL0031	No	00945	SULFATE, TOTAL (MG/L AS SO4)	11/05/92-07/12/95	2	4	
BADL0001	Yes	00946	SULFATE, DISSOLVED (MG/L AS SO4)	06/29/78-06/29/78	0	1	
BADL0003	No	00946	SULFATE, DISSOLVED (MG/L AS SO4)	12/19/77-06/08/78	0	2	
BADL0005	No	00950	FLUORIDE, DISSOLVED (MG/L AS F)	11/30/62-11/30/62	0	1	
BADL0008	No	00950	FLUORIDE, DISSOLVED (MG/L AS F)	11/12/92-07/11/95	2	4	
BADL0011	Yes	00950	FLUORIDE, DISSOLVED (MG/L AS F)	08/13/64-09/18/67	3	101	T
BADL0012	No	00950	FLUORIDE, DISSOLVED (MG/L AS F)	11/30/62-11/30/62	0	1	
BADL0013	No	00950	FLUORIDE, DISSOLVED (MG/L AS F)	09/23/92-06/21/95	2	4	
BADL0014	Yes	00950	FLUORIDE, DISSOLVED (MG/L AS F)	09/22/92-08/28/95	2	4	
BADL0015	No	00950	FLUORIDE, DISSOLVED (MG/L AS F)	09/25/92-08/28/95	2	4	
BADL0016	No	00950	FLUORIDE, DISSOLVED (MG/L AS F)	09/30/92-07/10/95	2	4	
BADL0017	Yes	00950	FLUORIDE, DISSOLVED (MG/L AS F)	09/21/92-09/21/92	0	1	
BADL0018	Yes	00950	FLUORIDE, DISSOLVED (MG/L AS F)	09/21/92-06/21/95	2	4	
BADL0019	No	00950	FLUORIDE, DISSOLVED (MG/L AS F)	10/08/92-08/29/95	2	4	
BADL0020	No	00950	FLUORIDE, DISSOLVED (MG/L AS F)	11/29/62-11/29/62	0	1	
BADL0021	Yes	00950	FLUORIDE, DISSOLVED (MG/L AS F)	10/01/92-07/13/95	2	5	
BADL0022	No	00950	FLUORIDE, DISSOLVED (MG/L AS F)	09/29/92-09/29/92	0	1	
BADL0024	Yes	00950	FLUORIDE, DISSOLVED (MG/L AS F)	09/30/92-09/30/92	0	1	
BADL0025	Yes	00950	FLUORIDE, DISSOLVED (MG/L AS F)	09/18/92-09/18/92	0	1	
BADL0026	No	00950	FLUORIDE, DISSOLVED (MG/L AS F)	09/07/93-08/30/94	0	2	
BADL0031	No	00950	FLUORIDE, DISSOLVED (MG/L AS F)	11/05/92-07/12/95	2	4	
BADL0005	No	00955	SILICA, DISSOLVED (MG/L AS SI02)	11/30/62-11/30/62	0	1	
BADL0008	No	00955	SILICA, DISSOLVED (MG/L AS SI02)	11/12/92-07/11/95	2	4	
BADL0011	Yes	00955	SILICA, DISSOLVED (MG/L AS SI02)	08/13/64-09/18/67	3	103	T
BADL0012	No	00955	SILICA, DISSOLVED (MG/L AS SI02)	11/30/62-11/30/62	0	1	
BADL0013	No	00955	SILICA, DISSOLVED (MG/L AS SI02)	09/23/92-06/21/95	2	4	
BADL0014	Yes	00955	SILICA, DISSOLVED (MG/L AS SI02)	09/22/92-08/28/95	2	4	
BADL0015	No	00955	SILICA, DISSOLVED (MG/L AS SI02)	09/25/92-08/28/95	2	4	
BADL0016	No	00955	SILICA, DISSOLVED (MG/L AS SI02)	09/30/92-07/10/95	2	4	
BADL0017	Yes	00955	SILICA, DISSOLVED (MG/L AS SI02)	09/21/92-09/21/92	0	1	
BADL0018	Yes	00955	SILICA, DISSOLVED (MG/L AS SI02)	09/21/92-06/21/95	2	4	
BADL0019	No	00955	SILICA, DISSOLVED (MG/L AS SI02)	10/08/92-08/29/95	2	4	
BADL0020	No	00955	SILICA, DISSOLVED (MG/L AS SI02)	11/29/62-11/29/62	0	1	
BADL0021	Yes	00955	SILICA, DISSOLVED (MG/L AS SI02)	10/01/92-07/13/95	2	5	
BADL0022	No	00955	SILICA, DISSOLVED (MG/L AS SI02)	09/29/92-09/29/92	0	1	
BADL0023	No	00955	SILICA, DISSOLVED (MG/L AS SI02)	06/01/71-08/21/72	1	3	
BADL0024	Yes	00955	SILICA, DISSOLVED (MG/L AS SI02)	09/30/92-09/30/92	0	1	
BADL0025	Yes	00955	SILICA, DISSOLVED (MG/L AS SI02)	09/18/92-09/18/92	0	1	
BADL0026	No	00955	SILICA, DISSOLVED (MG/L AS SI02)	09/07/93-08/30/94	0	2	
BADL0028	No	00955	SILICA, DISSOLVED (MG/L AS SI02)	06/03/71-06/03/71	0	1	
BADL0030	No	00955	SILICA, DISSOLVED (MG/L AS SI02)	11/04/92-11/04/92	0	1	
BADL0031	No	00955	SILICA, DISSOLVED (MG/L AS SI02)	11/05/92-07/12/95	2	4	
BADL0008	No	01000	ARSENIC, DISSOLVED (UG/L AS AS)	11/12/92-07/11/95	2	4	
BADL0013	No	01000	ARSENIC, DISSOLVED (UG/L AS AS)	09/23/92-06/21/95	2	4	

[1] T=Times Series Plot, A=Annual Plot, and S=Seasonal Plot

56

Station/Parameter Period of Record Tabulation
From 11/29/62 To 09/10/96

Station	In Park	Code	Name	Start - End	Years	Obs	Plots[1]
BADL0014	Yes	01000	ARSENIC, DISSOLVED (UG/L AS AS)	09/22/92-08/28/95	2	4	
BADL0015	No	01000	ARSENIC, DISSOLVED (UG/L AS AS)	09/25/92-08/28/95	2	4	
BADL0016	No	01000	ARSENIC, DISSOLVED (UG/L AS AS)	09/30/92-07/10/95	2	4	
BADL0017	Yes	01000	ARSENIC, DISSOLVED (UG/L AS AS)	09/21/92-09/21/92	0	1	
BADL0018	Yes	01000	ARSENIC, DISSOLVED (UG/L AS AS)	09/21/92-06/21/95	2	4	
BADL0019	No	01000	ARSENIC, DISSOLVED (UG/L AS AS)	10/08/92-08/29/95	2	4	
BADL0021	Yes	01000	ARSENIC, DISSOLVED (UG/L AS AS)	10/01/92-07/13/95	2	5	
BADL0022	No	01000	ARSENIC, DISSOLVED (UG/L AS AS)	09/29/92-09/29/92	0	1	
BADL0024	Yes	01000	ARSENIC, DISSOLVED (UG/L AS AS)	09/30/92-09/30/92	0	1	
BADL0025	Yes	01000	ARSENIC, DISSOLVED (UG/L AS AS)	09/18/92-09/18/92	0	1	
BADL0026	No	01000	ARSENIC, DISSOLVED (UG/L AS AS)	09/07/93-08/30/94	0	2	
BADL0027	No	01000	ARSENIC, DISSOLVED (UG/L AS AS)	05/06/88-09/08/94	6	6	
BADL0030	No	01000	ARSENIC, DISSOLVED (UG/L AS AS)	11/04/92-11/04/92	0	1	
BADL0031	No	01000	ARSENIC, DISSOLVED (UG/L AS AS)	11/05/92-07/12/95	2	4	
BADL0009	No	01002	ARSENIC, TOTAL (UG/L AS AS)	04/23/79-04/23/79	0	1	
BADL0010	No	01002	ARSENIC, TOTAL (UG/L AS AS)	04/18/79-04/18/79	0	1	
BADL0026	No	01002	ARSENIC, TOTAL (UG/L AS AS)	09/07/93-08/30/94	0	2	
BADL0008	No	01005	BARIUM, DISSOLVED (UG/L AS BA)	11/12/92-07/11/95	2	4	
BADL0013	No	01005	BARIUM, DISSOLVED (UG/L AS BA)	09/23/92-06/21/95	2	4	
BADL0014	Yes	01005	BARIUM, DISSOLVED (UG/L AS BA)	09/22/92-08/28/95	2	4	
BADL0015	No	01005	BARIUM, DISSOLVED (UG/L AS BA)	09/25/92-08/28/95	2	4	
BADL0016	No	01005	BARIUM, DISSOLVED (UG/L AS BA)	09/30/92-07/10/95	2	4	
BADL0017	Yes	01005	BARIUM, DISSOLVED (UG/L AS BA)	09/21/92-09/21/92	0	1	
BADL0018	Yes	01005	BARIUM, DISSOLVED (UG/L AS BA)	09/21/92-06/21/95	2	4	
BADL0019	No	01005	BARIUM, DISSOLVED (UG/L AS BA)	10/08/92-08/29/95	2	4	
BADL0021	Yes	01005	BARIUM, DISSOLVED (UG/L AS BA)	10/01/92-07/13/95	2	5	
BADL0022	No	01005	BARIUM, DISSOLVED (UG/L AS BA)	09/29/92-09/29/92	0	1	
BADL0024	Yes	01005	BARIUM, DISSOLVED (UG/L AS BA)	09/30/92-09/30/92	0	1	
BADL0025	Yes	01005	BARIUM, DISSOLVED (UG/L AS BA)	09/18/92-09/18/92	0	1	
BADL0026	No	01005	BARIUM, DISSOLVED (UG/L AS BA)	09/07/93-08/30/94	0	2	
BADL0030	No	01005	BARIUM, DISSOLVED (UG/L AS BA)	11/04/92-11/04/92	0	1	
BADL0031	No	01005	BARIUM, DISSOLVED (UG/L AS BA)	11/05/92-07/12/95	2	4	
BADL0009	No	01007	BARIUM, TOTAL (UG/L AS BA)	04/23/79-04/23/79	0	1	
BADL0010	No	01007	BARIUM, TOTAL (UG/L AS BA)	04/18/79-04/18/79	0	1	
BADL0008	No	01010	BERYLLIUM, DISSOLVED (UG/L AS BE)	11/12/92-07/11/95	2	4	
BADL0013	No	01010	BERYLLIUM, DISSOLVED (UG/L AS BE)	09/23/92-06/21/95	2	4	
BADL0014	Yes	01010	BERYLLIUM, DISSOLVED (UG/L AS BE)	09/22/92-08/28/95	2	4	
BADL0015	No	01010	BERYLLIUM, DISSOLVED (UG/L AS BE)	09/25/92-08/28/95	2	4	
BADL0016	No	01010	BERYLLIUM, DISSOLVED (UG/L AS BE)	09/30/92-07/10/95	2	4	
BADL0017	Yes	01010	BERYLLIUM, DISSOLVED (UG/L AS BE)	09/21/92-09/21/92	0	1	
BADL0018	Yes	01010	BERYLLIUM, DISSOLVED (UG/L AS BE)	09/21/92-06/21/95	2	4	
BADL0019	No	01010	BERYLLIUM, DISSOLVED (UG/L AS BE)	10/08/92-08/29/95	2	4	
BADL0021	Yes	01010	BERYLLIUM, DISSOLVED (UG/L AS BE)	10/01/92-07/13/95	2	5	
BADL0022	No	01010	BERYLLIUM, DISSOLVED (UG/L AS BE)	09/29/92-09/29/92	0	1	
BADL0024	Yes	01010	BERYLLIUM, DISSOLVED (UG/L AS BE)	09/30/92-09/30/92	0	1	
BADL0025	Yes	01010	BERYLLIUM, DISSOLVED (UG/L AS BE)	09/18/92-09/18/92	0	1	
BADL0026	No	01010	BERYLLIUM, DISSOLVED (UG/L AS BE)	09/07/93-08/30/94	0	2	
BADL0030	No	01010	BERYLLIUM, DISSOLVED (UG/L AS BE)	11/04/92-11/04/92	0	1	
BADL0031	No	01010	BERYLLIUM, DISSOLVED (UG/L AS BE)	11/05/92-07/12/95	2	4	
BADL0009	No	01012	BERYLLIUM, TOTAL (UG/L AS BE)	04/23/79-04/23/79	0	1	
BADL0010	No	01012	BERYLLIUM, TOTAL (UG/L AS BE)	04/18/79-04/18/79	0	1	
BADL0005	No	01020	BORON, DISSOLVED (UG/L AS B)	11/30/62-11/30/62	0	1	
BADL0008	No	01020	BORON, DISSOLVED (UG/L AS B)	11/12/92-07/11/95	2	4	
BADL0011	Yes	01020	BORON, DISSOLVED (UG/L AS B)	08/13/64-09/18/67	3	101	T
BADL0012	No	01020	BORON, DISSOLVED (UG/L AS B)	11/30/62-11/30/62	0	1	
BADL0013	No	01020	BORON, DISSOLVED (UG/L AS B)	09/23/92-06/21/95	2	4	
BADL0014	Yes	01020	BORON, DISSOLVED (UG/L AS B)	09/22/92-08/28/95	2	4	
BADL0015	No	01020	BORON, DISSOLVED (UG/L AS B)	09/25/92-08/28/95	2	4	
BADL0016	No	01020	BORON, DISSOLVED (UG/L AS B)	09/30/92-07/10/95	2	4	
BADL0017	Yes	01020	BORON, DISSOLVED (UG/L AS B)	09/21/92-09/21/92	0	1	
BADL0018	Yes	01020	BORON, DISSOLVED (UG/L AS B)	09/21/92-06/21/95	2	4	
BADL0019	No	01020	BORON, DISSOLVED (UG/L AS B)	10/08/92-08/29/95	2	4	
BADL0020	No	01020	BORON, DISSOLVED (UG/L AS B)	11/29/62-11/29/62	0	1	
BADL0021	Yes	01020	BORON, DISSOLVED (UG/L AS B)	10/01/92-07/13/95	2	5	
BADL0022	No	01020	BORON, DISSOLVED (UG/L AS B)	09/29/92-09/29/92	0	1	
BADL0025	Yes	01020	BORON, DISSOLVED (UG/L AS B)	09/18/92-09/18/92	0	1	
BADL0026	No	01020	BORON, DISSOLVED (UG/L AS B)	09/07/93-08/30/94	0	2	
BADL0027	No	01020	BORON, DISSOLVED (UG/L AS B)	05/06/88-09/08/94	6	6	
BADL0030	No	01020	BORON, DISSOLVED (UG/L AS B)	11/04/92-11/04/92	0	1	
BADL0031	No	01020	BORON, DISSOLVED (UG/L AS B)	11/05/92-07/12/95	2	4	
BADL0009	No	01022	BORON, TOTAL (UG/L AS B)	04/23/79-04/23/79	0	1	
BADL0010	No	01022	BORON, TOTAL (UG/L AS B)	04/18/79-04/18/79	0	1	

[1]T=Times Series Plot, A=Annual Plot, and S=Seasonal Plot

Station/Parameter Period of Record Tabulation
From 11/29/62 To 09/10/96

Station	In Park	Code	Name	Start - End	Years	Obs	Plots[1]
BADL0008	No	01025	CADMIUM, DISSOLVED (UG/L AS CD)	11/12/92-07/11/95	2	4	
BADL0013	No	01025	CADMIUM, DISSOLVED (UG/L AS CD)	09/23/92-06/21/95	2	4	
BADL0014	Yes	01025	CADMIUM, DISSOLVED (UG/L AS CD)	09/22/92-08/28/95	2	4	
BADL0015	No	01025	CADMIUM, DISSOLVED (UG/L AS CD)	09/25/92-08/28/95	2	4	
BADL0016	No	01025	CADMIUM, DISSOLVED (UG/L AS CD)	09/30/92-07/10/95	2	4	
BADL0017	Yes	01025	CADMIUM, DISSOLVED (UG/L AS CD)	09/21/92-09/21/92	0	1	
BADL0018	Yes	01025	CADMIUM, DISSOLVED (UG/L AS CD)	09/21/92-06/21/95	2	4	
BADL0019	No	01025	CADMIUM, DISSOLVED (UG/L AS CD)	10/08/92-08/29/95	2	4	
BADL0021	Yes	01025	CADMIUM, DISSOLVED (UG/L AS CD)	10/01/92-07/13/95	2	5	
BADL0022	No	01025	CADMIUM, DISSOLVED (UG/L AS CD)	09/29/92-09/29/92	0	1	
BADL0024	Yes	01025	CADMIUM, DISSOLVED (UG/L AS CD)	09/30/92-09/30/92	0	1	
BADL0025	Yes	01025	CADMIUM, DISSOLVED (UG/L AS CD)	09/18/92-09/18/92	0	1	
BADL0026	No	01025	CADMIUM, DISSOLVED (UG/L AS CD)	09/07/93-08/30/94	0	2	
BADL0027	No	01025	CADMIUM, DISSOLVED (UG/L AS CD)	05/06/88-09/08/94	6	6	
BADL0030	No	01025	CADMIUM, DISSOLVED (UG/L AS CD)	11/04/92-11/04/92	0	1	
BADL0031	No	01025	CADMIUM, DISSOLVED (UG/L AS CD)	11/05/92-07/12/95	2	4	
BADL0008	No	01027	CADMIUM, TOTAL (UG/L AS CD)	11/12/92-07/11/95	2	4	
BADL0013	No	01027	CADMIUM, TOTAL (UG/L AS CD)	09/23/92-06/21/95	2	3	
BADL0014	Yes	01027	CADMIUM, TOTAL (UG/L AS CD)	09/22/92-08/28/95	2	3	
BADL0015	No	01027	CADMIUM, TOTAL (UG/L AS CD)	09/25/92-08/28/95	2	3	
BADL0016	No	01027	CADMIUM, TOTAL (UG/L AS CD)	09/30/92-07/10/95	2	3	
BADL0017	Yes	01027	CADMIUM, TOTAL (UG/L AS CD)	09/21/92-09/21/92	0	1	
BADL0018	Yes	01027	CADMIUM, TOTAL (UG/L AS CD)	09/21/92-06/21/95	2	3	
BADL0019	No	01027	CADMIUM, TOTAL (UG/L AS CD)	10/08/92-08/29/95	2	3	
BADL0021	Yes	01027	CADMIUM, TOTAL (UG/L AS CD)	10/01/92-07/13/95	2	4	
BADL0022	No	01027	CADMIUM, TOTAL (UG/L AS CD)	09/29/92-09/29/92	0	1	
BADL0024	Yes	01027	CADMIUM, TOTAL (UG/L AS CD)	09/30/92-09/30/92	0	1	
BADL0025	Yes	01027	CADMIUM, TOTAL (UG/L AS CD)	09/18/92-09/18/92	0	1	
BADL0030	No	01027	CADMIUM, TOTAL (UG/L AS CD)	11/04/92-11/04/92	0	1	
BADL0031	No	01027	CADMIUM, TOTAL (UG/L AS CD)	11/05/92-07/12/95	2	4	
BADL0008	No	01030	CHROMIUM, DISSOLVED (UG/L AS CR)	11/12/92-07/11/95	2	4	
BADL0013	No	01030	CHROMIUM, DISSOLVED (UG/L AS CR)	09/23/92-06/21/95	2	4	
BADL0014	Yes	01030	CHROMIUM, DISSOLVED (UG/L AS CR)	09/22/92-08/28/95	2	4	
BADL0015	No	01030	CHROMIUM, DISSOLVED (UG/L AS CR)	09/25/92-08/28/95	2	4	
BADL0016	No	01030	CHROMIUM, DISSOLVED (UG/L AS CR)	09/30/92-07/10/95	2	4	
BADL0017	Yes	01030	CHROMIUM, DISSOLVED (UG/L AS CR)	09/21/92-09/21/92	0	1	
BADL0018	Yes	01030	CHROMIUM, DISSOLVED (UG/L AS CR)	09/21/92-06/21/95	2	4	
BADL0019	No	01030	CHROMIUM, DISSOLVED (UG/L AS CR)	10/08/92-08/29/95	2	4	
BADL0021	Yes	01030	CHROMIUM, DISSOLVED (UG/L AS CR)	10/01/92-07/13/95	2	5	
BADL0022	No	01030	CHROMIUM, DISSOLVED (UG/L AS CR)	09/29/92-09/29/92	0	1	
BADL0024	Yes	01030	CHROMIUM, DISSOLVED (UG/L AS CR)	09/30/92-09/30/92	0	1	
BADL0025	Yes	01030	CHROMIUM, DISSOLVED (UG/L AS CR)	09/18/92-09/18/92	0	1	
BADL0026	No	01030	CHROMIUM, DISSOLVED (UG/L AS CR)	09/07/93-08/30/94	0	2	
BADL0027	No	01030	CHROMIUM, DISSOLVED (UG/L AS CR)	05/06/88-09/08/94	6	6	
BADL0030	No	01030	CHROMIUM, DISSOLVED (UG/L AS CR)	11/04/92-11/04/92	0	1	
BADL0031	No	01030	CHROMIUM, DISSOLVED (UG/L AS CR)	11/05/92-07/12/95	2	4	
BADL0009	No	01034	CHROMIUM, TOTAL (UG/L AS CR)	04/23/79-04/23/79	0	1	
BADL0010	No	01034	CHROMIUM, TOTAL (UG/L AS CR)	04/18/79-04/18/79	0	1	
BADL0011	Yes	01034	CHROMIUM, TOTAL (UG/L AS CR)	01/31/67-01/31/67	0	1	
BADL0008	No	01035	COBALT, DISSOLVED (UG/L AS CO)	11/12/92-07/11/95	2	4	
BADL0013	No	01035	COBALT, DISSOLVED (UG/L AS CO)	09/23/92-06/21/95	2	4	
BADL0014	Yes	01035	COBALT, DISSOLVED (UG/L AS CO)	09/22/92-08/28/95	2	4	
BADL0015	No	01035	COBALT, DISSOLVED (UG/L AS CO)	09/25/92-08/28/95	2	4	
BADL0016	No	01035	COBALT, DISSOLVED (UG/L AS CO)	09/30/92-07/10/95	2	4	
BADL0017	Yes	01035	COBALT, DISSOLVED (UG/L AS CO)	09/21/92-09/21/92	0	1	
BADL0018	Yes	01035	COBALT, DISSOLVED (UG/L AS CO)	09/21/92-06/21/95	2	4	
BADL0019	No	01035	COBALT, DISSOLVED (UG/L AS CO)	10/08/92-08/29/95	2	4	
BADL0021	Yes	01035	COBALT, DISSOLVED (UG/L AS CO)	10/01/92-07/13/95	2	5	
BADL0022	No	01035	COBALT, DISSOLVED (UG/L AS CO)	09/29/92-09/29/92	0	1	
BADL0024	Yes	01035	COBALT, DISSOLVED (UG/L AS CO)	09/30/92-09/30/92	0	1	
BADL0025	Yes	01035	COBALT, DISSOLVED (UG/L AS CO)	09/18/92-09/18/92	0	1	
BADL0026	No	01035	COBALT, DISSOLVED (UG/L AS CO)	08/30/94-08/30/94	0	1	
BADL0030	No	01035	COBALT, DISSOLVED (UG/L AS CO)	11/04/92-11/04/92	0	1	
BADL0031	No	01035	COBALT, DISSOLVED (UG/L AS CO)	11/05/92-07/12/95	2	4	
BADL0009	No	01037	COBALT, TOTAL (UG/L AS CO)	04/23/79-04/23/79	0	1	
BADL0010	No	01037	COBALT, TOTAL (UG/L AS CO)	04/18/79-04/18/79	0	1	
BADL0008	No	01040	COPPER, DISSOLVED (UG/L AS CU)	11/12/92-07/11/95	2	4	
BADL0011	Yes	01040	COPPER, DISSOLVED (UG/L AS CU)	01/31/67-01/31/67	0	1	
BADL0013	No	01040	COPPER, DISSOLVED (UG/L AS CU)	09/23/92-06/21/95	2	4	
BADL0014	Yes	01040	COPPER, DISSOLVED (UG/L AS CU)	09/22/92-08/28/95	2	4	
BADL0015	No	01040	COPPER, DISSOLVED (UG/L AS CU)	09/25/92-08/28/95	2	4	
BADL0016	No	01040	COPPER, DISSOLVED (UG/L AS CU)	09/30/92-07/10/95	2	4	

[1]T=Times Series Plot, A=Annual Plot, and S=Seasonal Plot

Station/Parameter Period of Record Tabulation
From 11/29/62 To 09/10/96

Station	In Park	Code	Name	Start - End	Years	Obs	Plots[1]
BADL0017	Yes	01040	COPPER, DISSOLVED (UG/L AS CU)	09/21/92-09/21/92	0	1	
BADL0018	Yes	01040	COPPER, DISSOLVED (UG/L AS CU)	09/21/92-06/21/95	2	4	
BADL0019	No	01040	COPPER, DISSOLVED (UG/L AS CU)	10/08/92-08/29/95	2	4	
BADL0021	Yes	01040	COPPER, DISSOLVED (UG/L AS CU)	10/01/92-07/13/95	2	5	
BADL0022	No	01040	COPPER, DISSOLVED (UG/L AS CU)	09/29/92-09/29/92	0	1	
BADL0024	Yes	01040	COPPER, DISSOLVED (UG/L AS CU)	09/30/92-09/30/92	0	1	
BADL0025	Yes	01040	COPPER, DISSOLVED (UG/L AS CU)	09/18/92-09/18/92	0	1	
BADL0026	No	01040	COPPER, DISSOLVED (UG/L AS CU)	09/07/93-08/30/94	0	2	
BADL0027	No	01040	COPPER, DISSOLVED (UG/L AS CU)	05/06/88-09/08/94	6	6	
BADL0030	No	01040	COPPER, DISSOLVED (UG/L AS CU)	11/04/92-11/04/92	0	1	
BADL0031	No	01040	COPPER, DISSOLVED (UG/L AS CU)	11/05/92-07/12/95	2	4	
BADL0009	No	01042	COPPER, TOTAL (UG/L AS CU)	04/23/79-04/23/79	0	1	
BADL0010	No	01042	COPPER, TOTAL (UG/L AS CU)	04/18/79-04/18/79	0	1	
BADL0009	No	01045	IRON, TOTAL (UG/L AS FE)	04/23/79-04/23/79	0	1	
BADL0010	No	01045	IRON, TOTAL (UG/L AS FE)	04/18/79-04/18/79	0	1	
BADL0023	No	01045	IRON, TOTAL (UG/L AS FE)	06/01/71-08/21/72	1	3	
BADL0026	No	01045	IRON, TOTAL (UG/L AS FE)	09/07/93-08/30/94	0	2	
BADL0028	No	01045	IRON, TOTAL (UG/L AS FE)	06/03/71-06/03/71	0	1	
BADL0001	Yes	01046	IRON, DISSOLVED (UG/L AS FE)	06/29/78-06/29/78	0	1	
BADL0008	No	01046	IRON, DISSOLVED (UG/L AS FE)	11/12/92-07/11/95	2	4	
BADL0013	No	01046	IRON, DISSOLVED (UG/L AS FE)	09/23/92-06/21/95	2	4	
BADL0014	Yes	01046	IRON, DISSOLVED (UG/L AS FE)	09/22/92-08/28/95	2	4	
BADL0015	No	01046	IRON, DISSOLVED (UG/L AS FE)	09/25/92-08/28/95	2	4	
BADL0016	No	01046	IRON, DISSOLVED (UG/L AS FE)	09/30/92-07/10/95	2	4	
BADL0017	Yes	01046	IRON, DISSOLVED (UG/L AS FE)	09/21/92-09/21/92	0	1	
BADL0018	Yes	01046	IRON, DISSOLVED (UG/L AS FE)	09/21/92-06/21/95	2	4	
BADL0019	No	01046	IRON, DISSOLVED (UG/L AS FE)	10/08/92-08/29/95	2	4	
BADL0021	Yes	01046	IRON, DISSOLVED (UG/L AS FE)	10/01/92-07/13/95	2	5	
BADL0022	No	01046	IRON, DISSOLVED (UG/L AS FE)	09/29/92-09/29/92	0	1	
BADL0024	Yes	01046	IRON, DISSOLVED (UG/L AS FE)	09/30/92-09/30/92	0	1	
BADL0025	Yes	01046	IRON, DISSOLVED (UG/L AS FE)	09/18/92-09/18/92	0	1	
BADL0026	No	01046	IRON, DISSOLVED (UG/L AS FE)	09/07/93-08/30/94	0	2	
BADL0030	No	01046	IRON, DISSOLVED (UG/L AS FE)	11/04/92-11/04/92	0	1	
BADL0031	No	01046	IRON, DISSOLVED (UG/L AS FE)	11/05/92-07/12/95	2	4	
BADL0008	No	01049	LEAD, DISSOLVED (UG/L AS PB)	11/12/92-07/11/95	2	4	
BADL0013	No	01049	LEAD, DISSOLVED (UG/L AS PB)	09/23/92-06/21/95	2	4	
BADL0014	Yes	01049	LEAD, DISSOLVED (UG/L AS PB)	09/22/92-08/28/95	2	4	
BADL0015	No	01049	LEAD, DISSOLVED (UG/L AS PB)	09/25/92-08/28/95	2	4	
BADL0016	No	01049	LEAD, DISSOLVED (UG/L AS PB)	09/30/92-07/10/95	2	4	
BADL0017	Yes	01049	LEAD, DISSOLVED (UG/L AS PB)	09/21/92-09/21/92	0	1	
BADL0018	Yes	01049	LEAD, DISSOLVED (UG/L AS PB)	09/21/92-06/21/95	2	4	
BADL0019	No	01049	LEAD, DISSOLVED (UG/L AS PB)	10/08/92-08/29/95	2	4	
BADL0021	Yes	01049	LEAD, DISSOLVED (UG/L AS PB)	10/01/92-07/13/95	2	5	
BADL0022	No	01049	LEAD, DISSOLVED (UG/L AS PB)	09/29/92-09/29/92	0	1	
BADL0024	Yes	01049	LEAD, DISSOLVED (UG/L AS PB)	09/30/92-09/30/92	0	1	
BADL0025	Yes	01049	LEAD, DISSOLVED (UG/L AS PB)	09/18/92-09/18/92	0	1	
BADL0026	No	01049	LEAD, DISSOLVED (UG/L AS PB)	09/07/93-08/30/94	0	2	
BADL0027	No	01049	LEAD, DISSOLVED (UG/L AS PB)	05/06/88-09/08/94	6	6	
BADL0030	No	01049	LEAD, DISSOLVED (UG/L AS PB)	11/04/92-11/04/92	0	1	
BADL0031	No	01049	LEAD, DISSOLVED (UG/L AS PB)	11/05/92-07/12/95	2	4	
BADL0009	No	01055	MANGANESE, TOTAL (UG/L AS MN)	04/23/79-04/23/79	0	1	
BADL0010	No	01055	MANGANESE, TOTAL (UG/L AS MN)	04/18/79-04/18/79	0	1	
BADL0011	Yes	01055	MANGANESE, TOTAL (UG/L AS MN)	08/13/64-07/22/67	2	34	T
BADL0023	No	01055	MANGANESE, TOTAL (UG/L AS MN)	06/01/71-08/21/72	1	3	
BADL0028	No	01055	MANGANESE, TOTAL (UG/L AS MN)	06/03/71-06/03/71	0	1	
BADL0001	Yes	01056	MANGANESE, DISSOLVED (UG/L AS MN)	06/29/78-06/29/78	0	1	
BADL0008	No	01056	MANGANESE, DISSOLVED (UG/L AS MN)	11/12/92-07/11/95	2	4	
BADL0013	No	01056	MANGANESE, DISSOLVED (UG/L AS MN)	09/23/92-06/21/95	2	4	
BADL0014	Yes	01056	MANGANESE, DISSOLVED (UG/L AS MN)	09/22/92-08/28/95	2	4	
BADL0015	No	01056	MANGANESE, DISSOLVED (UG/L AS MN)	09/25/92-08/28/95	2	4	
BADL0016	No	01056	MANGANESE, DISSOLVED (UG/L AS MN)	09/30/92-07/10/95	2	4	
BADL0017	Yes	01056	MANGANESE, DISSOLVED (UG/L AS MN)	09/21/92-09/21/92	0	1	
BADL0018	Yes	01056	MANGANESE, DISSOLVED (UG/L AS MN)	09/21/92-06/21/95	2	4	
BADL0019	No	01056	MANGANESE, DISSOLVED (UG/L AS MN)	10/08/92-08/29/95	2	4	
BADL0021	Yes	01056	MANGANESE, DISSOLVED (UG/L AS MN)	10/01/92-07/13/95	2	5	
BADL0022	No	01056	MANGANESE, DISSOLVED (UG/L AS MN)	09/29/92-09/29/92	0	1	
BADL0024	Yes	01056	MANGANESE, DISSOLVED (UG/L AS MN)	09/30/92-09/30/92	0	1	
BADL0025	Yes	01056	MANGANESE, DISSOLVED (UG/L AS MN)	09/18/92-09/18/92	0	1	
BADL0026	No	01056	MANGANESE, DISSOLVED (UG/L AS MN)	09/07/93-08/30/94	0	2	
BADL0031	No	01056	MANGANESE, DISSOLVED (UG/L AS MN)	11/05/92-07/12/95	2	4	
BADL0008	No	01057	THALLIUM, DISSOLVED (UG/L AS TL)	05/17/95-07/11/95	0	2	
BADL0013	No	01057	THALLIUM, DISSOLVED (UG/L AS TL)	04/27/95-06/21/95	0	2	

[1]T=Times Series Plot, A=Annual Plot, and S=Seasonal Plot

Station/Parameter Period of Record Tabulation
From 11/29/62 To 09/10/96

Station	In Park	Code	Name	Start - End	Years	Obs	Plots[1]
BADL0014	Yes	01057	THALLIUM, DISSOLVED (UG/L AS TL)	04/24/95-08/28/95	0	2	
BADL0015	No	01057	THALLIUM, DISSOLVED (UG/L AS TL)	05/11/95-08/28/95	0	2	
BADL0016	No	01057	THALLIUM, DISSOLVED (UG/L AS TL)	04/14/95-07/10/95	0	2	
BADL0018	Yes	01057	THALLIUM, DISSOLVED (UG/L AS TL)	04/27/95-06/21/95	0	2	
BADL0019	No	01057	THALLIUM, DISSOLVED (UG/L AS TL)	05/12/95-08/29/95	0	2	
BADL0021	Yes	01057	THALLIUM, DISSOLVED (UG/L AS TL)	05/16/95-07/13/95	0	2	
BADL0031	No	01057	THALLIUM, DISSOLVED (UG/L AS TL)	05/16/95-07/12/95	0	2	
BADL0008	No	01060	MOLYBDENUM, DISSOLVED (UG/L AS MO)	11/12/92-07/11/95	2	4	
BADL0013	No	01060	MOLYBDENUM, DISSOLVED (UG/L AS MO)	09/23/92-06/21/95	2	4	
BADL0014	Yes	01060	MOLYBDENUM, DISSOLVED (UG/L AS MO)	09/22/92-08/28/95	2	4	
BADL0015	No	01060	MOLYBDENUM, DISSOLVED (UG/L AS MO)	09/25/92-08/28/95	2	4	
BADL0016	No	01060	MOLYBDENUM, DISSOLVED (UG/L AS MO)	09/30/92-07/10/95	2	4	
BADL0017	Yes	01060	MOLYBDENUM, DISSOLVED (UG/L AS MO)	09/21/92-09/21/92	0	1	
BADL0018	Yes	01060	MOLYBDENUM, DISSOLVED (UG/L AS MO)	09/21/92-06/21/95	2	4	
BADL0019	No	01060	MOLYBDENUM, DISSOLVED (UG/L AS MO)	10/08/92-08/29/95	2	4	
BADL0021	Yes	01060	MOLYBDENUM, DISSOLVED (UG/L AS MO)	10/01/92-07/13/95	2	5	
BADL0022	No	01060	MOLYBDENUM, DISSOLVED (UG/L AS MO)	09/29/92-09/29/92	0	1	
BADL0024	Yes	01060	MOLYBDENUM, DISSOLVED (UG/L AS MO)	09/30/92-09/30/92	0	1	
BADL0025	Yes	01060	MOLYBDENUM, DISSOLVED (UG/L AS MO)	09/18/92-09/18/92	0	1	
BADL0026	No	01060	MOLYBDENUM, DISSOLVED (UG/L AS MO)	08/30/94-08/30/94	0	1	
BADL0027	No	01060	MOLYBDENUM, DISSOLVED (UG/L AS MO)	05/06/88-09/08/94	6	6	
BADL0030	No	01060	MOLYBDENUM, DISSOLVED (UG/L AS MO)	11/04/92-11/04/92	0	1	
BADL0031	No	01060	MOLYBDENUM, DISSOLVED (UG/L AS MO)	11/05/92-07/12/95	2	4	
BADL0009	No	01062	MOLYBDENUM, TOTAL (UG/L AS MO)	04/23/79-04/23/79	0	1	
BADL0010	No	01062	MOLYBDENUM, TOTAL (UG/L AS MO)	04/18/79-04/18/79	0	1	
BADL0008	No	01065	NICKEL, DISSOLVED (UG/L AS NI)	11/12/92-07/11/95	2	4	
BADL0013	No	01065	NICKEL, DISSOLVED (UG/L AS NI)	09/23/92-06/21/95	2	4	
BADL0014	Yes	01065	NICKEL, DISSOLVED (UG/L AS NI)	09/22/92-08/28/95	2	4	
BADL0015	No	01065	NICKEL, DISSOLVED (UG/L AS NI)	09/25/92-08/28/95	2	4	
BADL0016	No	01065	NICKEL, DISSOLVED (UG/L AS NI)	09/30/92-07/10/95	2	4	
BADL0017	Yes	01065	NICKEL, DISSOLVED (UG/L AS NI)	09/21/92-09/21/92	0	1	
BADL0018	Yes	01065	NICKEL, DISSOLVED (UG/L AS NI)	09/21/92-06/21/95	2	4	
BADL0019	No	01065	NICKEL, DISSOLVED (UG/L AS NI)	10/08/92-08/29/95	2	4	
BADL0021	Yes	01065	NICKEL, DISSOLVED (UG/L AS NI)	10/01/92-07/13/95	2	5	
BADL0022	No	01065	NICKEL, DISSOLVED (UG/L AS NI)	09/29/92-09/29/92	0	1	
BADL0024	Yes	01065	NICKEL, DISSOLVED (UG/L AS NI)	09/30/92-09/30/92	0	1	
BADL0025	Yes	01065	NICKEL, DISSOLVED (UG/L AS NI)	09/18/92-09/18/92	0	1	
BADL0026	No	01065	NICKEL, DISSOLVED (UG/L AS NI)	08/30/94-08/30/94	0	1	
BADL0030	No	01065	NICKEL, DISSOLVED (UG/L AS NI)	11/04/92-11/04/92	0	1	
BADL0031	No	01065	NICKEL, DISSOLVED (UG/L AS NI)	11/05/92-07/12/95	2	4	
BADL0009	No	01067	NICKEL, TOTAL (UG/L AS NI)	04/23/79-04/23/79	0	1	
BADL0010	No	01067	NICKEL, TOTAL (UG/L AS NI)	04/18/79-04/18/79	0	1	
BADL0008	No	01075	SILVER, DISSOLVED (UG/L AS AG)	11/12/92-07/11/95	2	4	
BADL0013	No	01075	SILVER, DISSOLVED (UG/L AS AG)	09/23/92-06/21/95	2	4	
BADL0014	Yes	01075	SILVER, DISSOLVED (UG/L AS AG)	09/22/92-08/28/95	2	4	
BADL0015	No	01075	SILVER, DISSOLVED (UG/L AS AG)	09/25/92-08/28/95	2	4	
BADL0016	No	01075	SILVER, DISSOLVED (UG/L AS AG)	09/30/92-07/10/95	2	4	
BADL0017	Yes	01075	SILVER, DISSOLVED (UG/L AS AG)	09/21/92-09/21/92	0	1	
BADL0018	Yes	01075	SILVER, DISSOLVED (UG/L AS AG)	09/21/92-06/21/95	2	4	
BADL0019	No	01075	SILVER, DISSOLVED (UG/L AS AG)	10/08/92-08/29/95	2	4	
BADL0021	Yes	01075	SILVER, DISSOLVED (UG/L AS AG)	10/01/92-07/13/95	2	5	
BADL0022	No	01075	SILVER, DISSOLVED (UG/L AS AG)	09/29/92-09/29/92	0	1	
BADL0024	Yes	01075	SILVER, DISSOLVED (UG/L AS AG)	09/30/92-09/30/92	0	1	
BADL0025	Yes	01075	SILVER, DISSOLVED (UG/L AS AG)	09/18/92-09/18/92	0	1	
BADL0026	No	01075	SILVER, DISSOLVED (UG/L AS AG)	08/30/94-08/30/94	0	1	
BADL0030	No	01075	SILVER, DISSOLVED (UG/L AS AG)	11/04/92-11/04/92	0	1	
BADL0031	No	01075	SILVER, DISSOLVED (UG/L AS AG)	11/05/92-07/12/95	2	4	
BADL0009	No	01077	SILVER, TOTAL (UG/L AS AG)	04/23/79-04/23/79	0	1	
BADL0010	No	01077	SILVER, TOTAL (UG/L AS AG)	04/18/79-04/18/79	0	1	
BADL0008	No	01080	STRONTIUM, DISSOLVED (UG/L AS SR)	11/12/92-07/11/95	2	4	
BADL0013	No	01080	STRONTIUM, DISSOLVED (UG/L AS SR)	09/23/92-06/21/95	2	4	
BADL0014	Yes	01080	STRONTIUM, DISSOLVED (UG/L AS SR)	09/22/92-08/28/95	2	4	
BADL0015	No	01080	STRONTIUM, DISSOLVED (UG/L AS SR)	09/25/92-08/28/95	2	4	
BADL0016	No	01080	STRONTIUM, DISSOLVED (UG/L AS SR)	09/30/92-07/10/95	2	4	
BADL0017	Yes	01080	STRONTIUM, DISSOLVED (UG/L AS SR)	09/21/92-09/21/92	0	1	
BADL0018	Yes	01080	STRONTIUM, DISSOLVED (UG/L AS SR)	09/21/92-06/21/95	2	4	
BADL0019	No	01080	STRONTIUM, DISSOLVED (UG/L AS SR)	10/08/92-08/29/95	2	4	
BADL0021	Yes	01080	STRONTIUM, DISSOLVED (UG/L AS SR)	10/01/92-07/13/95	2	5	
BADL0022	No	01080	STRONTIUM, DISSOLVED (UG/L AS SR)	09/29/92-09/29/92	0	1	
BADL0024	Yes	01080	STRONTIUM, DISSOLVED (UG/L AS SR)	09/30/92-09/30/92	0	1	
BADL0025	Yes	01080	STRONTIUM, DISSOLVED (UG/L AS SR)	09/18/92-09/18/92	0	1	
BADL0026	No	01080	STRONTIUM, DISSOLVED (UG/L AS SR)	08/30/94-08/30/94	0	1	

[1]T=Times Series Plot, A=Annual Plot, and S=Seasonal Plot

Station	In Park	Code	Name	Start - End	Years	Obs	Plots[1]
BADL0030	No	01080	STRONTIUM, DISSOLVED (UG/L AS SR)	11/04/92-11/04/92	0	1	
BADL0031	No	01080	STRONTIUM, DISSOLVED (UG/L AS SR)	11/05/92-07/12/95	2	4	
BADL0009	No	01082	STRONTIUM, TOTAL (UG/L AS SR)	04/23/79-04/23/79	0	1	
BADL0010	No	01082	STRONTIUM, TOTAL (UG/L AS SR)	04/18/79-04/18/79	0	1	
BADL0008	No	01085	VANADIUM, DISSOLVED (UG/L AS V)	11/12/92-07/11/95	2	4	
BADL0013	No	01085	VANADIUM, DISSOLVED (UG/L AS V)	09/23/92-06/21/95	2	4	
BADL0014	Yes	01085	VANADIUM, DISSOLVED (UG/L AS V)	09/22/92-08/28/95	2	4	
BADL0015	No	01085	VANADIUM, DISSOLVED (UG/L AS V)	09/25/92-08/28/95	2	4	
BADL0016	No	01085	VANADIUM, DISSOLVED (UG/L AS V)	09/30/92-07/10/95	2	4	
BADL0017	Yes	01085	VANADIUM, DISSOLVED (UG/L AS V)	09/21/92-09/21/92	0	1	
BADL0018	Yes	01085	VANADIUM, DISSOLVED (UG/L AS V)	09/21/92-06/21/95	2	4	
BADL0019	No	01085	VANADIUM, DISSOLVED (UG/L AS V)	10/08/92-08/29/95	2	4	
BADL0021	Yes	01085	VANADIUM, DISSOLVED (UG/L AS V)	10/01/92-07/13/95	2	5	
BADL0022	No	01085	VANADIUM, DISSOLVED (UG/L AS V)	09/29/92-09/29/92	0	1	
BADL0024	Yes	01085	VANADIUM, DISSOLVED (UG/L AS V)	09/30/92-09/30/92	0	1	
BADL0025	Yes	01085	VANADIUM, DISSOLVED (UG/L AS V)	09/18/92-09/18/92	0	1	
BADL0026	No	01085	VANADIUM, DISSOLVED (UG/L AS V)	08/30/94-08/30/94	0	1	
BADL0027	No	01085	VANADIUM, DISSOLVED (UG/L AS V)	05/06/88-09/08/94	6	6	
BADL0030	No	01085	VANADIUM, DISSOLVED (UG/L AS V)	11/04/92-11/04/92	0	1	
BADL0031	No	01085	VANADIUM, DISSOLVED (UG/L AS V)	11/05/92-07/12/95	2	4	
BADL0009	No	01087	VANADIUM, TOTAL (UG/L AS V)	04/23/79-04/23/79	0	1	
BADL0010	No	01087	VANADIUM, TOTAL (UG/L AS V)	04/18/79-04/18/79	0	1	
BADL0008	No	01090	ZINC, DISSOLVED (UG/L AS ZN)	11/12/92-07/11/95	2	4	
BADL0011	Yes	01090	ZINC, DISSOLVED (UG/L AS ZN)	01/31/67-01/31/67	0	1	
BADL0013	No	01090	ZINC, DISSOLVED (UG/L AS ZN)	09/23/92-06/21/95	2	4	
BADL0014	Yes	01090	ZINC, DISSOLVED (UG/L AS ZN)	09/22/92-08/28/95	2	4	
BADL0015	No	01090	ZINC, DISSOLVED (UG/L AS ZN)	09/25/92-08/28/95	2	4	
BADL0016	No	01090	ZINC, DISSOLVED (UG/L AS ZN)	09/30/92-07/10/95	2	4	
BADL0017	Yes	01090	ZINC, DISSOLVED (UG/L AS ZN)	09/21/92-09/21/92	0	1	
BADL0018	Yes	01090	ZINC, DISSOLVED (UG/L AS ZN)	09/21/92-06/21/95	2	4	
BADL0019	No	01090	ZINC, DISSOLVED (UG/L AS ZN)	10/08/92-08/29/95	2	4	
BADL0021	Yes	01090	ZINC, DISSOLVED (UG/L AS ZN)	10/01/92-07/13/95	2	5	
BADL0022	No	01090	ZINC, DISSOLVED (UG/L AS ZN)	09/29/92-09/29/92	0	1	
BADL0024	Yes	01090	ZINC, DISSOLVED (UG/L AS ZN)	09/30/92-09/30/92	0	1	
BADL0025	Yes	01090	ZINC, DISSOLVED (UG/L AS ZN)	09/18/92-09/18/92	0	1	
BADL0026	No	01090	ZINC, DISSOLVED (UG/L AS ZN)	09/07/93-08/30/94	0	2	
BADL0027	No	01090	ZINC, DISSOLVED (UG/L AS ZN)	05/06/88-09/08/94	6	6	
BADL0030	No	01090	ZINC, DISSOLVED (UG/L AS ZN)	11/04/92-11/04/92	0	1	
BADL0031	No	01090	ZINC, DISSOLVED (UG/L AS ZN)	11/05/92-07/12/95	2	4	
BADL0009	No	01092	ZINC, TOTAL (UG/L AS ZN)	04/23/79-04/23/79	0	1	
BADL0010	No	01092	ZINC, TOTAL (UG/L AS ZN)	04/18/79-04/18/79	0	1	
BADL0026	No	01092	ZINC, TOTAL (UG/L AS ZN)	09/07/93-08/30/94	0	2	
BADL0026	No	01095	ANTIMONY, DISSOLVED (UG/L AS SB)	09/07/93-08/30/94	0	2	
BADL0009	No	01105	ALUMINUM, TOTAL (UG/L AS AL)	04/23/79-04/23/79	0	1	
BADL0010	No	01105	ALUMINUM, TOTAL (UG/L AS AL)	04/18/79-04/18/79	0	1	
BADL0027	No	01106	ALUMINUM, DISSOLVED (UG/L AS AL)	04/20/94-09/08/94	0	2	
BADL0009	No	01112	CERIUM, TOTAL (UG/L AS CE)	04/23/79-04/23/79	0	1	
BADL0010	No	01112	CERIUM, TOTAL (UG/L AS CE)	04/18/79-04/18/79	0	1	
BADL0008	No	01130	LITHIUM, DISSOLVED (UG/L AS LI)	11/12/92-07/11/95	2	4	
BADL0013	No	01130	LITHIUM, DISSOLVED (UG/L AS LI)	09/23/92-06/21/95	2	4	
BADL0014	Yes	01130	LITHIUM, DISSOLVED (UG/L AS LI)	09/22/92-08/28/95	2	4	
BADL0015	No	01130	LITHIUM, DISSOLVED (UG/L AS LI)	09/25/92-08/28/95	2	4	
BADL0016	No	01130	LITHIUM, DISSOLVED (UG/L AS LI)	09/30/92-07/10/95	2	4	
BADL0017	Yes	01130	LITHIUM, DISSOLVED (UG/L AS LI)	09/21/92-09/21/92	0	1	
BADL0018	Yes	01130	LITHIUM, DISSOLVED (UG/L AS LI)	09/21/92-06/21/95	2	4	
BADL0019	No	01130	LITHIUM, DISSOLVED (UG/L AS LI)	10/08/92-08/29/95	2	4	
BADL0021	Yes	01130	LITHIUM, DISSOLVED (UG/L AS LI)	10/01/92-07/13/95	2	5	
BADL0022	No	01130	LITHIUM, DISSOLVED (UG/L AS LI)	09/29/92-09/29/92	0	1	
BADL0024	Yes	01130	LITHIUM, DISSOLVED (UG/L AS LI)	09/30/92-09/30/92	0	1	
BADL0025	Yes	01130	LITHIUM, DISSOLVED (UG/L AS LI)	09/18/92-09/18/92	0	1	
BADL0026	No	01130	LITHIUM, DISSOLVED (UG/L AS LI)	09/07/93-08/30/94	0	2	
BADL0030	No	01130	LITHIUM, DISSOLVED (UG/L AS LI)	11/04/92-11/04/92	0	1	
BADL0031	No	01130	LITHIUM, DISSOLVED (UG/L AS LI)	11/05/92-07/12/95	2	4	
BADL0009	No	01132	LITHIUM, TOTAL (UG/L AS LI)	04/23/79-04/23/79	0	1	
BADL0010	No	01132	LITHIUM, TOTAL (UG/L AS LI)	04/18/79-04/18/79	0	1	
BADL0009	No	01142	SILICON, TOTAL (UG/L AS SI)	04/23/79-04/23/79	0	1	
BADL0010	No	01142	SILICON, TOTAL (UG/L AS SI)	04/18/79-04/18/79	0	1	
BADL0008	No	01145	SELENIUM, DISSOLVED (UG/L AS SE)	11/12/92-07/11/95	2	4	
BADL0013	No	01145	SELENIUM, DISSOLVED (UG/L AS SE)	09/23/92-06/21/95	2	4	
BADL0014	Yes	01145	SELENIUM, DISSOLVED (UG/L AS SE)	09/22/92-08/28/95	2	4	
BADL0015	No	01145	SELENIUM, DISSOLVED (UG/L AS SE)	09/25/92-08/28/95	2	4	
BADL0016	No	01145	SELENIUM, DISSOLVED (UG/L AS SE)	04/14/95-07/10/95	0	2	

[1] T=Times Series Plot, A=Annual Plot, and S=Seasonal Plot

Station	In Park	Code	Name	Start - End	Years	Obs	Plots[1]
BADL0019	No	01145	SELENIUM, DISSOLVED (UG/L AS SE)	10/08/92-08/29/95	2	4	
BADL0021	Yes	01145	SELENIUM, DISSOLVED (UG/L AS SE)	10/01/92-07/13/95	2	5	
BADL0022	No	01145	SELENIUM, DISSOLVED (UG/L AS SE)	09/29/92-09/29/92	0	1	
BADL0023	No	01145	SELENIUM, DISSOLVED (UG/L AS SE)	06/01/71-08/21/72	1	3	
BADL0024	Yes	01145	SELENIUM, DISSOLVED (UG/L AS SE)	09/30/92-09/30/92	0	1	
BADL0025	Yes	01145	SELENIUM, DISSOLVED (UG/L AS SE)	09/18/92-09/18/92	0	1	
BADL0026	No	01145	SELENIUM, DISSOLVED (UG/L AS SE)	09/07/93-08/30/94	0	2	
BADL0027	No	01145	SELENIUM, DISSOLVED (UG/L AS SE)	05/06/88-09/08/94	6	6	
BADL0028	No	01145	SELENIUM, DISSOLVED (UG/L AS SE)	06/03/71-06/03/71	0	1	
BADL0030	No	01145	SELENIUM, DISSOLVED (UG/L AS SE)	11/04/92-11/04/92	0	1	
BADL0031	No	01145	SELENIUM, DISSOLVED (UG/L AS SE)	11/05/92-07/12/95	2	4	
BADL0009	No	01147	SELENIUM, TOTAL (UG/L AS SE)	04/23/79-04/23/79	0	1	
BADL0010	No	01147	SELENIUM, TOTAL (UG/L AS SE)	04/18/79-04/18/79	0	1	
BADL0026	No	01147	SELENIUM, TOTAL (UG/L AS SE)	09/07/93-08/30/94	0	2	
BADL0009	No	01152	TITANIUM, TOTAL (UG/L AS TI)	04/23/79-04/23/79	0	1	
BADL0010	No	01152	TITANIUM, TOTAL (UG/L AS TI)	04/18/79-04/18/79	0	1	
BADL0009	No	01189	SCANDIUM, TOTAL (UG/L AS SC)	04/23/79-04/23/79	0	1	
BADL0010	No	01189	SCANDIUM, TOTAL (UG/L AS SC)	04/18/79-04/18/79	0	1	
BADL0009	No	01203	YTTRIUM, TOTAL (UG/L AS Y)	04/23/79-04/23/79	0	1	
BADL0010	No	01203	YTTRIUM, TOTAL (UG/L AS Y)	04/18/79-04/18/79	0	1	
BADL0009	No	01239	NIOBIUM, TOTAL UG/L	04/23/79-04/23/79	0	1	
BADL0010	No	01239	NIOBIUM, TOTAL UG/L	04/18/79-04/18/79	0	1	
BADL0008	No	03515	BETA, DISSOLVED GROSS, AS CS-137, PC/L	08/02/94-08/02/94	0	1	
BADL0013	No	03515	BETA, DISSOLVED GROSS, AS CS-137, PC/L	07/26/94-07/26/94	0	1	
BADL0014	Yes	03515	BETA, DISSOLVED GROSS, AS CS-137, PC/L	07/12/94-07/12/94	0	1	
BADL0015	No	03515	BETA, DISSOLVED GROSS, AS CS-137, PC/L	07/13/94-07/13/94	0	1	
BADL0016	No	03515	BETA, DISSOLVED GROSS, AS CS-137, PC/L	07/27/94-07/27/94	0	1	
BADL0018	Yes	03515	BETA, DISSOLVED GROSS, AS CS-137, PC/L	07/14/94-07/14/94	0	1	
BADL0019	No	03515	BETA, DISSOLVED GROSS, AS CS-137, PC/L	07/27/94-07/27/94	0	1	
BADL0021	Yes	03515	BETA, DISSOLVED GROSS, AS CS-137, PC/L	07/28/94-07/28/94	0	2	
BADL0026	No	03515	BETA, DISSOLVED GROSS, AS CS-137, PC/L	09/07/93-08/30/94	0	2	
BADL0031	No	03515	BETA, DISSOLVED GROSS, AS CS-137, PC/L	08/03/94-08/03/94	0	1	
BADL0008	No	04126	ALPHA, DISSOLVED, WATER (AS TH-230) PCI/L	08/02/94-08/02/94	0	1	
BADL0013	No	04126	ALPHA, DISSOLVED, WATER (AS TH-230) PCI/L	07/26/94-07/26/94	0	1	
BADL0014	Yes	04126	ALPHA, DISSOLVED, WATER (AS TH-230) PCI/L	07/12/94-07/12/94	0	1	
BADL0015	No	04126	ALPHA, DISSOLVED, WATER (AS TH-230) PCI/L	07/13/94-07/13/94	0	1	
BADL0016	No	04126	ALPHA, DISSOLVED, WATER (AS TH-230) PCI/L	07/27/94-07/27/94	0	1	
BADL0018	Yes	04126	ALPHA, DISSOLVED, WATER (AS TH-230) PCI/L	07/14/94-07/14/94	0	1	
BADL0019	No	04126	ALPHA, DISSOLVED, WATER (AS TH-230) PCI/L	07/27/94-07/27/94	0	1	
BADL0021	Yes	04126	ALPHA, DISSOLVED, WATER (AS TH-230) PCI/L	07/28/94-07/28/94	0	2	
BADL0026	No	04126	ALPHA, DISSOLVED, WATER (AS TH-230) PCI/L	09/07/93-08/30/94	0	2	
BADL0031	No	04126	ALPHA, DISSOLVED, WATER (AS TH-230) PCI/L	08/03/94-08/03/94	0	1	
BADL0008	No	09510	RADIUM 226, DISSOLVED, PLANCHET COUNT	08/02/94-08/02/94	0	1	
BADL0013	No	09510	RADIUM 226, DISSOLVED, PLANCHET COUNT	07/26/94-07/26/94	0	1	
BADL0014	Yes	09510	RADIUM 226, DISSOLVED, PLANCHET COUNT	07/12/94-07/12/94	0	1	
BADL0015	No	09510	RADIUM 226, DISSOLVED, PLANCHET COUNT	07/13/94-07/13/94	0	1	
BADL0016	No	09510	RADIUM 226, DISSOLVED, PLANCHET COUNT	07/27/94-07/27/94	0	1	
BADL0018	Yes	09510	RADIUM 226, DISSOLVED, PLANCHET COUNT	07/14/94-07/14/94	0	1	
BADL0019	No	09510	RADIUM 226, DISSOLVED, PLANCHET COUNT	07/27/94-07/27/94	0	1	
BADL0021	Yes	09510	RADIUM 226, DISSOLVED, PLANCHET COUNT	07/28/94-07/28/94	0	2	
BADL0031	No	09510	RADIUM 226, DISSOLVED, PLANCHET COUNT	08/03/94-08/03/94	0	1	
BADL0008	No	22703	URANIUM, NATURAL, DISSOLVED	08/02/94-08/02/94	0	1	
BADL0013	No	22703	URANIUM, NATURAL, DISSOLVED	07/26/94-07/26/94	0	1	
BADL0014	Yes	22703	URANIUM, NATURAL, DISSOLVED	07/12/94-07/12/94	0	1	
BADL0015	No	22703	URANIUM, NATURAL, DISSOLVED	07/13/94-07/13/94	0	1	
BADL0016	No	22703	URANIUM, NATURAL, DISSOLVED	07/27/94-07/27/94	0	1	
BADL0018	Yes	22703	URANIUM, NATURAL, DISSOLVED	07/14/94-07/14/94	0	1	
BADL0019	No	22703	URANIUM, NATURAL, DISSOLVED	07/27/94-07/27/94	0	1	
BADL0021	Yes	22703	URANIUM, NATURAL, DISSOLVED	07/28/94-07/28/94	0	2	
BADL0026	No	22703	URANIUM, NATURAL, DISSOLVED	09/07/93-08/30/94	0	2	
BADL0027	No	22703	URANIUM, NATURAL, DISSOLVED	05/06/88-09/08/94	6	6	
BADL0031	No	22703	URANIUM, NATURAL, DISSOLVED	08/03/94-08/03/94	0	1	
BADL0008	No	31501	COLIFORM,TOT,MEMBRANE FILTER,IMMED M-ENDO MED,35C	08/02/94-05/17/95	0	2	
BADL0013	No	31501	COLIFORM,TOT,MEMBRANE FILTER,IMMED M-ENDO MED,35C	07/26/94-04/27/95	0	2	
BADL0014	Yes	31501	COLIFORM,TOT,MEMBRANE FILTER,IMMED M-ENDO MED,35C	07/12/94-07/12/94	0	1	
BADL0015	No	31501	COLIFORM,TOT,MEMBRANE FILTER,IMMED M-ENDO MED,35C	07/13/94-05/11/95	0	2	
BADL0016	No	31501	COLIFORM,TOT,MEMBRANE FILTER,IMMED M-ENDO MED,35C	07/27/94-04/14/95	0	2	
BADL0018	Yes	31501	COLIFORM,TOT,MEMBRANE FILTER,IMMED M-ENDO MED,35C	07/14/94-04/27/95	0	2	
BADL0019	No	31501	COLIFORM,TOT,MEMBRANE FILTER,IMMED M-ENDO MED,35C	07/27/94-05/12/95	0	2	
BADL0021	Yes	31501	COLIFORM,TOT,MEMBRANE FILTER,IMMED M-ENDO MED,35C	07/28/94-05/16/95	0	2	
BADL0031	No	31501	COLIFORM,TOT,MEMBRANE FILTER,IMMED M-ENDO MED,35C	08/03/94-05/16/95	0	2	
BADL0008	No	31503	COLIFORM,TOT,MEMBR FILTER,DELAYED,M-ENDO MED,35 C	07/11/95-07/11/95	0	1	

[1]T=Times Series Plot, A=Annual Plot, and S=Seasonal Plot

Station	In Park	Code	Name	Start - End	Years	Obs	Plots[1]
BADL0013	No	31503	COLIFORM,TOT,MEMBR FILTER,DELAYED,M-ENDO MED,35 C	06/21/95-06/21/95	0	1	
BADL0016	No	31503	COLIFORM,TOT,MEMBR FILTER,DELAYED,M-ENDO MED,35 C	07/10/95-07/10/95	0	1	
BADL0018	Yes	31503	COLIFORM,TOT,MEMBR FILTER,DELAYED,M-ENDO MED,35 C	06/21/95-06/21/95	0	1	
BADL0021	Yes	31503	COLIFORM,TOT,MEMBR FILTER,DELAYED,M-ENDO MED,35 C	07/13/95-07/13/95	0	1	
BADL0031	No	31503	COLIFORM,TOT,MEMBR FILTER,DELAYED,M-ENDO MED,35 C	07/12/95-07/12/95	0	1	
BADL0014	Yes	31504	COLIFORM,TOT,MEMBR FILTER,IMMED,LES ENDO AGAR,35C	08/28/95-08/28/95	0	1	
BADL0015	No	31504	COLIFORM,TOT,MEMBR FILTER,IMMED,LES ENDO AGAR,35C	08/28/95-08/28/95	0	1	
BADL0019	No	31504	COLIFORM,TOT,MEMBR FILTER,IMMED,LES ENDO AGAR,35C	08/29/95-08/29/95	0	1	
BADL0002	No	31616	FECAL COLIFORM,MEMBR FILTER,M-FC BROTH,44 5 C	06/08/78-09/19/78	0	2	
BADL0003	No	31616	FECAL COLIFORM,MEMBR FILTER,M-FC BROTH,44 5 C	03/15/78-09/19/78	0	3	
BADL0004	No	31616	FECAL COLIFORM,MEMBR FILTER,M-FC BROTH,44 5 C	06/08/78-09/19/78	0	2	
BADL0008	No	31625	FECAL COLIFORM, MF,M-FC, 0 7 UM	08/02/94-07/11/95	0	3	
BADL0013	No	31625	FECAL COLIFORM, MF,M-FC, 0 7 UM	07/26/94-06/21/95	0	3	
BADL0014	Yes	31625	FECAL COLIFORM, MF,M-FC, 0 7 UM	07/12/94-08/28/95	1	2	
BADL0015	No	31625	FECAL COLIFORM, MF,M-FC, 0 7 UM	07/13/94-08/28/95	1	3	
BADL0016	No	31625	FECAL COLIFORM, MF,M-FC, 0 7 UM	07/27/94-04/14/95	0	2	
BADL0018	Yes	31625	FECAL COLIFORM, MF,M-FC, 0 7 UM	07/14/94-06/21/95	0	3	
BADL0019	No	31625	FECAL COLIFORM, MF,M-FC, 0 7 UM	07/27/94-08/29/95	1	3	
BADL0021	Yes	31625	FECAL COLIFORM, MF,M-FC, 0 7 UM	07/28/94-07/13/95	0	3	
BADL0026	No	31625	FECAL COLIFORM, MF,M-FC, 0 7 UM	09/07/93-08/30/94	0	2	
BADL0031	No	31625	FECAL COLIFORM, MF,M-FC, 0 7 UM	08/03/94-07/12/95	0	3	
BADL0008	No	31633	E COLI,THERMOTOL,MF,M-TEC,IN SITU UREASE #/100ML	08/02/94-07/11/95	0	3	
BADL0013	No	31633	E COLI,THERMOTOL,MF,M-TEC,IN SITU UREASE #/100ML	07/26/94-06/21/95	0	3	
BADL0014	Yes	31633	E COLI,THERMOTOL,MF,M-TEC,IN SITU UREASE #/100ML	07/12/94-08/28/95	1	2	
BADL0015	No	31633	E COLI,THERMOTOL,MF,M-TEC,IN SITU UREASE #/100ML	07/13/94-08/28/95	1	3	
BADL0016	No	31633	E COLI,THERMOTOL,MF,M-TEC,IN SITU UREASE #/100ML	07/27/94-04/14/95	0	2	
BADL0018	Yes	31633	E COLI,THERMOTOL,MF,M-TEC,IN SITU UREASE #/100ML	07/14/94-06/21/95	0	3	
BADL0019	No	31633	E COLI,THERMOTOL,MF,M-TEC,IN SITU UREASE #/100ML	07/27/94-08/29/95	1	3	
BADL0021	Yes	31633	E COLI,THERMOTOL,MF,M-TEC,IN SITU UREASE #/100ML	07/28/94-07/13/95	0	3	
BADL0031	No	31633	E COLI,THERMOTOL,MF,M-TEC,IN SITU UREASE #/100ML	08/03/94-07/12/95	0	3	
BADL0008	No	31673	FECAL STREPTOCOCCI, MBR FILT,KF AGAR,35C,48HR	08/02/94-07/11/95	0	3	
BADL0013	No	31673	FECAL STREPTOCOCCI, MBR FILT,KF AGAR,35C,48HR	07/26/94-06/21/95	0	3	
BADL0014	Yes	31673	FECAL STREPTOCOCCI, MBR FILT,KF AGAR,35C,48HR	07/12/94-08/28/95	1	2	
BADL0015	No	31673	FECAL STREPTOCOCCI, MBR FILT,KF AGAR,35C,48HR	07/13/94-08/28/95	1	3	
BADL0016	No	31673	FECAL STREPTOCOCCI, MBR FILT,KF AGAR,35C,48HR	07/27/94-04/14/95	0	2	
BADL0018	Yes	31673	FECAL STREPTOCOCCI, MBR FILT,KF AGAR,35C,48HR	07/14/94-06/21/95	0	3	
BADL0019	No	31673	FECAL STREPTOCOCCI, MBR FILT,KF AGAR,35C,48HR	07/27/94-08/29/95	1	3	
BADL0021	Yes	31673	FECAL STREPTOCOCCI, MBR FILT,KF AGAR,35C,48HR	07/28/94-07/13/95	0	3	
BADL0026	No	31673	FECAL STREPTOCOCCI, MBR FILT,KF AGAR,35C,48HR	09/07/93-08/30/94	0	2	
BADL0031	No	31673	FECAL STREPTOCOCCI, MBR FILT,KF AGAR,35C,48HR	08/03/94-07/12/95	0	3	
BADL0011	Yes	32730	PHENOLICS, TOTAL, RECOVERABLE (UG/L)	10/01/65-09/29/66	0	45	
BADL0026	No	32730	PHENOLICS, TOTAL, RECOVERABLE (UG/L)	09/07/93-08/30/94	0	2	
BADL0027	No	39024	PROPAZINE,COULSON CONDUCTIVITY,WATER SAMPL(UG/L)	05/06/88-10/31/88	0	4	
BADL0027	No	39030	TREFLAN, MICROCOULOMETRIC,WATER SAMPLE (UG/L)	05/06/88-10/31/88	0	4	
BADL0027	No	39051	METHOMYL IN WHOLE WATER (UG/L)	05/06/88-10/31/88	0	4	
BADL0027	No	39052	PROPHAM IN WHOLE WATER (UG/L)	05/06/88-10/31/88	0	4	
BADL0027	No	39054	SIMETRYNE IN WHOLE WATER (UG/L)	05/06/88-10/31/88	0	4	
BADL0027	No	39055	SIMAZINE IN WHOLE WATER (UG/L)	05/06/88-10/31/88	0	4	
BADL0027	No	39056	PROMETONE IN WHOLE WATER (UG/L)	05/06/88-10/31/88	0	4	
BADL0027	No	39057	PROMETRYNE IN WHOLE WATER (UG/L)	05/06/88-10/31/88	0	4	
BADL0008	No	39086	ALKALINITY,WATER,DISS,INCR TIT,FIELD,AS CACO3,MG/L	07/11/95-07/11/95	0	1	
BADL0013	No	39086	ALKALINITY,WATER,DISS,INCR TIT,FIELD,AS CACO3,MG/L	06/21/95-06/21/95	0	1	
BADL0016	No	39086	ALKALINITY,WATER,DISS,INCR TIT,FIELD,AS CACO3,MG/L	07/10/95-07/10/95	0	1	
BADL0018	Yes	39086	ALKALINITY,WATER,DISS,INCR TIT,FIELD,AS CACO3,MG/L	06/21/95-06/21/95	0	1	
BADL0021	Yes	39086	ALKALINITY,WATER,DISS,INCR TIT,FIELD,AS CACO3,MG/L	07/13/95-07/13/95	0	1	
BADL0027	No	39086	ALKALINITY,WATER,DISS,INCR TIT,FIELD,AS CACO3,MG/L	10/31/88-04/20/94	5	2	
BADL0031	No	39086	ALKALINITY,WATER,DISS,INCR TIT,FIELD,AS CACO3,MG/L	07/12/95-07/12/95	0	1	
BADL0027	No	39630	ATRAZINE(AATREX) IN WHOLE WATER SAMPLE (UG/L)	05/06/88-10/31/88	0	4	
BADL0027	No	39750	SEVIN IN WHOLE WATER SAMPLE (UG/L)	05/06/88-10/31/88	0	4	
BADL0005	No	70300	RESIDUE,TOTAL FILTRABLE (DRIED AT 180C),MG/L	11/30/62-11/30/62	0	1	
BADL0006	No	70300	RESIDUE,TOTAL FILTRABLE (DRIED AT 180C),MG/L	07/12/89-08/16/89	0	2	
BADL0007	No	70300	RESIDUE,TOTAL FILTRABLE (DRIED AT 180C),MG/L	07/12/89-08/16/89	0	2	
BADL0008	No	70300	RESIDUE,TOTAL FILTRABLE (DRIED AT 180C),MG/L	11/12/92-07/11/95	2	4	
BADL0011	Yes	70300	RESIDUE,TOTAL FILTRABLE (DRIED AT 180C),MG/L	08/13/64-09/18/67	3	101	T
BADL0012	No	70300	RESIDUE,TOTAL FILTRABLE (DRIED AT 180C),MG/L	11/30/62-11/30/62	0	1	
BADL0013	No	70300	RESIDUE,TOTAL FILTRABLE (DRIED AT 180C),MG/L	09/23/92-06/21/95	2	4	
BADL0014	Yes	70300	RESIDUE,TOTAL FILTRABLE (DRIED AT 180C),MG/L	09/22/92-08/28/95	2	4	
BADL0015	No	70300	RESIDUE,TOTAL FILTRABLE (DRIED AT 180C),MG/L	09/25/92-08/28/95	2	4	
BADL0016	No	70300	RESIDUE,TOTAL FILTRABLE (DRIED AT 180C),MG/L	09/30/92-07/10/95	2	4	
BADL0017	Yes	70300	RESIDUE,TOTAL FILTRABLE (DRIED AT 180C),MG/L	09/21/92-09/21/92	0	1	
BADL0018	Yes	70300	RESIDUE,TOTAL FILTRABLE (DRIED AT 180C),MG/L	09/21/92-06/21/95	2	4	
BADL0019	No	70300	RESIDUE,TOTAL FILTRABLE (DRIED AT 180C),MG/L	10/08/92-08/29/95	2	4	

[1]T=Times Series Plot, A=Annual Plot, and S=Seasonal Plot

Station/Parameter Period of Record Tabulation
From 11/29/62 To 09/10/96

Station	In Park	Code	Name	Start - End	Years	Obs	Plots[1]
BADL0020	No	70300	RESIDUE,TOTAL FILTRABLE (DRIED AT 180C),MG/L	11/29/62-11/29/62	0	1	
BADL0021	Yes	70300	RESIDUE,TOTAL FILTRABLE (DRIED AT 180C),MG/L	10/01/92-07/13/95	2	5	
BADL0022	No	70300	RESIDUE,TOTAL FILTRABLE (DRIED AT 180C),MG/L	09/29/92-09/29/92	0	1	
BADL0024	Yes	70300	RESIDUE,TOTAL FILTRABLE (DRIED AT 180C),MG/L	09/30/92-09/30/92	0	1	
BADL0025	Yes	70300	RESIDUE,TOTAL FILTRABLE (DRIED AT 180C),MG/L	09/18/92-09/18/92	0	1	
BADL0026	No	70300	RESIDUE,TOTAL FILTRABLE (DRIED AT 180C),MG/L	09/07/93-08/30/94	0	2	
BADL0027	No	70300	RESIDUE,TOTAL FILTRABLE (DRIED AT 180C),MG/L	05/06/88-09/08/94	6	6	
BADL0030	No	70300	RESIDUE,TOTAL FILTRABLE (DRIED AT 180C),MG/L	11/04/92-11/04/92	0	1	
BADL0031	No	70300	RESIDUE,TOTAL FILTRABLE (DRIED AT 180C),MG/L	11/05/92-07/12/95	2	4	
BADL0005	No	70301	SOLIDS, DISSOLVED-SUM OF CONSTITUENTS (MG/L)	11/30/62-11/30/62	0	1	
BADL0012	No	70301	SOLIDS, DISSOLVED-SUM OF CONSTITUENTS (MG/L)	11/30/62-11/30/62	0	1	
BADL0020	No	70301	SOLIDS, DISSOLVED-SUM OF CONSTITUENTS (MG/L)	11/29/62-11/29/62	0	1	
BADL0011	Yes	70302	SOLIDS, DISSOLVED-TONS PER DAY	08/13/64-09/18/67	3	101	T
BADL0011	Yes	70303	SOLIDS, DISSOLVED-TONS PER ACRE-FT	08/13/64-09/18/67	3	101	T
BADL0011	Yes	70331	SUSPENDED SED SIEVE DIAMETER,% FINER THAN 062MM	11/04/64-08/16/67	2	21	T
BADL0011	Yes	70337	SUS SED FALL DIA(DISTLD WATER)%FINER THAN 002MM	11/04/64-07/24/67	2	31	T
BADL0011	Yes	70338	SUS SED FALL DIA(DISTLD WATER)%FINER THAN 004MM	11/04/64-07/24/67	2	31	T
BADL0011	Yes	70339	SUS SED FALL DIA(DISTLD WATER)%FINER THAN 008MM	11/04/64-08/17/66	1	16	
BADL0011	Yes	70340	SUS SED FALL DIA(DISTLD WATER)%FINER THAN 016MM	11/04/64-07/24/67	2	30	T
BADL0011	Yes	70341	SUS SED FALL DIA(DISTLD WATER)%FINER THAN 031MM	11/04/64-08/17/66	1	15	
BADL0011	Yes	70342	SUS SED FALL DIA(DISTLD WATER)%FINER THAN 062MM	05/15/65-06/22/67	2	17	T
BADL0011	Yes	70343	SUS SED FALL DIA(DISTLD WATER)%FINER THAN 125MM	05/15/65-06/22/67	2	10	T
BADL0011	Yes	70344	SUS SED FALL DIA(DISTLD WATER)%FINER THAN 250MM	05/15/65-06/22/67	2	7	
BADL0011	Yes	70345	SUS SED FALL DIA(DISTLD WATER)%FINER THAN 500MM	05/15/65-06/22/67	2	5	
BADL0011	Yes	70346	SUS SED FALL DIA(DISTLD WATER)%FINER THAN 1 00MM	03/14/66-03/14/66	0	1	
BADL0011	Yes	70347	SUS SED FALL DIA(DISTLD WATER)%FINER THAN 2 00MM	03/14/66-03/14/66	0	1	
BADL0002	No	70505	PHOSPHATE,TOTAL,COLORIMETRIC METHOD (MG/L AS P)	06/08/78-09/19/78	0	2	
BADL0003	No	70505	PHOSPHATE,TOTAL,COLORIMETRIC METHOD (MG/L AS P)	12/19/77-09/19/78	0	4	
BADL0004	No	70505	PHOSPHATE,TOTAL,COLORIMETRIC METHOD (MG/L AS P)	06/08/78-09/19/78	0	2	
BADL0002	No	70507	PHOSPHORUS,IN TOTAL ORTHOPHOSPHATE (MG/L AS P)	09/19/78-09/19/78	0	1	
BADL0003	No	70507	PHOSPHORUS,IN TOTAL ORTHOPHOSPHATE (MG/L AS P)	09/19/78-09/19/78	0	1	
BADL0004	No	70507	PHOSPHORUS,IN TOTAL ORTHOPHOSPHATE (MG/L AS P)	09/19/78-09/19/78	0	1	
BADL0015	No	71832	HYDROXIDE,INCREMENTAL TITRATION,(OH) FIELD MG/L	08/28/95-08/28/95	0	1	
BADL0011	Yes	71846	NITROGEN, AMMONIA, DISSOLVED (MG/L AS NH4)	11/04/64-09/28/65	0	18	
BADL0005	No	71851	NITRATE NITROGEN, DISSOLVED (MG/L AS NO3)	11/30/62-11/30/62	0	1	
BADL0011	Yes	71851	NITRATE NITROGEN, DISSOLVED (MG/L AS NO3)	08/13/64-09/18/67	3	101	T
BADL0012	No	71851	NITRATE NITROGEN, DISSOLVED (MG/L AS NO3)	11/30/62-11/30/62	0	1	
BADL0020	No	71851	NITRATE NITROGEN, DISSOLVED (MG/L AS NO3)	11/29/62-11/29/62	0	1	
BADL0011	Yes	71856	NITRITE NITROGEN, DISSOLVED (MG/L AS NO2)	11/04/64-09/28/65	0	14	
BADL0011	Yes	71870	BROMIDE (MG/L AS BR)	08/13/64-09/28/65	1	19	
BADL0005	No	71883	MANGANESE, TOTAL ELEMENTAL (UG/L AS MN)	11/30/62-11/30/62	0	1	
BADL0012	No	71883	MANGANESE, TOTAL ELEMENTAL (UG/L AS MN)	11/30/62-11/30/62	0	1	
BADL0020	No	71883	MANGANESE, TOTAL ELEMENTAL (UG/L AS MN)	11/29/62-11/29/62	0	1	
BADL0005	No	71885	IRON (UG/L AS FE)	11/30/62-11/30/62	0	1	
BADL0011	Yes	71885	IRON (UG/L AS FE)	08/13/64-09/18/67	3	56	T
BADL0012	No	71885	IRON (UG/L AS FE)	11/30/62-11/30/62	0	1	
BADL0020	No	71885	IRON (UG/L AS FE)	11/29/62-11/29/62	0	1	
BADL0008	No	71890	MERCURY, DISSOLVED (UG/L AS HG)	11/12/92-07/11/95	2	4	
BADL0013	No	71890	MERCURY, DISSOLVED (UG/L AS HG)	09/23/92-06/21/95	2	4	
BADL0014	Yes	71890	MERCURY, DISSOLVED (UG/L AS HG)	09/22/92-08/28/95	2	4	
BADL0015	No	71890	MERCURY, DISSOLVED (UG/L AS HG)	09/25/92-08/28/95	2	4	
BADL0016	No	71890	MERCURY, DISSOLVED (UG/L AS HG)	09/30/92-07/10/95	2	4	
BADL0017	Yes	71890	MERCURY, DISSOLVED (UG/L AS HG)	09/21/92-09/21/92	0	1	
BADL0018	Yes	71890	MERCURY, DISSOLVED (UG/L AS HG)	09/21/92-06/21/95	2	4	
BADL0019	No	71890	MERCURY, DISSOLVED (UG/L AS HG)	10/08/92-08/29/95	2	4	
BADL0021	Yes	71890	MERCURY, DISSOLVED (UG/L AS HG)	10/01/92-07/13/95	2	5	
BADL0022	No	71890	MERCURY, DISSOLVED (UG/L AS HG)	09/29/92-09/29/92	0	1	
BADL0024	Yes	71890	MERCURY, DISSOLVED (UG/L AS HG)	09/30/92-09/30/92	0	1	
BADL0025	Yes	71890	MERCURY, DISSOLVED (UG/L AS HG)	09/18/92-09/18/92	0	1	
BADL0026	No	71890	MERCURY, DISSOLVED (UG/L AS HG)	09/07/93-08/30/94	0	2	
BADL0027	No	71890	MERCURY, DISSOLVED (UG/L AS HG)	05/06/88-09/08/94	6	6	
BADL0030	No	71890	MERCURY, DISSOLVED (UG/L AS HG)	11/04/92-11/04/92	0	1	
BADL0031	No	71890	MERCURY, DISSOLVED (UG/L AS HG)	11/05/92-07/12/95	2	4	
BADL0005	No	72000	ELEVATION OF LAND SURFACE DATUM (FT ABOVE MSL)	11/30/62-11/30/62	0	1	
BADL0012	No	72000	ELEVATION OF LAND SURFACE DATUM (FT ABOVE MSL)	11/30/62-11/30/62	0	1	
BADL0020	No	72000	ELEVATION OF LAND SURFACE DATUM (FT ABOVE MSL)	11/29/62-11/29/62	0	1	
BADL0008	No	75986	ALPHA GROSS,1 SIGMA PRC EST AS NAT U,DISS,WTR UG/L	08/02/94-08/02/94	0	1	
BADL0013	No	75986	ALPHA GROSS,1 SIGMA PRC EST AS NAT U,DISS,WTR UG/L	07/26/94-07/26/94	0	1	
BADL0014	Yes	75986	ALPHA GROSS,1 SIGMA PRC EST AS NAT U,DISS,WTR UG/L	07/12/94-07/12/94	0	1	
BADL0015	No	75986	ALPHA GROSS,1 SIGMA PRC EST AS NAT U,DISS,WTR UG/L	07/13/94-07/13/94	0	1	
BADL0016	No	75986	ALPHA GROSS,1 SIGMA PRC EST AS NAT U,DISS,WTR UG/L	07/27/94-07/27/94	0	1	
BADL0018	Yes	75986	ALPHA GROSS,1 SIGMA PRC EST AS NAT U,DISS,WTR UG/L	07/14/94-07/14/94	0	1	

[1]T=Times Series Plot, A=Annual Plot, and S=Seasonal Plot

Station	In Park	Code	Name	Start - End	Years	Obs	Plots[1]
BADL0019	No	75986	ALPHA GROSS,1 SIGMA PRC EST AS NAT U,DISS,WTR UG/L	07/27/94-07/27/94	0	1	
BADL0021	Yes	75986	ALPHA GROSS,1 SIGMA PRC EST AS NAT U,DISS,WTR UG/L	07/28/94-07/28/94	0	2	
BADL0026	No	75986	ALPHA GROSS,1 SIGMA PRC EST AS NAT U,DISS,WTR UG/L	09/07/93-08/30/94	0	2	
BADL0031	No	75986	ALPHA GROSS,1 SIGMA PRC EST AS NAT U,DISS,WTR UG/L	08/03/94-08/03/94	0	1	
BADL0008	No	75987	ALPHA GROSS,DISS,1 SIGMA PRC EST AS TH230,WTR PC/L	08/02/94-08/02/94	0	1	
BADL0013	No	75987	ALPHA GROSS,DISS,1 SIGMA PRC EST AS TH230,WTR PC/L	07/26/94-07/26/94	0	1	
BADL0014	Yes	75987	ALPHA GROSS,DISS,1 SIGMA PRC EST AS TH230,WTR PC/L	07/12/94-07/12/94	0	1	
BADL0015	No	75987	ALPHA GROSS,DISS,1 SIGMA PRC EST AS TH230,WTR PC/L	07/13/94-07/13/94	0	1	
BADL0016	No	75987	ALPHA GROSS,DISS,1 SIGMA PRC EST AS TH230,WTR PC/L	07/27/94-07/27/94	0	1	
BADL0018	Yes	75987	ALPHA GROSS,DISS,1 SIGMA PRC EST AS TH230,WTR PC/L	07/14/94-07/14/94	0	1	
BADL0019	No	75987	ALPHA GROSS,DISS,1 SIGMA PRC EST AS TH230,WTR PC/L	07/27/94-07/27/94	0	1	
BADL0021	Yes	75987	ALPHA GROSS,DISS,1 SIGMA PRC EST AS TH230,WTR PC/L	07/28/94-07/28/94	0	2	
BADL0026	No	75987	ALPHA GROSS,DISS,1 SIGMA PRC EST AS TH230,WTR PC/L	09/07/93-08/30/94	0	2	
BADL0031	No	75987	ALPHA GROSS,DISS,1 SIGMA PRC EST AS TH230,WTR PC/L	08/03/94-08/03/94	0	1	
BADL0008	No	75988	BETA GROSS,DISS,1 SIGMA PRC EST AS SR90/Y90 PC/L	08/02/94-08/02/94	0	1	
BADL0013	No	75988	BETA GROSS,DISS,1 SIGMA PRC EST AS SR90/Y90 PC/L	07/26/94-07/26/94	0	1	
BADL0014	Yes	75988	BETA GROSS,DISS,1 SIGMA PRC EST AS SR90/Y90 PC/L	07/12/94-07/12/94	0	1	
BADL0015	No	75988	BETA GROSS,DISS,1 SIGMA PRC EST AS SR90/Y90 PC/L	07/13/94-07/13/94	0	1	
BADL0016	No	75988	BETA GROSS,DISS,1 SIGMA PRC EST AS SR90/Y90 PC/L	07/27/94-07/27/94	0	1	
BADL0018	Yes	75988	BETA GROSS,DISS,1 SIGMA PRC EST AS SR90/Y90 PC/L	07/14/94-07/14/94	0	1	
BADL0019	No	75988	BETA GROSS,DISS,1 SIGMA PRC EST AS SR90/Y90 PC/L	07/27/94-07/27/94	0	1	
BADL0021	Yes	75988	BETA GROSS,DISS,1 SIGMA PRC EST AS SR90/Y90 PC/L	07/28/94-07/28/94	0	2	
BADL0026	No	75988	BETA GROSS,DISS,1 SIGMA PRC EST AS SR90/Y90 PC/L	09/07/93-08/30/94	0	2	
BADL0031	No	75988	BETA GROSS,DISS,1 SIGMA PRC EST AS SR90/Y90 PC/L	08/03/94-08/03/94	0	1	
BADL0008	No	75989	BETA GROSS,1 SIGMA PRC EST AS CS-137,DISS,WTR PC/L	08/02/94-08/02/94	0	1	
BADL0013	No	75989	BETA GROSS,1 SIGMA PRC EST AS CS-137,DISS,WTR PC/L	07/26/94-07/26/94	0	1	
BADL0014	Yes	75989	BETA GROSS,1 SIGMA PRC EST AS CS-137,DISS,WTR PC/L	07/12/94-07/12/94	0	1	
BADL0015	No	75989	BETA GROSS,1 SIGMA PRC EST AS CS-137,DISS,WTR PC/L	07/13/94-07/13/94	0	1	
BADL0016	No	75989	BETA GROSS,1 SIGMA PRC EST AS CS-137,DISS,WTR PC/L	07/27/94-07/27/94	0	1	
BADL0018	Yes	75989	BETA GROSS,1 SIGMA PRC EST AS CS-137,DISS,WTR PC/L	07/14/94-07/14/94	0	1	
BADL0019	No	75989	BETA GROSS,1 SIGMA PRC EST AS CS-137,DISS,WTR PC/L	07/27/94-07/27/94	0	1	
BADL0021	Yes	75989	BETA GROSS,1 SIGMA PRC EST AS CS-137,DISS,WTR PC/L	07/28/94-07/28/94	0	2	
BADL0026	No	75989	BETA GROSS,1 SIGMA PRC EST AS CS-137,DISS,WTR PC/L	09/07/93-08/30/94	0	2	
BADL0031	No	75989	BETA GROSS,1 SIGMA PRC EST AS CS-137,DISS,WTR PC/L	08/03/94-08/03/94	0	1	
BADL0008	No	75990	URANIUM,NATURAL,1 SIGMA PRC EST,DISS,WATER UG/L	08/02/94-08/02/94	0	1	
BADL0013	No	75990	URANIUM,NATURAL,1 SIGMA PRC EST,DISS,WATER UG/L	07/26/94-07/26/94	0	1	
BADL0014	Yes	75990	URANIUM,NATURAL,1 SIGMA PRC EST,DISS,WATER UG/L	07/12/94-07/12/94	0	1	
BADL0015	No	75990	URANIUM,NATURAL,1 SIGMA PRC EST,DISS,WATER UG/L	07/13/94-07/13/94	0	1	
BADL0016	No	75990	URANIUM,NATURAL,1 SIGMA PRC EST,DISS,WATER UG/L	07/27/94-07/27/94	0	1	
BADL0018	Yes	75990	URANIUM,NATURAL,1 SIGMA PRC EST,DISS,WATER UG/L	07/14/94-07/14/94	0	1	
BADL0019	No	75990	URANIUM,NATURAL,1 SIGMA PRC EST,DISS,WATER UG/L	07/27/94-07/27/94	0	1	
BADL0021	Yes	75990	URANIUM,NATURAL,1 SIGMA PRC EST,DISS,WATER UG/L	07/28/94-07/28/94	0	2	
BADL0026	No	75990	URANIUM,NATURAL,1 SIGMA PRC EST,DISS,WATER UG/L	09/07/93-08/30/94	0	2	
BADL0027	No	75990	URANIUM,NATURAL,1 SIGMA PRC EST,DISS,WATER UG/L	04/20/94-09/08/94	0	2	
BADL0031	No	75990	URANIUM,NATURAL,1 SIGMA PRC EST,DISS,WATER UG/L	08/03/94-08/03/94	0	1	
BADL0008	No	76001	RADIUM 226,1 SIGMA PRC EST,DISSOLVED,WATER PC/L	08/02/94-08/02/94	0	1	
BADL0013	No	76001	RADIUM 226,1 SIGMA PRC EST,DISSOLVED,WATER PC/L	07/26/94-07/26/94	0	1	
BADL0014	Yes	76001	RADIUM 226,1 SIGMA PRC EST,DISSOLVED,WATER PC/L	07/12/94-07/12/94	0	1	
BADL0015	No	76001	RADIUM 226,1 SIGMA PRC EST,DISSOLVED,WATER PC/L	07/13/94-07/13/94	0	1	
BADL0016	No	76001	RADIUM 226,1 SIGMA PRC EST,DISSOLVED,WATER PC/L	07/27/94-07/27/94	0	1	
BADL0018	Yes	76001	RADIUM 226,1 SIGMA PRC EST,DISSOLVED,WATER PC/L	07/14/94-07/14/94	0	1	
BADL0019	No	76001	RADIUM 226,1 SIGMA PRC EST,DISSOLVED,WATER PC/L	07/27/94-07/27/94	0	1	
BADL0021	Yes	76001	RADIUM 226,1 SIGMA PRC EST,DISSOLVED,WATER PC/L	07/28/94-07/28/94	0	2	
BADL0031	No	76001	RADIUM 226,1 SIGMA PRC EST,DISSOLVED,WATER PC/L	08/03/94-08/03/94	0	1	
BADL0027	No	77825	ALACHLOR WHOLE WATER,UG/L	05/06/88-10/31/88	0	4	
BADL0008	No	80030	ALPHA,DISSOLVED GROSS,AS URANIUM-NATURAL,UG/L	08/02/94-08/02/94	0	1	
BADL0013	No	80030	ALPHA,DISSOLVED GROSS,AS URANIUM-NATURAL,UG/L	07/26/94-07/26/94	0	1	
BADL0014	Yes	80030	ALPHA,DISSOLVED GROSS,AS URANIUM-NATURAL,UG/L	07/12/94-07/12/94	0	1	
BADL0015	No	80030	ALPHA,DISSOLVED GROSS,AS URANIUM-NATURAL,UG/L	07/13/94-07/13/94	0	1	
BADL0016	No	80030	ALPHA,DISSOLVED GROSS,AS URANIUM-NATURAL,UG/L	07/27/94-07/27/94	0	1	
BADL0018	Yes	80030	ALPHA,DISSOLVED GROSS,AS URANIUM-NATURAL,UG/L	07/14/94-07/14/94	0	1	
BADL0019	No	80030	ALPHA,DISSOLVED GROSS,AS URANIUM-NATURAL,UG/L	07/27/94-07/27/94	0	1	
BADL0021	Yes	80030	ALPHA,DISSOLVED GROSS,AS URANIUM-NATURAL,UG/L	07/28/94-07/28/94	0	2	
BADL0026	No	80030	ALPHA,DISSOLVED GROSS,AS URANIUM-NATURAL,UG/L	09/07/93-08/30/94	0	2	
BADL0031	No	80030	ALPHA,DISSOLVED GROSS,AS URANIUM-NATURAL,UG/L	08/03/94-08/03/94	0	1	
BADL0008	No	80050	BETA,DISSOLVED GROSS,AS SR-Y-90, PC/L	08/02/94-08/02/94	0	1	
BADL0013	No	80050	BETA,DISSOLVED GROSS,AS SR-Y-90, PC/L	07/26/94-07/26/94	0	1	
BADL0014	Yes	80050	BETA,DISSOLVED GROSS,AS SR-Y-90, PC/L	07/12/94-07/12/94	0	1	
BADL0015	No	80050	BETA,DISSOLVED GROSS,AS SR-Y-90, PC/L	07/13/94-07/13/94	0	1	
BADL0016	No	80050	BETA,DISSOLVED GROSS,AS SR-Y-90, PC/L	07/27/94-07/27/94	0	1	
BADL0018	Yes	80050	BETA,DISSOLVED GROSS,AS SR-Y-90, PC/L	07/14/94-07/14/94	0	1	
BADL0019	No	80050	BETA,DISSOLVED GROSS,AS SR-Y-90, PC/L	07/27/94-07/27/94	0	1	

[1] T=Times Series Plot, A=Annual Plot, and S=Seasonal Plot

Station/Parameter Period of Record Tabulation
From 11/29/62 To 09/10/96

Station	In Park	Code	Name	Start - End	Years	Obs	Plots[1]
BADL0021	Yes	80050	BETA,DISSOLVED GROSS,AS SR-Y-90, PC/L	07/28/94-07/28/94	0	2	
BADL0026	No	80050	BETA,DISSOLVED GROSS,AS SR-Y-90, PC/L	09/07/93-08/30/94	0	2	
BADL0031	No	80050	BETA,DISSOLVED GROSS,AS SR-Y-90, PC/L	08/03/94-08/03/94	0	1	
BADL0011	Yes	80154	SUSP SEDIMENT CONCENTRATION-EVAP AT 110C (MG/L)	08/13/64-09/19/67	3	86	T
BADL0026	No	80154	SUSP SEDIMENT CONCENTRATION-EVAP AT 110C (MG/L)	08/30/94-08/30/94	0	1	
BADL0011	Yes	80155	SUSPENDED SEDIMENT DISCHARGE (TONS/DAY)	08/13/64-09/19/67	3	86	T
BADL0011	Yes	80158	BED MATERIAL FALL DIAMETER, % FINER THAN 062MM	07/27/65-07/27/65	0	1	
BADL0011	Yes	80159	BED MATERIAL FALL DIAMETER, % FINER THAN 125MM	07/27/65-07/27/65	0	1	
BADL0011	Yes	80160	BED MATERIAL FALL DIAMETER, % FINER THAN 250MM	07/27/65-09/08/66	1	2	
BADL0011	Yes	80161	BED MATERIAL FALL DIAMETER, % FINER THAN 500MM	07/27/65-09/08/66	1	2	
BADL0011	Yes	80162	BED MATERIAL FALL DIAMETER, % FINER THAN 1 00MM	07/27/65-09/08/66	1	2	
BADL0011	Yes	80163	BED MATERIAL FALL DIAMETER, % FINER THAN 2 00MM	07/27/65-09/08/66	1	2	
BADL0011	Yes	80170	BED MATERIAL SIEVE DIAMETER,% FINER THAN 4 00MM	07/27/65-09/08/66	1	3	
BADL0011	Yes	80171	BED MATERIAL SIEVE DIAMETER,% FINER THAN 8 00MM	07/27/65-09/08/66	1	3	
BADL0011	Yes	80172	BED MATERIAL SIEVE DIAMETER,% FINER THAN 16 0MM	07/27/65-09/08/66	1	3	
BADL0011	Yes	80173	BED MATERIAL SIEVE DIAMETER,% FINER THAN 32 0MM	07/27/65-09/08/66	1	3	
BADL0027	No	81757	CYANAZINE IN THE WHOLE WATER SAMPLE UG/L	05/06/88-10/31/88	0	4	
BADL0009	No	82033	MAGNESIUM - TOTAL UG/L(AS MG)	04/23/79-04/23/79	0	1	
BADL0010	No	82033	MAGNESIUM - TOTAL UG/L(AS MG)	04/18/79-04/18/79	0	1	
BADL0027	No	82184	AMETRYNE (GESAPAX OR EVIK) TOTAL UG/L	05/06/88-10/31/88	0	4	
BADL0001	Yes	82233	SILICON (SI) TOTAL IN WATER MG/L AS (SIO2)	06/29/78-06/29/78	0	1	
BADL0009	No	82364	THORIUM, TOTAL IN WATER UG/L	04/23/79-04/23/79	0	1	
BADL0010	No	82364	THORIUM, TOTAL IN WATER UG/L	04/18/79-04/18/79	0	1	
BADL0027	No	82611	METRIBUZIN, WHOLE WATER, TOTAL RECOVERABLE UG/L	05/06/88-10/31/88	0	4	
BADL0027	No	82612	METOLACHLOR, WHOLE WATER, TOTAL RECOVERABLE UG/L	05/06/88-10/31/88	0	4	
BADL0005	No	84000	GEOLOGIC AGE CODE (SEE USGS CATALOG)	11/30/62-11/30/62	0	1	
BADL0012	No	84000	GEOLOGIC AGE CODE (SEE USGS CATALOG)	11/30/62-11/30/62	0	1	
BADL0013	No	84000	GEOLOGIC AGE CODE (SEE USGS CATALOG)	07/26/94-06/21/95	0	3	
BADL0014	Yes	84000	GEOLOGIC AGE CODE (SEE USGS CATALOG)	07/12/94-07/12/94	0	1	
BADL0015	No	84000	GEOLOGIC AGE CODE (SEE USGS CATALOG)	07/13/94-08/28/95	1	2	
BADL0016	No	84000	GEOLOGIC AGE CODE (SEE USGS CATALOG)	07/10/95-07/10/95	0	1	
BADL0019	No	84000	GEOLOGIC AGE CODE (SEE USGS CATALOG)	10/08/92-05/12/95	2	3	
BADL0020	No	84000	GEOLOGIC AGE CODE (SEE USGS CATALOG)	11/29/62-11/29/62	0	1	
BADL0021	Yes	84000	GEOLOGIC AGE CODE (SEE USGS CATALOG)	07/28/94-05/16/95	0	2	
BADL0031	No	84000	GEOLOGIC AGE CODE (SEE USGS CATALOG)	11/05/92-11/05/92	0	1	
BADL0005	No	84001	AQUIFER NAME CODE (SEE USGS CATALOG)	11/30/62-11/30/62	0	1	
BADL0012	No	84001	AQUIFER NAME CODE (SEE USGS CATALOG)	11/30/62-11/30/62	0	1	
BADL0013	No	84001	AQUIFER NAME CODE (SEE USGS CATALOG)	07/26/94-06/21/95	0	3	
BADL0014	Yes	84001	AQUIFER NAME CODE (SEE USGS CATALOG)	07/12/94-07/12/94	0	1	
BADL0015	No	84001	AQUIFER NAME CODE (SEE USGS CATALOG)	07/13/94-08/28/95	1	2	
BADL0016	No	84001	AQUIFER NAME CODE (SEE USGS CATALOG)	07/10/95-07/10/95	0	1	
BADL0019	No	84001	AQUIFER NAME CODE (SEE USGS CATALOG)	10/08/92-05/12/95	2	3	
BADL0020	No	84001	AQUIFER NAME CODE (SEE USGS CATALOG)	11/29/62-11/29/62	0	1	
BADL0021	Yes	84001	AQUIFER NAME CODE (SEE USGS CATALOG)	07/28/94-05/16/95	0	2	
BADL0031	No	84001	AQUIFER NAME CODE (SEE USGS CATALOG)	11/05/92-11/05/92	0	1	

[1]T=Time Series Plot, A=Annual Plot, S=Seasonal Plot

Station-By-Station Results

Station Inventory for Station: BADL0001

NPS Station ID: BADL0001
Location: SPRING NW OF BADLANDS NP HEADQUARTERS
Station Type: /TYPA/AMBNT/SPRING
RMI-Indexes:
RMI-Miles:
HUC: 10140102
Major Basin: MISSOURI RIVER
Minor Basin: SOUTH CENTRAL MISSOURI
RF1 Index: 10140102
RF3 Index: 10140201028200 00
Description:

LAT/LON: 43 755142/-101 944310

Depth of Water: 0
Elevation: 2443

RF1 Mile Point: 0 000
RF3 Mile Point: 0 00

Agency: 11NPSWRD
FIPS State/County: 46071 SOUTH DAKOTA/JACKSON
STORET Station ID(s): BADL_SDSM_1
Within Park Boundary: Yes

Aquifer:
Water Body Id:
ECO Region:
Distance from RF1: 2 90
Distance from RF3: 0 06

On/Off RF1:
On/Off RF3:

Date Created: 12/20/97

THE SITE IS LOCATED ON THE COTTONWOOD SW SOUTH DAKOTA JACKSON COUNTY 7 MINUTE SERIES (TOPOGRAPHIC) QUADRANGLE. THE SITE IS LOCATED AT A SPRING NORTHWEST OF THE NEARBY BADLANDS NATIONAL PARK (BADL) HEADQUARTERS. DATA ARE FROM A 1978 "REPORT OF ANALYSIS" DATA SHEET SUBMITTED TO THE PARK SUPERINTENDENT BY THE SOUTH DAKOTA SCHOOL OF MINES AND TECHNOLOGY. LATITUDE AND LONGITUDE COORDINATE VALUES WERE OBTAINED BY DIGITIZING A POINT ON THE QUADRANGLE MAP MATCHING THE SITE DESCRIPTION IN THE REPORT. FOR MORE INFORMATION CONTACT RESOURCE MANAGEMENT AT BADLANDS NATIONAL PARK; P O BOX 6 INTERIOR SD 57750 TEL 605-433-5361. DATA PROCESSED AND UPLOADED TO STORET BY J. CHRIS ECHOHAWK, NATIONAL PARK SERVICE WATER RESOURCES DIVISION; 1201 OAK RIDGE DR SUITE 250; FORT COLLINS CO 80525 TEL 970-225-3516 FAX 970-225-9965

Parameter Inventory for Station: BADL0001

Parameter	Period of Record	Obs	Median	Mean	Maximum	Minimum	Variance	Std Dev	10th	25th	75th	90th
00403	PH, LAB, STANDARD UNITS SU	1	7 2	7 2	7 2	7 2	0	0	**	**	**	**
00403	CONVERTED PH, LAB, STANDARD UNITS	1	7 2	7 2	7 2	7 2	0	0	**	**	**	**
00403	MICRO EQUIVALENTS/LITER OF H+ COMPUTED FROM PH	1	0 063	0 063	0 063	0 063	0	0	**	**	**	**
00411	ALKALINITY, METHYL ORANGE MG/L	1	344	344	344	344	0	0	**	**	**	**
00520	RESIDUE, VOLATILE FILTRABLE (MG/L)	1	542	542	542	542	0	0	**	**	**	**
00535	RESIDUE, VOLATILE NONFILTRABLE (MG/L)	1	##	0 1	0 1	0 1	0	0	**	**	**	**
00618	NITRATE NITROGEN, DISSOLVED (MG/L AS N)	1	0 03	0 03	0 03	0 03	0	0	**	**	**	**
00650	PHOSPHATE, TOTAL (MG/L AS PO4)	1	0 09	0 09	0 09	0 09	0	0	**	**	**	**
00660	PHOSPHATE, ORTHO (MG/L AS PO4)	1	0 06	0 06	0 06	0 06	0	0	**	**	**	**
00900	HARDNESS, TOTAL (MG/L AS CACO3)	1	128	128	128	128	0	0	**	**	**	**
00901	HARDNESS, CARBONATE (MG/L AS CACO3)	1	111	111	111	111	0	0	**	**	**	**
00915	CALCIUM, DISSOLVED (MG/L AS CA)	1	44	44	44	44	0	0	**	**	**	**
00925	MAGNESIUM, DISSOLVED (MG/L AS MG)	1	4	4	4	4	0	0	**	**	**	**
00930	SODIUM, DISSOLVED (MG/L AS NA)	1	100	100	100	100	0	0	**	**	**	**
00935	POTASSIUM, DISSOLVED (MG/L AS K)	1	17	17	17	17	0	0	**	**	**	**
00941	CHLORIDE, DISSOLVED IN WATER MG/L	1	10	10	10	10	0	0	**	**	**	**
00946	SULFATE, DISSOLVED (MG/L AS SO4)	1	63	63	63	63	0	0	**	**	**	**
01046	IRON, DISSOLVED (UG/L AS FE)	1	0 1	0 1	0 1	0 1	0	0	**	**	**	**
01056	MANGANESE, DISSOLVED (UG/L AS MN)	1	## 0 025	0 025	0 025	0 025	0	0	**	**	**	**
82233	SILICON (SI) TOTAL IN WATER MG/L AS (SIO2)	1	27	27	27	27	0	0	**	**	**	**

** - Less than 9 observations   ## - Computed with 50% or more of the total observations as values that were half the detection limit   p - Has a corresponding time series plot

69

EPA Water Quality Criteria Analysis for Station: BADL0001

Parameter		Std Type	Std Value	Total Obs	Exceed Standard	Prop Exceeding	10/01-1/31			2/01-4/14			4/15-6/30			7/01-9/30		
							Obs	Exceed	Prop	Obs	Exceed	Prop	Obs	Exceed	Prop	Obs	Exceed	Prop
00403	PH, LAB	Other-Hi Lim	9	1	0	0.00							1	0	0.00			
		Other-Lo Lim	6 5															
00618	NITRATE NITROGEN, DISSOLVED AS N	Drinking Water	10	1	0	0.00							1	0	0.00			
00941	CHLORIDE, DISSOLVED IN WATER	Fresh Acute	860	1	0	0.00							1	0	0.00			
		Drinking Water	250	1	0	0.00							1	0	0.00			
00946	SULFATE, DISSOLVED (AS SO4)	Drinking Water	250	1	0	0.00							1	0	0.00			

& - Below detection limit observations, for which half the detection limit exceeded the criterion, were excluded from the criterion comparison for this parameter

70

NPS Station ID: BADL0002
Location: PIKE IN PENNINGTON COUNTY
Station Type: /TYPA/AMBNT/LAKE
RMI-Indexes:
RMI-Miles:
HUC: 10140102
Major Basin: MISSOURI RIVER
Minor Basin: BAD RIVER
RF1 Index: 10140102067
RF3 Index: 10140102140000 68
Description:
PIKE-SECTION 24, T 2S, R 17, LATITUDE 43 51 45 6, LONGITUDE 102 02 18 4, PENNINGTON COUNTY COMPOSITE SAMPLE

LAT/LON: 43 862670/-102 038448
Agency: 21SDAK01
FIPS State/County: 46103 SOUTH DAKOTA/PENNINGTON
STORET Station ID(s): 46PE33
Within Park Boundary: No

Depth of Water: 0
Elevation: 0
Aquifer:
Water Body Id:
ECO Region:
Distance from RF1: 0 00
Distance from RF3: 0 02
RF1 Mile Point: 2 390
RF3 Mile Point: 0 68

Date Created: 11/10/77
On/Off RF1: OFF
On/Off RF3:

Parameter Inventory for Station: BADL0002

Parameter		Period of Record	Obs	Median	Mean	Maximum	Minimum	Variance	Std Dev	10th	25th	75th	90th
00010	TEMPERATURE, WATER (DEGREES CENTIGRADE)	06/08/78-06/08/78	2	21.7	21.7	21.7	21.7	0	0	**	**	**	**
00011	TEMPERATURE, WATER (DEGREES FAHRENHEIT)	06/08/78-09/19/78	2	64.5	64.5	71	58	84.5	9.192	**	**	**	**
00020	TEMPERATURE, AIR (DEGREES CENTIGRADE)	06/08/78-06/08/78	1	24.4	24.4	24.4	24.4		0	**	**	**	**
00021	TEMPERATURE, AIR (DEGREES FAHRENHEIT)	06/08/78-09/19/78	2	64	64	76	52	288	16.971	**	**	**	**
00095	SPECIFIC CONDUCTANCE (UMHOS/CM @ 25C)	06/08/78-09/19/78	2	255	255	290	220	2450	49.497	**	**	**	**
00300	OXYGEN, DISSOLVED MG/L	06/08/78-09/19/78	2	10.75	10.75	11	10.5	0.125	0.354	**	**	**	**
00400	PH (STANDARD UNITS)	06/08/78-09/19/78	2	7.8	7.8	8.3	7.3	0.5	0.707	**	**	**	**
00400	CONVERTED PH (STANDARD UNITS)	06/08/78-09/19/78	2	7.56	7.56	8.3	7.3	0.616	0.785	**	**	**	**
00400	MICRO EQUIVALENTS/LITER OF H+ COMPUTED FROM PH	06/08/78-09/19/78	2	0.028	0.028	0.05	0.005	0.001	0.032	**	**	**	**
00403	PH, LAB, STANDARD UNITS SU	06/08/78-09/19/78	2	9	9	9.3	8.7	0.18	0.424	**	**	**	**
00403	CONVERTED PH, LAB, STANDARD UNITS	06/08/78-09/19/78	2	8.904	8.904	9.3	8.7	0.199	0.446	**	**	**	**
00403	MICRO EQUIVALENTS/LITER OF H+ COMPUTED FROM PH	06/08/78-09/19/78	2	0.001	0.001	0.002	0.001	0.001	0.001	**	**	**	**
00410	ALKALINITY, TOTAL (MG/L AS CACO3)	06/08/78-09/19/78	2	137.5	137.5	156	119	684.5	26.163	**	**	**	**
00515	RESIDUE, TOTAL FILTRABLE (DRIED AT 105C),MG/L	06/08/78-09/19/78	2	214	214	243	185	1682	41.012	**	**	**	**
00530	RESIDUE, TOTAL NONFILTRABLE (MG/L)	06/08/78-09/19/78	2	12.5	12.5	20	5	112.5	10.607	**	**	**	**
00610	NITROGEN, AMMONIA, TOTAL (MG/L AS N)	06/08/78-09/19/78	2 ##	0.01	0.01	0.01	0.01	0	0	**	**	**	**
00613	NITRITE NITROGEN, DISSOLVED (MG/L AS N)	06/08/78-09/19/78	2 ##	0.008	0.008	0.01	0.005	0	0.004	**	**	**	**
00625	NITROGEN, KJELDAHL, TOTAL, (MG/L AS N)	06/08/78-09/19/78	2 ##	1.17	1.17	1.38	0.96	0.088	0.297	**	**	**	**
00630	NITRITE PLUS NITRATE, TOTAL 1 DET (MG/L AS N)	06/08/78-09/19/78	1 ##	0.075	0.075	0.1	0.05	0.001	0.035	**	**	**	**
00671	PHOSPHORUS, DISSOLVED ORTHOPHOSPHATE (MG/L AS P)	06/08/78-06/08/78	1 ##	0.002	0.002	0.002	0.002	0	0	**	**	**	**
00940	CHLORIDE, TOTAL IN WATER MG/L	06/08/78-09/19/78	2	2	2	2	2	0	0	**	**	**	**
00945	SULFATE, TOTAL (MG/L AS SO4)	06/08/78-09/19/78	2 ##	3.25	3.25	4	2.5	1.125	1.061	**	**	**	**
31616	FECAL COLIFORM,MEMBR FILTER,M-FC BROTH,44 5 C	06/08/78-09/19/78	2	8	8	13	3	50	7.071	**	**	**	**
31616	LOG FECAL COLIFORM,MEMBR FILTER,M-FC BROTH,44 5 C	06/08/78-09/19/78	2	0.796	0.796	1.114	0.477	0.203	0.45	**	**	**	**
31616	GM FECAL COLIFORM,MEMBR FILTER,M-FC BROTH,44 5 C	GEOMETRIC MEAN =			6.245								
70507	PHOSPHORUS,IN TOTAL ORTHOPHOSPHATE (MG/L AS P)	09/19/78-09/19/78	1	0.017	0.017	0.017	0.017	0	0	**	**	**	**

** - Less than 9 observations   ## - Computed with 50% or more of the total observations as values that were half the detection limit   p - Has a corresponding time series plot

EPA Water Quality Criteria Analysis for Station: BADL0002

Parameter		Std Type	Std Value	Total Obs	Exceed Standard	Prop Exceeding	10/01-1/31 Obs	Exceed	Prop	2/01-4/14 Obs	Exceed	Prop	4/15-6/30 Obs	Exceed	Prop	7/01-9/30 Obs	Exceed	Prop
00300	OXYGEN, DISSOLVED	Other-Lo Lim	4	2	0	0.00							1	0	0.00	1	0	0.00
00400	PH	Other-Hi Lim	9	2	0	0.00							1	0	0.00	1	0	0.00
		Other-Lo Lim	6 5	2	0	0.00							1	0	0.00	1	0	0.00
00403	PH, LAB	Other-Hi Lim	9	2	1	0.50							1	1	1 00	1	0	0.00
		Other-Lo Lim	6 5	2	0	0.00							1	0	0.00	1	0	0.00
00613	NITRITE NITROGEN, DISSOLVED AS N	Drinking Water	1	2	0	0.00							1	0	0.00	1	0	0.00
00630	NITRITE PLUS NITRATE, TOTAL 1 DET	Drinking Water	10	2	0	0.00							1	0	0.00	1	0	0.00
00940	CHLORIDE, TOTAL IN WATER	Fresh Acute	860	2	0	0.00							1	0	0.00	1	0	0.00
00945	SULFATE, TOTAL (AS SO4)	Drinking Water	250	2	0	0.00							1	0	0.00	1	0	0.00
		Drinking Water	250	2	0	0.00							1	0	0.00	1	0	0.00
31616	FECAL COLIFORM, MEMBRANE FILTER, BROTH	Other-Hi Lim	200	2	0	0.00							1	0	0.00	1	0	0.00

& - Below detection limit observations, for which half the detection limit exceeded the criterion, were excluded from the criterion comparison for this parameter

72

Station Inventory for Station: BADL0003

NPS Station ID: BADL0003
Location: N WHITE WATER IN PENN COUNTY
Station Type: /TYPA/AMBNT/LAKE
RMI-Indexes:
RMI-Miles:
HUC: 10140102
Major Basin: MISSOURI RIVER
Minor Basin: BAD RIVER
RF1 Index: 10140102067
RF3 Index: 10120111060902 34
Description
N WHITE WATER-SECTION 35, T 25 , R 17, LATITUDE 43 50 00 0, LONGITUDE 102 03 10 9, PENNINGTON COUNTY COMPOSITE SAMPLE

LAT/LON: 43 833338,-102 053031

Agency: 21SDAK01
FIPS State/County: 46103 SOUTH DAKOTA/PENNINGTON
STORET Station IDs: 46PE29
Within Park Boundary: No

Aquifer:
Water Body Id:
ECO Region:
Distance from RF1: 0 00
Distance from RF3: 0 07

Depth of Water: 0
Elevation: 0

RF1 Mile Point: 4 860
RF3 Mile Point: 2 33

Date Created: 11/10/77

On/Off RF1: OFF
On/Off RF3:

Parameter Inventory for Station: BADL0003

Parameter	Period of Record	Obs	Median	Mean	Maximum	Minimum	Variance	Std Dev	10th	25th	75th	90th
00010 TEMPERATURE, WATER (DEGREES CENTIGRADE)	12/19/77-06/08/78	3	2 2	7 4	20	0	120 28	10 967	**	**	**	**
00011 TEMPERATURE, WATER (DEGREES FAHRENHEIT)	12/19/77-09/19/78	4	47 5	48 75	68	32	306 25	17 5	**	**	**	**
00020 TEMPERATURE, AIR (DEGREES CENTIGRADE)	12/19/77-06/08/78	3	-3 3	8 133	24 4	-3 3	209 343	14 469	**	**	**	**
00021 TEMPERATURE, AIR (DEGREES FAHRENHEIT)	12/19/77-09/19/78	4	46	48 5	76	26	467 667	21 626	**	**	**	**
00095 SPECIFIC CONDUCTANCE (UMHOS/CM @ 25C)	12/19/77-09/19/78	4	390	425	820	100	90433 333	300 721	**	**	**	**
00300 OXYGEN, DISSOLVED MG/L	12/19/77-09/19/78	4	9 95	10 225	12 5	8 5	2 969	1 723	**	**	**	**
00400 PH (STANDARD UNITS)	12/19/77-09/19/78	4	7 7	7 765	8	7 2	0 117	0 342	**	**	**	**
00400 CONVERTED PH (STANDARD UNITS)	12/19/77-09/19/78	4	7 689	7 545	8	7 2	0 131	0 362	**	**	**	**
00400 MICRO EQUIVALENTS/LITER OF H+ COMPUTED FROM PH	12/19/77-09/19/78	4	0 02	0 029	0 063	0 01	0 001	0 024	**	**	**	**
00403 PH, LAB, STANDARD UNITS SU	12/19/77-09/19/78	4	8 5	8 55	9	8 2	0 117	0 342	**	**	**	**
00403 CONVERTED PH, LAB, STANDARD UNITS	12/19/77-09/19/78	4	8 489	8 462	8 55	8 2	0 127	0 356	**	**	**	**
00403 MICRO EQUIVALENTS/LITER OF H+ COMPUTED FROM PH	12/19/77-09/19/78	4	0 003	0 003	0 006	0 001		0 002	**	**	**	**
00410 ALKALINITY, TOTAL (MG/L AS CACO3)	12/19/77-09/19/78	4	198	208 25	397	40	22107 583	148 686	**	**	**	**
00500 RESIDUE, TOTAL (MG/L)	12/19/77-09/19/78	4	544 5	523 25	789	215	58462 917	241 791	**	**	**	**
00515 RESIDUE, TOTAL FILTRABLE (DRIED AT 105C),MG/L	12/19/77-09/19/78	4	249	267 5	489	83	29731 667	172 429	**	**	**	**
00530 RESIDUE, TOTAL NONFILTRABLE (MG/L)	12/19/77-09/19/78	4	238 5	255 75	414	132	16172 25	127 17	**	**	**	**
00608 NITROGEN, AMMONIA, DISSOLVED (MG/L AS N)	03/15/78-09/19/78	2	0 385	0 385	0 6	0 17	0 092	0 304	**	**	**	**
00610 NITROGEN, AMMONIA, TOTAL (MG/L AS N)	12/19/77-09/19/78	4 ##	0 07	0 07	0 09	0 05	0 001	0 028	**	**	**	**
00613 NITRITE NITROGEN, DISSOLVED (MG/L AS N)	12/19/77-09/19/78	4	0 008	0 01	0 02	0 005	0	0 007	**	**	**	**
00625 NITROGEN, KJELDAHL, TOTAL (MG/L AS N)	12/19/77-09/19/78	4	1 565	1 685	2 38	1 23	0 298	0 546	**	**	**	**
00630 NITRITE PLUS NITRATE, TOTAL 1 DET (MG/L AS N)	12/19/77-09/19/78	4	0 15	0 175	0 3	0 1	0 009	0 096	**	**	**	**
00671 PHOSPHORUS, DISSOLVED ORTHOPHOSPHATE (MG/L AS P)	12/19/77-06/08/78	3	0 169	0 148	0 172	0 104	0 001	0 038	**	**	**	**
00940 CHLORIDE,TOTAL IN WATER MG/L	12/19/77-09/19/78	4	4	6 75	17	2	48 25	6 946	**	**	**	**
00945 SULFATE, TOTAL (MG/L AS SO4)	03/15/78-09/19/78	2	9 5	9 5	17	2	112 5	10 607	**	**	**	**
00946 SULFATE, DISSOLVED (MG/L AS SO4)	12/19/77-06/08/78	2	28	28	40	16	288	16 971	**	**	**	**
31616 FECAL COLIFORM,MEMBR FILTER,M-FC BROTH,44 5 C	12/19/77-06/08/78	3	40	46 667	67	33	322 333	17 954	**	**	**	**
31616 LOG FECAL COLIFORM,MEMBR FILTER,M-FC BROTH,44 5 C	03/15/78-09/19/78	3	1 602	1 649	1 826	1 519	0 025	0 159	**	**	**	**
31616 GM FECAL COLIFORM,MEMBR FILTER,M-FC BROTH,44 5 C	GEOMETRIC MEAN =			44 554								
70507 PHOSPHORUS,IN TOTAL ORTHOPHOSPHATE (MG/L AS P)	09/19/78-09/19/78	1	0 01	0 01	0 01	0 01	0	0	**	**	**	**

** - Less than 9 observations   ## - Computed with 50% or more of the total observations as values that were half the detection limit   p - Has a corresponding time series plot

73

EPA Water Quality Criteria Analysis for Station: BADL0003

Parameter		Std Type	Std Value	Total Obs	Exceed Standard	Prop Exceeding	10/01-1/31 Obs	Exceed	Prop	2/01-4/14 Obs	Exceed	Prop	4/15-6/30 Obs	Exceed	Prop	7/01-9/30 Obs	Exceed	Prop
00300	OXYGEN, DISSOLVED	Other-Lo Lim	4	4	0	0.00	1	0	0.00	1	0	0.00	1	0	0.00	1	0	0.00
00400	PH	Other-Hi Lim	9	4	0	0.00	1	0	0.00	1	0	0.00	1	0	0.00	1	0	0.00
		Other-Lo Lim	6.5	4	0	0.00	1	0	0.00	1	0	0.00	1	0	0.00	1	0	0.00
00403	PH, LAB	Other-Hi Lim	9	4	1	0.25	1	0	0.00	1	0	0.00	1	0	0.00	1	1	1.00
		Other-Lo Lim	6.5	4	0	0.00	1	0	0.00	1	0	0.00	1	0	0.00	1	0	0.00
00613	NITRITE NITROGEN, DISSOLVED AS N	Drinking Water	1	4	0	0.00	1	0	0.00	1	0	0.00	1	0	0.00	1	0	0.00
00630	NITRITE PLUS NITRATE, TOTAL 1 DET	Drinking Water	10	4	0	0.00	1	0	0.00	1	0	0.00	1	0	0.00	1	0	0.00
00940	CHLORIDE,TOTAL IN WATER	Fresh Acute	860	4	0	0.00	1	0	0.00	1	0	0.00	1	0	0.00	1	0	0.00
		Drinking Water	250	4	0	0.00	1	0	0.00	1	0	0.00	1	0	0.00	1	0	0.00
00945	SULFATE, TOTAL (AS SO4)	Drinking Water	250	2	0	0.00							1	0	0.00	1	0	0.00
00946	SULFATE, DISSOLVED (AS SO4)	Drinking Water	250	2	0	0.00							1	0	0.00	1	0	0.00
31616	FECAL COLIFORM, MEMBRANE FILTER, BROTH	Other-Hi Lim	200	3	0	0.00	1	0	0.00	1	0	0.00				1	0	0.00

& - Below detection limit observations, for which half the detection limit exceeded the criterion, were excluded from the criterion comparison for this parameter

Station Inventory for Station: BADL0004

Date Created: 11/10/77

NPS Station ID: BADL0004
Location: MISSLE ALLOTMENT IN PENN COUNTY
Station Type: /TYPA/AMBNT/LAKE
RMI-Indexes:
HUC: 10140102
Major Basin: MISSOURI RIVER
Minor Basin: BAD RIVER
RF1 Index: 10140102067
RF3 Index: 10120109031100 00
Description:
MISSLE ALLOTMENT-SECTION 28, T 25, R 17, LATITUDE 43 50 29 6, LONGITUDE 102 05 40 9, PENNINGTON COUNTY COMPOSITE SAMPLE

LAT/LON: 43 841560/-102 094698

Agency: 21SDAK01
FIPS State/County: 46103 SOUTH DAKOTA/PENNINGTON
STORET Station ID(s): 46PE25
Within Park Boundary: No

Depth of Water: 0
Elevation: 0
RF1 Mile Point: 5 630
RF3 Mile Point: 1 07

Aquifer:
Water Body Id:
ECO Region:
Distance from RF1: 0 00
Distance from RF3: 0 13

On/Off RF1: ON
On/Off RF3:

Parameter Inventory for Station: BADL0004

Parameter		Period of Record	Obs	Median	Mean	Maximum	Minimum	Variance	Std Dev	10th	25th	75th	90th
00010	TEMPERATURE, WATER (DEGREES CENTIGRADE)	06/08/78-06/08/78	1	22 2	22 2	22 2	22 2	0	0	**	**	**	**
00011	TEMPERATURE, WATER (DEGREES FAHRENHEIT)	06/08/78-09/19/78	2	64 5	64 5	72	57	112 5	10 607	**	**	**	**
00020	TEMPERATURE, AIR (DEGREES CENTIGRADE)	06/08/78-06/08/78	2	25	25	25	25	0	0	**	**	**	**
00021	TEMPERATURE, AIR (DEGREES FAHRENHEIT)	06/08/78-09/19/78	2	64 5	64 5	77	52	312 5	17 678	**	**	**	**
00095	SPECIFIC CONDUCTANCE (UMHOS/CM @ 25C)	06/08/78-09/19/78	2	192 5	192 5	220	165	1512 5	38 891	**	**	**	**
00300	OXYGEN, DISSOLVED MG/L	06/08/78-09/19/78	2	9 25	9 25	10	8 5	1 125	1 061	**	**	**	**
00400	PH (STANDARD UNITS)	06/08/78-09/19/78	2	7 4	7 4	7 4	7 4	0	0	**	**	**	**
00400	CONVERTED PH (STANDARD UNITS)	06/08/78-09/19/78	2	7 4	7 4	7 4	7 4	0	0	**	**	**	**
00400	MICRO EQUIVALENTS/LITER OF H+ COMPUTED FROM PH	06/08/78-09/19/78	2	0 04	0 04	0 04	0 04	0	0	**	**	**	**
00403	PH, LAB, STANDARD UNITS SU	06/08/78-09/19/78	2	8 75	8 75	9 2	8 3	0 405	0 636	**	**	**	**
00403	CONVERTED PH, LAB, STANDARD UNITS	06/08/78-09/19/78	2	8 55	8 55	9 2	8 3	0 485	0 697	**	**	**	**
00403	MICRO EQUIVALENTS/LITER OF H+ COMPUTED FROM PH	06/08/78-09/19/78	2	0 003	0 003	0 005	0 001	0	0 003	**	**	**	**
00410	ALKALINITY, TOTAL (MG/L AS CACO3)	06/08/78-09/19/78	2	104	104	119	89	450	21 213	**	**	**	**
00500	RESIDUE, TOTAL (MG/L)	06/08/78-09/19/78	2	154 5	154 5	165	144	220 5	14 849	**	**	**	**
00515	RESIDUE, TOTAL FILTRABLE (DRIED AT 105C),MG/L	06/08/78-09/19/78	2	133 5	133 5	142	125	144 5	12 021	**	**	**	**
00530	RESIDUE, TOTAL NONFILTRABLE (MG/L)	06/08/78-09/19/78	2	21	21	23	19	8	2 828	**	**	**	**
00610	NITROGEN, AMMONIA, TOTAL (MG/L AS N)	06/08/78-09/19/78	2 ##	0 01	0 01	0 01	0 01	0	0	**	**	**	**
00613	NITRITE NITROGEN, DISSOLVED (MG/L AS N)	06/08/78-09/19/78	2 ##	0 005	0 005	0 005	0 005	0	0	**	**	**	**
00625	NITROGEN, KJELDAHL, TOTAL, (MG/L AS N)	06/08/78-09/19/78	2	1 09	1 09	1 22	0 96	0 034	0 184	**	**	**	**
00630	NITRITE PLUS NITRATE, TOTAL 1 DET (MG/L AS N)	06/08/78-09/19/78	2	0 1	0 1	0 1	0 1	0	0	**	**	**	**
00671	PHOSPHORUS, DISSOLVED ORTHOPHOSPHATE (MG/L AS P)	06/08/78-06/08/78	1	0 005	0 005	0 005	0 005	0	0	**	**	**	**
00940	CHLORIDE,TOTAL IN WATER MG/L	06/08/78-09/19/78	2 ##	1 3	1 3	2	0 6	0 98	0 99	**	**	**	**
00945	SULFATE, TOTAL (MG/L AS SO4)	06/08/78-09/19/78	2 ##	2 25	2 25	2 5	2	0 125	0 354	**	**	**	**
31616	FECAL COLIFORM,MEMBR FILTER,M-FC BROTH,44 5 C	06/08/78-09/19/78	2	58 5	58 5	100	17	3444 5	58 69	**	**	**	**
31616	LOG FECAL COLIFORM,MEMBR FILTER,M-FC BROTH,44 5 C	06/08/78-09/19/78	2	1 615	1 615	2	1 23	0 296	0 544	**	**	**	**
31616	GM FECAL COLIFORM,MEMBR FILTER,M-FC BROTH,44 5 C	GEOMETRIC MEAN =			41 231						**		**
70507	PHOSPHORUS,IN TOTAL ORTHOPHOSPHATE (MG/L AS P)	09/19/78-09/19/78	1 ##	0 003	0 003	0 003	0 003	0	0	**	**	**	**

** - Less than 9 observations   ## - Computed with 50% or more of the total observations as values that were half the detection limit   p - Has a corresponding time series plot

EPA Water Quality Criteria Analysis for Station: BADL0004

Parameter	Std Type	Std Value	Total Obs	Exceed Standard	Prop Exceeding	10/01-1/31 Obs	Exceed	Prop	2/01-4/14 Obs	Exceed	Prop	4/15-6/30 Obs	Exceed	Prop	7/01-9/30 Obs	Exceed	Prop
00300 OXYGEN, DISSOLVED	Other-Lo Lim	4	2	0	0.00							1	0	0.00	1	0	0.00
00400 PH	Other-Hi Lim	9	2	0	0.00							1	0	0.00	1	0	0.00
	Other-Lo Lim	6 5	2	0	0.00							1	0	0.00	1	0	0.00
00403 PH, LAB	Other-Hi Lim	9	2	1	0.50							1	1	1.00	1	0	0.00
	Other-Lo Lim	6 5	2	0	0.00							1	0	0.00	1	0	0.00
00613 NITRITE NITROGEN, DISSOLVED AS N	Drinking Water	1	2	0	0.00							1	0	0.00	1	0	0.00
00630 NITRITE PLUS NITRATE, TOTAL 1 DET	Drinking Water	10	2	0	0.00							1	0	0.00	1	0	0.00
00940 CHLORIDE, TOTAL IN WATER	Fresh Acute	860	2	0	0.00							1	0	0.00	1	0	0.00
00945 SULFATE, TOTAL (AS SO4)	Drinking Water	250	2	0	0.00							1	0	0.00	1	0	0.00
	Drinking Water	250	2	0	0.00							1	0	0.00	1	0	0.00
31616 FECAL COLIFORM, MEMBRANE FILTER, BROTH	Other-Hi Lim	200	2	0	0.00							1	0	0.00	1	0	0.00

& - Below detection limit observations, for which half the detection limit exceeded the criterion, were excluded from the criterion comparison for this parameter

76

Station Inventory for Station: BADL0005

NPS Station ID: BADL0005
Location: 42N42W 2BB JOHN POURIER
Station Type: /TYPA/AMBNT/SPRING
RMI-Indexes:
HUC: 10140201
Major Basin:
Minor Basin:
RF1 Index: 10140201
RF3 Index: 101402012317708 64
Description:

LAT/LON: 43 605559/-102 275003

Agency: 112WRD
FIPS State/County: 46113 SOUTH DAKOTA/SHANNON
STORET Station ID(s): 4336201021630001
Within Park Boundary: No

Depth of Water: 0
Elevation: 0

RF1 Mile Point: 0 000
RF3 Mile Point: 9 36

Aquifer:
Water Body Id:
ECO Region:
Distance from RF1: 7 20
Distance from RF3: 0 21

On/Off RF1:
On/Off RF3:

Date Created: 01/16/79

Parameter Inventory for Station: BADL0005

Parameter		Period of Record	Obs	Median	Mean	Maximum	Minimum	Variance	Std Dev	10th	25th	75th	90th
00095	SPECIFIC CONDUCTANCE (UMHOS/CM @ 25C)	11/30/62-11/30/62	1	473	473	473	473	0	0	**	**	**	**
00400	PH (STANDARD UNITS)	11/30/62-11/30/62	1	7 8	7 8	7 8	7 8	0	0	**	**	**	**
00400	CONVERTED PH (STANDARD UNITS)	11/30/62-11/30/62	1	7 8	7 8	7 8	7 8	0	0	**	**	**	**
00400	MICRO EQUIVALENTS/LITER OF H+ COMPUTED FROM PH	11/30/62-11/30/62	1	0 016	0 016	0 016	0 016	0	0	**	**	**	**
00405	CARBON DIOXIDE (MG/L AS CO2)	11/30/62-11/30/62	1	7 2	7 2	7 2	7 2	0	0	**	**	**	**
00410	ALKALINITY, TOTAL (MG/L AS CACO3)	11/30/62-11/30/62	1	232	232	232	232	0	0	**	**	**	**
00440	BICARBONATE ION (MG/L AS HCO3)	11/30/62-11/30/62	1	283	283	283	283	0	0	**	**	**	**
00445	CARBONATE ION (MG/L AS CO3)	11/30/62-11/30/62	1	0	0	0	0	0	0	**	**	**	**
00618	NITRATE NITROGEN, DISSOLVED (MG/L AS N)	11/30/62-11/30/62	1	1 2	1 2	1 2	1 2	0	0	**	**	**	**
00900	HARDNESS, TOTAL (MG/L AS CACO3)	11/30/62-11/30/62	1	158	158	158	158	0	0	**	**	**	**
00902	HARDNESS, NON-CARBONATE (MG/L AS CACO3)	11/30/62-11/30/62	1	0	0	0	0	0	0	**	**	**	**
00915	CALCIUM, DISSOLVED (MG/L AS CA)	11/30/62-11/30/62	1	53	53	53	53	0	0	**	**	**	**
00925	MAGNESIUM, DISSOLVED (MG/L AS MG)	11/30/62-11/30/62	1	63	63	63	63	0	0	**	**	**	**
00930	SODIUM, DISSOLVED (MG/L AS NA)	11/30/62-11/30/62	1	39	39	39	39	0	0	**	**	**	**
00931	SODIUM ADSORPTION RATIO	11/30/62-11/30/62	1	1 3	1 3	1 3	1 3	0	0	**	**	**	**
00932	SODIUM PERCENT	11/30/62-11/30/62	1	33	33	33	33	0	0	**	**	**	**
00935	POTASSIUM, DISSOLVED (MG/L AS K)	11/30/62-11/30/62	1	11	11	11	11	0	0	**	**	**	**
00940	CHLORIDE, TOTAL IN WATER MG/L	11/30/62-11/30/62	1	2	2	2	2	0	0	**	**	**	**
00945	SULFATE, TOTAL (MG/L AS SO4)	11/30/62-11/30/62	1	15	15	15	15	0	0	**	**	**	**
00950	FLUORIDE, DISSOLVED (MG/L AS F)	11/30/62-11/30/62	1	0 4	0 4	0 4	0 4	0	0	**	**	**	**
00955	SILICA, DISSOLVED (MG/L AS SI02)	11/30/62-11/30/62	1	47	47	47	47	0	0	**	**	**	**
01020	BORON, DISSOLVED (UG/L AS B)	11/30/62-11/30/62	1	160	160	160	160	0	0	**	**	**	**
70300	RESIDUE, TOTAL FILTRABLE (DRIED AT 180C),MG/L	11/30/62-11/30/62	1	331	331	331	331	0	0	**	**	**	**
70301	SOLIDS, DISSOLVED-SUM OF CONSTITUENTS (MG/L)	11/30/62-11/30/62	1	318	318	318	318	0	0	**	**	**	**
71851	NITRATE NITROGEN, DISSOLVED (MG/L AS NO3)	11/30/62-11/30/62	1	5 2	5 2	5 2	5 2	0	0	**	**	**	**
71883	MANGANESE, TOTAL ELEMENTAL (UG/L AS MN)	11/30/62-11/30/62	1	10	10	10	10	0	0	**	**	**	**
71885	IRON (UG/L AS FE)	11/30/62-11/30/62	1	40	40	40	40	0	0	**	**	**	**

** - Less than 9 observations   ## - Computed with 50% or more of the total observations as values that were half the detection limit   p - Has a corresponding time series plot

EPA Water Quality Criteria Analysis for Station: BADL0005

Parameter		Std Type	Std Value	Total Obs	Exceed Standard	Prop Exceeding	Obs	10/01-1/31 Exceed	Prop	Obs	2/01-4/14 Exceed	Prop	Obs	4/15-6/30 Exceed	Prop	Obs	7/01-9/30 Exceed	Prop
00400	PH	Other-Hi Lim	9	1	0	0 00	0	0 00											
		Other-Lo Lim	6 5	1	0	0 00	0	0 00											

& - Below detection limit observations, for which half the detection limit exceeded the criterion, were excluded from the criterion comparison for this parameter

EPA Water Quality Criteria Analysis for Station: BADL0005

Parameter	Std Type	Std Value	Total Obs	Exceed Standard	Prop Exceeding	10/01-1/31			2/01-4/14			4/15-6/30			7/01-9/30		
						Obs	Exceed	Prop	Obs	Exceed	Prop	Obs	Exceed	Prop	Obs	Exceed	Prop
00618	NITRATE NITROGEN, DISSOLVED AS N	Drinking Water	10	1	0	0.00	1	0	0.00								
00940	CHLORIDE,TOTAL IN WATER	Fresh Acute	860	1	0	0.00	1	0	0.00								
00945	SULFATE, TOTAL (AS SO4)	Drinking Water	250	1	0	0.00	1	0	0.00								
00950	FLUORIDE, DISSOLVED AS F	Drinking Water	4	1	0	0.00	1	0	0.00								
71851	NITRATE NITROGEN, DISSOLVED (AS NO3)	Drinking Water	44	1	0	0.00	1	0	0.00								

& - Below detection limit observations, for which half the detection limit exceeded the criterion, were excluded from the criterion comparison for this parameter

78

Station Inventory for Station: BADL0006

NPS Station ID: BADL0006
Location: NEW WALL-E PENNINGTON CO
Station Type: /TYPA/AMBNT/LAKE
RMI-Indexes:
HUC: 10120111
Major Basin: MISSOURI RIVER BASIN
Minor Basin: S CENTRAL MISSOURI RIVER BASIN
RF1 Index: 10120111
RF3 Index: 10120109006301 37
Description:
COMPOSITE SAMPLE OF 3 INLAKE SITES - SURFACE   AND BOTTOM SAMPLES

LAT/LON: 43 977504/-102 277781

Agency: 21SDLASS
FIPS State/County: 46103 SOUTH DAKOTA/PENNINGTON
STORET Station ID(s): 46LAS58S
Within Park Boundary: No

Aquifer:
Water Body Id:
ECO Region:
Distance from RF1: 0 60
Distance from RF3: 0 04

Depth of Water: 17
Elevation: 0

RF1 Mile Point: 0 000
RF3 Mile Point: 1 36

Date Created: 08/10/91

On/Off RF1:
On/Off RF3:

Parameter Inventory for Station: BADL0006

Parameter		Period of Record	Obs	Median	Mean	Maximum	Minimum	Variance	Std Dev	10th	25th	75th	90th
00010	TEMPERATURE, WATER (DEGREES CENTIGRADE)	07/12/89-08/16/89	2	23 75	23 75	24	23 5	0 125	0 354	**	**	**	**
00021	TEMPERATURE, AIR (DEGREES FAHRENHEIT)	07/12/89-08/16/89	2	84 5	84 5	89	80	40 5	6 364	**	**	**	**
00078	TRANSPARENCY, SECCHI DISC (METERS)	07/12/89-08/16/89	2	0 785	0 785	0 91	0 66	0	0 177	**	**	**	**
00095	SPECIFIC CONDUCTANCE (UMHOS/CM @ 25C)	08/16/89-08/16/89	1	525	525	525	525	0	0	**	**	**	**
00300	OXYGEN, DISSOLVED MG/L	07/12/89-08/16/89	2	9 55	9 55	11 6	7 5	8 405	2 899	**	**	**	**
00400	PH (STANDARD UNITS)	07/12/89-08/16/89	2	8 7	8 7	8 8	8 6	0 02	0 141	**	**	**	**
00400	MICRO EQUIVALENTS/LITER OF H+ COMPUTED FROM PH	07/12/89-08/16/89	2	0 002	0 002	0 003	0 002	0	0 001	**	**	**	**
00403	PH, LAB, STANDARD UNITS SU	07/12/89-08/16/89	2	8 5	8 5	8 7	8 3	0 08	0 283	**	**	**	**
00403	CONVERTED PH LAB, STANDARD UNITS	07/12/89-08/16/89	2	8 455	8 455	8 7	8 3	0 084	0 29	**	**	**	**
00403	MICRO EQUIVALENTS/LITER OF H+ COMPUTED FROM PH	07/12/89-08/16/89	2	0 004	0 004	0 005	0 002	0	0 002	**	**	**	**
00410	ALKALINITY, TOTAL (MG/L AS CACO3)	07/12/89-08/16/89	2	222	222	224	220	8	2 828	**	**	**	**
00415	ALKALINITY, PHENOLPHTHALEIN (MG/L)	07/12/89-08/16/89	2	4 5	4 5	6	3	4 5	2 121	**	**	**	**
00500	RESIDUE, TOTAL (MG/L)	07/12/89-08/16/89	2	382	382	386	378	32	5 657	**	**	**	**
00530	RESIDUE, TOTAL NONFILTRABLE (MG/L)	07/12/89-08/16/89	2	21	21	32	10	242	15 556	**	**	**	**
00610	NITROGEN, AMMONIA, TOTAL (MG/L AS N)	07/12/89-08/16/89	2	0 17	0 17	0 3	0 04	0 034	0 184	**	**	**	**
00625	NITROGEN, KJELDAHL, TOTAL, (MG/L AS N)	07/12/89-08/16/89	2	1 335	1 335	1 37	1 3	0 002	0 049	**	**	**	**
00630	NITRITE PLUS NITRATE, TOTAL 1 DET (MG/L AS N)	07/12/89-08/16/89	2 ##	0 05	0 05	0 05	0 05	0	0	**	**	**	**
00665	PHOSPHORUS, TOTAL (MG/L AS P)	07/12/89-08/16/89	2	0 105	0 105	0 142	0 068	0 003	0 052	**	**	**	**
00671	PHOSPHORUS, DISSOLVED ORTHOPHOSPHATE (MG/L AS P)	07/12/89-08/16/89	2 ##	0 017	0 017	0 032	0 003	0	0 021	**	**	**	**
70300	RESIDUE, TOTAL FILTRABLE (DRIED AT 180C),MG/L	07/12/89-08/16/89	2	361	361	376	346	450	21 213	**	**	**	**

** - Less than 9 observations   ## - Computed with 50% or more of the total observations as values that were half the detection limit   p - Has a corresponding time series plot

EPA Water Quality Criteria Analysis for Station: BADL0006

Parameter		Std Type	Std Value	Total Obs	Exceed Standard	Prop Exceeding	10/01-1/31 Obs	10/01-1/31 Exceed	10/01-1/31 Prop	2/01-4/14 Obs	2/01-4/14 Exceed	2/01-4/14 Prop	4/15-6/30 Obs	4/15-6/30 Exceed	4/15-6/30 Prop	7/01-9/30 Obs	7/01-9/30 Exceed	7/01-9/30 Prop
00300	OXYGEN, DISSOLVED	Other-Lo Lim	4	2	0	0.00										2	0	0 00
00400	PH	Other-Hi Lim	9	2	0	0.00										2	0	0 00
		Other-Lo Lim	6 5	2	0	0.00										2	0	0 00
00403	PH, LAB	Other-Hi Lim	9	2	0	0.00										2	0	0 00
		Other-Lo Lim	6 5	2	0	0.00										2	0	0 00

& - Below detection limit observations, for which half the detection limit exceeded the criterion, were excluded from the criterion comparison for this parameter

79

EPA Water Quality Criteria Analysis for Station: BADL0006

Parameter	Std Type	Std Value	Total Obs	Exceed Standard	Prop Exceeding	10/01-1/31			2/01-4/14			4/15-6/30			7/01-9/30		
						Obs	Exceed	Prop	Obs	Exceed	Prop	Obs	Exceed	Prop	Obs	Exceed	Prop
00630 NITRITE PLUS NITRATE, TOTAL 1 DET	Drinking Water	10	2	0	0 00										2	0	0 00

& - Below detection limit observations, for which half the detection limit exceeded the criterion, were excluded from the criterion comparison for this parameter

80

Station Inventory for Station: BADL0007

Date Created: 08/17/91

NPS Station ID: BADL0007
Location: NEW WALL LAKE-E PENNINGTON CO
Station Type: /TYPA/AMBNT/LAKE
RMI-Indexes:
RMI-Miles:
HUC: 10120111
Major Basin: MISSOURI RIVER BASIN
Minor Basin: S CENTRAL MISSOURI RIVER BASIN
RF1 Index: 10120111
RF3 Index: 101201110870001 62
Description:
COMPOSITE SAMPLE OF 3 INLAKE SITES - SURFACE   AND BOTTOM SAMPLES

LAT/LON: 43 977504/-102 277781

Agency: 21SDLASS
FIPS State/County: 46103 SOUTH DAKOTA/PENNINGTON
STORET Station ID(s) 46LAS58B
Within Park Boundary: No

Aquifer:
Water Body Id:
ECO Region:
Distance from RF1: 11 70
Distance from RF3: 0 07

Depth of Water: 17
Elevation: 0

RF1 Mile Point: 0 000
RF3 Mile Point: 1 61

On/Off RF1:
On/Off RF3:

Parameter Inventory for Station: BADL0007

Parameter		Period of Record	Obs	Median	Mean	Maximum	Minimum	Variance	Std Dev	10th	25th	75th	90th
00010	TEMPERATURE, WATER (DEGREES CENTIGRADE)	07/12/89-08/16/89	2	23 7	23 7	24	23 4	0 18	0 424	**	**	**	**
00021	TEMPERATURE, AIR (DEGREES FAHRENHEIT)	07/12/89-08/16/89	2	84 5	84 5	89	80	40 5	6 364	**	**	**	**
00095	SPECIFIC CONDUCTANCE (UMHOS/CM @ 25C)	08/16/89-08/16/89	1	520	520	520	520	0	0	**	**	**	**
00300	OXYGEN, DISSOLVED MG/L	07/12/89-08/16/89	2	6 1	6 1	7 1	5 1	2	1 414	**	**	**	**
00400	PH (STANDARD UNITS)	07/12/89-08/16/89	2	8 6	8 6	8 8	8 4	0 08	0 283	**	**	**	**
00400	CONVERTED PH (STANDARD UNITS)	07/12/89-08/16/89	2	8 555	8 555	8 8	8 4	0 084	0 29	**	**	**	**
00400	MICRO EQUIVALENTS/LITER OF H+ COMPUTED FROM PH	07/12/89-08/16/89	2	0 003	0 003	0 004	0 002	0	0 002	**	**	**	**
00403	PH, LAB, STANDARD UNITS SU	07/12/89-08/16/89	2	8 7	8 7	8 8	8 6	0 02	0 141	**	**	**	**
00403	CONVERTED PH, LAB, STANDARD UNITS	07/12/89-08/16/89	2	8 689	8 689	8 8	8 6	0 02	0 142	**	**	**	**
00403	MICRO EQUIVALENTS/LITER OF H+ COMPUTED FROM PH	07/12/89-08/16/89	2	0 002	0 002	0 003	0 002	0	0 001	**	**	**	**
00410	ALKALINITY, TOTAL (MG/L AS CACO3)	07/12/89-08/16/89	2	218 5	218 5	224	213	60 5	7 778	**	**	**	**
00415	ALKALINITY, PHENOLPHTHALEIN (MG/L)	07/12/89-08/16/89	2	7	7	8	6	2	1 414	**	**	**	**
00500	RESIDUE, TOTAL (MG/L)	07/12/89-08/16/89	2	381	381	396	366	450	21 213	**	**	**	**
00530	RESIDUE, TOTAL NONFILTRABLE (MG/L)	07/12/89-08/16/89	2	14	14	18	10	32	5 657	**	**	**	**
00610	NITROGEN, AMMONIA, TOTAL (MG/L AS N)	07/12/89-08/16/89	2	0 065	0 065	0 09	0 04	0 001	0 035	**	**	**	**
00625	NITROGEN, KJELDAHL, TOTAL, (MG/L AS N)	07/12/89-08/16/89	2	1 12	1 12	1 15	1 09	0 002	0 042	**	**	**	**
00630	NITRITE PLUS NITRATE, TOTAL 1 DET (MG/L AS N)	07/12/89-08/16/89	2 ##	0 05	0 05	0 05	0 05	0	0	**	**	**	**
00665	PHOSPHORUS, TOTAL (MG/L AS P)	07/12/89-08/16/89	2 ##	0 095	0 095	0 095	0 095	0	0	**	**	**	**
00671	PHOSPHORUS, DISSOLVED ORTHOPHOSPHATE (MG/L AS P)	07/12/89-08/16/89	2	0 017	0 017	0 032	0 003	0	0 021	**	**	**	**
70300	RESIDUE, TOTAL FILTRABLE (DRIED AT 180C), MG/L	07/12/89-08/16/89	2	372	372	388	356	512	22 627	**	**	**	**

** - Less than 9 observations   ## - Computed with 50% or more of the total observations as values that were half the detection limit   p - Has a corresponding time series plot

EPA Water Quality Criteria Analysis for Station: BADL0007

Parameter		Std Type	Std Value	Total Obs	Exceed Standard	Prop Exceeding	------10/01-1/31------			------2/01-4/14------			------4/15-6/30------			------7/01-9/30------		
							Obs	Exceed	Prop	Obs	Exceed	Prop	Obs	Exceed	Prop	Obs	Exceed	Prop
00300	OXYGEN, DISSOLVED	Other-Lo Lim	4	2	0	0 00										2	0	0 00
00400	PH	Other-Hi Lim	9	2	0	0 00										2	0	0 00
		Other-Lo Lim	6 5	2	0	0 00										2	0	0 00
00403	PH, LAB	Other-Hi Lim	9	2	0	0 00										2	0	0 00
		Other-Lo Lim	6 5	2	0	0 00										2	0	0 00
00630	NITRITE PLUS NITRATE, TOTAL 1 DET	Drinking Water	10	2	0	0 00										2	0	0 00

& - Below detection limit observations, for which half the detection limit exceeded the criterion, were excluded from the criterion comparison for this parameter

Station Inventory for Station: BADL0008

NPS Station ID: BADL0008
Location: 42N42W 2DACA
Station Type: /TYPA/AMBNT/SPRING
RMI-Indexes:
HUC: 10140201
Major Basin:
Minor Basin:
RF1 Index: 10140201
RF3 Index: 10140201028200 00
Description:

LAT/LON: 43 639448/-102 279448

Depth of Water: 0
Elevation: 0

RF1 Mile Point: 0 000
RF3 Mile Point: 0 00

Agency: 112WRD
FIPS State/County: 46113 SOUTH DAKOTA/SHANNON
STORET Station ID(s): 4338221 02164601
Within Park Boundary: No

Aquifer:
Water Body Id:
ECO Region:
Distance from RF1: 2 90
Distance from RF3: 0 06

Date Created: 05/01/93

On/Off RF1:
On/Off RF3:

Parameter Inventory for Station: BADL0008

Parameter	(name)	Period of Record	Obs	Median	Mean	Maximum	Minimum	Variance	Std Dev	10th	25th	75th	90th
00010	TEMPERATURE, WATER (DEGREES CENTIGRADE)	11/12/92-07/11/95	4	15 4	14 55	17 4	10	11 797	3 435	**	**	**	**
00020	TEMPERATURE, AIR (DEGREES CENTIGRADE)	11/12/92-07/11/95	4	21	20 5	28	12	67	8 185	**	**	**	**
00025	BAROMETRIC PRESSURE (MM OF HG)	11/12/92-08/02/94	2	673 5	673 5	674	673	0 5	0 707	**	**	**	**
00061	FLOW, STREAM, INSTANTANEOUS CFS	11/12/92-07/11/95	4	0 065	0 065	0 08	0 05	0	0 013	**	**	**	**
00065	STAGE, STREAM (FEET)	08/02/94-07/11/95	2	0 27	0 27	0 28	0 26	0	0 014	**	**	**	**
00095	SPECIFIC CONDUCTANCE (UMHOS/CM @ 25C)	11/12/92-07/11/95	4	427	430 25	445	422	112 917	10 626	**	**	**	**
00300	OXYGEN, DISSOLVED MG/L	11/12/92-07/11/95	3	8 2	7 867	8 6	6 8	0 893	0 945	**	**	**	**
00400	PH (STANDARD UNITS)	11/12/92-07/11/95	4	8 385	8 41	8 6	8 26	0 022	0 147	**	**	**	**
00400	CONVERTED PH (STANDARD UNITS)	11/12/95-07/11/95	4	8 384	8 392	8 61	8 26	0 022	0 149	**	**	**	**
00400	MICRO EQUIVALENTS/LITER OF H+ COMPUTED FROM PH	11/12/92-07/11/95	4	0 004	0 004	0 005	0 002	0	0 001	**	**	**	**
00403	PH, LAB, STANDARD UNITS SU	11/12/92-07/11/95	4	7 95	7 85	8	7 5	0 057	0 238	**	**	**	**
00403	CONVERTED PH, LAB, STANDARD UNITS	11/12/92-07/11/95	4	7 947	7 794	8	7 5	0 061	0 247	**	**	**	**
00403	MICRO EQUIVALENTS/LITER OF H+ COMPUTED FROM PH	11/12/92-07/11/95	4	0 011	0 016	0 032	0 01	0	0 01	**	**	**	**
00410	ALKALINITY, TOTAL (MG/L AS CACO3)	11/12/92-08/02/94	2	226	226	247	205	882	29 698	**	**	**	**
00440	BICARBONATE ION (MG/L AS HCO3)	08/02/94-08/02/94	1	250	250	250	250			**	**	**	**
00452	CARBONATE WATER DISS.INCR TIT, FIELD, AS CO3, MG/L	07/11/95-07/11/95	1	2	2	2		0	0	**	**	**	**
00453	BICARBONATE WATER DISS,INCR TIT,FIELD,AS HCO3,MG/L	07/11/95-07/11/95	1	195	195	195	195	0	0	**	**	**	**
00608	NITROGEN, AMMONIA, DISSOLVED (MG/L AS N)	08/02/94-07/11/95	2 ##	0 03	0 033	0 05	0 02	0	0 015	**	**	**	**
00613	NITRITE NITROGEN, DISSOLVED (MG/L AS N)	05/17/95-07/11/95	2 ##	0 005	0 005	0 005	0 005	0	0	**	**	**	**
00623	NITROGEN, KJELDAHL, DISSOLVED (MG/L AS N)	08/02/94-07/11/95	3 ##	0 1	0 167	0 3	0 1	0 013	0 115	**	**	**	**
00631	NITRITE PLUS NITRATE, DISS 1 DET (MG/L AS N)	08/02/94-07/11/95	3	1 2	1 067	1 3	0 7	0 103	0 321	**	**	**	**
00666	PHOSPHORUS, DISSOLVED (MG/L AS P)	08/02/94-07/11/95	3	0 03	0 033	0 039	0 03	0	0 005	**	**	**	**
00671	PHOSPHORUS, DISSOLVED ORTHOPHOSPHATE (MG/L AS P)	05/17/95-07/11/95	2	0 035	0 035	0 04	0 03	0	0 007	**	**	**	**
00915	CALCIUM, DISSOLVED (MG/L AS CA)	11/12/92-07/11/95	4	46	46 5	51	43	11 667	3 416	**	**	**	**
00925	MAGNESIUM, DISSOLVED (MG/L AS MG)	11/12/92-07/11/95	4	5 2	5 3	5 8	5	0 127	0 356	**	**	**	**
00930	SODIUM, DISSOLVED (MG/L AS NA)	11/12/92-07/11/95	4	34 5	35	38	33	6	2 449	**	**	**	**
00935	POTASSIUM, DISSOLVED (MG/L AS K)	11/12/92-07/11/95	4	11	11	11	11	0	0	**	**	**	**
00940	CHLORIDE,TOTAL IN WATER MG/L	11/12/92-07/11/95	4	11	11	11	11	0	0	**	**	**	**
00945	SULFATE, TOTAL (MG/L AS SO4)	11/12/92-07/11/95	4	10 5	10 25	11	9	0 917	0 957	**	**	**	**
00950	FLUORIDE, DISSOLVED (MG/L AS F)	11/12/92-07/11/95	4	0 3	0 3	0 4	0 2	0 007	0 082	**	**	**	**
00955	SILICA, DISSOLVED (MG/L AS SI02)	11/12/92-07/11/95	4	46 5	46 5	48	45	1 667	1 291	**	**	**	**
01000	ARSENIC, DISSOLVED (UG/L AS AS)	11/12/92-07/11/95	4	6	6	7	5	0 667	0 816	**	**	**	**
01005	BARIUM, DISSOLVED (UG/L AS BA)	11/12/92-07/11/95	4	175	180	200	170	200	14 142	**	**	**	**
01010	BERYLLIUM, DISSOLVED (UG/L AS BE)	11/12/92-07/11/95	4 ##	0 25	0 25	0 25	0 25		0	**	**	**	**
01020	BORON, DISSOLVED (UG/L AS B)	11/12/92-07/11/95	4	55	55	60	50	33 333	5 774	**	**	**	**
01025	CADMIUM, DISSOLVED (UG/L AS CD)	11/12/92-07/11/95	4 ##	0 5	0 875	2	0 5	0 563	0 75	**	**	**	**
01027	CADMIUM, TOTAL (UG/L AS CD)	11/12/92-07/11/95	4 ##	0 5	0 5	0 5	0 5	0	0	**	**	**	**
01030	CHROMIUM, DISSOLVED (UG/L AS CR)	11/12/92-07/11/95	4 ##	2 5	2 5	2 5	2 5	0	0	**	**	**	**

** - Less than 9 observations   ## - Computed with 50% or more of the total observations as values that were half the detection limit   p - Has a corresponding time series plot

Parameter Inventory for Station: BADL0008

Parameter		Period of Record	Obs	Median	Mean	Maximum	Minimum	Variance	Std Dev	10th	25th	75th	90th
01035	COBALT, DISSOLVED (UG/L AS CO)	11/12/92-07/11/95	4 ##	15	15	15	15	0	0	**	**	**	**
01040	COPPER, DISSOLVED (UG/L AS CU)	11/12/92-07/11/95	4 ##	5	5	5	5	0	0	**	**	**	**
01046	IRON, DISSOLVED (UG/L AS FE)	11/12/92-07/11/95	4	15.5	14.25	21	5	58.25	7.632	**	**	**	**
01049	LEAD, DISSOLVED (UG/L AS PB)	11/12/92-07/11/95	4	5	6.25	10	5	6.25	2.5	**	**	**	**
01056	MANGANESE, DISSOLVED (UG/L AS MN)	11/12/92-07/11/95	4	5	6	11	3	14.667	3.83	**	**	**	**
01057	THALLIUM, DISSOLVED (UG/L AS TL)	05/17/95-07/11/95	2 ##	0.25	0.25	0.25	0.25	0	0		**	**	**
01060	MOLYBDENUM, DISSOLVED (UG/L AS MO)	11/12/92-07/11/95	4 ##	5	5	5	5	0	0	**	**	**	**
01065	NICKEL, DISSOLVED (UG/L AS NI)	11/12/92-07/11/95	4 ##	5	5	5	5	0	0	**	**	**	**
01075	SILVER, DISSOLVED (UG/L AS AG)	11/12/92-07/11/95	4	0.5	0.875	2	0.5	0.563	0.75	**	**	**	**
01080	STRONTIUM, DISSOLVED (UG/L AS SR)	11/12/92-07/11/95	4	500	500	520	480	266.667	16.33	**	**	**	**
01085	VANADIUM, DISSOLVED (UG/L AS V)	11/12/92-07/11/95	4	6	5.5	7	3	3	1.732	**	**	**	**
01090	ZINC, DISSOLVED (UG/L AS ZN)	11/12/92-07/11/95	4 ##	1.5	1.5	1.5	1.5	0	0	**	**	**	**
01130	LITHIUM, DISSOLVED (UG/L AS LI)	11/12/92-07/11/95	4	26.5	26.75	30	24	6.25	2.5	**	**	**	**
01145	SELENIUM, DISSOLVED (UG/L AS SE)	11/12/92-07/11/95	4	15	15	15	15	0	0	**	**	**	**
03515	BETA, DISSOLVED GROSS, AS CS-137, PC/L	08/02/94-08/02/94	1	6.8	6.8	6.8	6.8	0	0		**	**	**
04126	ALPHA, DISSOLVED, WATER (AS TH-230) PC/L	08/02/94-08/02/94	1	0.2	0.2	0.2	0.2	0	0		**	**	**
09510	RADIUM 226, DISSOLVED, PLANCHET COUNT	08/02/94-08/02/94	1	0.2	0.2	0.2	0.2	0	0		**	**	**
22703	URANIUM, NATURAL, DISSOLVED	08/02/94-08/02/94	1	72	72	72	72	0	0		**	**	**
31501	COLIFORM,TOT.MEMBRANE FILTER,IMMED M-ENDO MED,35C	08/02/94-05/17/95	1	407	407	407	407	0	0		**	**	**
31501	LOG COLIFORM,TOT.MEMBRANE FILTER,IMMED M-ENDO MED,	08/02/94-05/17/95	1	2.61	2.61	2.61	2.61	0	0			**	**
31501	GM COLIFORM,TOT.MEMBRANE FILTER,IMMED M-ENDO MED,3	GEOMETRIC MEAN =		407									
31503	COLIFORM,TOT,MEMBR FILTER,DELAYED,M-ENDO MED,35 C	07/11/95-07/11/95	1	10800	10800	10800	10800	0	0		**	**	**
31503	LOG COLIFORM,TOT,MEMBR FILTER,DELAYED,M-ENDO MED,3	07/11/95-07/11/95	1	4.033	4.033	4.033	4.033	0	0		**	**	**
31503	GM COLIFORM,TOT,MEMBR FILTER,DELAYED,M-ENDO MED,35	GEOMETRIC MEAN =		10800									
31625	FECAL COLIFORM, MF,M-FC, 0.7 UM	08/02/94-07/11/95	2	453.5	453.5	700	207	121524.5	348.604	**	**	**	**
31625	LOG FECAL COLIFORM, MF,M-FC, 0.7 UM	08/02/94-07/11/95	2	2.581	2.581	2.845	2.316	0.14	0.374	**	**	**	**
31625	GM FECAL COLIFORM, MF,M-FC, 0.7 UM	GEOMETRIC MEAN =		380.657									
31633	E COLI,THERMOTOL,MF,M-TEC,IN SITU UREASE #/100ML	08/02/94-07/11/95	2	485	485	920	50	378450	615.183	**	**	**	**
31633	LOG E COLI,THERMOTOL,MF,M-TEC,IN SITU UREASE #/100	08/02/94-07/11/95	2	2.331	2.331	2.964	1.699	0.8	0.894	**	**	**	**
31633	GM E COLI,THERMOTOL,MF,M-TEC,IN SITU UREASE #/100M	GEOMETRIC MEAN =		214.476									
31673	FECAL STREPTOCOCCI, MBR FILT,KF AGAR,35C,48HR	08/02/94-07/11/95	3	2240	2611	4900	693	4527943	2127.896	**	**	**	**
31673	LOG FECAL STREPTOCOCCI, MBR FILT,KF AGAR,35C,48HR	08/02/94-07/11/95	3	3.35	3.294	3.69	2.841	0.183	0.428	**	**	**	**
31673	GM FECAL STREPTOCOCCI, MBR FILT,KF AGAR,35C,48HR	GEOMETRIC MEAN =		1966.644									
39086	ALKALINITY,WATER,DISS,INCR TIT,FIELD,AS CACO3,MG/L	07/11/95-07/11/95	1	200	200	200	200	0	0	**	**	**	**
70300	RESIDUE,TOTAL FILTRABLE (DRIED AT 180C),MG/L	11/12/92-07/11/95	4 ##	282	289	314	278	284	16.852	**	**	**	**
71890	MERCURY, DISSOLVED (UG/L AS HG)	08/02/94-08/02/94	1	0.05	0.05	0.05	0.05	0	0		**	**	**
75986	ALPHA GROSS,1 SIGMA PRC EST AS NAT U,DISS,WTR UG/L	08/02/94-08/02/94	1	3.95	3.95	3.95	3.95	0	0		**	**	**
75987	ALPHA GROSS,DISS,1 SIGMA PRC EST AS TH230,WTR PC/L	08/02/94-08/02/94	1	2.92	2.92	2.92	2.92	0	0		**	**	**
75988	BETA GROSS,DISS,1 SIGMA PRC EST AS SR90/Y90 PC/L	08/02/94-08/02/94	1	1.73	1.73	1.73	1.73	0	0		**	**	**
75989	BETA GROSS,1 SIGMA PRC EST AS CS-137,DISS,WTR PC/L	08/02/94-08/02/94	1	2.3	2.3	2.3	2.3	0	0		**	**	**
75990	URANIUM,NATURAL,1 SIGMA PRC EST,DISS,WATER UG/L	08/02/94-08/02/94	1	2.2	2.2	2.2	2.2	0	0		**	**	**
76001	RADIUM 226,1 SIGMA PRC EST,DISSOLVED,WATER PC/L	08/02/94-08/02/94	1	0.146	0.146	0.146	0.146	0	0		**	**	**
80030	ALPHA,DISSOLVED GROSS,AS URANIUM-NATURAL,UG/L	08/02/94-08/02/94	1	9.1	9.1	9.1	9.1	0	0		**	**	**
80050	BETA,DISSOLVED GROSS,AS SR-Y-90, PC/L	08/02/94-08/02/94	1	11	11	11	11	0	0		**	**	**

** - Less than 9 observations   ## - Computed with 50% or more of the total observations as values that were half the detection limit   p - Has a corresponding time series plot

EPA Water Quality Criteria Analysis for Station: BADL0008

Parameter		Std Type	Std Value	Total Obs	Exceed Standard	Prop Exceeding	10/01-1/31 Obs	Exceed	Prop	2/01-4/14 Obs	Exceed	Prop	4/15-6/30 Obs	Exceed	Prop	7/01-9/30 Obs	Exceed	Prop
00300	OXYGEN, DISSOLVED	Other-Lo Lim	4	3	0	0.00	-			-			-			2	0	0.00
00400	PH	Other-Hi Lim	9	4	0	0.00	-			0	0.00		0	0.00		2	0	0.00
		Other-Lo Lim	6.5	4	0	0.00	-			0	0.00		0	0.00		2	0	0.00
00403	PH, LAB	Other-Hi Lim	9	4	0	0.00	-			0	0.00		0	0.00		2	0	0.00
		Other-Lo Lim	6.5	4	0	0.00	-			0	0.00		0	0.00		2	0	0.00
00613	NITRITE NITROGEN, DISSOLVED AS N	Drinking Water	1	2	0	0.00	-									1	0	0.00
00631	NITRITE PLUS NITRATE, DISS 1 DET	Drinking Water	10	3	0	0.00	-									2	0	0.00
00940	CHLORIDE,TOTAL IN WATER	Drinking Water	250	4	0	0.00	0	0.00								2	0	0.00

| Fresh Acute | 860 |

& - Below detection limit observations, for which half the detection limit exceeded the criterion, were excluded from the criterion comparison for this parameter

EPA Water Quality Criteria Analysis for Station: BADL0008

Parameter	Std Type	Std Value	Total Obs	Exceed Standard	Prop Exceeding	10/01-1/31			2/01-4/14			4/15-6/30			7/01-9/30		
						Obs	Exceed	Prop	Obs	Exceed	Prop	Obs	Exceed	Prop	Obs	Exceed	Prop
00945 SULFATE, TOTAL (AS SO4)	Drinking Water	250	4	0	0.00	1	0	0.00				1	0	0.00	2	0	0.00
00950 FLUORIDE, DISSOLVED AS F	Drinking Water	4	4	0	0.00	1	0	0.00				1	0	0.00	2	0	0.00
01000 ARSENIC, DISSOLVED	Fresh Acute	360	4	0	0.00	1	0	0.00				1	0	0.00	2	0	0.00
	Drinking Water	50	4	0	0.00	1	0	0.00				1	0	0.00	2	0	0.00
01005 BARIUM, DISSOLVED	Drinking Water	2000	4	0	0.00	1	0	0.00				1	0	0.00	2	0	0.00
01010 BERYLLIUM, DISSOLVED	Fresh Acute	130	4	0	0.00	1	0	0.00				1	0	0.00	2	0	0.00
	Drinking Water	4	4	0	0.00	1	0	0.00				1	0	0.00	2	0	0.00
01025 CADMIUM, DISSOLVED	Fresh Acute	3.9	4	0	0.00	1	0	0.00				1	0	0.00	2	0	0.00
	Drinking Water	5	4	0	0.00	1	0	0.00				1	0	0.00	2	0	0.00
01027 CADMIUM, TOTAL	Fresh Acute	3.9	4	0	0.00	1	0	0.00				1	0	0.00	2	0	0.00
	Drinking Water	5	4	0	0.00	1	0	0.00				1	0	0.00	2	0	0.00
01030 CHROMIUM, DISSOLVED	Drinking Water	100	4	0	0.00	1	0	0.00				1	0	0.00	2	0	0.00
01040 COPPER, DISSOLVED	Fresh Acute	18	4	0	0.00	1	0	0.00				1	0	0.00	2	0	0.00
	Drinking Water	1300	4	0	0.00	1	0	0.00				1	0	0.00	2	0	0.00
01049 LEAD, DISSOLVED	Fresh Acute	82	4	0	0.00	1	0	0.00				1	0	0.00	2	0	0.00
	Drinking Water	15	4	0	0.00	1	0	0.00				1	0	0.00	2	0	0.00
01057 THALLIUM, DISSOLVED	Fresh Acute	1400	2	0	0.00	1	0	0.00							1	0	0.00
	Drinking Water	2	2	0	0.00	1	0	0.00							1	0	0.00
01065 NICKEL, DISSOLVED	Fresh Acute	1400	4	0	0.00	1	0	0.00				1	0	0.00	2	0	0.00
	Drinking Water	100	4	0	0.00	1	0	0.00				1	0	0.00	2	0	0.00
01075 SILVER, DISSOLVED	Fresh Acute	4.1	4	0	0.00	1	0	0.00				1	0	0.00	2	0	0.00
	Drinking Water	100	4	0	0.00	1	0	0.00				1	0	0.00	2	0	0.00
01090 ZINC, DISSOLVED	Fresh Acute	120	4	0	0.00	1	0	0.00				1	0	0.00	2	0	0.00
	Drinking Water	5000	4	0	0.00	1	0	0.00				1	0	0.00	2	0	0.00
01145 SELENIUM, DISSOLVED	Fresh Acute	20	4	0	0.00	1	0	0.00				1	0	0.00	2	0	0.00
	Drinking Water	50	4	0	0.00	1	0	0.00				1	0	0.00	2	0	0.00
22703 URANIUM, NATURAL DISSOLVED	Drinking Water	20	1	0	0.00										1	0	0.00
31501 COLIFORM, TOTAL, MEMBRANE FILTER, IMMED	Other-Hi Lim	1000	1	0	0.00										1	0	0.00
31503 COLIFORM,TOT MEMBRANE FILTR,DELAY M-END	Other-Hi Lim	1000	1	1	1.00										1	1	1.00
31625 FECAL COLIFORM, MF	Other-Hi Lim	200	2	2	1.00										2	2	1.00
71890 MERCURY, DISSOLVED	Fresh Acute	2.4	4	0	0.00	1	0	0.00				1	0	0.00	2	0	0.00
	Drinking Water	2	4	0	0.00	1	0	0.00				1	0	0.00	2	0	0.00

& - Below detection limit observations, for which half the detection limit exceeded the criterion, were excluded from the criterion comparison for this parameter

Station Inventory for Station: BADL0009

NPS Station ID: BADL0009
Location: 402462
Station Type: /TYPA/AMBNT/SPRING
RMI-Indexes:
RMI-Miles:
HUC: 10140201
Major Basin: MISSOURI RIVER
Minor Basin: SOUTH CENTRAL MISSOURI
RF1 Index: 10140201
RF3 Index: 10140201028200 00

LAT/LON: 43.730004/-102.379003

Agency: 11NPSWRD
FIPS State/County: 46103 SOUTH DAKOTA/PENNINGTON
STORET Station ID(s): BADL_NURE_02 /6038749
Within Park Boundary: No

Date Created: 12/20/97

Depth of Water: 0
Elevation: 0

RF1 Mile Point: 0.000
RF3 Mile Point: 0.00

Aquifer:
Water Body Id:
ECO Region:
Distance from RF1: 2.90
Distance from RF3: 0.06

On/Off RF1:
On/Off RF3:

Description:
THE SITE IS LOCATED ON THE IMLAY SOUTH DAKOTA PENNINGTON COUNTY 7.5 MINUTE SERIES (TOPOGRAPHIC) QUADRANGLE. THE SITE IS AT A SPRING OUTSIDE OF BADLANDS NATIONAL PARK. DATA ARE FROM THE "U.S. GEOLOGICAL SURVEY NATIONAL GEOCHEMICAL DATABASE: NATIONAL URANIUM RESOURCE EVALUATION DATA FOR THE CONTERMINOUS UNITED STATES" 1994 CD-ROM BY J.D. HOFFMAN AND K. BUTTLEMAN (USGS DIGITAL DATA SERIES DDS-18-A). THE DATABASE INCLUDES SEDIMENT; SOIL, SURFACE WATER; AND GROUND WATER DATA. THE "UNIQID" FIELD ENTRY WAS USED TO CREATE THE SECONDARY STATION NAME. THE "ORNLID" FIELD ENTRY (OAK RIDGE NATIONAL LABORATORY SAMPLE NUMBER) WAS USED TO CREATE THE STATION LOCATION. THE SAMPLES WERE ANALYZED BY OAK RIDGE NATIONAL LABORATORY. DATA PROCESSED AND UPLOADED TO STORET BY J. CHRIS ECHOHAWK, NATIONAL PARK SERVICE WATER RESOURCES DIVISION;
1201 OAK RIDGE DRIVE SUITE 250; FORT COLLINS CO 80525   TEL 970-225-3516; FAX 970-225-9965

Parameter Inventory for Station: BADL0009

Parameter		Period of Record	Obs	Median	Mean	Maximum	Minimum	Variance	Std Dev	10th	25th	75th	90th
00010	TEMPERATURE, WATER (DEGREES CENTIGRADE)	04/23/79-04/23/79	1	13	13	13	13	0	0	**	**	**	**
00400	PH (STANDARD UNITS)	04/23/79-04/23/79	1	7.4	7.4	7.4	7.4	0	0	**	**	**	**
00400	CONVERTED PH (STANDARD UNITS)	04/23/79-04/23/79	1	7.4	7.4	7.4	7.4	0	0	**	**	**	**
00400	MICRO EQUIVALENTS/LITER OF H+ COMPUTED FROM PH	04/23/79-04/23/79	1	0.04	0.04	0.04	0.04	0	0	**	**	**	**
00410	ALKALINITY, TOTAL (MG/L AS CACO3)	04/23/79-04/23/79	2	400	400	400	400	0	0	**	**	**	**
00665	PHOSPHORUS, TOTAL (MG/L AS P)	04/23/79-04/23/79	1##	0.02	0.02	0.02	0.02	0	0	**	**	**	**
00916	CALCIUM, TOTAL (MG/L AS CA)	04/23/79-04/23/79	1	52.7	52.7	52.7	52.7	0	0	**	**	**	**
00929	SODIUM, TOTAL (MG/L AS NA)	04/23/79-04/23/79	1	123.3	123.3	123.3	123.3	0	0	**	**	**	**
00937	POTASSIUM, TOTAL (MG/L AS K)	04/23/79-04/23/79	1	7	7	7	7	0	0	**	**	**	**
00945	SULFATE, TOTAL (MG/L AS SO4)	04/23/79-04/23/79	1	30	30	30	30	0	0	**	**	**	**
01002	ARSENIC, TOTAL (UG/L AS AS)	04/23/79-04/23/79	1	2	2	2	2	0	0	**	**	**	**
01007	BARIUM, TOTAL (UG/L AS BA)	04/23/79-04/23/79	1	72	72	72	72	0	0	**	**	**	**
01012	BERYLLIUM, TOTAL (UG/L AS BE)	04/23/79-04/23/79	1##	0.5	0.5	0.5	0.5	0	0	**	**	**	**
01022	BORON, TOTAL (UG/L AS B)	04/23/79-04/23/79	1##	760	760	760	760	0	0	**	**	**	**
01034	CHROMIUM, TOTAL (UG/L AS CR)	04/23/79-04/23/79	1##	2	2	2	2	0	0	**	**	**	**
01037	COBALT, TOTAL (UG/L AS CO)	04/23/79-04/23/79	1##	1	1	1	1	0	0	**	**	**	**
01042	COPPER, TOTAL (UG/L AS CU)	04/23/79-04/23/79	1##	1	1	1	1	0	0	**	**	**	**
01045	IRON, TOTAL (UG/L AS FE)	04/23/79-04/23/79	1##	5	5	5	5	0	0	**	**	**	**
01055	MANGANESE, TOTAL (UG/L AS MN)	04/23/79-04/23/79	1##	9	9	9	9	0	0	**	**	**	**
01062	MOLYBDENUM, TOTAL (UG/L AS MO)	04/23/79-04/23/79	1	2	2	2	2	0	0	**	**	**	**
01067	NICKEL, TOTAL (UG/L AS NI)	04/23/79-04/23/79	1##	2	2	2	2	0	0	**	**	**	**
01077	SILVER, TOTAL (UG/L AS AG)	04/23/79-04/23/79	1							**	**	**	**

** - Less than 9 observations   ## - Computed with 50% or more of the total observations as values that were half the detection limit   p - Has a corresponding time series plot

Parameter Inventory for Station: BADL0009

Parameter		Period of Record	Obs	Median	Mean	Maximum	Minimum	Variance	Std Dev	10th	25th	75th	90th
01082	STRONTIUM, TOTAL (UG/L AS SR)	04/23/79-04/23/79	1	320	320	320	320	0	0	**	**	**	**
01087	VANADIUM, TOTAL (UG/L AS V)	04/23/79-04/23/79	1	6	6	6	6	0	0	**	**	**	**
01092	ZINC, TOTAL (UG/L AS ZN)	04/23/79-04/23/79	1	18	18	18	18	0	0	**	**	**	**
01105	ALUMINUM, TOTAL (UG/L AS AL)	04/23/79-04/23/79	1	37	37	37	37	0	0	**	**	**	**
01112	CERIUM, TOTAL (UG/L AS CE)	04/23/79-04/23/79	##	15	15	15	15	0	0	**	**	**	**
01132	LITHIUM, TOTAL (UG/L AS LI)	04/23/79-04/23/79	1	54	54	54	54	0	0	**	**	**	**
01142	SILICON, TOTAL (UG/L AS SI)	04/23/79-04/23/79	1	9699	9699	9699	9699	0	0	**	**	**	**
01147	SELENIUM, TOTAL (UG/L AS SE)	04/23/79-04/23/79	1	0.6	0.6	0.6	0.6	0	0	**	**	**	**
01152	TITANIUM, TOTAL (UG/L AS TI)	04/23/79-04/23/79	1	2	2	2	2	0	0	**	**	**	**
01189	SCANDIUM, TOTAL (UG/L AS SC)	04/23/79-04/23/79	1	0.5	0.5	0.5	0.5	0	0	**	**	**	**
01203	YTTRIUM, TOTAL (UG/L AS Y)	04/23/79-04/23/79	##	0.5	0.5	0.5	0.5	0	0	**	**	**	**
01239	NIOBIUM, TOTAL UG/L	04/23/79-04/23/79	##	4	4	4	4	0	0	**	**	**	**
82033	MAGNESIUM - TOTAL UG/L(AS MG)	04/23/79-04/23/79	1	3399	3399	3399	3399	0	0	**	**	**	**
82364	THORIUM, TOTAL IN WATER UG/L	04/23/79-04/23/79	1	7	7	7	7	0	0	**	**	**	**

** - Less than 9 observations   ## - Computed with 50% or more of the total observations as values that were half the detection limit   p - Has a corresponding time series plot

EPA Water Quality Criteria Analysis for Station: BADL0009

Parameter		Std Type	Std Value	Total Obs	Exceed Standard	Prop Exceeding	10/01-1/31 Obs	Exceed	Prop	2/01-4/14 Obs	Exceed	Prop	4/15-6/30 Obs	Exceed	Prop	7/01-9/30 Obs	Exceed	Prop
00400	PH	Other-Hi Lim	9	1	0	0.00							1	0	0.00			
		Other-Lo Lim	6.5	1	0	0.00							1	0	0.00			
00945	SULFATE, TOTAL (AS SO4)	Drinking Water	250	1	0	0.00							1	0	0.00			
01002	ARSENIC, TOTAL	Fresh Acute	360	1	0	0.00							1	0	0.00			
		Drinking Water	50	1	0	0.00							1	0	0.00			
01007	BARIUM, TOTAL	Drinking Water	2000	1	0	0.00							1	0	0.00			
01012	BERYLLIUM, TOTAL	Fresh Acute	130	1	0	0.00							1	0	0.00			
		Drinking Water	4	1	0	0.00							1	0	0.00			
01034	CHROMIUM, TOTAL	Drinking Water	100	1	0	0.00							1	0	0.00			
01042	COPPER, TOTAL	Fresh Acute	18	1	0	0.00							1	0	0.00			
		Drinking Water	1300	1	0	0.00							1	0	0.00			
01067	NICKEL, TOTAL	Fresh Acute	1400	1	0	0.00							1	0	0.00			
		Drinking Water	100	1	0	0.00							1	0	0.00			
01077	SILVER, TOTAL	Fresh Acute	4.1	1	0	0.00							1	0	0.00			
		Drinking Water	100	1	0	0.00							1	0	0.00			
01092	ZINC, TOTAL	Fresh Acute	120	1	0	0.00							1	0	0.00			
		Drinking Water	5000	1	0	0.00							1	0	0.00			
01147	SELENIUM, TOTAL	Fresh Acute	20	1	0	0.00							1	0	0.00			
		Drinking Water	50	1	0	0.00							1	0	0.00			

& - Below detection limit observations, for which half the detection limit exceeded the criterion, were excluded from the criterion comparison for this parameter

Station Inventory for Station: BADL0010

NPS Station ID: BADL0010
Location: 401905
Station Type: /TYPA/AMBNT/SPRING
RMI-Indexes:
RMI-Miles:
HUC: 10140201
Major Basin: MISSOURI RIVER
Minor Basin: SOUTH CENTRAL MISSOURI
RF1 Index: 10140201
RF3 Index: 1014020102820000

LAT/LON: 43.738003/-102.403003

Depth of Water: 0
Elevation: 0

RF1 Mile Point: 0.000
RF3 Mile Point: 0.00

Agency: 11NPSWRD
FIPS State/County: 46103 SOUTH DAKOTA/PENNINGTON
STORET Station ID(s): BADL_NURE_01 /6038355
Within Park Boundary: No

Aquifer:
Water Body Id:
ECO Region:
Distance from RF1: 2.90
Distance from RF3: 0.06

Date Created: 12/20/97

On/Off RF1:
On/Off RF3:

Description:
THE SITE IS LOCATED ON THE IMLAY SOUTH DAKOTA PENNINGTON COUNTY 7.5 MINUTE SERIES (TOPOGRAPHIC) QUADRANGLE. THE SITE IS AT A SPRING OUTSIDE OF BADLANDS NATIONAL PARK. DATA ARE FROM THE "U.S. GEOLOGICAL SURVEY NATIONAL GEOCHEMICAL DATABASE: NATIONAL URANIUM RESOURCE EVALUATION DATA FOR THE CONTERMINOUS UNITED STATES" 1994 CD-ROM BY J.D. HOFFMAN AND K. BUTTLEMAN (USGS DIGITAL DATA SERIES DDS-18-A). THE DATABASE INCLUDES SEDIMENT, SOIL, SURFACE WATER, AND GROUND WATER DATA. THE "UNIQID" FIELD ENTRY WAS USED TO CREATE THE SECONDARY STATION NAME. THE "ORNLID" FIELD ENTRY (OAK RIDGE NATIONAL LABORATORY SAMPLE NUMBER) WAS USED TO CREATE THE STATION LOCATION. THE SAMPLES WERE ANALYZED BY OAK RIDGE NATIONAL LABORATORY. DATA PROCESSED AND UPLOADED TO STORET BY J. CHRIS ECHOHAWK; NATIONAL PARK SERVICE WATER RESOURCES DIVISION; 1201 OAK RIDGE DRIVE SUITE 250; FORT COLLINS CO 80525   TEL 970-225-3516; FAX 970-225-9965

Parameter Inventory for Station: BADL0010

Parameter		Period of Record	Obs	Median	Mean	Maximum	Minimum	Variance	Std Dev	10th	25th	75th	90th
00010	TEMPERATURE, WATER (DEGREES CENTIGRADE)	04/18/79-04/18/79	1	12.4	12.4	12.4	12.4	0	0	**	**	**	**
00400	PH (STANDARD UNITS)	04/18/79-04/18/79	1	7	7	7	7	0	0	**	**	**	**
00400	CONVERTED PH (STANDARD UNITS)	04/18/79-04/18/79	1	7	7	7	7	0	0	**	**	**	**
00400	MICRO EQUIVALENTS/LITER OF H+ COMPUTED FROM PH	04/18/79-04/18/79	2	0.1	0.1	0.1	0.1	0	0	**	**	**	**
00410	ALKALINITY, TOTAL (MG/L AS CACO3)	04/18/79-04/18/79	2	325	325	334	316	162	12.728	**	**	**	**
00665	PHOSPHORUS, TOTAL (MG/L AS P)	04/18/79-04/18/79	1 ##	0.02	0.02	0.02	0.02	0	0	**	**	**	**
00916	CALCIUM, TOTAL (MG/L AS CA)	04/18/79-04/18/79	1	38.3	38.3	38.3	38.3	0	0	**	**	**	**
00929	SODIUM, TOTAL (MG/L AS NA)	04/18/79-04/18/79	1	113.6	113.6	113.6	113.6	0	0	**	**	**	**
00937	POTASSIUM, TOTAL (MG/L AS K)	04/18/79-04/18/79	1	6.67	6.67	6.67	6.67	0	0	**	**	**	**
00945	SULFATE, TOTAL (MG/L AS SO4)	04/18/79-04/18/79	1	96	96	96	96	0	0	**	**	**	**
01002	ARSENIC, TOTAL (UG/L AS AS)	04/18/79-04/18/79	1	2	2	2	2	0	0	**	**	**	**
01007	BARIUM, TOTAL (UG/L AS BA)	04/18/79-04/18/79	1	36	36	36	36	0	0	**	**	**	**
01012	BERYLLIUM, TOTAL (UG/L AS BE)	04/18/79-04/18/79	1 ##	0.5	0.5	0.5	0.5	0	0	**	**	**	**
01022	BORON, TOTAL (UG/L AS B)	04/18/79-04/18/79	1 ##	760	760	760	760	0	0	**	**	**	**
01034	CHROMIUM, TOTAL (UG/L AS CR)	04/18/79-04/18/79	1 ##	2	2	2	2	0	0	**	**	**	**
01037	COBALT, TOTAL (UG/L AS CO)	04/18/79-04/18/79	1 ##	1	1	1	1	0	0	**	**	**	**
01042	COPPER, TOTAL (UG/L AS CU)	04/18/79-04/18/79	1 ##	1	1	1	1	0	0	**	**	**	**
01045	IRON, TOTAL (UG/L AS FE)	04/18/79-04/18/79	1 ##	5	5	5	5	0	0	**	**	**	**
01055	MANGANESE, TOTAL (UG/L AS MN)	04/18/79-04/18/79	1 ##	1	1	1	1	0	0	**	**	**	**
01062	MOLYBDENUM, TOTAL (UG/L AS MO)	04/18/79-04/18/79	1	6	6	6	6	0	0	**	**	**	**
01067	NICKEL, TOTAL (UG/L AS NI)	04/18/79-04/18/79	1	2	2	2	2	0	0	**	**	**	**
01077	SILVER, TOTAL (UG/L AS AG)	04/18/79-04/18/79	1	2	2	2	2	0	0	**	**	**	**

** - Less than 9 observations   ## - Computed with 50% or more of the total observations as values that were half the detection limit   p - Has a corresponding time series plot

Parameter Inventory for Station: BADL0010

Parameter		Period of Record	Obs	Median	Mean	Maximum	Minimum	Variance	Std Dev	10th	25th	75th	90th
01082	STRONTIUM, TOTAL (UG/L AS SR)	04/18/79-04/18/79	1	258	258	258	258	0	0	**	**	**	**
01087	VANADIUM, TOTAL (UG/L AS V)	04/18/79-04/18/79	1##	2	2	2	2	0	0	**	**	**	**
01092	ZINC, TOTAL (UG/L AS ZN)	04/18/79-04/18/79	1	90	90	90	90	0	0	**	**	**	**
01105	ALUMINUM, TOTAL (UG/L AS AL)	04/18/79-04/18/79	1	13	13	13	13	0	0	**	**	**	**
01112	CERIUM, TOTAL (UG/L AS CE)	04/18/79-04/18/79	1##	15	15	15	15	0	0	**	**	**	**
01132	LITHIUM, TOTAL (UG/L AS LI)	04/18/79-04/18/79	1	56	56	56	56	0	0	**	**	**	**
01142	SILICON, TOTAL (UG/L AS SI)	04/18/79-04/18/79	1	12500	12500	12500	12500	0	0	**	**	**	**
01147	SELENIUM, TOTAL (UG/L AS SE)	04/18/79-04/18/79	1	0.4	0.4	0.4	0.4	0	0	**	**	**	**
01152	TITANIUM, TOTAL (UG/L AS TI)	04/18/79-04/18/79	1	2	2	2	2	0	0	**	**	**	**
01189	SCANDIUM, TOTAL (UG/L AS SC)	04/18/79-04/18/79	1##	0.5	0.5	0.5	0.5	0	0	**	**	**	**
01203	YTTRIUM, TOTAL (UG/L AS Y)	04/18/79-04/18/79	1##	0.5	0.5	0.5	0.5	0	0	**	**	**	**
01239	NIOBIUM, TOTAL UG/L	04/18/79-04/18/79	1	4	4	4	4	0	0	**	**	**	**
82033	MAGNESIUM - TOTAL UG/L(AS MG)	04/18/79-04/18/79	1	1699	1699	1699	1699	0	0	**	**	**	**
82364	THORIUM, TOTAL IN WATER UG/L	04/18/79-04/18/79	1##	2.5	2.5	2.5	2.5	0	0	**	**	**	**

** - Less than 9 observations   ## - Computed with 50% or more of the total observations as values that were half the detection limit   p - Has a corresponding time series plot

EPA Water Quality Criteria Analysis for Station: BADL0010

Parameter		Std Type	Std Value	Total Obs	Exceed Standard	Prop Exceeding	10/01-1/31 Obs	Exceed	Prop	2/01-4/14 Obs	Exceed	Prop	4/15-6/30 Obs	Exceed	Prop	7/01-9/30 Obs	Exceed	Prop
00400	PH	Other-Hi Lim	9	1	0	0.00							1	0	0.00			
		Other-Lo Lim	6.5															
00945	SULFATE, TOTAL (AS SO4)	Drinking Water	250	1	0	0.00							1	0	0.00			
01002	ARSENIC, TOTAL	Fresh Acute	360	1	0	0.00							1	0	0.00			
		Drinking Water	50															
01007	BARIUM, TOTAL	Drinking Water	2000	1	0	0.00							1	0	0.00			
01012	BERYLLIUM, TOTAL	Fresh Acute	130	1	0	0.00							1	0	0.00			
		Drinking Water	4															
01034	CHROMIUM, TOTAL	Fresh Acute	100	1	0	0.00							1	0	0.00			
01042	COPPER, TOTAL	Fresh Acute	18	1	0	0.00							1	0	0.00			
		Drinking Water	1300															
01067	NICKEL, TOTAL	Fresh Acute	1400	1	0	0.00							1	0	0.00			
		Drinking Water	100															
01077	SILVER, TOTAL	Fresh Acute	4.1	1	0	0.00							1	0	0.00			
		Drinking Water	100															
01092	ZINC, TOTAL	Fresh Acute	120	1	0	0.00							1	0	0.00			
		Drinking Water	5000															
01147	SELENIUM, TOTAL	Fresh Acute	20	1	0	0.00							1	0	0.00			
		Drinking Water	50															

& - Below detection limit observations, for which half the detection limit exceeded the criterion, were excluded from the criterion comparison for this parameter

Station Inventory for Station: BADL0011

NPS Station ID: BADL0011
Location: WHITE R NEAR ROCKYFORD SD
Station Type: /TYPA/AMBNT/STREAM
RMI-Miles:
HUC: 10140201
Major Basin:
Minor Basin:
RF1 Index: 10140201
RF3 Index: 10140201005500 00
Description:

LAT/LON: 43.514448,-102.491671

Agency: 112WRD
FIPS State/County: 46113 SOUTH DAKOTA/SHANNON
STORET Station ID(s): 06446200
Within Park Boundary: Yes

Depth of Water: 0
Elevation: 0

Aquifer:
Water Body Id:
ECO Region:
Distance from RF1: 0.00
Distance from RF3: 0.02

RF1 Mile Point: 0.000
RF3 Mile Point: 3.08

Date Created: / /

On/Off RF1:
On/Off RF3:

Parameter Inventory for Station: BADL0011

Parameter	Period of Record	Obs	Median	Mean	Maximum	Minimum	Variance	Std Dev	10th	25th	75th	90th
00010 TEMPERATURE, WATER (DEGREES CENTIGRADE)	08/13/64-09/19/67	105	14.4	13.113	32.2	0	88.882	9.428	0	3.9	20	25.6
00060p FLOW, STREAM, MEAN DAILY CFS	08/13/64-09/19/67	122	38	344.274	7743	0.2	933822.138	966.345	4	12	154	889.9
00080p COLOR (PLATINUM-COBALT UNITS)	08/13/64-09/18/67	17	15	18	60	3	221.875	14.895	3	6	27	42.4
00095p SPECIFIC CONDUCTANCE (UMHOS/CM @ 25C)	08/13/64-09/18/67	36	667	699.167	1120	364	32628.2	180.633	491.9	569	805	994.3
00300 OXYGEN, DISSOLVED MG/L	01/31/67-07/22/67	2	10.45	10.45	11.8	9.1	3.645	1.909	**	**	**	**
00400p PH (STANDARD UNITS)	08/13/64-09/18/67	36	7.9	7.869	8.3	7.3	0.072	0.268	7.5	7.625	8.075	8.2
00400p CONVERTED PH (STANDARD UNITS)	08/13/64-09/18/67	36	7.9	7.785	8.3	7.3	0.079	0.282	7.5	7.625	8.075	8.2
00400p MICRO EQUIVALENTS/LITER OF H+ COMPUTED FROM PH	08/13/64-09/18/67	36	0.013	0.016	0.05	0.005	0	0.011	0.006	0.008	0.024	0.032
00410 ALKALINITY, TOTAL (MG/L AS CACO3)	10/01/65-09/18/67	17	217	201.765	270	132	1747.941	41.808	140.8	169.5	233	265.2
00440p BICARBONATE ION (MG/L AS HCO3)	08/13/64-09/18/67	36	261.5	256.972	354	161	2446.999	49.467	189.9	218	286.5	332.3
00445p CARBONATE ION (MG/L AS CO3)	08/13/64-09/18/67	36	0	0.528	11	0	4.999	2.236	0	0	0	0
00650p PHOSPHATE, TOTAL (MG/L AS PO4)	11/04/64-07/22/67	8	0.275	0.334	0.86	0.016	0.048	0.218	**	**	**	**
00900p HARDNESS, TOTAL (MG/L AS CACO3)	08/13/64-09/18/67	36	124	138.917	309	20	5847.736	76.47	26.8	88.25	210.25	240
00902p HARDNESS, NON-CARBONATE (MG/L AS CACO3)	08/13/64-09/18/67	36	0	10.111	149	0	969.302	31.134	0	0	0	46.4
00915p CALCIUM, DISSOLVED (MG/L AS CA)	08/13/64-09/18/67	36	41	44.847	105	3.2	565.404	23.778	8.92	29.25	64.25	72.2
00925p MAGNESIUM, DISSOLVED (MG/L AS MG)	08/13/64-09/18/67	36	4.8	6.519	20	0.6	24.7	4.97	1.3	2.625	9.95	15
00930p SODIUM, DISSOLVED (MG/L AS NA)	08/13/64-09/18/67	36	101	99.278	194	32	1393.406	37.328	55.2	68	126.75	150.9
00931p SODIUM ADSORPTION RATIO	08/13/64-09/18/67	36	3.1	4.664	15	1.2	12.158	3.487	1.84	2.325	5.875	10.3
00932 SODIUM, PERCENT	10/01/65-09/18/67	17	51	55.412	92	31	280.007	16.733	39	43	67	89.6
00935p POTASSIUM, DISSOLVED (MG/L AS K)	08/13/64-09/18/67	36	10	9.933	16	5.5	6.252	2.5	6.71	7.725	12	13.3
00940p CHLORIDE,TOTAL IN WATER MG/L	08/13/64-09/18/67	36	6	6.889	14	2	8.559	2.926	3.7	5	8	11.3
00945p SULFATE, TOTAL (MG/L AS SO4)	08/13/64-09/18/67	36	106.5	138.389	352	36	7776.93	88.187	54	78.5	167	305.1
00950p FLUORIDE, DISSOLVED (MG/L AS F)	08/13/64-09/18/67	36	0.5	0.508	0.8	0.2	0.019	0.136	0.37	0.4	0.6	0.7
00955p SILICA, DISSOLVED (MG/L AS SI02)	08/13/64-09/18/67	36	32	35.278	58	20	111.806	10.574	23.8	27.25	42	52.6
01020p BORON, DISSOLVED (UG/L AS B)	08/13/64-09/18/67	36	125	161.667	820	60	17260	131.377	70	85	190	289
01034 CHROMIUM, TOTAL (UG/L AS CR)	01/31/67-01/31/67	1	0	0	0	0	0	0	**	**	**	**
01040 COPPER, DISSOLVED (UG/L AS CU)	01/31/67-01/31/67	1	80	80	80	80	0	0	**	**	**	**
01055p MANGANESE, TOTAL (UG/L AS MN)	08/13/64-07/22/67	27	30	34.444	170	0	1541.026	39.256	0	0	40	92
01090 ZINC, DISSOLVED (UG/L AS ZN)	01/31/67-01/31/67											
32730 PHENOLICS, TOTAL, RECOVERABLE (UG/L)	10/01/65-09/29/66	8	7900	7487.5	8300	4000	2021250	1421.707	**	**	**	**
70300p RESIDUE,TOTAL FILTRABLE (DRIED AT 180C),MG/L	08/13/64-09/18/67	36	462	488.861	806	274	16389.894	128.023	341.5	404	557.75	704.8
70302p SOLIDS, DISSOLVED-TONS PER DAY	08/13/64-09/18/67	36	40.2	375.922	7230	0.3	1474486.427	1214.284	3.464	13.85	168.5	752.9
70303p SOLIDS, DISSOLVED-TONS PER ACRE-FT	08/13/64-09/18/67	36	0.63	0.664	1.1	0.37	0.031	0.175	0.464	0.553	0.758	0.959
70331p SUSPENDED SED SIEVE DIAMETER,% FINER THAN .062MM	11/04/64-08/16/67	21	100	95.143	100	17	324.529	18.015	92.4	98.5	100	100
70337p SUS SED FALL DIA(DISTLD WATER)%FINER THAN .002MM	11/04/64-07/24/67	31	58	58.935	93	18	457.529	21.39	26	42	78	86
70338p SUS SED FALL DIA(DISTLD WATER)%FINER THAN .004MM	11/04/64-07/24/67	31	71	68.871	97	22	481.116	21.934	36.8	49	88	91.8
70339 SUS SED FALL DIA(DISTLD WATER)%FINER THAN .008MM	11/04/64-08/17/66	16	94	77.625	98	25	627.317	25.046	29.2	57.75	95.75	96.6
70340p SUS SED FALL DIA(DISTLD WATER)%FINER THAN .016MM	11/04/64-07/24/67	30	86.5	77.833	99	0	634.213	25.184	40.2	65	98	99

** - Less than 9 observations   ## - Computed with 50% or more of the total observations as values that were half the detection limit   p - Has a corresponding time series plot

Parameter Inventory for Station: BADL0011

Parameter	Period of Record	Obs	Median	Maximum	Mean	Variance	Minimum	Std Dev	10th	25th	75th	90th
70341 SUS SED FALL DIA(DISTLD WATER)%FINER THAN 031MM	11/04/64-08/17/66	15	97	100	85.933	365.781	37	19.125	48.4	72	99	100
70342p SUS SED FALL DIA(DISTLD WATER)%FINER THAN 062MM	05/15/65-06/22/67	17	99	100	94.588	117.507	55	10.84	81.4	95	100	100
70343p SUS SED FALL DIA(DISTLD WATER)%FINER THAN 125MM	05/15/65-06/22/67	10	98	100	94.3	164.9	58	12.841	61.7	97.25	100	100
70344 SUS SED FALL DIA(DISTLD WATER)%FINER THAN 250MM	05/15/65-06/22/67	7	99	100	95.143	123.143	70	11.097	**	**	**	**
70345 SUS SED FALL DIA(DISTLD WATER)%FINER THAN 500MM	05/15/65-06/22/67	5	100	100	96.6	57.8	83	7.603	**	**	**	**
70346 SUS SED FALL DIA(DISTLD WATER)%FINER THAN 1 00MM	03/14/66-03/14/66	1	91	91	91	0	91	0	**	**	**	**
70347 SUS SED FALL DIA(DISTLD WATER)%FINER THAN 2 00MM	03/14/66-03/14/66	1	98	98	98	0	98	0	**	**	**	**
71846 NITROGEN, AMMONIA, DISSOLVED (MG/L AS NH4)	11/04/64-09/28/65	18	2.05	2.9	2.167	0.168	1.6	0.41	1.69	1.875	2.525	2.9
71851p NITRATE NITROGEN, DISSOLVED (MG/L AS NO3)	08/13/64-09/18/67	36	0.5	8.4	1.297	3.969	0.1	1.992	0.17	0.2	1.3	4.62
71856 NITRITE NITROGEN, DISSOLVED (MG/L AS NO2)	11/04/64-09/28/65	14	2.5	9	3.357	5.17	1	2.274		2	4.5	7.5
71870 BROMIDE (MG/L AS BR)	08/13/64-09/28/65	19	0.54	0.91	0.616	0.034	0.35	0.184	0.37	0.48	0.8	0.88
71885p IRON (UG/L AS FE)	08/13/64-09/18/67	30	70	500	97.333	10344.368	20	101.707	30	37.5	110	200
80154p SUSP SEDIMENT CONCENTRATION-EVAP AT 110C (MG/L)	08/13/64-09/19/67	86	1735	38600	5047.488	71312180.3	46	8444.654	136.1	236.5	5115	14020
80155p SUSPENDED SEDIMENT DISCHARGE (TONS/DAY)	08/13/64-09/19/67	86	214.5	199000	7977.48	1237807.893	0.0961523	24803.988	1.77	12.5	2737.5	26820
80158 BED MATERIAL FALL DIAMETER, % FINER THAN 062MM	07/27/65-07/27/65	1	2	2	2	0	2		**	**	**	**
80159 BED MATERIAL FALL DIAMETER, % FINER THAN 125MM	07/27/65-07/27/65	1	2	2	2	0	2	0	**	**	**	**
80160 BED MATERIAL FALL DIAMETER, % FINER THAN 250MM	07/27/65-09/08/66	2	3.5	7	3.5	24.5	0	4.95	**	**	**	**
80161 BED MATERIAL FALL DIAMETER, % FINER THAN 500MM	07/27/65-09/08/66	2	15.5	23	15.5	112.5	8	10.607	**	**	**	**
80162 BED MATERIAL FALL DIAMETER, % FINER THAN 1 00MM	07/27/65-09/08/66	2	20.5	25	20.5	40.5	16	6.364	**	**	**	**
80163 BED MATERIAL FALL DIAMETER, % FINER THAN 2 00MM	07/27/65-09/08/66	2	29.5	41	29.5	264.5	18	16.263	**	**	**	**
80170 BED MATERIAL SIEVE DIAMETER,% FINER THAN 4 00MM	07/27/65-09/08/66	3	21	51	24	657	0	25.632	**	**	**	**
80171 BED MATERIAL SIEVE DIAMETER,% FINER THAN 8 00MM	07/27/65-09/08/66	3	27	66	32.333	982.333	4	31.342	**	**	**	**
80172 BED MATERIAL SIEVE DIAMETER,% FINER THAN 16 0MM	07/27/65-09/08/66	3	61	80	66	151	57	12.288	**	**	**	**
80173 BED MATERIAL SIEVE DIAMETER,% FINER THAN 32 0MM	07/27/65-09/08/66	3	100	100	100	0	100	0	**	**	**	**

** - Less than 9 observations   ## - Computed with 50% or more of the total observations as values that were half the detection limit   p - Has a corresponding time series plot

EPA Water Quality Criteria Analysis for Station: BADL0011

Parameter	Std Type	Std Value	Total Obs	Exceed Standard	Prop Exceeding	10/01-1/31 Obs	Exceed	Prop	2/01-4/14 Obs	Exceed	Prop	4/15-6/30 Obs	Exceed	Prop	7/01-9/30 Obs	Exceed	Prop
00300 OXYGEN, DISSOLVED	Other-Lo Lim	4	2	0	0.00	1	0	0.00	0	0	0.00	1	0	0.00	0	0	0.00
00400 PH	Other-Hi Lim	9	36	0	0.00	10	0	0.00	6	0	0.00	10	0	0.00	10	0	0.00
	Other-Lo Lim	6.5	36	0	0.00	10	0	0.00	6	0	0.00	10	0	0.00	10	0	0.00
00940 CHLORIDE,TOTAL IN WATER	Fresh Acute	860	36	0	0.00	10	0	0.00	6	0	0.00	10	0	0.00	10	0	0.00
	Drinking Water	250	36	0	0.00	10	0	0.00	6	0	0.00	10	0	0.00	10	0	0.00
00945 SULFATE, TOTAL (AS SO4)	Drinking Water	250	36	5	0.14	10	0	0.00	6	0	0.00	10	2	0.20	10	3	0.30
00950 FLUORIDE, DISSOLVED AS F	Drinking Water	4	36	0	0.00	10	0	0.00	6	0	0.00	10	0	0.00	10	0	0.00
01034 CHROMIUM, TOTAL	Drinking Water	100	1	0	0.00	1	0	0.00									
01040 COPPER, DISSOLVED	Fresh Acute	18	1	1	1.00	1	1	1.00									
	Drinking Water	1300	1	0	0.00	1	0	0.00									
01090 ZINC, DISSOLVED	Fresh Acute	120	1	0	0.00	1	0	0.00									
	Drinking Water	5000	1	0	0.00	1	0	0.00									
71851 NITRATE NITROGEN, DISSOLVED (AS NO3)	Drinking Water	44	36	0	0.00	10	0	0.00	6	0	0.00	10	0	0.00	10	0	0.00
71856 NITRITE NITROGEN, DISSOLVED (AS NO2)	Drinking Water	3.3	14	5	0.36	4	2	0.50	1	0	0.00	2	0	0.00	7	3	0.43

& - Below detection limit observations, for which half the detection limit exceeded the criterion, were excluded from the criterion comparison for this parameter

Station: BADL0011 Parameter Code: 00060

FLOW, STREAM, MEAN DAILY

(X 1000)

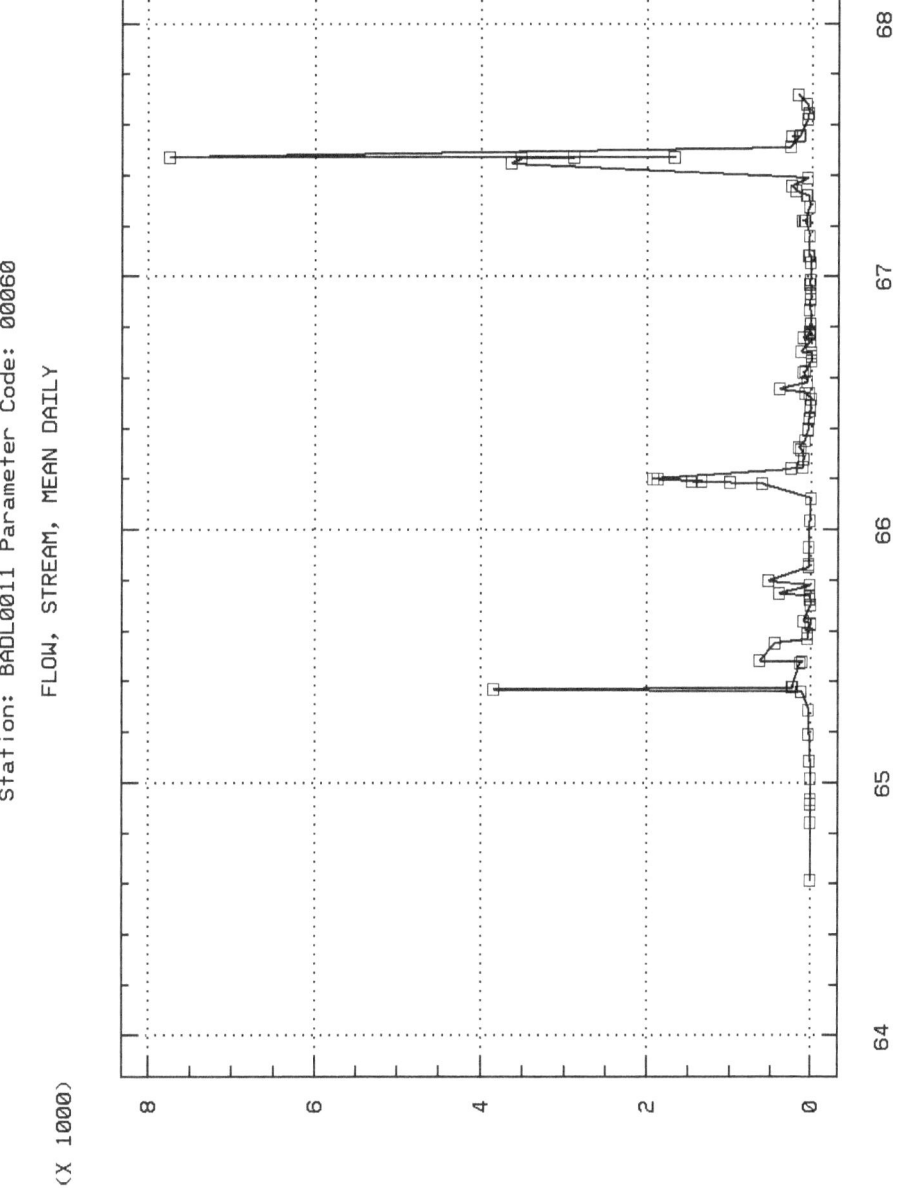

Sample Date (Years)

STREAM FLOW CFS

WHITE R NEAR ROCKYFORD SD

91

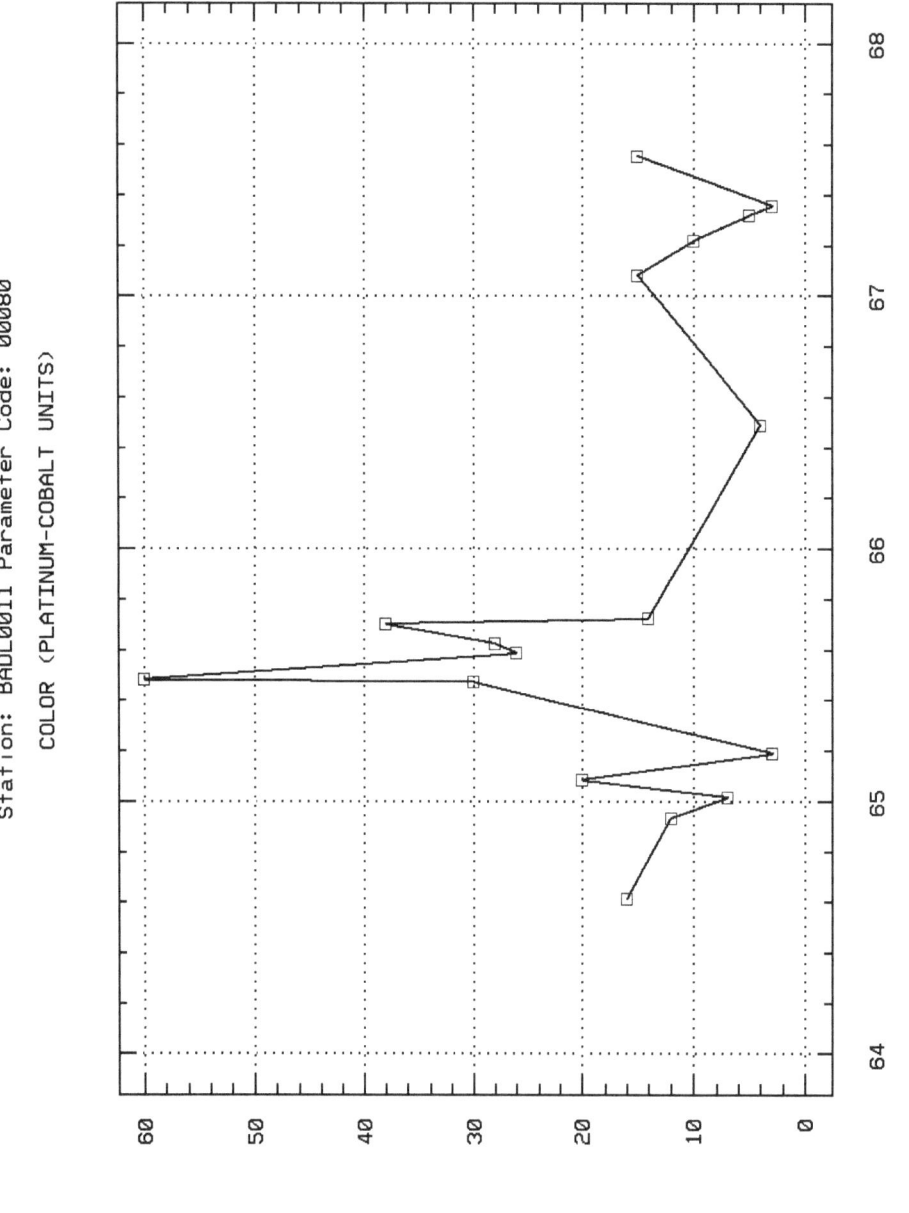

Station: BADL0011 Parameter Code: 00080

COLOR (PLATINUM-COBALT UNITS)

Sample Date (Years)

COLOR PT-CO UNITS

92

WHITE R NEAR ROCKYFORD SD

Station: BADL0011 Parameter Code: 00095

SPECIFIC CONDUCTANCE (UMHOS/CM @ 25C)

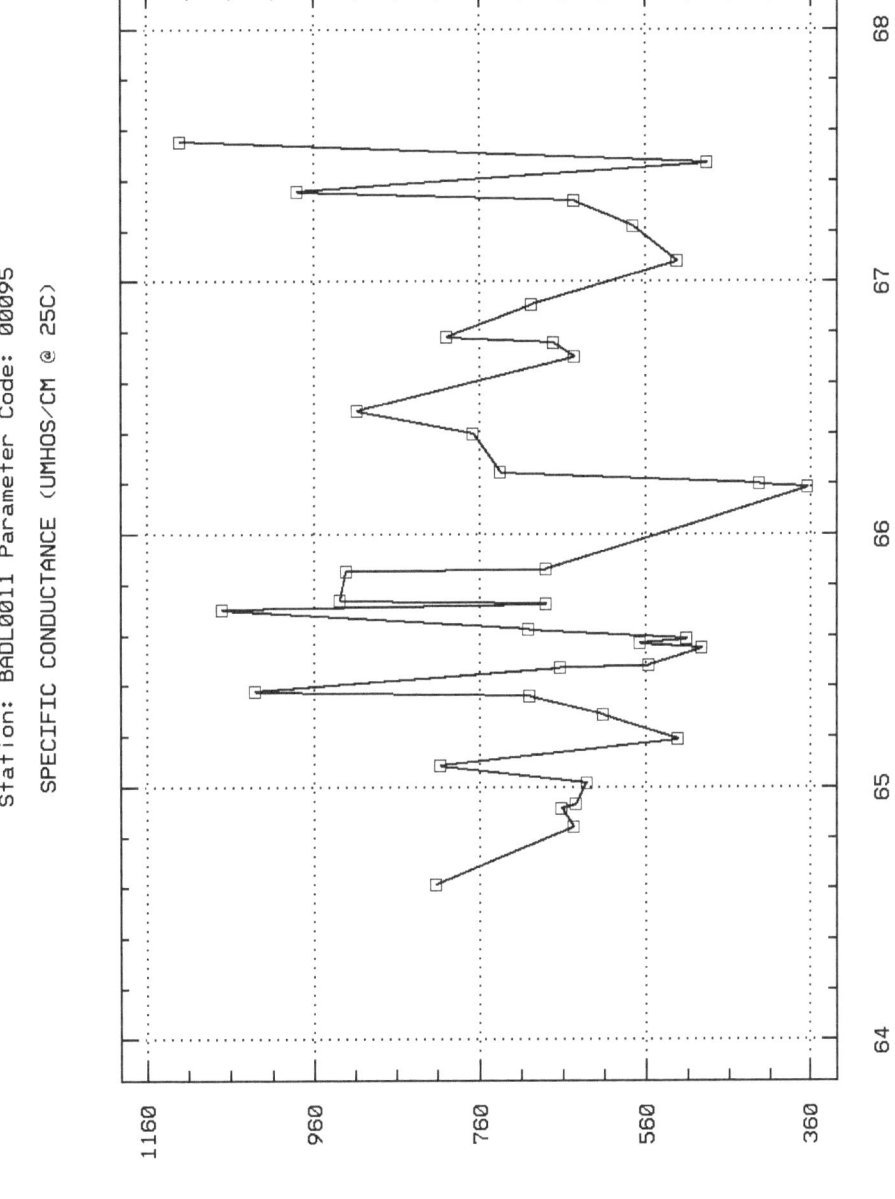

Sample Date (Years)

CNDUCTVY AT 25C MICROMHO

WHITE R NEAR ROCKYFORD SD

93

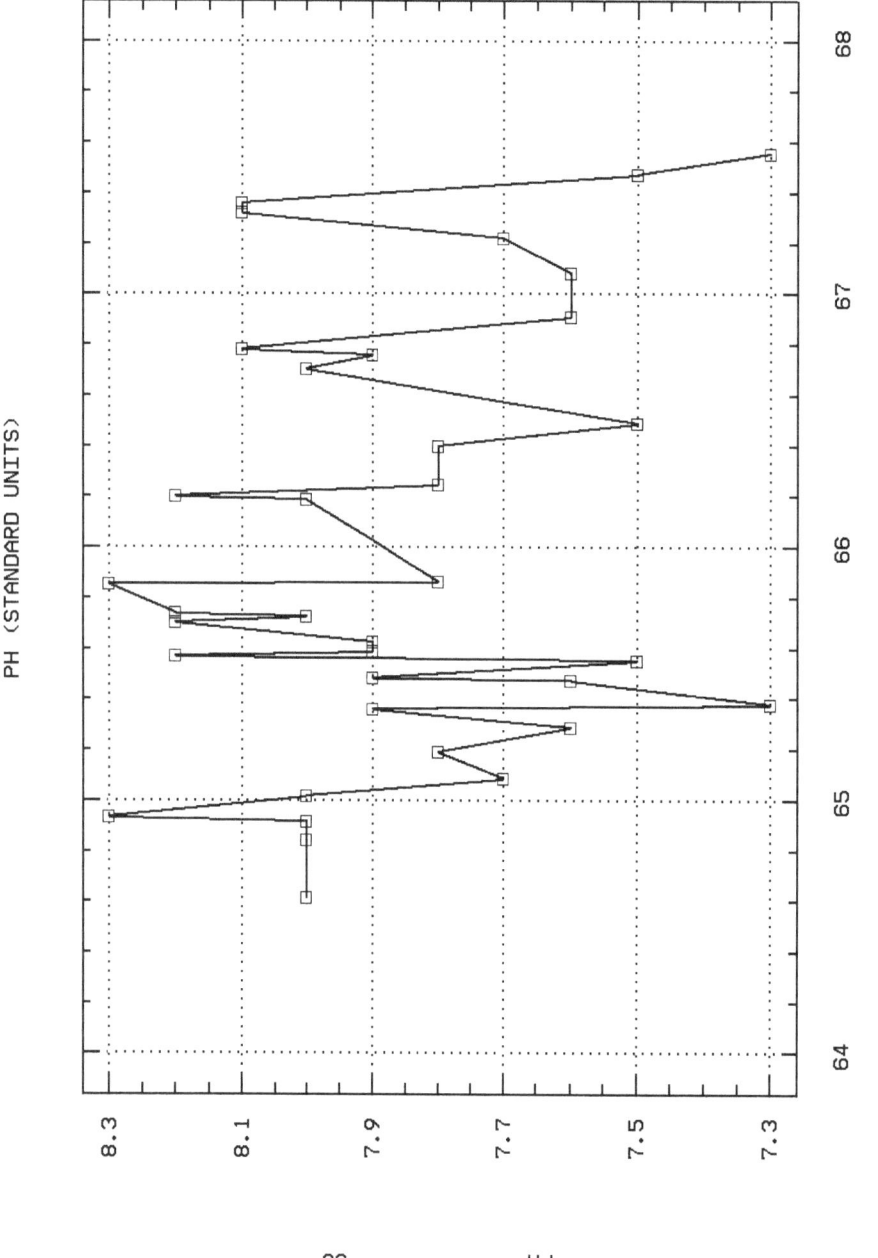

Station: BADL0011 Parameter Code: 00400

PH (STANDARD UNITS)

Sample Date (Years)

94

WHITE R NEAR ROCKYFORD SD

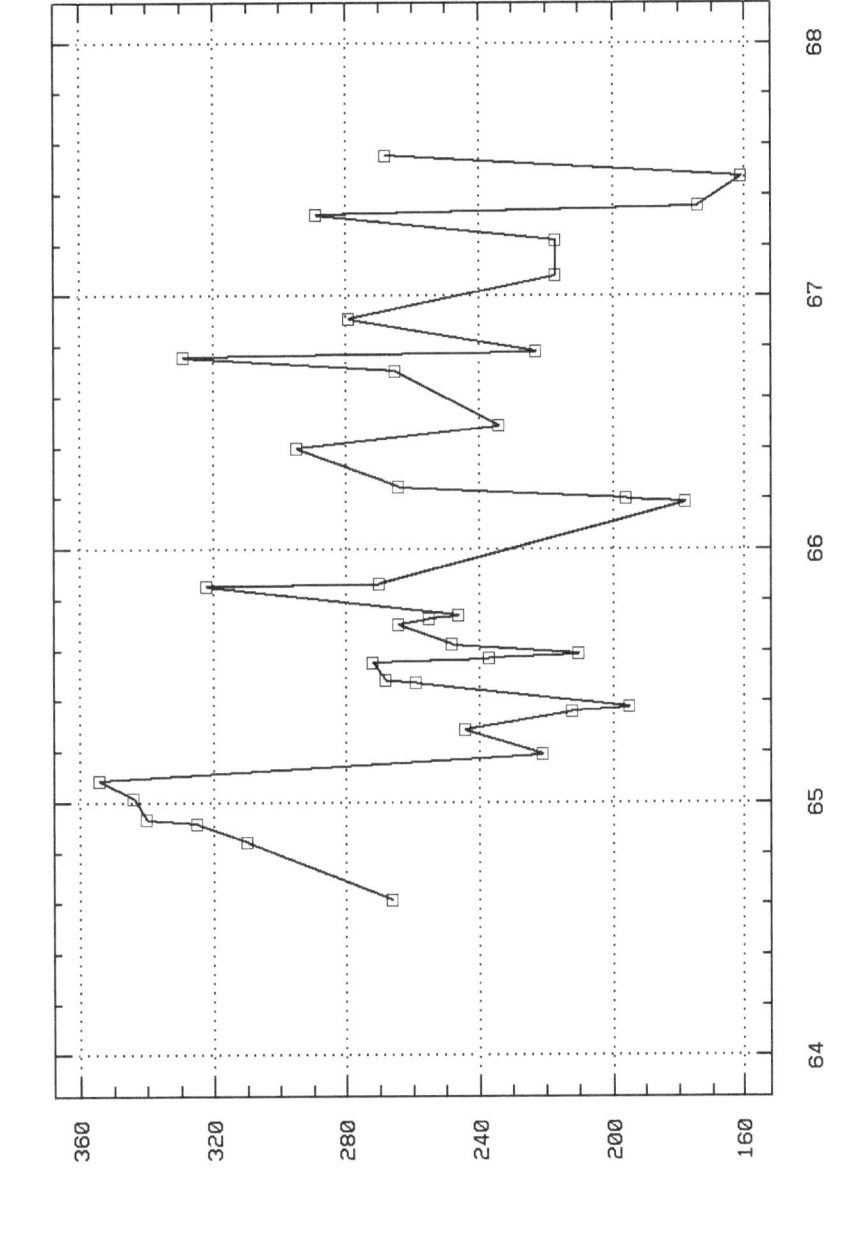

Station: BADL0011 Parameter Code: 00440

BICARBONATE ION (MG/L AS HCO3)

Sample Date (Years)

HCO3 ION HCO3   MG/L

WHITE R NEAR ROCKYFORD SD

95

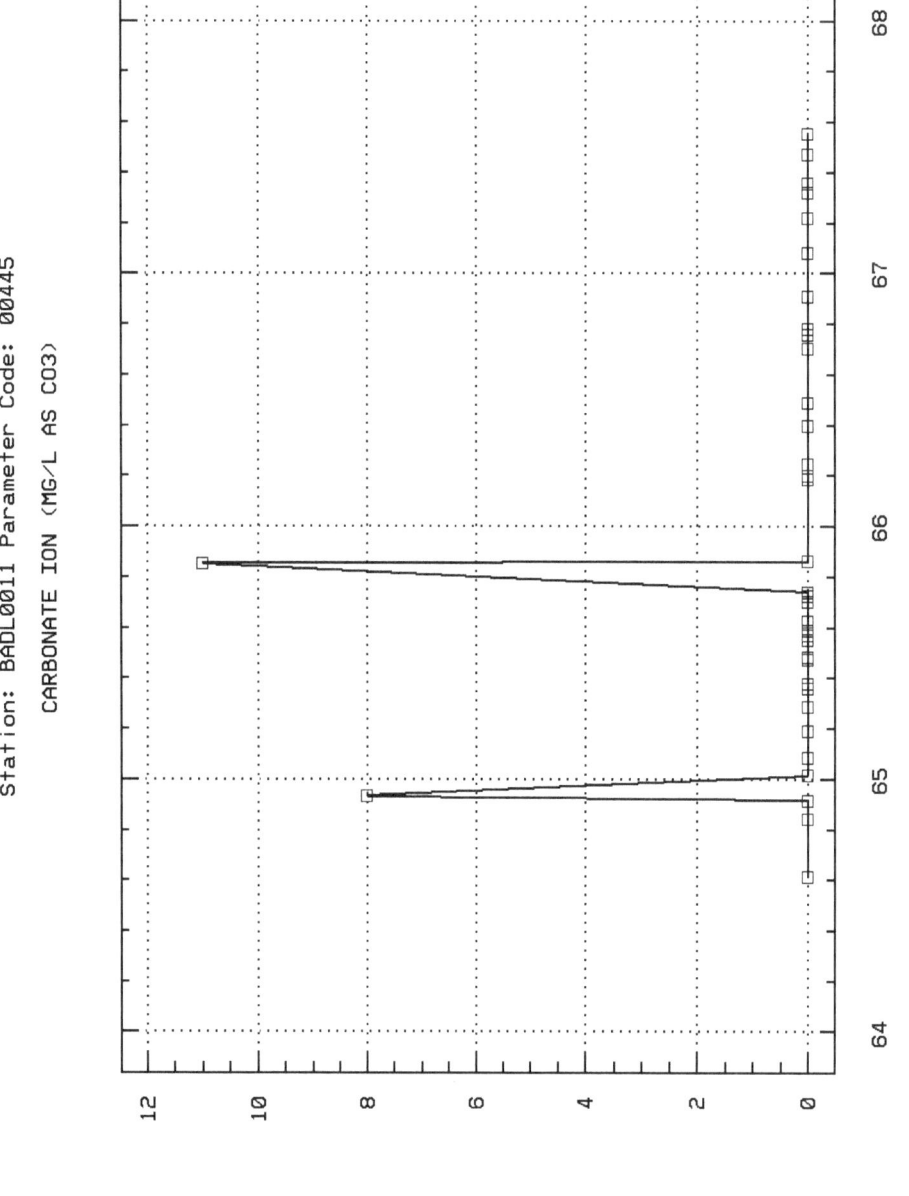

Station: BADL0011 Parameter Code: 00445

CARBONATE ION (MG/L AS CO3)

Sample Date (Years)

CO3 ION CO3   MG/L

WHITE R NEAR ROCKYFORD SD

96

Station: BADL0011 Parameter Code: 00650
PHOSPHATE, TOTAL (MG/L AS PO4)

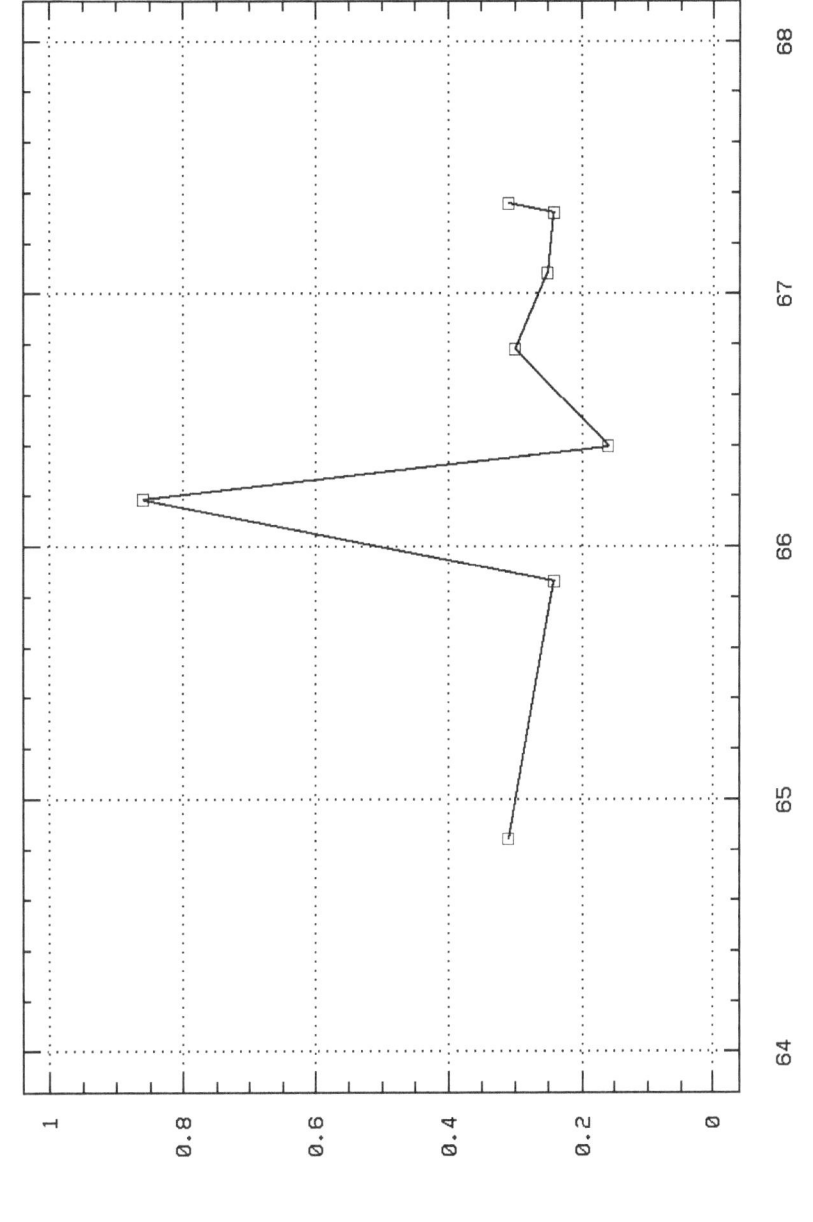

Sample Date (Years)

T PO4   PO4   MG/L

97

WHITE R NEAR ROCKYFORD SD

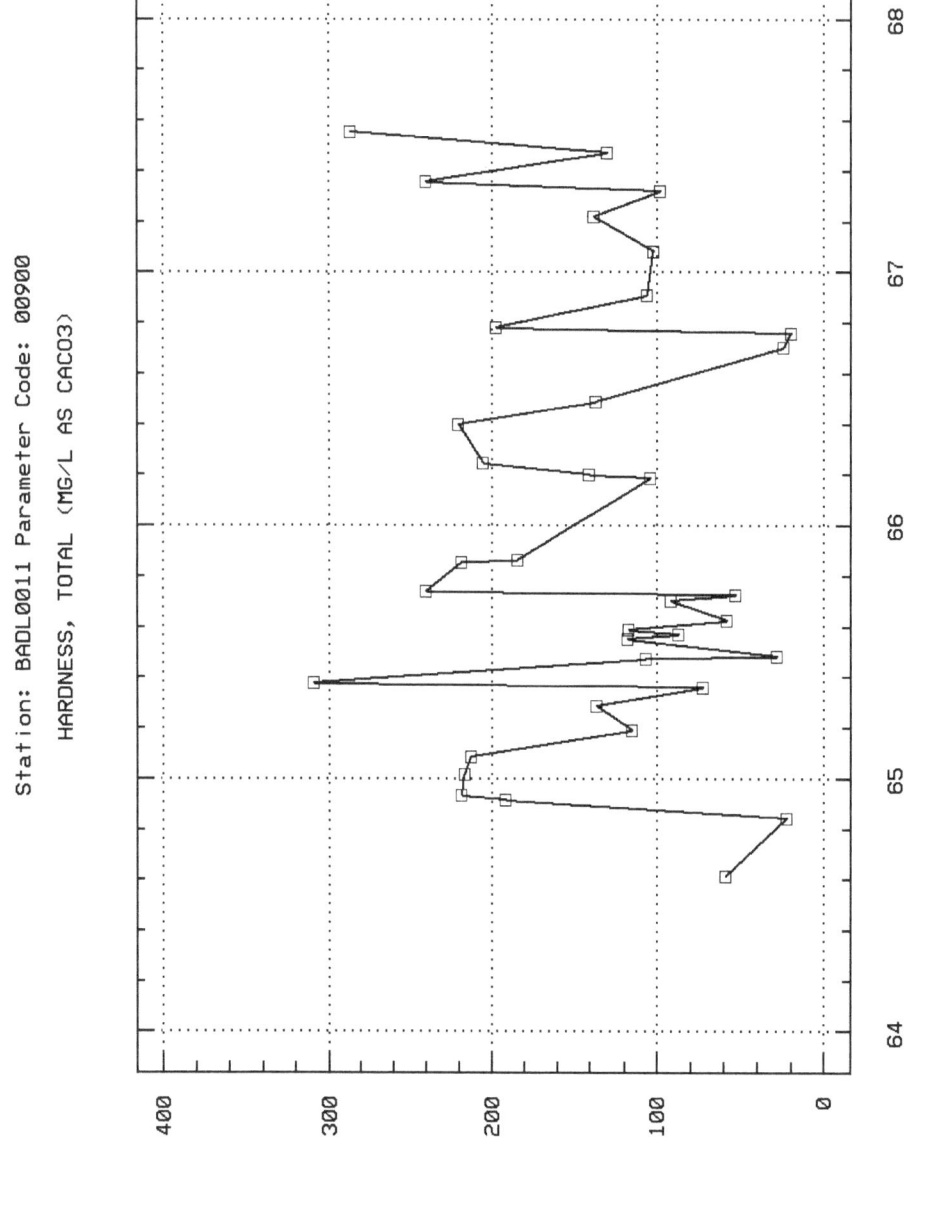

Station: BADL0011 Parameter Code: 00900

HARDNESS, TOTAL (MG/L AS CACO3)

Sample Date (Years)

TOT HARD CACO3   MG/L

98

WHITE R NEAR ROCKYFORD SD

Station: BADL0011 Parameter Code: 00902

HARDNESS, NON-CARBONATE (MG/L AS CACO3)

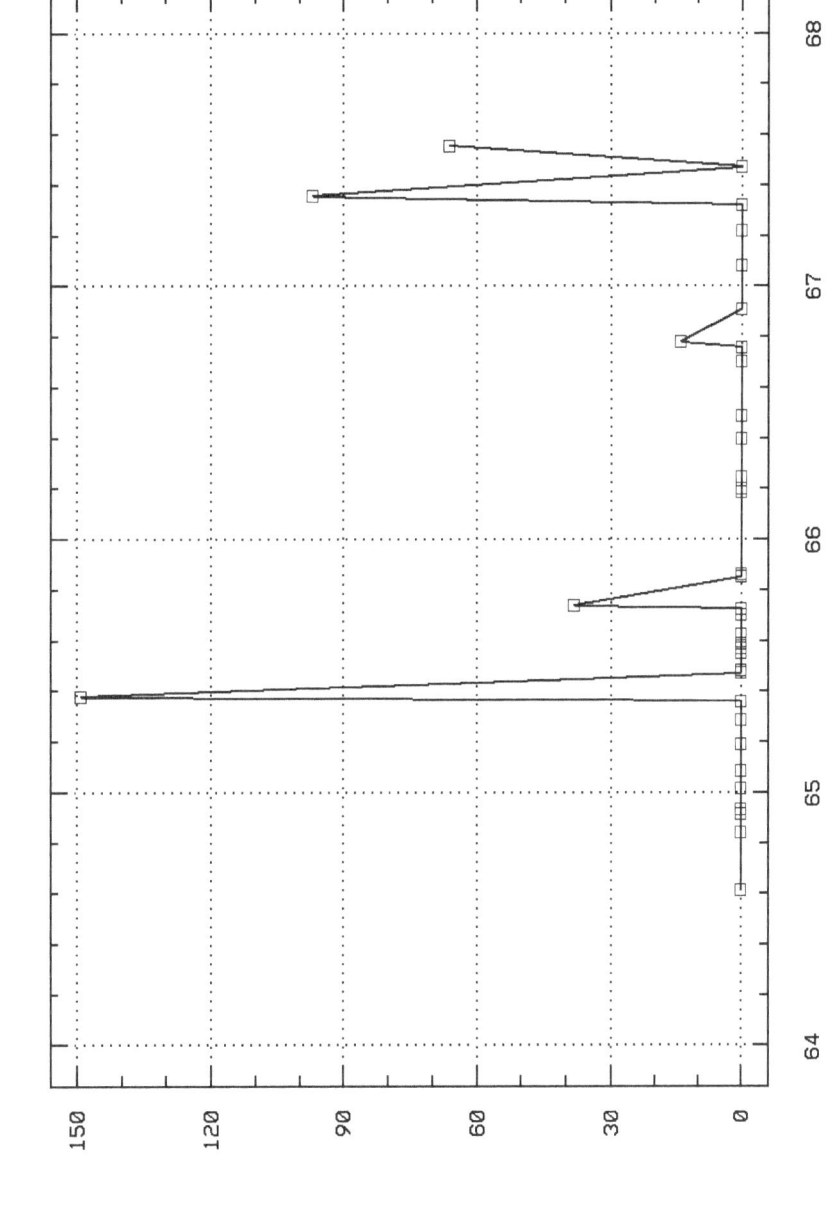

Sample Date (Years)

NC HARD CACO3 MG/L

WHITE R NEAR ROCKYFORD SD

99

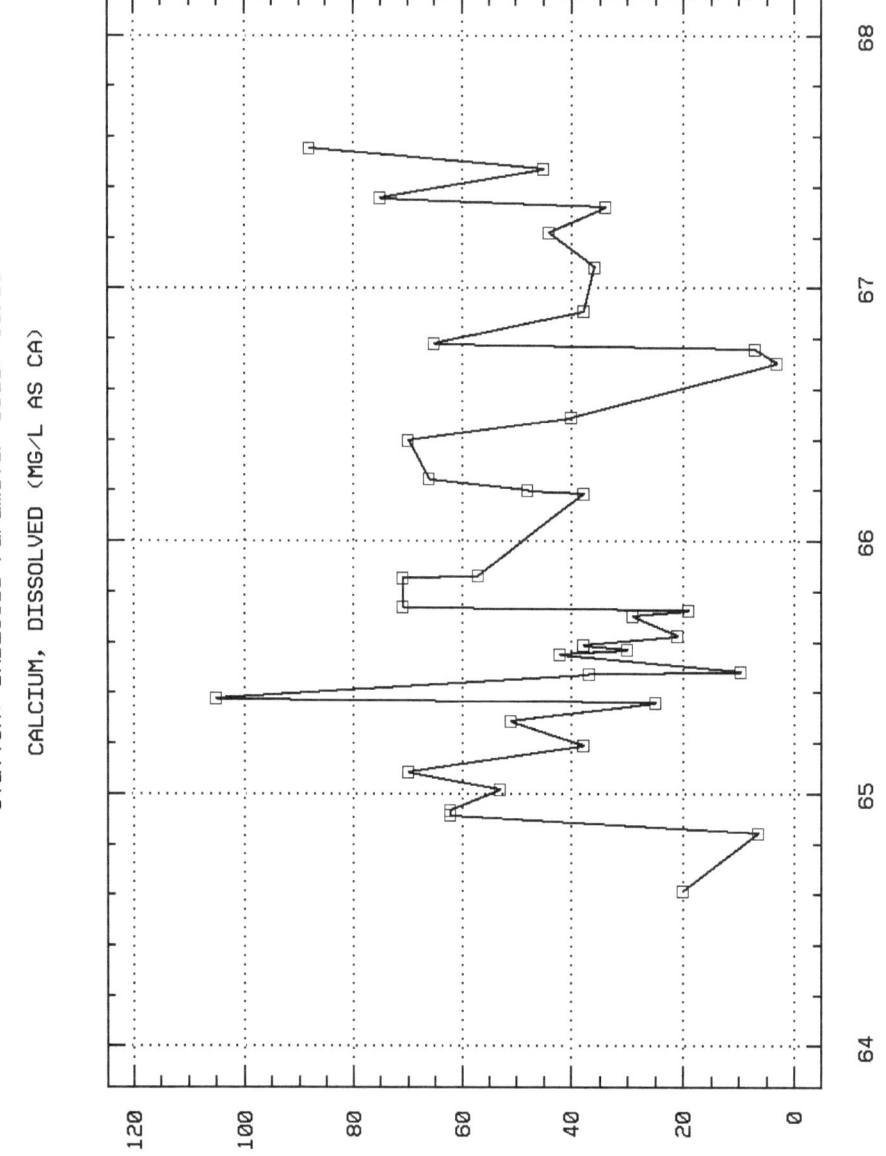

Station: BADL0011 Parameter Code: 00915
CALCIUM, DISSOLVED (MG/L AS CA)

Sample Date (Years)

CALCIUM CA,DISS   MG/L

100

WHITE R NEAR ROCKYFORD SD

Station: BADL0011 Parameter Code: 00925
MAGNESIUM, DISSOLVED (MG/L AS MG)

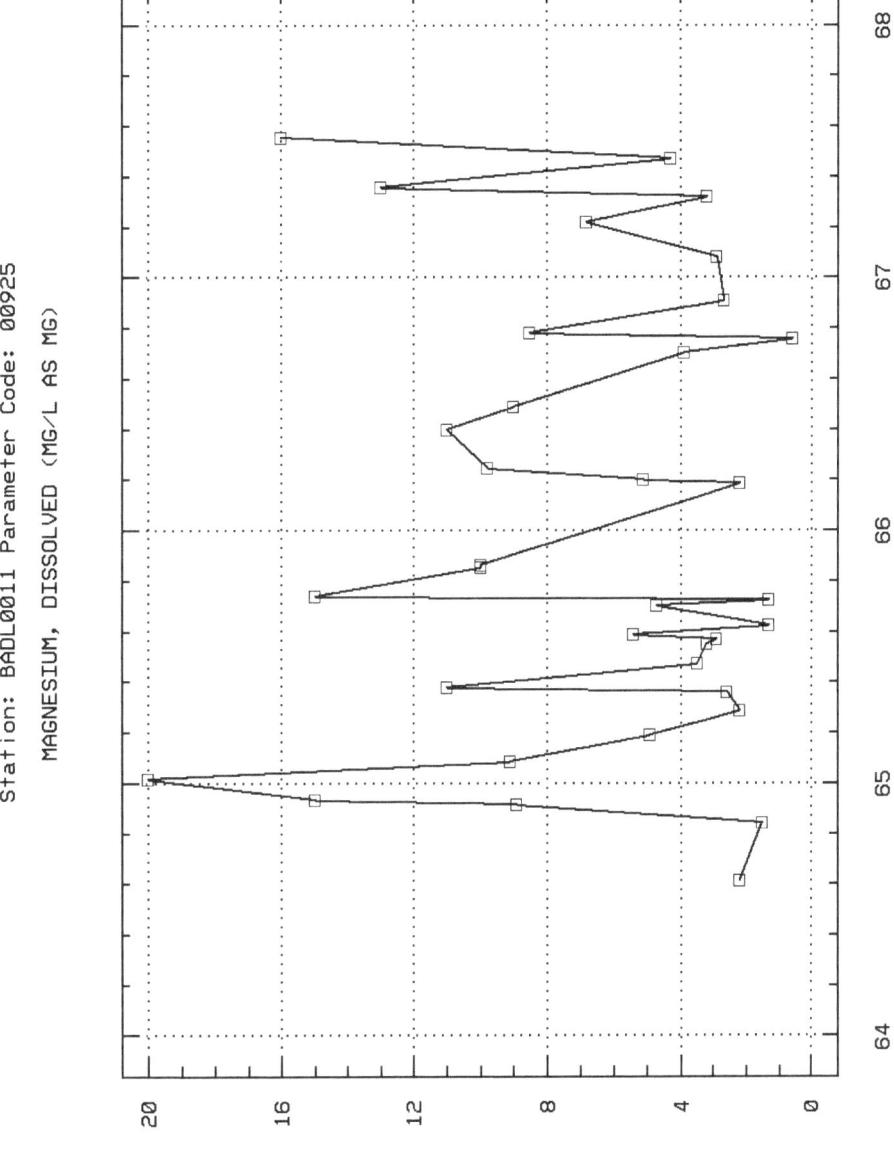

Sample Date (Years)

MGNSIUM MG,DISS   MG/L

WHITE R NEAR ROCKYFORD SD

101

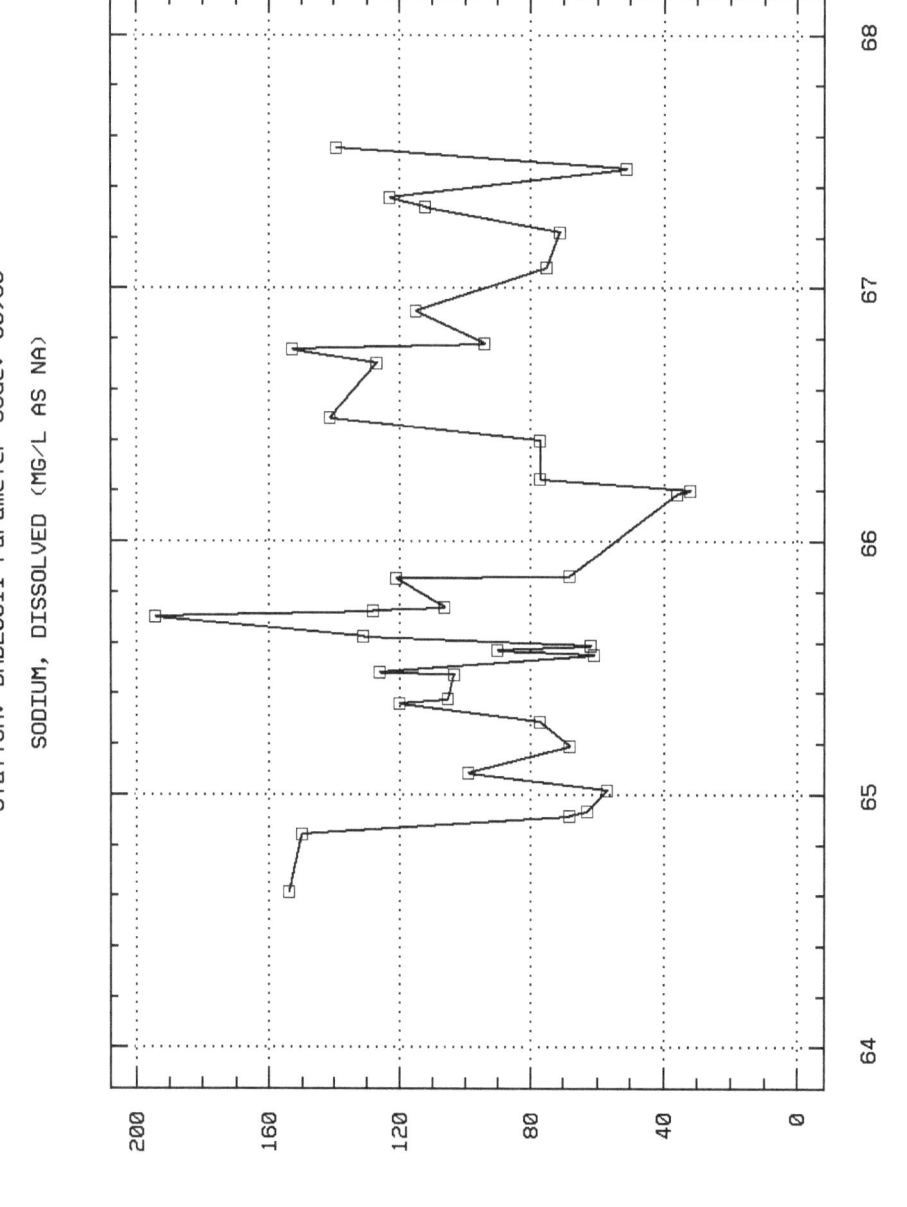

Station: BADL0011 Parameter Code: 00930

SODIUM, DISSOLVED (MG/L AS NA)

Sample Date (Years)

SODIUM NA,DISS   MG/L

WHITE R NEAR ROCKYFORD SD

102

Station: BADL0011 Parameter Code: 00931

SODIUM ADSORPTION RATIO

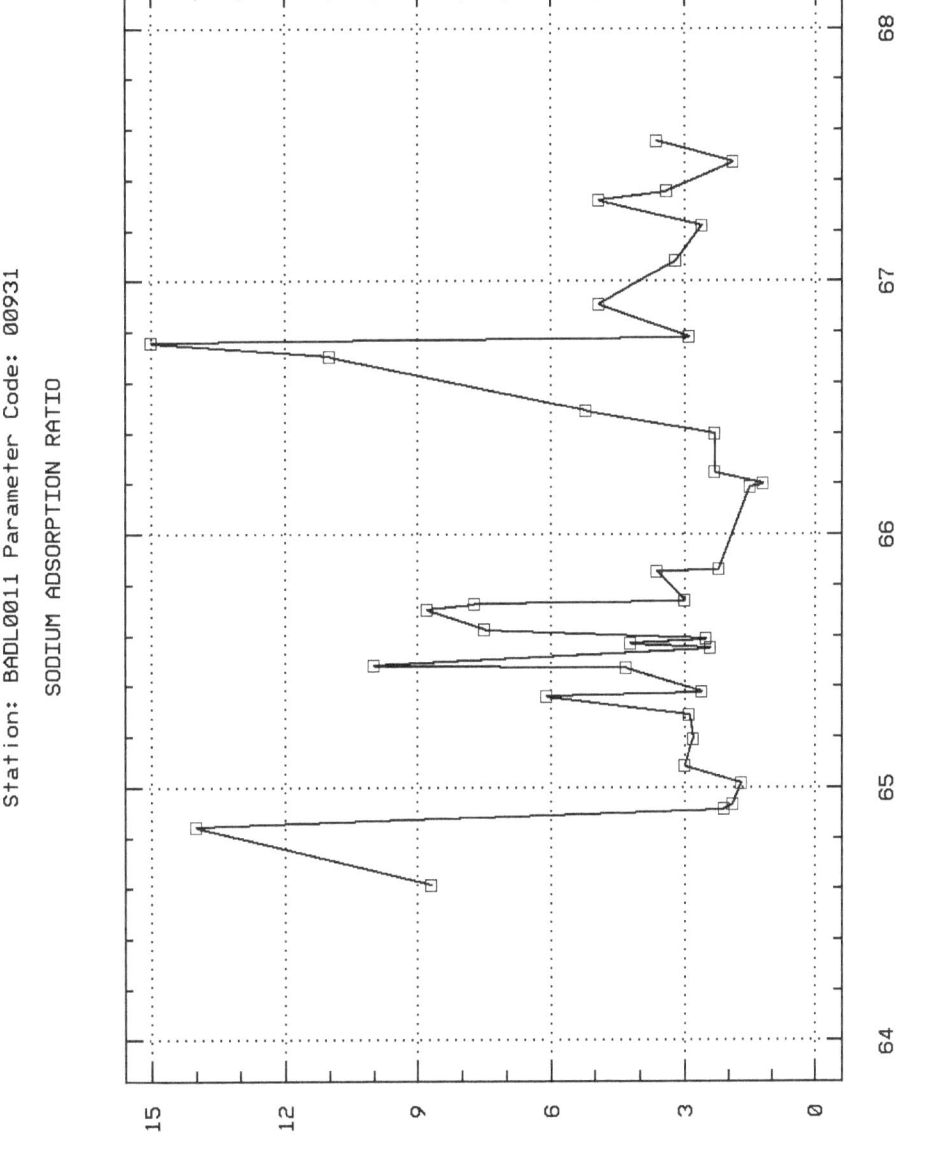

SODIUM ADSBTION RATIO

Sample Date (Years)

WHITE R NEAR ROCKYFORD SD

103

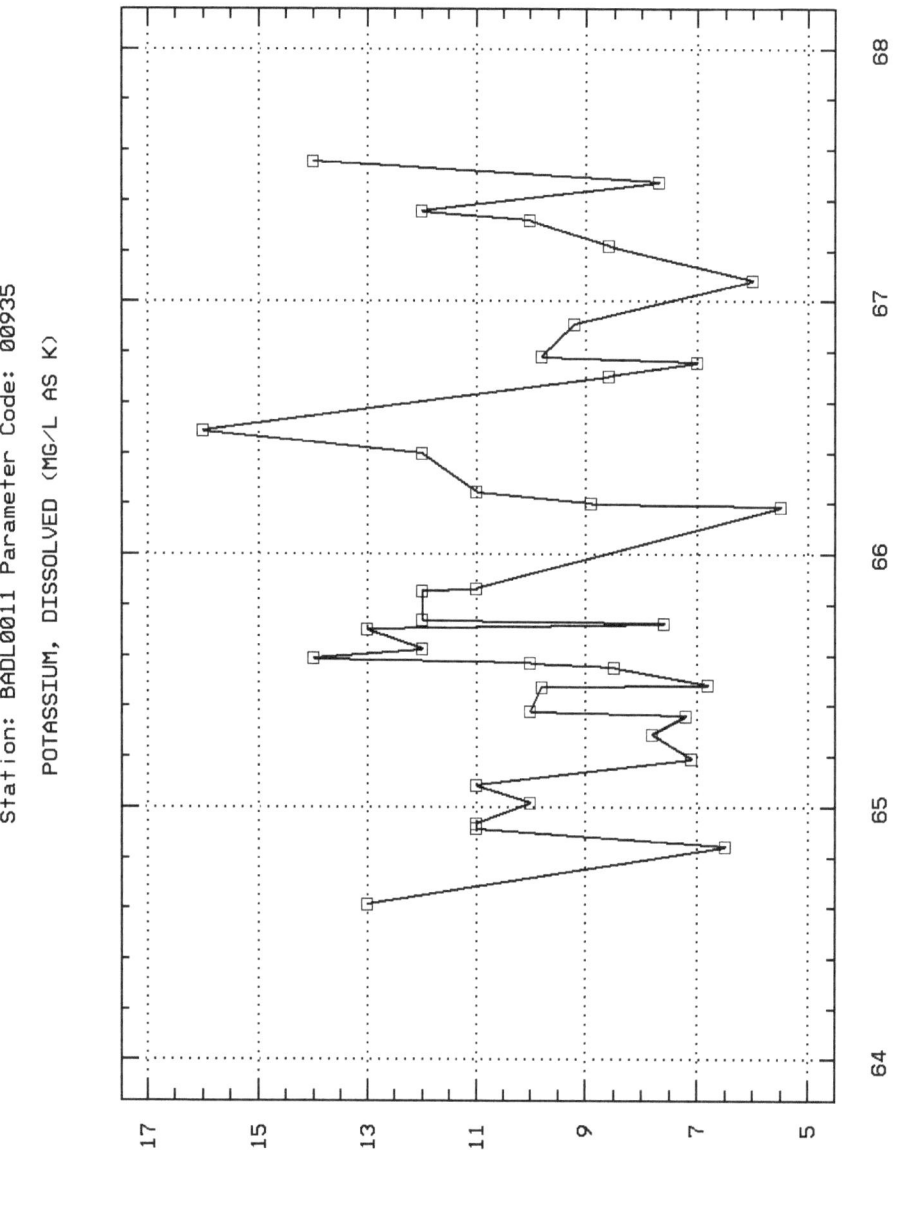

Station: BADL0011 Parameter Code: 00935

POTASSIUM, DISSOLVED (MG/L AS K)

Sample Date (Years)

PTSSIUM K,DISS MG/L

WHITE R NEAR ROCKYFORD SD

104

Station: BADL0011 Parameter Code: 00940

CHLORIDE, TOTAL IN WATER

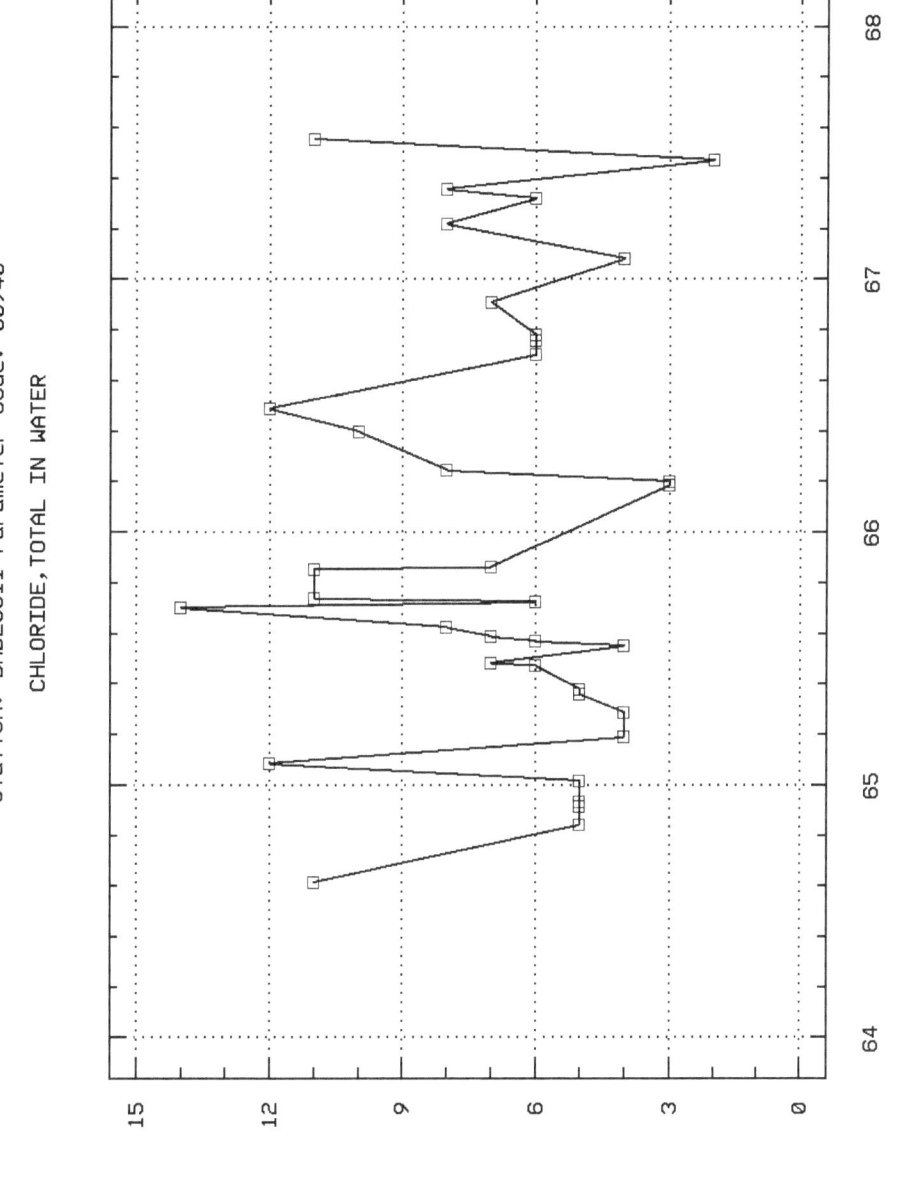

WHITE R NEAR ROCKYFORD SD

105

Station: BADL0011 Parameter Code: 00945

SULFATE, TOTAL (MG/L AS SO4)

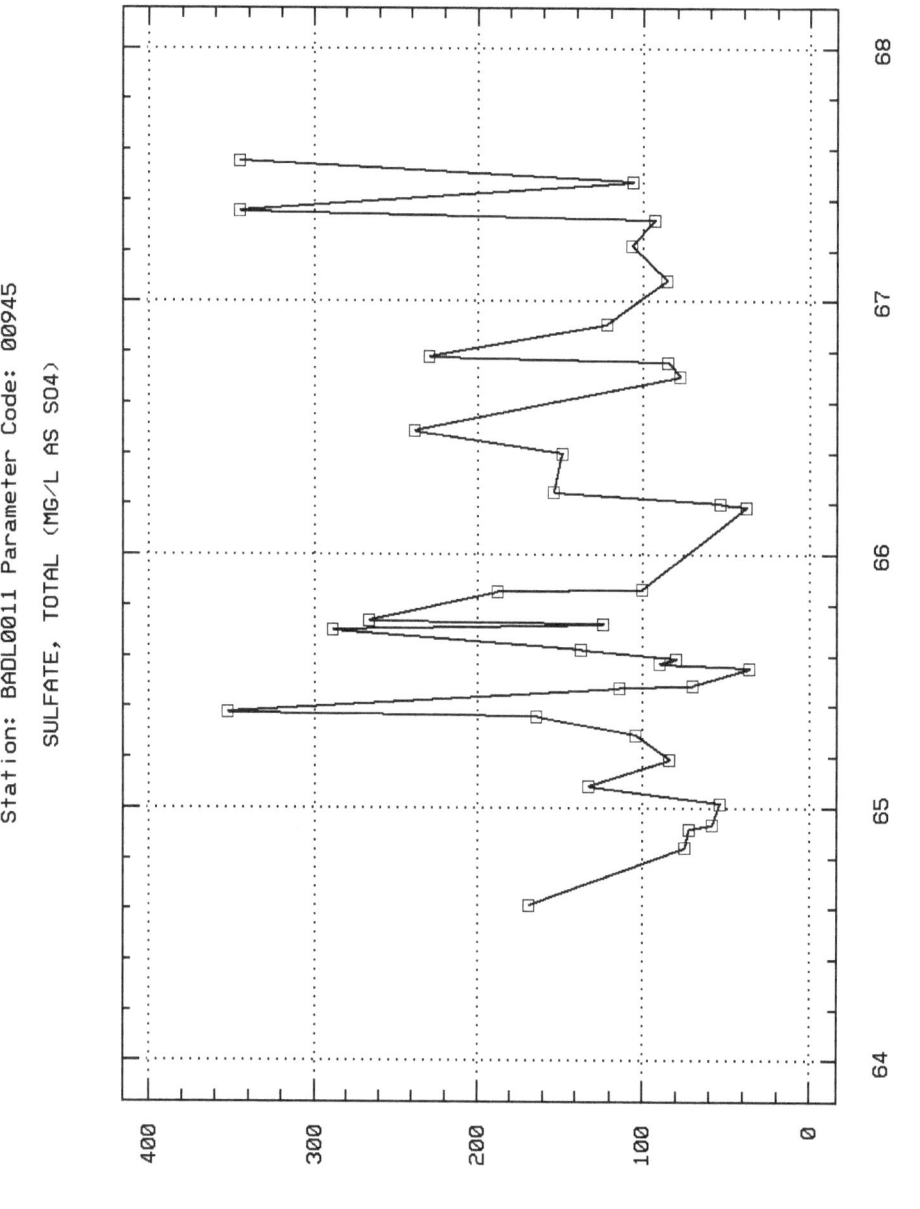

Sample Date (Years)

SULFATE SO4-TOT   MG/L

WHITE R NEAR ROCKYFORD SD

106

Station: BADL0011 Parameter Code: 00950

FLUORIDE, DISSOLVED (MG/L AS F)

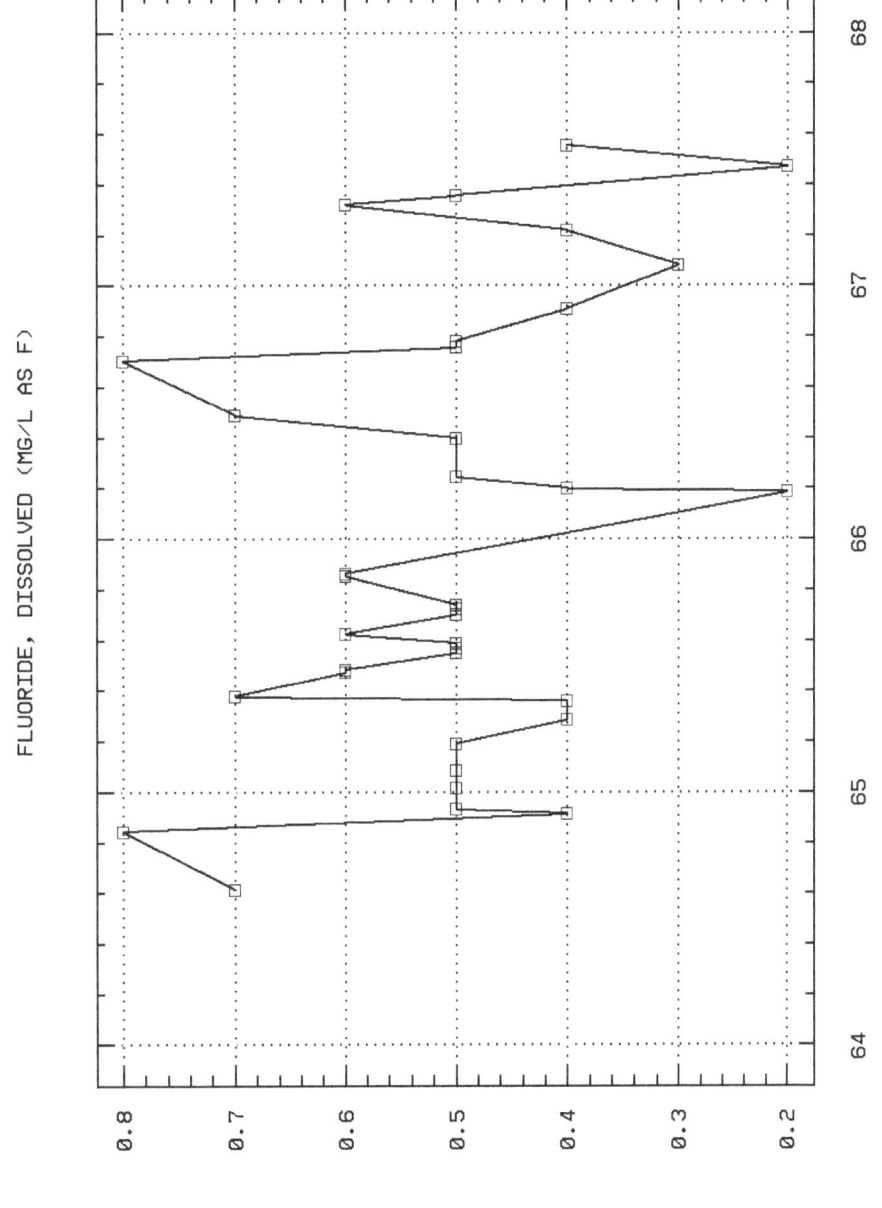

Sample Date (Years)

FLUORIDE F,DISS   MG/L

107

WHITE R NEAR ROCKYFORD SD

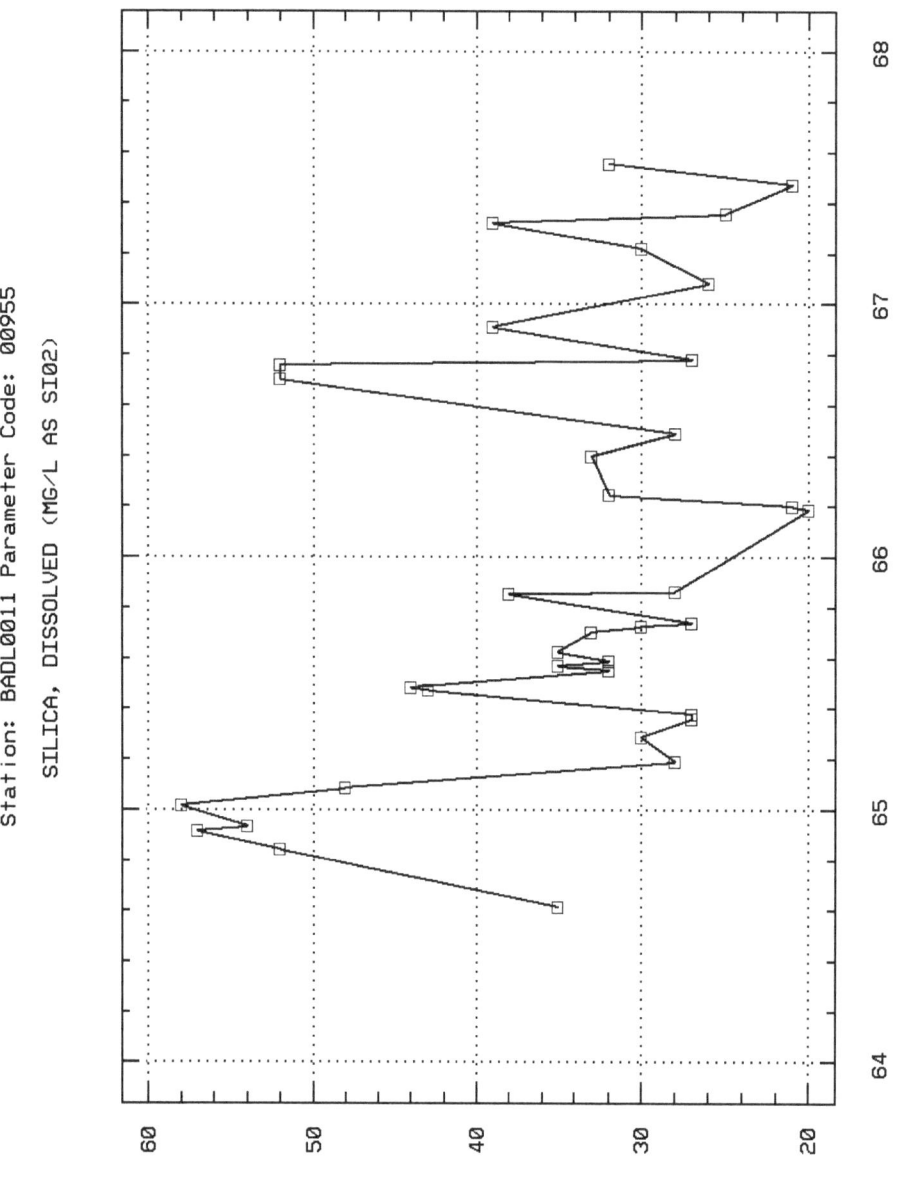

Station: BADL0011 Parameter Code: 00955

SILICA, DISSOLVED (MG/L AS SIO2)

Sample Date (Years)

SILICA DISSOLVED MG/L

108

WHITE R NEAR ROCKYFORD SD

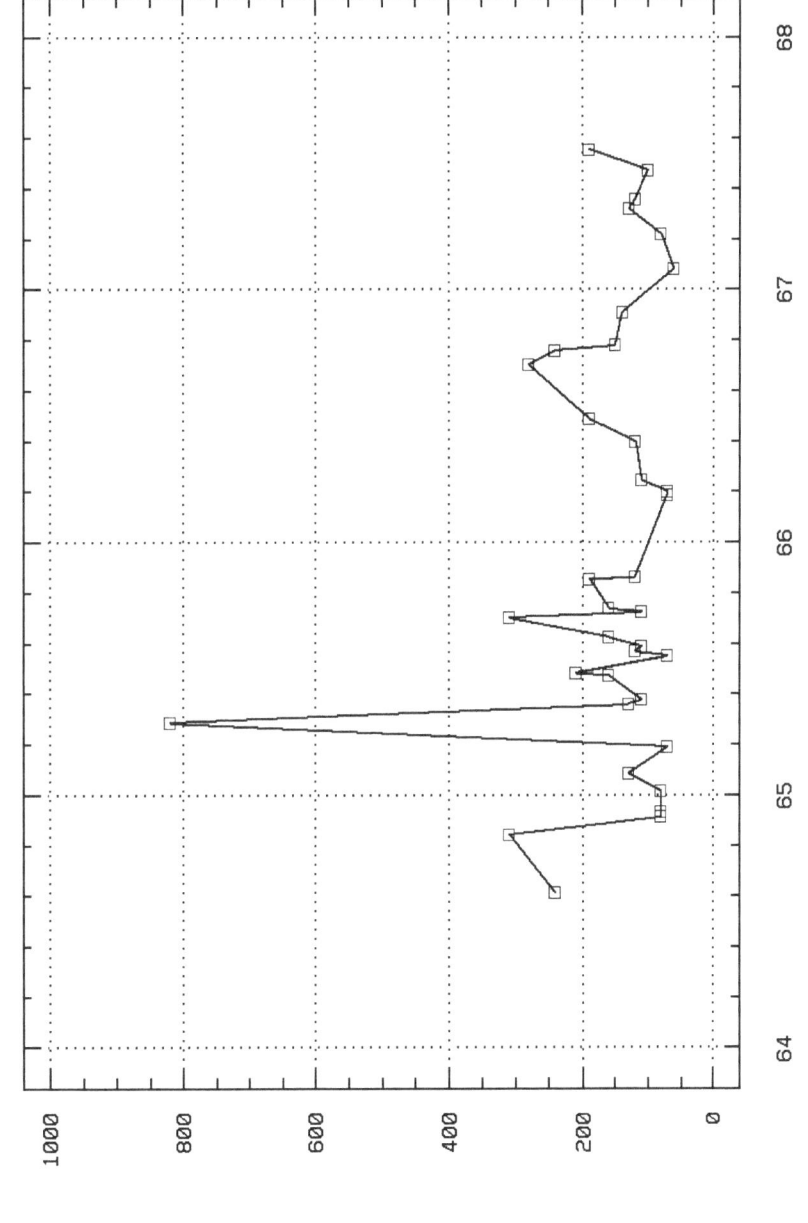

Station: BADL0011 Parameter Code: 01020

BORON, DISSOLVED (UG/L AS B)

BORON B,DISS UG/L

Sample Date (Years)

WHITE R NEAR ROCKYFORD SD

109

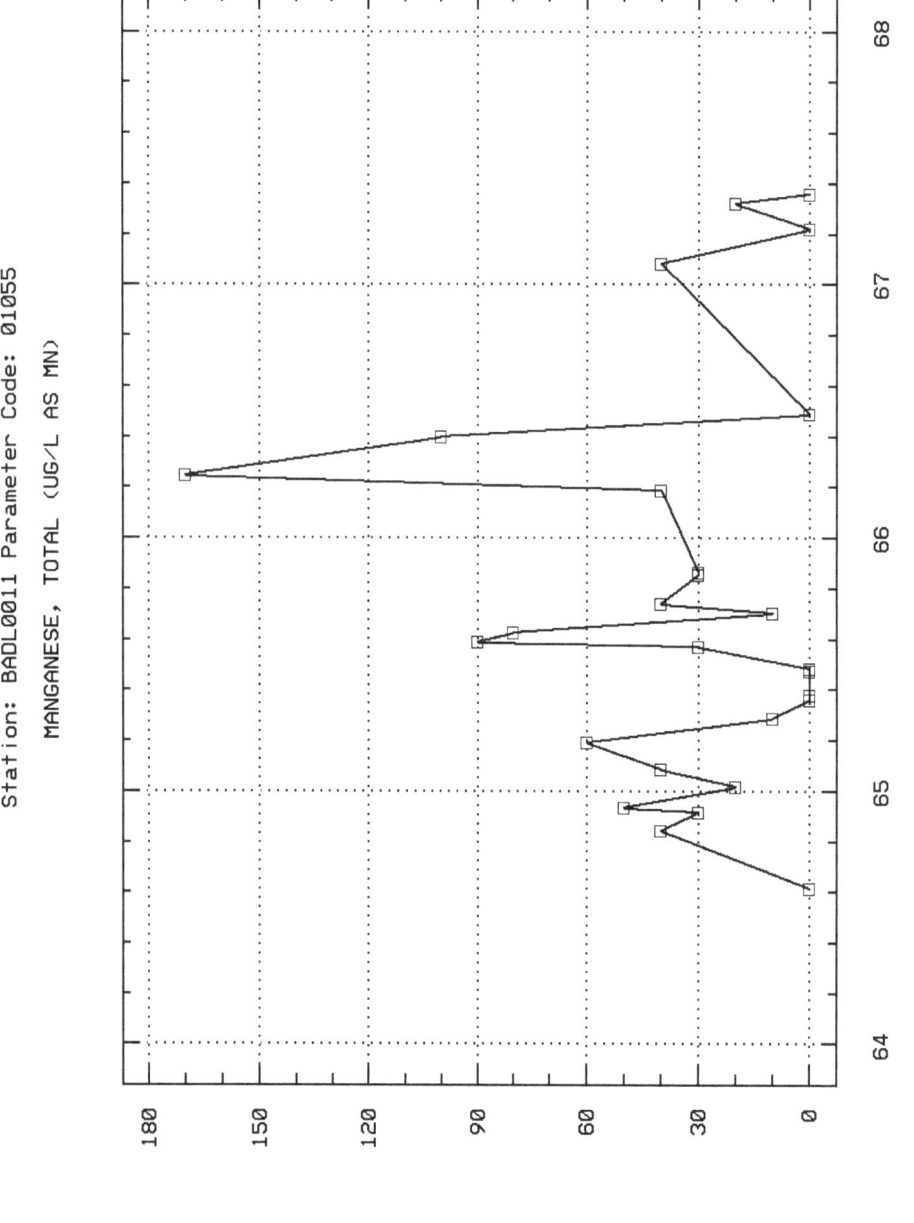

Station: BADL0011 Parameter Code: 01055

MANGANESE, TOTAL (UG/L AS MN)

Sample Date (Years)

MANGNESE MN UG/L

WHITE R NEAR ROCKYFORD SD

110

Station: BADL0011 Parameter Code: 70300
RESIDUE,TOTAL FILTRABLE (DRIED AT 180C)

RESIDUE DISS-180C   MG/L

Sample Date (Years)

WHITE R NEAR ROCKYFORD SD

111

Station: BADL0011 Parameter Code: 70302

SOLIDS, DISSOLVED-TONS PER DAY

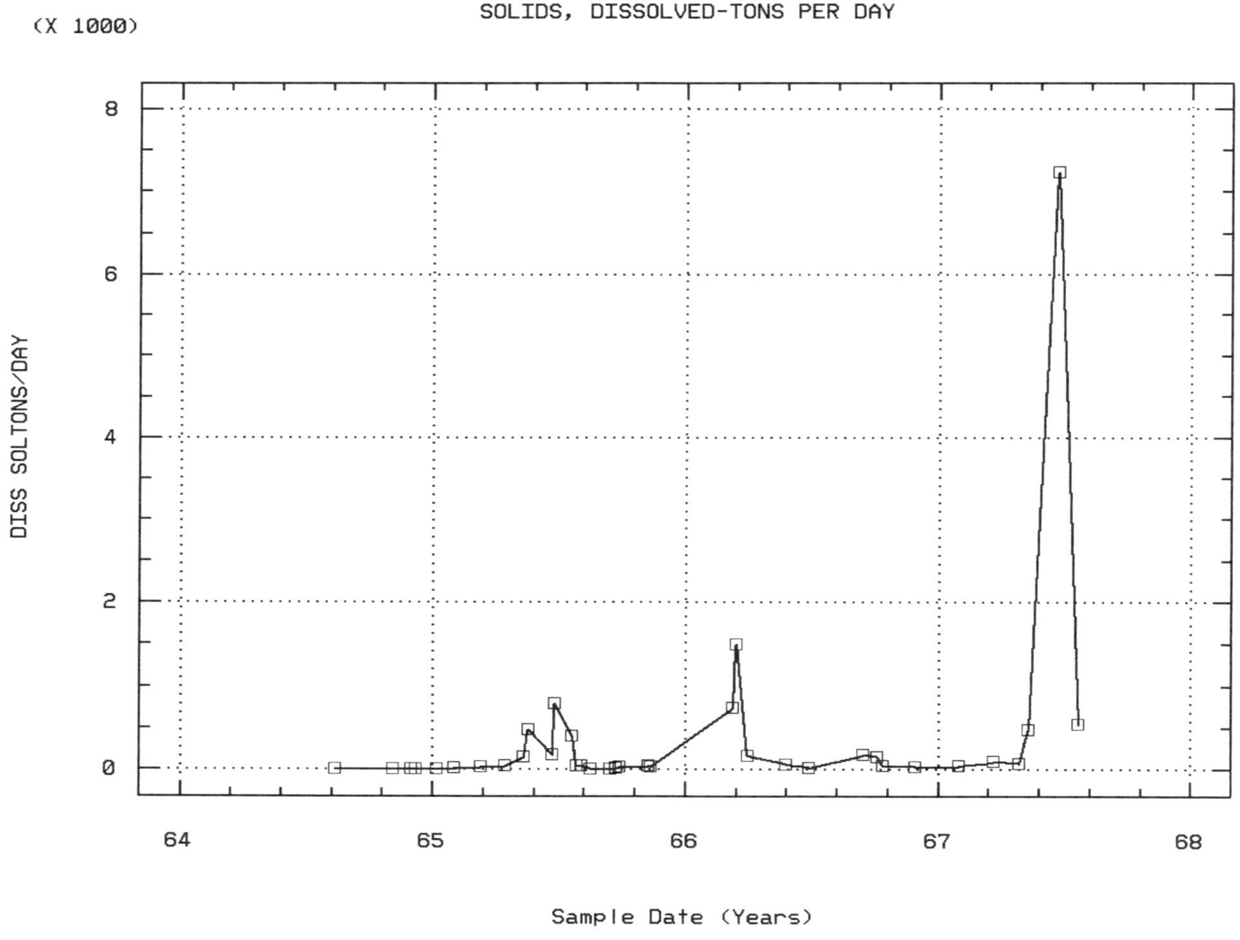
(X 1000)
DISS SOLTONS/DAY
8
6
4
2
0
64
65
66
67
68
Sample Date (Years)

WHITE R NEAR ROCKYFORD SD

Station: BADL0011 Parameter Code: 70303

SOLIDS, DISSOLVED-TONS PER ACRE-FT

(X 0.01)

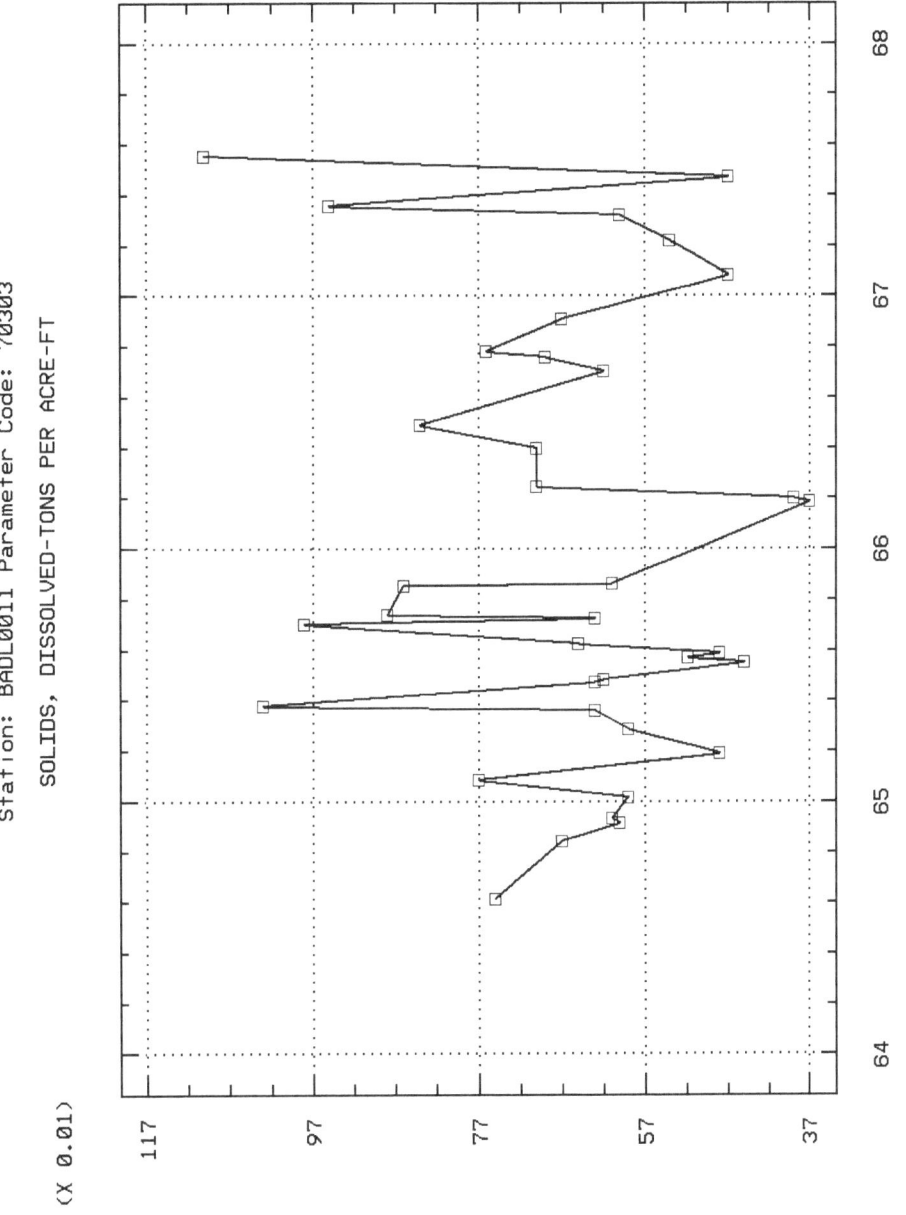

WHITE R NEAR ROCKYFORD SD

113

Station: BADL0011 Parameter Code: 70331
SUSPENDED SED SIEVE DIAMETER,% FINER TH

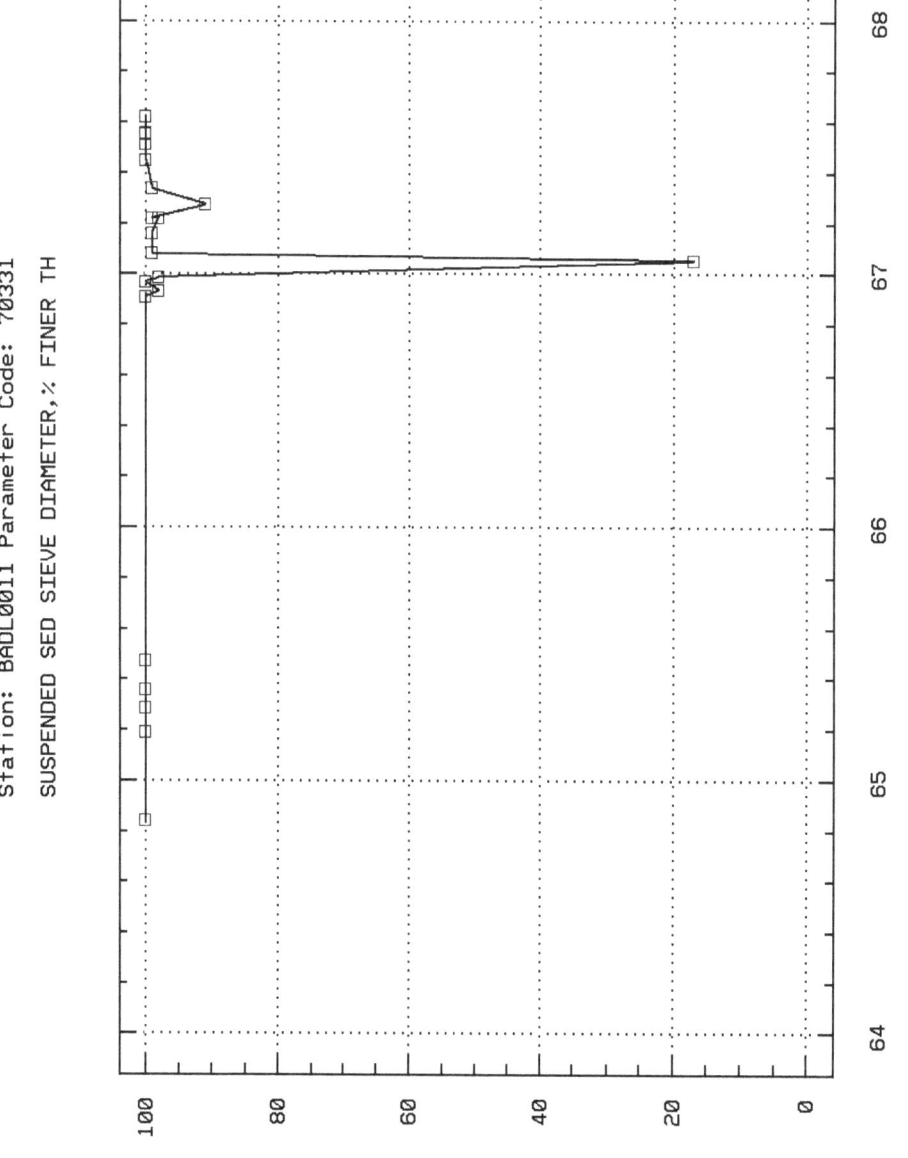

Sample Date (Years)

SUSP SEDPARTSIZE%.062MM

114

WHITE R NEAR ROCKYFORD SD

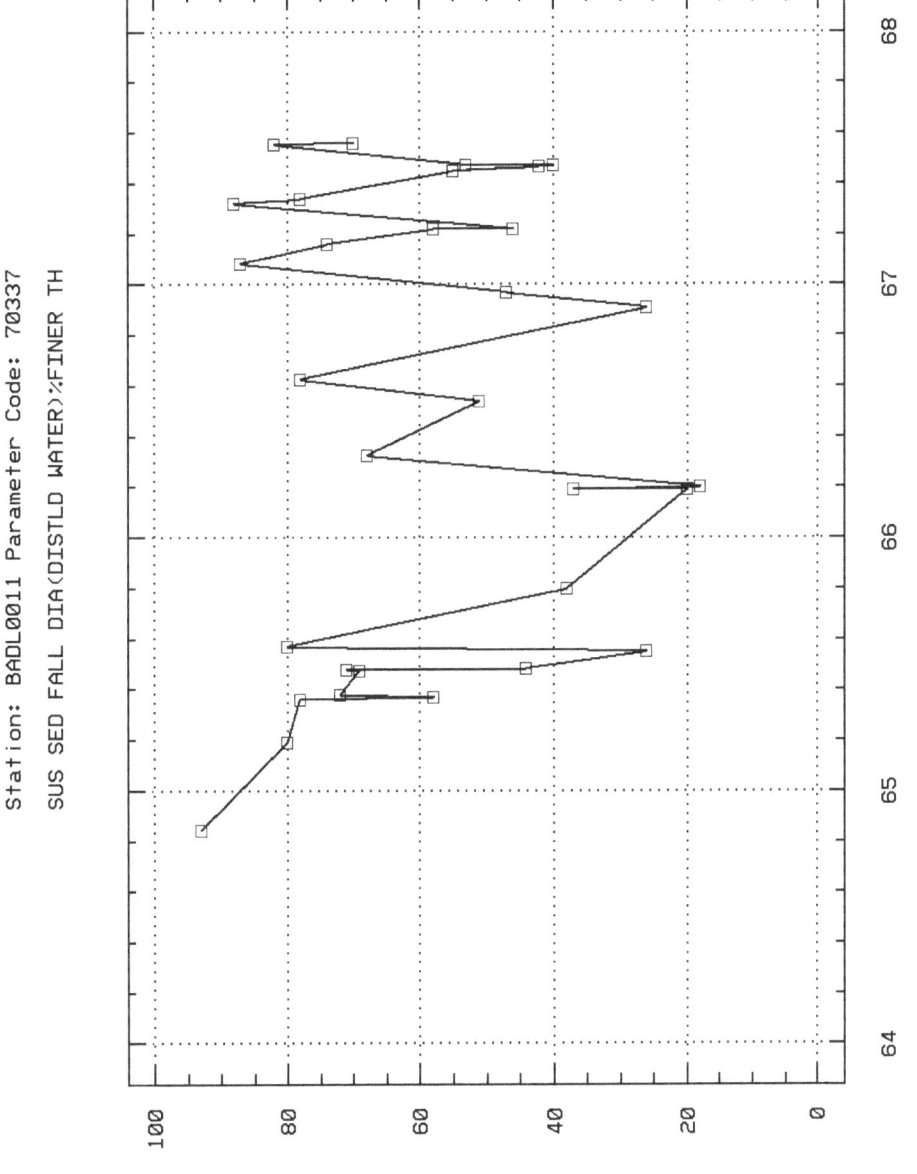

Station: BADL0011 Parameter Code: 70337

SUS SED FALL DIA(DISTLD WATER)%FINER TH

Sample Date (Years)

SUSP SEDPARTSIZE%.002MM

WHITE R NEAR ROCKYFORD SD

115

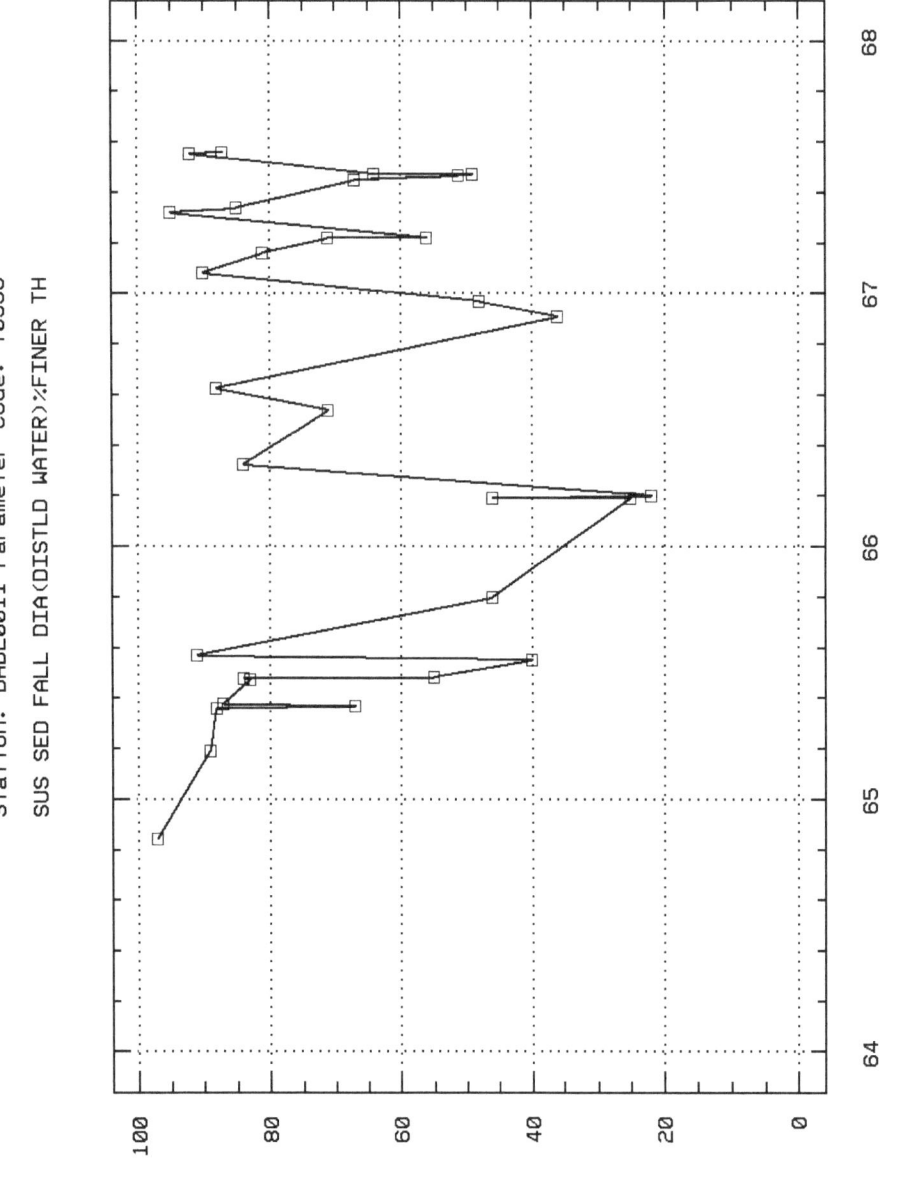

Station: BADL0011 Parameter Code: 70338

SUS SED FALL DIA(DISTLD WATER)%FINER TH

Sample Date (Years)

SUSP SEDPARTSIZE%<.004MM

WHITE R NEAR ROCKYFORD SD

116

Station: BADL0011 Parameter Code: 70340

SUS SED FALL DIA(DISTLD WATER)%FINER TH

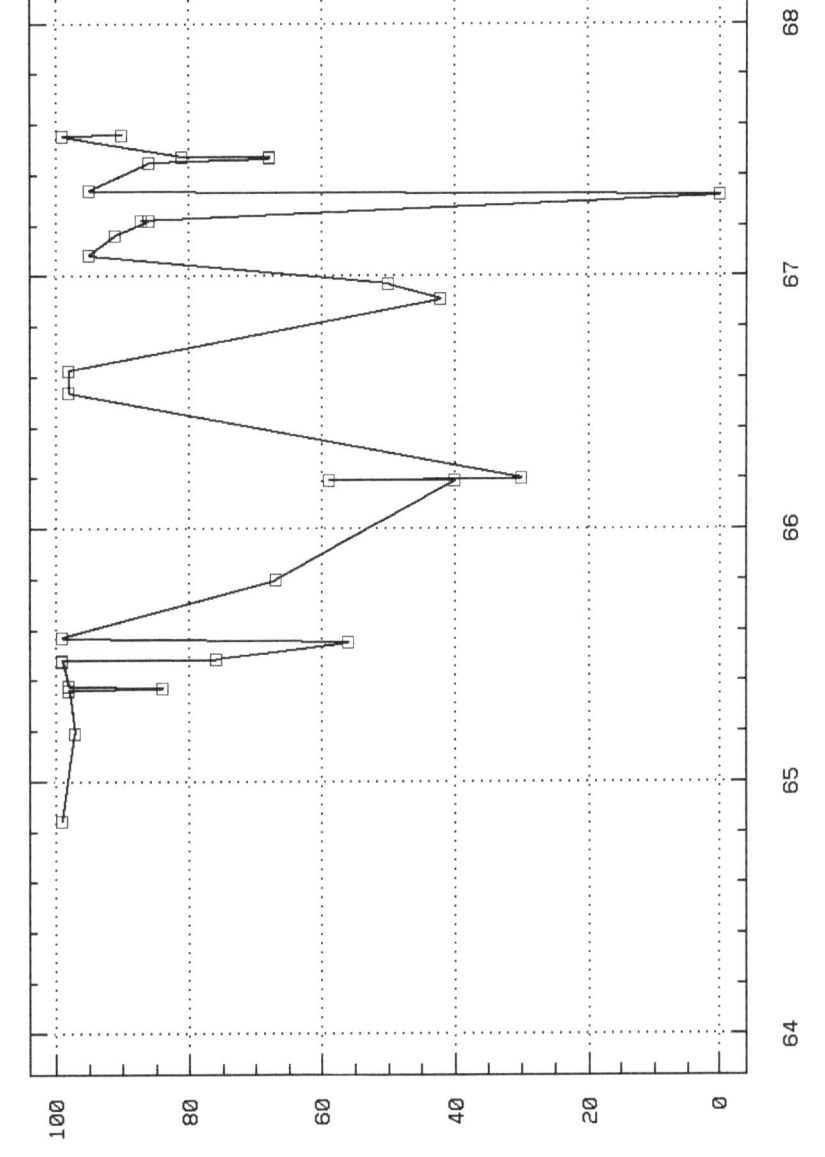

Sample Date (Years)

117

WHITE R NEAR ROCKYFORD SD

Station: BADL0011 Parameter Code: 70342

SUS SED FALL DIA(DISTLD WATER)%FINER TH

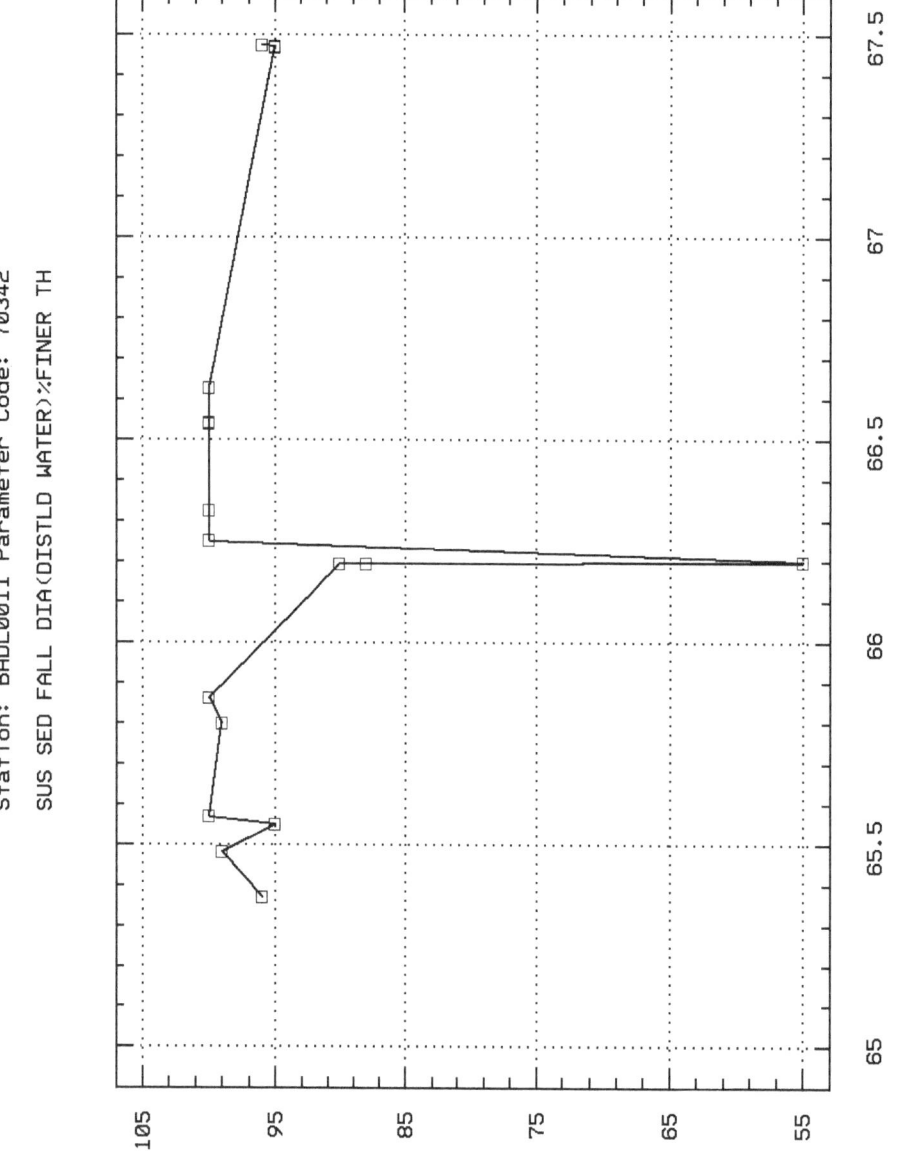

Sample Date (Years)

SUSP SEDPARTSIZE%<.062MM

118

WHITE R NEAR ROCKYFORD SD

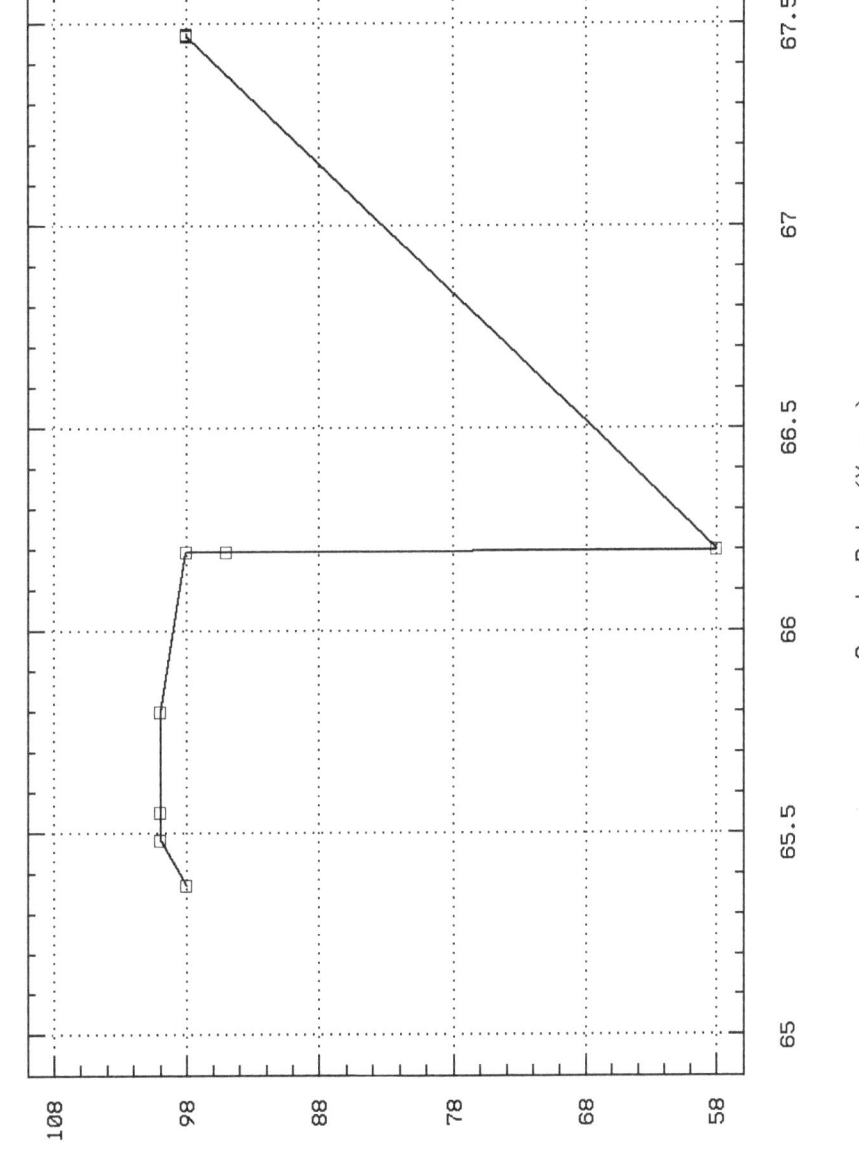

Station: BADL0011 Parameter Code: 70343

SUS SED FALL DIA(DISTLD WATER)%FINER TH

Sample Date (Years)

SUSP SEDPARTSIZE%.125MM

WHITE R NEAR ROCKYFORD SD

119

Station: BADL0011 Parameter Code: 71851
NITRATE NITROGEN, DISSOLVED (MG/L AS NO

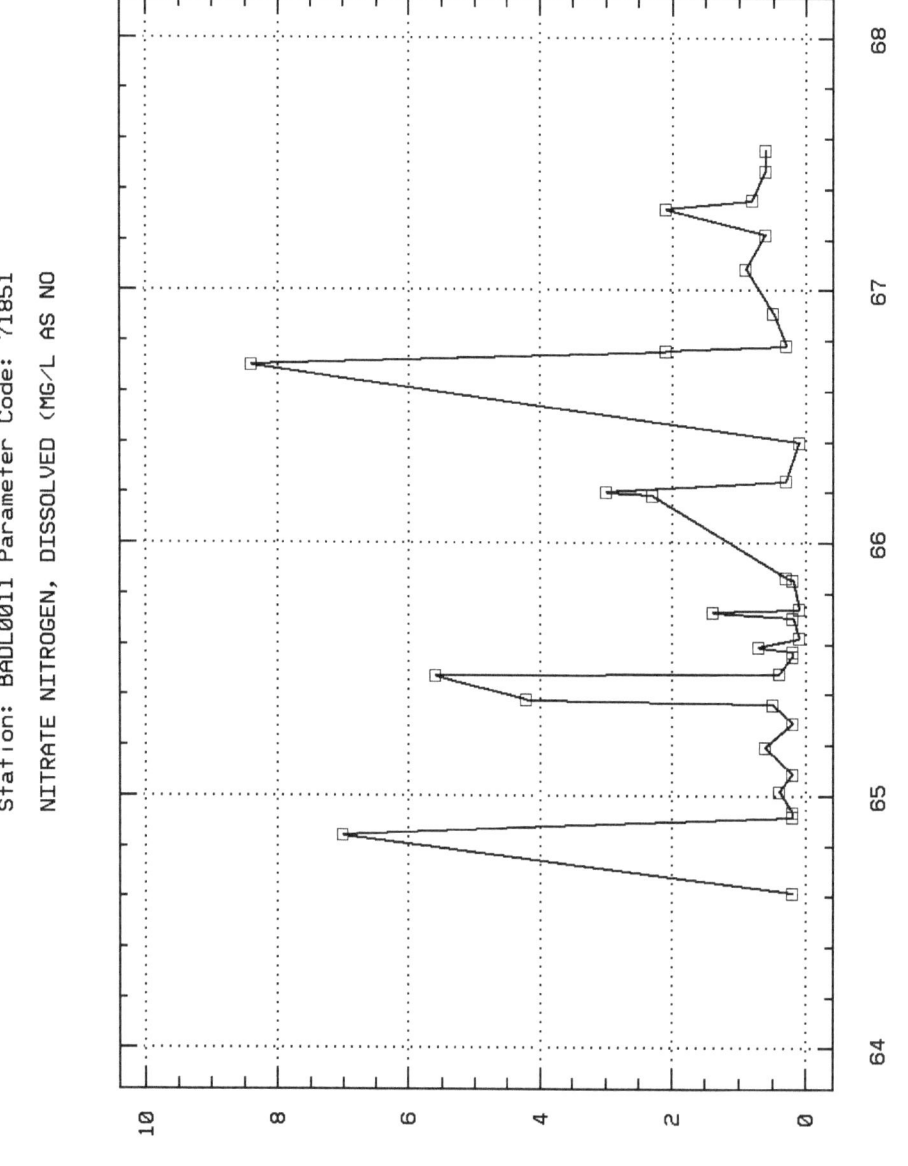

Sample Date (Years)

NITRATE DISS-NO3 MG/L

120

WHITE R NEAR ROCKYFORD SD

Station: BADL0011 Parameter Code: 71885

IRON (UG/L AS FE)

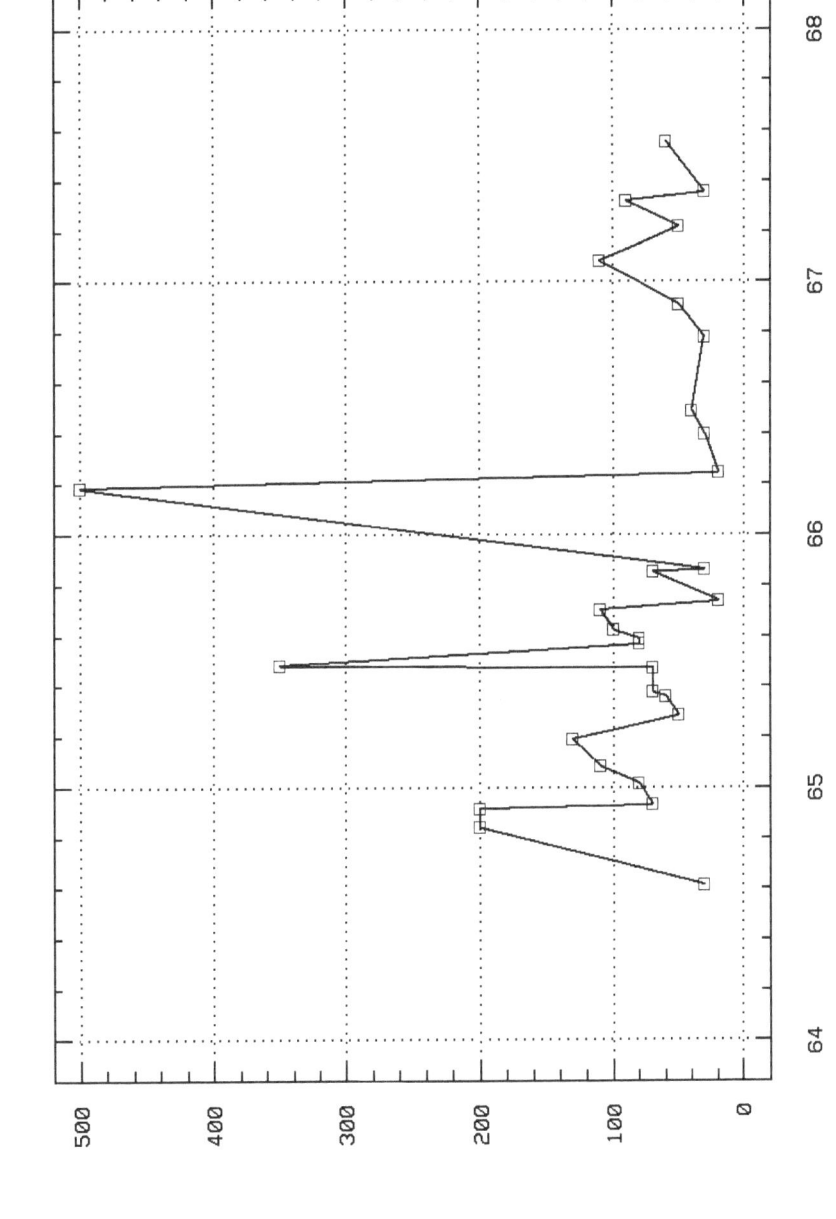

Sample Date (Years)

IRON   FE   UG/L

121

WHITE R NEAR ROCKYFORD SD

Station: BADL0011 Parameter Code: 80154

SUSP. SEDIMENT CONCENTRATION-EVAP. AT 1

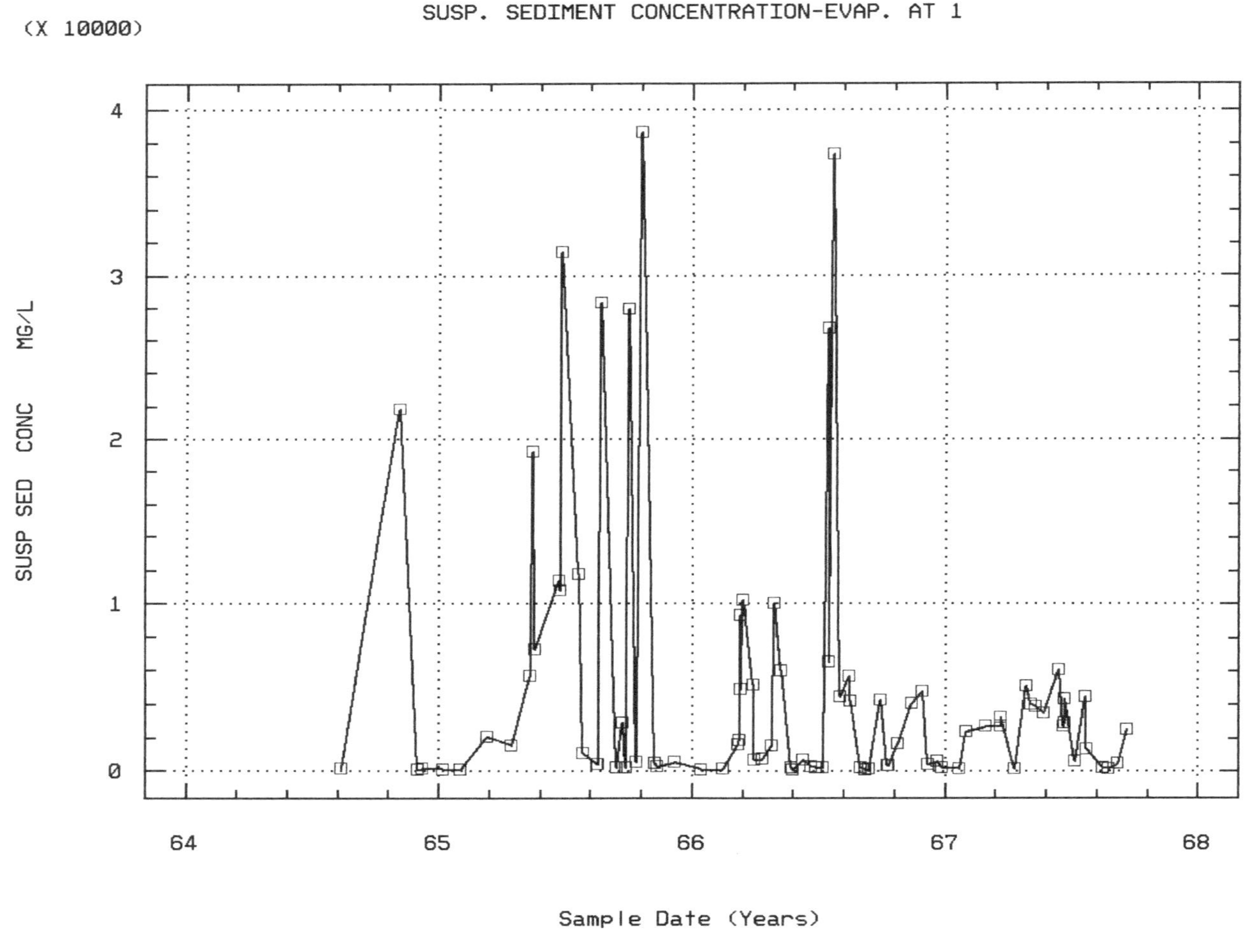
(X 10000)
SUSP SED CONC MG/L
4
3
2
1
0
64
65
66
67
68
Sample Date (Years)

WHITE R NEAR ROCKYFORD SD

Station: BADL0011 Parameter Code: 80155

SUSPENDED SEDIMENT DISCHARGE (TONS/DAY)

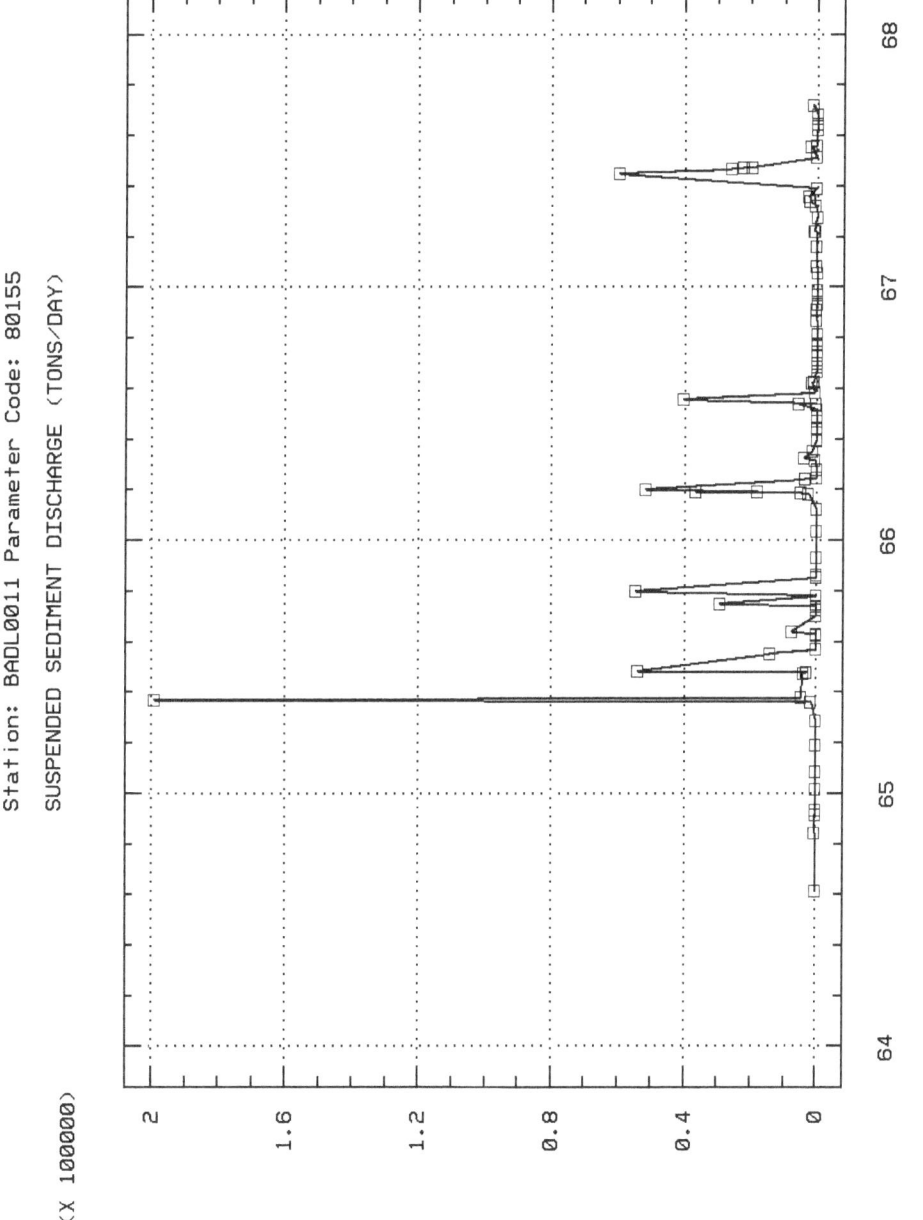

Sample Date (Years)

SUSP SEDDISCHARGTONS/DAY

(X 100000)

WHITE R NEAR ROCKYFORD SD

123

Station Inventory for Station: BADL0012

NPS Station ID: BADL0012
Location: 43N44W35BAD
Station Type: /TYPA/AMBNT/SPRING
RMI-Indexes:
RMI-Miles:
HUC: 10140201
Major Basin:
Minor Basin:
RF1 Index: 10140201
RF3 Index: 10120109103000 00
Description:

LAT/LON: 43 654448/-102 526948

Agency: 112WRD
FIPS State/County: 46113 SOUTH DAKOTA/SHANNON
STORET Station ID(s) 433916102313701
Within Park Boundary: No

Depth of Water: 0
Elevation: 0

Aquifer:
Water Body Id:
ECO Region:
Distance from RF1: 7 60
Distance from RF3: 0 25

Date Created: 01/16/79

On/Off RF1:
On/Off RF3:

Parameter Inventory for Station: BADL0012

Parameter		Period of Record	Obs	Median	Mean	Maximum	Minimum	Variance	Std Dev	10th	25th	75th	90th
00095	SPECIFIC CONDUCTANCE (UMHOS/CM @ 25C)	11/30/62-11/30/62	1	487	487	487	487	0	0	**	**	**	**
00400	PH (STANDARD UNITS)	11/30/62-11/30/62	1	7 5	7 5	7 5	7 5	0	0	**	**	**	**
00400	CONVERTED PH (STANDARD UNITS)	11/30/62-11/30/62	1	7 5	7 5	7 5	7 5	0	0	**	**	**	**
00400	MICRO EQUIVALENTS/LITER OF H+ COMPUTED FROM PH	11/30/62-11/30/62	1	0 032	0 032	0 032	0 032	0	0	**	**	**	**
00405	CARBON DIOXIDE (MG/L AS CO2)	11/30/62-11/30/62	1	14	14	14	14	0	0	**	**	**	**
00410	ALKALINITY, TOTAL (MG/L AS CACO3)	11/30/62-11/30/62	1	235	235	235	235	0	0	**	**	**	**
00440	BICARBONATE ION (MG/L AS HCO3)	11/30/62-11/30/62	1	286	286	286	286	0	0	**	**	**	**
00445	CARBONATE ION (MG/L AS CO3)	11/30/62-11/30/62	1	0	0	0	0	0	0	**	**	**	**
00618	NITRATE NITROGEN, DISSOLVED (MG/L AS N)	11/30/62-11/30/62	1	0 07	0 07	0 07	0 07	0	0	**	**	**	**
00900	HARDNESS, TOTAL (MG/L AS CACO3)	11/30/62-11/30/62	1	142	142	142	142	0	0	**	**	**	**
00902	HARDNESS, NON-CARBONATE (MG/L AS CACO3)	11/30/62-11/30/62	1	0	0	0	0	0	0	**	**	**	**
00915	CALCIUM, DISSOLVED (MG/L AS CA)	11/30/62-11/30/62	1	51	51	51	51	0	0	**	**	**	**
00925	MAGNESIUM, DISSOLVED (MG/L AS MG)	11/30/62-11/30/62	1	3 6	3 6	3 6	3 6	0	0	**	**	**	**
00930	SODIUM, DISSOLVED (MG/L AS NA)	11/30/62-11/30/62	1	55	55	55	55	0	0	**	**	**	**
00931	SODIUM ADSORPTION RATIO	11/30/62-11/30/62	1	2	2	2	2	0	0	**	**	**	**
00932	SODIUM, PERCENT	11/30/62-11/30/62	1	44	44	44	44	0	0	**	**	**	**
00935	POTASSIUM, DISSOLVED (MG/L AS K)	11/30/62-11/30/62	1	6	6	6	6	0	0	**	**	**	**
00940	CHLORIDE, TOTAL IN WATER MG/L	11/30/62-11/30/62	1	1	1	1	1	0	0	**	**	**	**
00945	SULFATE, TOTAL (MG/L AS SO4)	11/30/62-11/30/62	1	27	27	27	27	0	0	**	**	**	**
00950	FLUORIDE, DISSOLVED (MG/L AS F)	11/30/62-11/30/62	1	0 3	0 3	0 3	0 3	0	0	**	**	**	**
00955	SILICA, DISSOLVED (MG/L AS SI02)	11/30/62-11/30/62	1	33	33	33	33	0	0	**	**	**	**
01020	BORON, DISSOLVED (UG/L AS B)	11/30/62-11/30/62	1	50	50	50	50	0	0	**	**	**	**
70300	RESIDUE, TOTAL FILTRABLE (DRIED AT 180C) MG/L	11/30/62-11/30/62	1	322	322	322	322	0	0	**	**	**	**
70301	SOLIDS, DISSOLVED-SUM OF CONSTITUENTS (MG/L)	11/30/62-11/30/62	1	319	319	319	319	0	0	**	**	**	**
71851	NITRATE NITROGEN, DISSOLVED (MG/L AS NO3)	11/30/62-11/30/62	1	0 3	0 3	0 3	0 3	0	0	**	**	**	**
71883	MANGANESE, TOTAL ELEMENTAL (UG/L AS MN)	11/30/62-11/30/62	1	0	0	0	0	0	0	**	**	**	**
71885	IRON (UG/L AS FE)	11/30/62-11/30/62	1	70	70	70	70	0	0	**	**	**	**

** - Less than 9 observations   ## - Computed with 50% or more of the total observations as values that were half the detection limit   p - Has a corresponding time series plot

EPA Water Quality Criteria Analysis for Station: BADL0012

Parameter			Std Type	Std Value	Total Obs	Exceed Standard	Prop Exceeding	------10/01-1/31------			------2/01-4/14------			------4/15-6/30------			------7/01-9/30------		
								Obs	Exceed	Prop	Obs	Exceed	Prop	Obs	Exceed	Prop	Obs	Exceed	Prop
00400	PH		Other-Hi Lim	9	1	0	0 00	1	0	0 00									
			Other-Lo Lim	6 5	1	0	0 00	1	0	0 00									

& - Below detection limit observations, for which half the detection limit exceeded the criterion, were excluded from the criterion comparison for this parameter

EPA Water Quality Criteria Analysis for Station: BADL0012

Parameter	Std Type	Std Value	Total Obs	Exceed Standard	Prop Exceeding	10/01-1/31 Obs	Exceed	Prop	2/01-4/14 Obs	Exceed	Prop	4/15-6/30 Obs	Exceed	Prop	7/01-9/30 Obs	Exceed	Prop
00618 NITRATE NITROGEN, DISSOLVED AS N	Drinking Water	10	1	0	0.00	1	0	0.00									
00940 CHLORIDE,TOTAL IN WATER	Fresh Acute	860	1	0	0.00	1	0	0.00									
00945 SULFATE, TOTAL (AS SO4)	Drinking Water	250	1	0	0.00	1	0	0.00									
00950 FLUORIDE, DISSOLVED AS F	Drinking Water	250	1	0	0.00	1	0	0.00									
71851 NITRATE NITROGEN, DISSOLVED (AS NO3)	Drinking Water	44	1	0	0.00	1	0	0.00									

& - Below detection limit observations, for which half the detection limit exceeded the criterion comparison for this parameter

125

Station Inventory for Station: BADL0013

NPS Station ID: BADL0013
Location: 41N44W31CBCC
Station Type: /TYPA/AMBNT/SPRING
RMI-Indexes:
RMI-Miles:
HUC: 10140201
Major Basin:
Minor Basin:
RF1 Index: 101402011
RF3 Index: 1014020102820000
Description:

LAT/LON: 43.481392/-102.615282

Depth of Water: 0
Elevation: 0

RF1 Mile Point: 0.000
RF3 Mile Point: 0.00

Agency: 112WRD
FIPS State/County: 46113 SOUTH DAKOTA/SHANNON
STORET Station ID(s): 432853102365501
Within Park Boundary: No

Aquifer:
Water Body Id:
ECO Region:
Distance from RF1: 2.90
Distance from RF3: 0.06

Date Created: 05/01/93

On/Off RF1:
On/Off RF3:

Parameter Inventory for Station: BADL0013

Parameter		Period of Record	Obs	Median	Mean	Maximum	Minimum	Variance	Std Dev	10th	25th	75th	90th
00010	TEMPERATURE, WATER (DEGREES CENTIGRADE)	09/23/92-06/21/95	4	15.75	15.025	18.6	10	13.629	3.692	**	**	**	**
00020	TEMPERATURE, AIR (DEGREES CENTIGRADE)	09/23/92-06/21/95	4	22.5	20.5	28	9	65.667	8.103	**	**	**	**
00025	BAROMETRIC PRESSURE (MM OF HG)	09/23/92-09/23/92	1	685	685	685	685	0	0	**	**	**	**
00061	FLOW, STREAM, INSTANTANEOUS CFS	09/23/92-06/21/95	4	0.01	0.01	0.01	0.01	0	0	**	**	**	**
00095	SPECIFIC CONDUCTANCE (UMHOS/CM @ 25C)	09/23/92-06/21/95	4	494	494.5	497	493	3	1.732	**	**	**	**
00300	OXYGEN, DISSOLVED MG/L	07/26/94-06/21/95	2	5.75	5.75	5.8	5.7	0.005	0.071	**	**	**	**
00400	PH (STANDARD UNITS)	09/23/92-06/21/95	4	7.43	7.395	7.62	7.1	0.052	0.228	**	**	**	**
00400	CONVERTED PH (STANDARD UNITS)	09/23/92-06/21/95	4	7.421	7.348	7.62	7.1	0.055	0.235	**	**	**	**
00400	MICRO EQUIVALENTS/LITER OF H+ COMPUTED FROM PH	09/23/92-06/21/95	4	0.038	0.045	0.079	0.024	0.001	0.025	**	**	**	**
00403	PH, LAB, STANDARD UNITS SU	09/23/92-06/21/95	4	7.45	7.5	7.7	7.4	0.02	0.141	**	**	**	**
00403	CONVERTED PH, LAB, STANDARD UNITS	09/23/92-06/21/95	4	7.447	7.484	7.7	7.4	0.02	0.143	**	**	**	**
00403	MICRO EQUIVALENTS/LITER OF H+ COMPUTED FROM PH	09/23/92-06/21/95	4	0.036	0.033	0.04	0.02	0	0.009	**	**	**	**
00410	ALKALINITY, TOTAL (MG/L AS CACO3)	07/26/94-07/26/94	1	223	223	223	223	0	0	**	**	**	**
00440	BICARBONATE ION (MG/L AS HCO3)	07/26/94-07/26/94	1	272	272	272	272	0	0	**	**	**	**
00453	BICARBONATE,WATER,DISS,INCR TIT,FIELD,AS HCO3,MG/L	06/21/95-06/21/95	1	263	263	263	263	0	0	**	**	**	**
00608	NITROGEN, AMMONIA, DISSOLVED (MG/L AS N)	07/26/94-06/21/95	3 ##	0.01	0.013	0.02	0.008	0	0.007	**	**	**	**
00613	NITRITE NITROGEN, DISSOLVED (MG/L AS N)	04/27/95-06/21/95	2 ##	0.005	0.005	0.005	0.005	0	0	**	**	**	**
00623	NITROGEN, KJELDAHL, DISSOLVED (MG/L AS N)	07/26/94-06/21/95	3 ##	0.1	0.1	0.1	0.1	0	0	**	**	**	**
00631	NITRITE PLUS NITRATE, DISS 1 DET (MG/L AS N)	07/26/94-06/21/95	3	5	3.667	5	1	5.333	2.309	**	**	**	**
00666	PHOSPHORUS, DISSOLVED (MG/L AS P)	04/27/95-06/21/95	2	0.075	0.065	0.08	0.04	0	0.022	**	**	**	**
00671	PHOSPHORUS, DISSOLVED ORTHOPHOSPHATE (MG/L AS P)	04/27/95-06/21/95	2	0.07	0.07	0.07	0.07	0	0	**	**	**	**
00915	CALCIUM, DISSOLVED (MG/L AS CA)	09/23/92-06/21/95	4	45	44.25	45	42	2.25	1.5	**	**	**	**
00925	MAGNESIUM, DISSOLVED (MG/L AS MG)	09/23/92-06/21/95	4	2.3	2.275	2.3	2.2	0.002	0.05	**	**	**	**
00930	SODIUM, DISSOLVED (MG/L AS NA)	09/23/92-06/21/95	4	61	60.75	61	60	0.25	0.5	**	**	**	**
00935	POTASSIUM, DISSOLVED (MG/L AS K)	09/23/92-06/21/95	4	6.6	6.65	7.1	6.3	0.117	0.342	**	**	**	**
00940	CHLORIDE,TOTAL IN WATER MG/L	09/23/92-06/21/95	4	2	2	2	2	0	0	**	**	**	**
00945	SULFATE, TOTAL (MG/L AS SO4)	09/23/92-06/21/95	4	18	18	19	17	0.667	0.816	**	**	**	**
00950	FLUORIDE, DISSOLVED (MG/L AS F)	09/23/92-06/21/95	4	0.25	0.25	0.3	0.2	0.003	0.058	**	**	**	**
00955	SILICA, DISSOLVED (MG/L AS SI02)	09/23/92-06/21/95	4	41	40.25	41	38	2.25	1.5	**	**	**	**
01000	ARSENIC, DISSOLVED (UG/L AS AS)	09/23/92-06/21/95	4	4.5	4.5	5	4	0.333	0.577	**	**	**	**
01005	BARIUM, DISSOLVED (UG/L AS BA)	09/23/92-06/21/95	4	150	147.5	150	140	25	5	**	**	**	**
01010	BERYLLIUM, DISSOLVED (UG/L AS BE)	09/23/92-06/21/95	4 ##	0.25	0.25	0.25	0.25	0	0	**	**	**	**
01020	BORON, DISSOLVED (UG/L AS B)	09/23/92-06/21/95	4	80	77.5	80	70	25	5	**	**	**	**
01025	CADMIUM, DISSOLVED (UG/L AS CD)	09/23/92-06/21/95	4 ##	0.5	0.625	1	0.5	0.063	0.25	**	**	**	**
01027	CADMIUM, TOTAL (UG/L AS CD)	09/23/92-06/21/95	3 ##	0.5	0.5	0.5	0.5	0	0	**	**	**	**
01030	CHROMIUM, DISSOLVED (UG/L AS CR)	09/23/92-06/21/95	4 ##	2.5	2.5	2.5	2.5	0	0	**	**	**	**
01035	COBALT, DISSOLVED (UG/L AS CO)	09/23/92-06/21/95	4 ##	1.5	1.5	1.5	1.5	0	0	**	**	**	**
01040	COPPER, DISSOLVED (UG/L AS CU)	09/23/92-06/21/95	4 ##	5	5	5	5	0	0	**	**	**	**

** - Less than 9 observations   ## - Computed with 50% or more of the total observations as values that were half the detection limit   p - Has a corresponding time series plot

Parameter Inventory for Station: BADL0013

Parameter		Period of Record	Obs	Median	Mean	Maximum	Minimum	Variance	Std Dev	10th	25th	75th	90th
01046	IRON, DISSOLVED (UG/L AS FE)	09/23/92-06/21/95	4 ##	3.25	3.25	5	1.5	4.083	2.021	**	**	**	**
01049	LEAD, DISSOLVED (UG/L AS PB)	09/23/92-06/21/95	4 ##	7.5	7.5	7.5	5	8.333	2.887	**	**	**	**
01056	MANGANESE, DISSOLVED (UG/L AS MN)	09/23/92-06/21/95	4 ##	0.75	1	2	0.5	0.5	0.707	**	**	**	**
01057	THALLIUM, DISSOLVED (UG/L AS TL)	04/27/95-06/21/95	2 ##	0.25	0.25	0.25	0.25	0	0	**	**	**	**
01060	MOLYBDENUM, DISSOLVED (UG/L AS MO)	09/23/92-06/21/95	2 ##	5	5	5	5	0	0	**	**	**	**
01065	NICKEL, DISSOLVED (UG/L AS NI)	09/23/92-06/21/95	4 ##	5	5	5	5	0	0	**	**	**	**
01075	SILVER, DISSOLVED (UG/L AS AG)	09/23/92-06/21/95	4 ##	0.5	0.875	2	0.5	0.563	0.75	**	**	**	**
01080	STRONTIUM, DISSOLVED (UG/L AS SR)	09/23/92-06/21/95	4	180	177.5	180	170	25	5	**	**	**	**
01085	VANADIUM, DISSOLVED (UG/L AS V)	09/23/92-06/21/95	4	3	4	7	3	4	2	**	**	**	**
01090	ZINC, DISSOLVED (UG/L AS ZN)	09/23/92-06/21/95	4 ##	2.25	3	6	1.5	4.5	2.121	**	**	**	**
01130	LITHIUM, DISSOLVED (UG/L AS LI)	09/23/92-06/21/95	4	27.5	27.25	31	23	10.917	3.304	**	**	**	**
01145	SELENIUM, DISSOLVED (UG/L AS SE)	09/23/92-06/21/95	4	2.5	2.5	3	2	0.333	0.577	**	**	**	**
03515	BETA, DISSOLVED GROSS, AS CS-137_PC/L	07/26/94-07/26/94	1	10	10	10	10			**	**	**	**
04126	ALPHA, DISSOLVED, WATER (AS TH-230) PCI/L	07/26/94-07/26/94	1	5.2	5.2	5.2	5.2	0	0	**	**	**	**
09510	RADIUM 226, DISSOLVED, PLANCHET COUNT	07/26/94-07/26/94	1	0.2	0.2	0.2	0.2	0	0	**	**	**	**
22703	URANIUM, NATURAL, DISSOLVED	07/26/94-07/26/94	1	71	71	71	71	0	0	**	**	**	**
31501	COLIFORM,TOT.MEMBRANE FILTER,IMMED M-ENDO MED,35C	07/26/94-04/27/95	1	1600	1600	1600	1600			**	**	**	**
31501	LOG COLIFORM,TOT.MEMBRANE FILTER,IMMED M-ENDO MED,3	07/26/94-07/26/94	1	3.204	3.204	3.204	3.204	0	0	**	**	**	**
31501	GEOMETRIC MEAN = 1600												
31503	GM COLIFORM,TOT.MEMBR FILTER,DELAYED,M-ENDO MED,35 C	06/21/95-06/21/95	1	2	2	2	2	0	0		**		**
31503	LOG COLIFORM,TOT,MEMBR FILTER,DELAYED,M-ENDO MED,3	06/21/95-06/21/95	1	0.301	0.301	0.301	0.301	0	0	**	**	**	**
31503	GEOMETRIC MEAN = 2												
31625	FECAL COLIFORM, MF,M-FC, 0.7 UM	07/26/94-06/21/95	3 ##	2	1.667	2	1	0.333	0.577	**	**	**	**
31625	LOG FECAL COLIFORM, MF,M-FC, 0.7 UM	07/26/94-06/21/95	3 ##	0.301	0.201	0.301	0	0.03	0.174	**	**	**	**
31625	GEOMETRIC MEAN = 1.587												
31633	E COLI,THERMOTOL,MF,M-TEC,IN SITU UREASE #/100ML	07/26/94-06/21/95	3 ##	2	2.167	2.5	2	0.083	0.289	**	**	**	**
31633	LOG E COLI,THERMOTOL,MF,M-TEC,IN SITU UREASE #/100	07/26/94-06/21/95	3 ##	0.301	0.333	0.398	0.301	0.003	0.056	**	**	**	**
31633	GEOMETRIC MEAN = 2.154												
31673	FECAL STREPTOCOCCI, MBR FILT,KF AGAR,35C,48HR	07/26/94-06/21/95	3	14	28	68	2	1236	35.157	**	**	**	**
31673	LOG FECAL STREPTOCOCCI, MBR FILT,KF AGAR,35C,48HR	07/26/94-06/21/95	3	1.146	1.093	1.833	0.301	0.588	0.767	**	**	**	**
31673	GEOMETRIC MEAN = 12.394												
39086	ALKALINITY,WATER,DISS,INCR TIT,FIELD,AS CACO3,MG/L	06/21/95-06/21/95	1	216	216	216	216	0	0	**	**	**	**
70300	RESIDUE,TOTAL FILTRABLE (DRIED AT 180C),MG/L	09/23/92-06/21/95	4 ##	327	326	340	310	205.333	14.329	**	**	**	**
71890	MERCURY, DISSOLVED (UG/L AS HG)	09/23/92-06/21/95	4 ##	0.05	0.05	0.05	0.05	0	0	**	**	**	**
75986	ALPHA GROSS,1 SIGMA PRC EST AS NAT U,DISS,WTR UG/L	07/26/94-07/26/94	1	4.02	4.02	4.02	4.02	0	0	**	**	**	**
75987	ALPHA GROSS,DISS,1 SIGMA PRC EST AS TH230,WTR PC/L	07/26/94-07/26/94	1	2.9	2.9	2.9	2.9	0	0	**	**	**	**
75988	BETA GROSS,DISS,1 SIGMA PRC EST AS SR90/Y90 PC/L	07/26/94-07/26/94	1	1.39	1.39	1.39	1.39	0	0	**	**	**	**
75989	BETA GROSS,1 SIGMA PRC EST AS CS-137,DISS,WTR PC/L	07/26/94-07/26/94	1	1.86	1.86	1.86	1.86	0	0	**	**	**	**
75990	URANIUM,NATURAL,1 SIGMA PRC EST,DISS,WATER UG/L	07/26/94-07/26/94	1	2.2	2.2	2.2	2.2	0	0	**	**	**	**
76001	RADIUM 226,1 SIGMA PRC EST,DISSOLVED,WATER PC/L	07/26/94-07/26/94	1	0.151	0.151	0.151	0.151	0	0	**	**	**	**
80030	ALPHA,DISSOLVED GROSS,AS URANIUM-NATURAL,UG/L	07/26/94-07/26/94	1	7.3	7.3	7.3	7.3	0	0	**	**	**	**
80050	BETA,DISSOLVED GROSS,AS SR-Y-90, PC/L	07/26/94-07/26/94	1	7.5	7.5	7.5	7.5	0	0	**	**	**	**

** - Less than 9 observations   ## - Computed with 50% or more of the total observations as values that were half the detection limit   p - Has a corresponding time series plot

EPA Water Quality Criteria Analysis for Station: BADL0013

Parameter		Std Type	Std Value	Total Obs	Exceed Standard	Prop Exceeding	10/01-1/31 Obs	Exceed	Prop	2/01-4/14 Obs	Exceed	Prop	4/15-6/30 Obs	Exceed	Prop	7/01-9/30 Obs	Exceed	Prop
00300	OXYGEN, DISSOLVED	Other-Lo Lim	4	2	0	0.00							1	0	0.00	1	0	0.00
00400	PH	Other-Hi Lim	9	4	0	0.00							2	0	0.00	2	0	0.00
		Other-Lo Lim	6.5	4	0	0.00							2	0	0.00	2	0	0.00
00403	PH, LAB	Other-Hi Lim	9	4	0	0.00							2	0	0.00	2	0	0.00
		Other-Lo Lim	6.5	4	0	0.00							2	0	0.00	2	0	0.00
00613	NITRITE NITROGEN, DISSOLVED AS N	Drinking Water	1	2	0	0.00							1	0	0.00	1	0	0.00
00631	NITRITE PLUS NITRATE, DISS 1 DET	Drinking Water	10	3	0	0.00				1	0	0.00				2	0	0.00
00940	CHLORIDE,TOTAL IN WATER	Fresh Acute	860	4	0	0.00							2	0	0.00	2	0	0.00
00945	SULFATE, TOTAL (AS SO4)	Drinking Water	250	4	0	0.00							2	0	0.00	2	0	0.00
00950	FLUORIDE, DISSOLVED AS F	Drinking Water	4	4	0	0.00							2	0	0.00	2	0	0.00

& - Below detection limit observations, for which half the detection limit exceeded the criterion, were excluded from the criterion comparison for this parameter

127

EPA Water Quality Criteria Analysis for Station: BADL0013

Parameter	Std Type	Std Value	Total Obs	Exceed Standard	Prop Exceeding	10/01-1/31 Obs	10/01-1/31 Exceed	10/01-1/31 Prop	2/01-4/14 Obs	2/01-4/14 Exceed	2/01-4/14 Prop	4/15-6/30 Obs	4/15-6/30 Exceed	4/15-6/30 Prop	7/01-9/30 Obs	7/01-9/30 Exceed	7/01-9/30 Prop
01000 ARSENIC, DISSOLVED	Fresh Acute	360	4	0	0.00							2	0	0.00	2	0	0.00
	Drinking Water	50	4	0	0.00							2	0	0.00	2	0	0.00
01005 BARIUM, DISSOLVED	Drinking Water	2000	4	0	0.00							2	0	0.00	2	0	0.00
01010 BERYLLIUM, DISSOLVED	Fresh Acute	130	4	0	0.00							2	0	0.00	2	0	0.00
	Drinking Water	4	4	0	0.00							2	0	0.00	2	0	0.00
01025 CADMIUM, DISSOLVED	Fresh Acute	3.9	4	0	0.00							2	0	0.00	2	0	0.00
	Drinking Water	5	4	0	0.00							2	0	0.00	2	0	0.00
01027 CADMIUM, TOTAL	Fresh Acute	3.9	3	0	0.00							2	0	0.00	1	0	0.00
	Drinking Water	5	3	0	0.00							2	0	0.00	1	0	0.00
01030 CHROMIUM, DISSOLVED	Drinking Water	100	4	0	0.00							2	0	0.00	2	0	0.00
01040 COPPER, DISSOLVED	Fresh Acute	18	4	0	0.00							2	0	0.00	2	0	0.00
	Drinking Water	1300	4	0	0.00							2	0	0.00	2	0	0.00
01049 LEAD, DISSOLVED	Fresh Acute	82	4	0	0.00							2	0	0.00	2	0	0.00
	Drinking Water	15	4	0	0.00							2	0	0.00	2	0	0.00
01057 THALLIUM, DISSOLVED	Fresh Acute	1400	2	0	0.00							2	0	0.00			
	Drinking Water	2	2	0	0.00							2	0	0.00			
01065 NICKEL, DISSOLVED	Fresh Acute	1400	4	0	0.00							2	0	0.00	2	0	0.00
	Drinking Water	100	4	0	0.00							2	0	0.00	2	0	0.00
01075 SILVER, DISSOLVED	Fresh Acute	4.1	4	0	0.00							2	0	0.00	2	0	0.00
	Drinking Water	100	4	0	0.00							2	0	0.00	2	0	0.00
01090 ZINC, DISSOLVED	Fresh Acute	120	4	0	0.00							2	0	0.00	2	0	0.00
	Drinking Water	5000	4	0	0.00							2	0	0.00	2	0	0.00
01145 SELENIUM, DISSOLVED	Fresh Acute	20	4	0	0.00							2	0	0.00	2	0	0.00
	Drinking Water	50	4	0	0.00							2	0	0.00	2	0	0.00
22703 URANIUM, NATURAL DISSOLVED	Drinking Water	20	1	0	0.00										1	0	0.00
31501 COLIFORM, TOTAL, MEMBRANE FILTER, IMMED	Other-Hi Lim	1000	1	1	1.00										1	1	1.00
31503 COLIFORM,TOT,MEMBRANE FILTR,DELAY M-END	Other-Hi Lim	1000	1	1	1.00										1	1	1.00
31625 FECAL COLIFORM, MF	Other-Hi Lim	200	3	0	0.00							2	0	0.00	1	0	0.00
71890 MERCURY, DISSOLVED	Fresh Acute	2.4	4	0	0.00							2	0	0.00	2	0	0.00
	Drinking Water	2	4	0	0.00							2	0	0.00	2	0	0.00

&- Below detection limit observations, for which half the detection limit exceeded the criterion, were excluded from the criterion comparison for this parameter

Station Inventory for Station: BADL0014

NPS Station ID: BADL0014
Location: 41N45W24ACCA
Station Type: /TYPA/AMBNT/SPRING
RMI-Miles:
HUC: 10140201
Major Basin:
Minor Basin:
RF1 Index: 101402010
RF3 Index: 10140201028200 00
Description:

LAT/LON: 43 514448/-102 623338

Agency: 112WRD
FIPS State/County: 46113 SOUTH DAKOTA/SHANNON
STORET Station ID(s): 433052102372401
Within Park Boundary: Yes

Depth of Water: 0
Elevation: 0

Aquifer:
Water Body Id:
ECO Region:
Distance from RF1: 2 90
Distance from RF3: 0 06

RF1 Mile Point: 0 000
RF3 Mile Point: 0 00

Date Created: 05/01/93

On/Off RF1:
On/Off RF3:

Parameter Inventory for Station: BADL0014

Parameter		Period of Record	Obs	Median	Mean	Maximum	Minimum	Variance	Std Dev	10th	25th	75th	90th
00010	TEMPERATURE, WATER (DEGREES CENTIGRADE)	09/22/92-08/28/95	4	15 9	15 6	19 6	11	13 113	3 621	**	**	**	**
00020	TEMPERATURE, AIR (DEGREES CENTIGRADE)	09/22/92-08/28/95	4	24 25	22 375	26	15	25 229	5 023	**	**	**	**
00025	BAROMETRIC PRESSURE (MM OF HG)	09/22/92-09/22/92	1	686	686	686	686	0	0	**	**	**	**
00061	FLOW, STREAM, INSTANTANEOUS CFS	09/22/92-08/28/95	4	0 06	0 068	0 1	0 05	0 001	0 024	**	**	**	**
00065	STAGE, STREAM (FEET)	07/14/94-08/28/95	2	0 24	0 24	0 26	0 22	0 001	0 028	**	**	**	**
00095	SPECIFIC CONDUCTANCE (UMHOS/CM @ 25C)	09/22/92-08/28/95	4	373 5	374 25	390	360	158 917	12 606	**	**	**	**
00300	OXYGEN, DISSOLVED MG/L	07/12/94-08/28/95	2	7 75	7 75	7 9	7 6	0 045	0 212	**	**	**	**
00400	PH (STANDARD UNITS)	09/22/92-08/28/95	4	8 04	8 07	8 4	7 8	0 064	0 252	**	**	**	**
00400	CONVERTED PH (STANDARD UNITS)	09/22/92-08/28/95	4	8 036	8 019	8 4	7 8	0 067	0 259	**	**	**	**
00400	MICRO EQUIVALENTS/LITER OF H+ COMPUTED FROM PH	09/22/92-08/28/95	4	0 009	0 01	0 016	0 004	0	0 005	**	**	**	**
00403	PH, LAB, STANDARD UNITS SU	09/22/92-08/28/95	4	7 9	7 925	8 1	7 8	0 022	0 15	**	**	**	**
00403	CONVERTED PH, LAB, STANDARD UNITS	09/22/92-08/28/95	4	7 889	7 906	8 1	7 8	0 023	0 152	**	**	**	**
00403	MICRO EQUIVALENTS/LITER OF H+ COMPUTED FROM PH	09/22/92-08/28/95	4	0 013	0 012	0 016	0 008	0	0 004	**	**	**	**
00419	ALKALINITY,CARBONATE,INCREMENTAL TITR FIELD MG/L	08/28/95-08/28/95	1	172	172	172	172	0	0	**	**	**	**
00450	BICARBONATE,INCREMENTAL TITRATION (HCO3) FIELDMG/L	08/28/95-08/28/95	1	210	210	210	210	0	0	**	**	**	**
00608	NITROGEN, AMMONIA, DISSOLVED (MG/L AS N)	07/12/94-08/28/95	3	0 01	0 023	0 05	0 008	0 001	0 024	**	**	**	**
00613	NITRITE NITROGEN, DISSOLVED (MG/L AS N)	04/24/95-08/28/95	2	0 025	0 025	0 03	0 02	0	0 007	**	**	**	**
00623	NITROGEN, KJELDAHL, DISSOLVED (MG/L AS N)	07/12/94-08/28/95	3	0 2	0 2	0 3	0 1	0 01	0 1	**	**	**	**
00631	NITRITE PLUS NITRATE, DISS 1 DET (MG/L AS N)	07/12/94-08/28/95	3	2 4	2 4	2 7	2 1	0 09	0 3	**	**	**	**
00666	PHOSPHORUS, DISSOLVED (MG/L AS P)	04/24/95-08/28/95	3	0 02	0 023	0 029	0 02	0	0 005	**	**	**	**
00671	PHOSPHORUS, DISSOLVED ORTHOPHOSPHATE (MG/L AS P)	04/24/95-08/28/95	2	0 02	0 02	0 02	0 02	0	0	**	**	**	**
00915	CALCIUM, DISSOLVED (MG/L AS CA)	09/22/92-08/28/95	4	54	55	58	54	4	2	**	**	**	**
00925	MAGNESIUM, DISSOLVED (MG/L AS MG)	09/22/92-08/28/95	4	7 25	7 25	7 5	7	0 057	0 238	**	**	**	**
00930	SODIUM, DISSOLVED (MG/L AS NA)	09/22/92-08/28/95	4	9 5	9 525	10	9 1	0 143	0 377	**	**	**	**
00935	POTASSIUM, DISSOLVED (MG/L AS K)	09/22/92-08/28/95	4	6 45	6 4	6 6	6 1	0 047	0 216	**	**	**	**
00940	CHLORIDE,TOTAL IN WATER MG/L	09/22/92-08/28/95	4	0 95	0 85	1	0 5	0 057	0 238	**	**	**	**
00945	SULFATE, TOTAL (MG/L AS SO4)	09/22/92-08/28/95	4	8	8	9	7	0 667	0 816	**	**	**	**
00950	FLUORIDE, DISSOLVED (MG/L AS F)	09/22/92-08/28/95	4	0 3	0 275	0 4	0 1	0 016	0 126	**	**	**	**
00955	SILICA, DISSOLVED (MG/L AS SI02)	09/22/92-08/28/95	4	49 5	49 25	51	47	2 917	1 708	**	**	**	**
01000	ARSENIC, DISSOLVED (UG/L AS AS)	09/22/92-08/28/95	4	8 5	8 25	9	7	0 917	0 957	**	**	**	**
01005	BARIUM, DISSOLVED (UG/L AS BA)	09/22/92-08/28/95	4 ##	270	275	300	260	366 667	19 149	**	**	**	**
01010	BERYLLIUM, DISSOLVED (UG/L AS BE)	09/22/92-08/28/95	4 ##	0 25	0 25	0 25	0 25	0	0	**	**	**	**
01020	BORON, DISSOLVED (UG/L AS B)	09/22/92-08/28/95	4 ##	30	30	40	20	66 667	8 165	**	**	**	**
01025	CADMIUM, DISSOLVED (UG/L AS CD)	09/22/92-08/28/95	4 ##	0 5	0 875	2	0 5	0 563	075	**	**	**	**
01027	CADMIUM, TOTAL (UG/L AS CD)	09/22/92-08/28/95	3 ##	0 5	0 5	0 5	0 5	0	0	**	**	**	**
01030	CHROMIUM, DISSOLVED (UG/L AS CR)	09/22/92-08/28/95	4 ##	2 5	2 5	2 5	2 5	0	0	**	**	**	**
01035	COBALT, DISSOLVED (UG/L AS CO)	09/22/92-08/28/95	4 ##	1 5	2 125	4	1 5	1 563	1 25	**	**	**	**
01040	COPPER, DISSOLVED (UG/L AS CU)	09/22/92-08/28/95	4 ##	5	5	5	5	0	0	**	**	**	**

** - Less than 9 observations   ## - Computed with 50% or more of the total observations as values that were half the detection limit   p - Has a corresponding time series plot

129

Parameter Inventory for Station: BADL0014

Parameter		Period of Record	Obs	Median	Mean	Maximum	Minimum	Variance	Std Dev	10th	25th	75th	90th
01046	IRON, DISSOLVED (UG/L AS FE)	09/22/92-08/28/95	4	60 5	60	100	19	1431 333	37 833	**	**	**	**
01049	LEAD, DISSOLVED (UG/L AS PB)	09/22/92-08/28/95	4 ##	5	11 25	30	5	156 25	12 5	**	**	**	**
01056	MANGANESE, DISSOLVED (UG/L AS MN)	09/22/92-08/28/95	4	16	17 5	34	4	207	14 387	**	**	**	**
01057	THALLIUM, DISSOLVED (UG/L AS TL)	04/24/95-08/28/95	2 ##	0 25	0 25	0 25	0 25	0	0	**	**	**	**
01060	MOLYBDENUM, DISSOLVED (UG/L AS MO)	09/22/92-08/28/95	4 ##	5	8 75	20	5	56 25	7 5	**	**	**	**
01065	NICKEL, DISSOLVED (UG/L AS NI)	09/22/92-08/28/95	4 ##	5	5	5	5	0	0	**	**	**	**
01075	SILVER, DISSOLVED (UG/L AS AG)	09/22/92-08/28/95	4 ##	0 5	0 625	1	0 5	0 063	0 25	**	**	**	**
01080	STRONTIUM, DISSOLVED (UG/L AS SR)	09/22/92-08/28/95	4	340	340	350	330	133 333	11 547	**	**	**	**
01085	VANADIUM, DISSOLVED (UG/L AS V)	09/22/92-08/28/95	4	7 5	6 75	9	3	6 917	2 63	**	**	**	**
01090	ZINC, DISSOLVED (UG/L AS ZN)	09/22/92-08/28/95	4 ##	2 75	3	5	1 5	3 167	1 78	**	**	**	**
01130	LITHIUM, DISSOLVED (UG/L AS LI)	09/22/92-08/28/95	4	14	12 75	17	6	24 917	4 992	**	**	**	**
01145	SELENIUM, DISSOLVED (UG/L AS SE)	09/22/92-08/28/95	4	1 5	1 5	2	1	0 333	0 577	**	**	**	**
03515	BETA, DISSOLVED GROSS, AS CS-137, PC/L	07/12/94-07/12/94	1	7 5	7 5	7 5	7 5	0	0	**	**	**	**
04126	ALPHA, DISSOLVED, WATER (AS TH-230) PCI/L	07/12/94-07/12/94	1	17	17	17	17	0	0	**	**	**	**
09510	RADIUM 226, DISSOLVED, PLANCHET COUNT	07/12/94-07/12/94	1 ##	0 05	0 05	0 05	0 05	0	0	**	**	**	**
22703	URANIUM, NATURAL, DISSOLVED	07/12/94-07/12/94	1	2 1	2 1	2 1	2 1	0	0	**	**	**	**
31501	COLIFORM,TOT,MEMBRANE FILTER,IMMED M-ENDO MED,35C	07/12/94-07/12/94	1	82	82	82	82	0	0	**	**	**	**
31501	GM COLIFORM,TOT,MEMBRANE FILTER,IMMED M-ENDO MED,3	GEOMETRIC MEAN =			82								
31504	COLIFORM,TOT,MEMBR FILTER,IMMED,LES ENDO AGAR,35C	08/28/95-08/28/95	1	1200	1200	1200	1200	0	0	**	**	**	**
31504	LOG COLIFORM,TOT,MEMBR FILTER,IMMED,LES ENDO AGAR,	08/28/95-08/28/95	1	3 079	3 079	3 079	3 079	0	0	**	**	**	**
31504	GM COLIFORM,TOT,MEMBR FILTER,IMMED,LES ENDO AGAR,3	GEOMETRIC MEAN =			1200								
31625	FECAL COLIFORM, MF,M-FC, 0 7 UM	07/12/94-08/28/95	1	1200	1200	1200	1200	0	0	**	**	**	**
31625	LOG FECAL COLIFORM, MF,M-FC, 0 7 UM	07/12/94-08/28/95	1	3 079	3 079	3 079	3 079	0	0	**	**	**	**
31625	GM FECAL COLIFORM, MF,M-FC, 0 7 UM	GEOMETRIC MEAN =			1200								
31633	E COLI,THERMOTOL,MF,M-TEC,IN SITU UREASE #/100ML	07/12/94-08/28/95	2	825	825	1600	50	1201250	1096 016	**	**	**	**
31633	LOG E COLI,THERMOTOL,MF M-TEC,IN SITU UREASE #/100	07/12/94-08/28/95	2	2 452	2 452	3 204	1 699	1 133	1 064	**	**	**	**
31633	GM E COLI,THERMOTOL,MF M-TEC,IN SITU UREASE #/100M	GEOMETRIC MEAN =			282 843								
31673	FECAL STREPTOCOCCI, MBR FILT,KF AGAR,35C,48HR	07/12/94-08/28/95	2	2070	2070	2180	1960	24200	155 563	**	**	**	**
31673	LOG FECAL STREPTOCOCCI, MBR FILT,KF AGAR,35C,48HR	07/12/94-08/28/95	2	3 315	3 315	3 338	3 292	0 001	0 033	**	**	**	**
31673	GM FECAL STREPTOCOCCI, MBR FILT,KF AGAR,35C,48HR	GEOMETRIC MEAN =			2067 075								
70300	RESIDUE,TOTAL FILTRABLE (DRIED AT 180C),MG/L	09/22/92-08/28/95	4	249 5	249 25	256	242	52 917	7 274	**	**	**	**
71890	MERCURY, DISSOLVED (UG/L AS HG)	09/22/92-08/28/95	4 ##	0 05	0 05	0 05	0 05	0	0	**	**	**	**
75986	ALPHA GROSS,1 SIGMA PRC EST AS NAT U,DISS,WTR UG/L	07/12/94-07/12/94	1	2 12	2 12	2 12	2 12	0	0	**	**	**	**
75987	ALPHA GROSS,DISS,1 SIGMA PRC EST AS TH230,WTR PC/L	07/12/94-07/12/94	1	1 51	1 51	1 51	1 51	0	0	**	**	**	**
75988	BETA GROSS,DISS,1 SIGMA PRC EST AS SR90/Y90 PC/L	07/12/94-07/12/94	1	1 14	1 14	1 14	1 14	0	0	**	**	**	**
75989	BETA GROSS,1 SIGMA PRC EST AS CS-137,DISS,WTR PC/L	07/12/94-07/12/94	1	1 5	1 5	1 5	1 5	0	0	**	**	**	**
75990	URANIUM,NATURAL,1 SIGMA PRC EST,DISS,WATER UG/L	07/12/94-07/12/94	1	0 7	0 7	0 7	0 7	0	0	**	**	**	**
76001	RADIUM 226,1 SIGMA PRC EST,DISSOLVED,WATER PC/L	07/12/94-07/12/94	1	0 134	0 134	0 134	0 134	0	0	**	**	**	**
80030	ALPHA,DISSOLVED GROSS,AS URANIUM-NATURAL,UG/L	07/12/94-07/12/94	1	2 4	2 4	2 4	2 4	0	0	**	**	**	**
80050	BETA,DISSOLVED GROSS,AS SR-Y-90, PC/L	07/12/94-07/12/94	1	5 7	5 7	5 7	5 7	0	0	**	**	**	**

** - Less than 9 observations   ## - Computed with 50% or more of the total observations as values that were half the detection limit   p - Has a corresponding time series plot

EPA Water Quality Criteria Analysis for Station: BADL0014

Parameter		Std Type	Std Value	Total Obs	Exceed Standard	Prop Exceeding	10/01-1/31 Obs	Exceed	Prop	2/01-4/14 Obs	Exceed	Prop	4/15-6/30 Obs	Exceed	Prop	7/01-9/30 Obs	Exceed	Prop
00300	OXYGEN, DISSOLVED	Other-Lo Lim	4	2	0	0.00				1		0.00	0	0	0.00	1	0	0.00
00400	PH	Other-Hi Lim	9	4	0	0.00				1		0.00	0	0	0.00	3	0	0.00
00400	PH	Other-Lo Lim	6 5	4	0	0.00				1		0.00	0	0	0.00	3	0	0.00
00403	PH, LAB	Other-Hi Lim	9	4	0	0.00				1		0.00	0	0	0.00	3	0	0.00
00403	PH, LAB	Other-Lo Lim	6 5	4	0	0.00				1		0.00	0	0	0.00	3	0	0.00
00613	NITRITE NITROGEN, DISSOLVED AS N	Drinking Water	1	2	0	0.00				1		0.00	0	0	0.00	1	0	0.00
00631	NITRITE PLUS NITRATE, DISS 1 DET	Drinking Water	10	3	0	0.00				1		0.00	0	0	0.00	2	0	0.00
00940	CHLORIDE,TOTAL IN WATER	Fresh Acute	860	4	0	0.00				1		0.00	0	0	0.00	3	0	0.00
00945	SULFATE, TOTAL (AS SO4)	Drinking Water	250	4	0	0.00				1		0.00	0	0	0.00	3	0	0.00
00950	FLUORIDE, DISSOLVED AS F	Drinking Water	250	4	0	0.00				1		0.00	0	0	0.00	3	0	0.00
01000	ARSENIC, DISSOLVED	Fresh Acute	360	4	0	0.00				1		0.00	0	0	0.00	3	0	0.00
01000	ARSENIC, DISSOLVED	Drinking Water	50	4	0	0.00				1		0.00	0	0	0.00	3	0	0.00

& - Below detection limit observations, for which half the detection limit exceeded the criterion, were excluded from the criterion comparison for this parameter

EPA Water Quality Criteria Analysis for Station: BADL0014

Parameter		Std Type	Std Value	Total Obs	Exceed Standard	Prop Exceeding	10/01-1/31 Obs	Exceed	Prop	2/01-4/14 Obs	Exceed	Prop	4/15-6/30 Obs	Exceed	Prop	7/01-9/30 Obs	Exceed	Prop
01005	BARIUM, DISSOLVED	Drinking Water	2000	4	0	0.00							1	0	0.00	3	0	0.00
01010	BERYLLIUM, DISSOLVED	Fresh Acute	130	4	0	0.00							1	0	0.00	3	0	0.00
01025	CADMIUM, DISSOLVED	Drinking Water	4	4	0	0.00							1	0	0.00	3	0	0.00
		Fresh Acute	3.9	4	0	0.00							1	0	0.00	3	0	0.00
01027	CADMIUM, TOTAL	Drinking Water	5	3	0	0.00							1	0	0.00	2	0	0.00
		Fresh Acute	3.9	3	0	0.00							1	0	0.00	2	0	0.00
01030	CHROMIUM, DISSOLVED	Drinking Water	5	3	0	0.00							1	0	0.00	3	0	0.00
01040	COPPER, DISSOLVED	Drinking Water	100	4	0	0.00							1	0	0.00	3	0	0.00
		Fresh Acute	18	4	0	0.00							1	0	0.00	3	0	0.00
01049	LEAD, DISSOLVED	Drinking Water	1300	4	0	0.00							1	0	0.00	3	0	0.00
		Fresh Acute	82	4	1	0.25							1	1	1.00	3	0	0.00
01057	THALLIUM, DISSOLVED	Drinking Water	15	4	0	0.00							1	0	0.00	3	0	0.00
		Fresh Acute	1400	2	0	0.00							1	0	0.00	1	0	0.00
01065	NICKEL, DISSOLVED	Fresh Acute	2	2	0	0.00							1	0	0.00	1	0	0.00
		Drinking Water	1400	4	0	0.00							1	0	0.00	3	0	0.00
01075	SILVER, DISSOLVED	Fresh Acute	100	4	0	0.00							1	0	0.00	3	0	0.00
		Drinking Water	4.1	4	0	0.00							1	0	0.00	3	0	0.00
01090	ZINC, DISSOLVED	Drinking Water	100	4	0	0.00							1	0	0.00	3	0	0.00
		Fresh Acute	120	4	0	0.00							1	0	0.00	3	0	0.00
01145	SELENIUM, DISSOLVED	Drinking Water	5000	4	0	0.00							1	0	0.00	3	0	0.00
		Fresh Acute	20	4	0	0.00							1	0	0.00	3	0	0.00
22703	URANIUM, NATURAL DISSOLVED	Drinking Water	50	1	0	0.00										1	0	0.00
31501	COLIFORM, TOTAL, MEMBRANE FILTER,IMMED	Drinking Water	20	1	0	0.00										1	0	0.00
31504	COLIFORM,TOT,MEMBRANE FILTR,IMMED LES-E	Other-Hi Lim	1000	1	1	1.00										1	1	1.00
31625	FECAL COLIFORM, MF	Other-Hi Lim	1000	1	1	1.00										1	1	1.00
		Other-Hi Lim	200	1	1	1.00							1	0	0.00			
71890	MERCURY, DISSOLVED	Fresh Acute	2.4	4	0	0.00										3	0	0.00
		Drinking Water	2	4	0	0.00							1	0	0.00	3	0	0.00

& - Below detection limit observations, for which half the detection limit exceeded the criterion, were excluded from the criterion comparison for this parameter

131

Station Inventory for Station: BADL0015

NPS Station ID: BADL0015
Location: 41N45W23BBCB
Station Type: /TYPA/AMBNT/SPRING
RMI-Indexes:
RMI-Miles:
HUC: 10140201
Major Basin:
Minor Basin:
RF1 Index: 10140201
RF3 Index: 10140201028200 00
Description:

LAT/LON: 43.518059/-102.653892

Agency: 112WRD
FIPS State/County: 46113 SOUTH DAKOTA/SHANNON
STORET Station ID(s): 43310510239 1401
Within Park Boundary: No

Depth of Water: 0
Elevation: 0

Aquifer:
Water Body Id:
ECO Region:
Distance from RF1: 2.90
Distance from RF3: 0.06

RF1 Mile Point: 0.000
RF3 Mile Point: 0.00

Date Created: 05/01/93

On/Off RF1:
On/Off RF3:

Parameter Inventory for Station: BADL0015

Parameter	Period of Record	Obs	Median	Mean	Maximum	Minimum	Variance	Std Dev	10th	25th	75th	90th
00010 TEMPERATURE, WATER (DEGREES CENTIGRADE)	09/25/92-08/28/95	4	16.7	16.8	19.7	14.1	7.647	2.765	**	**	**	**
00020 TEMPERATURE, AIR (DEGREES CENTIGRADE)	09/25/92-08/28/95	4	17.65	18.825	24	16	12.789	3.576	**	**	**	**
00025 BAROMETRIC PRESSURE (MM OF HG)	09/25/92-09/25/92	1	673	673	673	673			**	**	**	**
00061 FLOW, STREAM, INSTANTANEOUS CFS	09/25/92-08/28/95	4	0.085	0.105	0.2	0.05	0.004	0.067	**	**	**	**
00065 STAGE, STREAM (FEET)	07/13/94-08/28/95	2	0.32	0.32	0.36	0.28	0.003	0.057	**	**	**	**
00095 SPECIFIC CONDUCTANCE (UMHOS/CM @ 25C)	07/13/94-08/28/95	3	486	523.667	608	477	5354.333	73.173	**	**	**	**
00300 OXYGEN, DISSOLVED MG/L	08/28/95-08/28/95	1	6.7	6.7	6.7	6.7		0.21	**	**	**	**
00400 PH (STANDARD UNITS)	09/25/92-08/28/95	4	7.915	7.993	8.3	7.84	0.044	0.21	**	**	**	**
00400 CONVERTED PH (STANDARD UNITS)	09/25/92-08/28/95	4	7.914	7.96	8.3	7.84	0.046	0.213	**	**	**	**
00400 MICRO EQUIVALENTS/LITER OF H+ COMPUTED FROM PH	09/25/92-08/28/95	4	0.012	0.011	0.014	0.005	0	0.004	**	**	**	**
00403 PH, LAB, STANDARD UNITS SU	09/25/92-08/28/95	4	7.9	7.9	8	7.8	0.007	0.082	**	**	**	**
00403 CONVERTED PH, LAB, STANDARD UNITS	09/25/92-08/28/95	4	7.9	7.894	8	7.8	0.007	0.082	**	**	**	**
00403 MICRO EQUIVALENTS/LITER OF H+ COMPUTED FROM PH	09/25/92-08/28/95	4	0.013	0.013	0.016	0.01	0	0.002	**	**	**	**
00419 ALKALINITY CARBONATE, INCREMENTAL TITR FIELD MG/L	08/28/95-08/28/95	1	278	278	278	278	0	0	**	**	**	**
00447 CARBONATE, INCREMENTAL TITRATION, (CO3) FIELD MG/L	08/28/95-08/28/95	1	0	0	0	0	0	0	**	**	**	**
00450 BICARBONATE, INCREMENTAL TITRATION, (HCO3) FIELD MG/L	08/28/95-08/28/95	1	339	339	339	339	0	0	**	**	**	**
00608 NITROGEN, AMMONIA, DISSOLVED (MG/L AS N)	07/13/94-08/28/95	3 ##	0.01	0.019	0.04	0.008	0	0.018	**	**	**	**
00613 NITRITE NITROGEN, DISSOLVED (MG/L AS N)	05/11/95-08/28/95	2 ##	0.005	0.005	0.005	0.005	0	0	**	**	**	**
00623 NITROGEN, KJELDAHL, DISSOLVED (MG/L AS N)	07/13/94-08/28/95	3	0.3	0.367	0.6	0.2	0.043	0.208	**	**	**	**
00631 NITRITE PLUS NITRATE, DISS 1 DET (MG/L AS N)	07/13/94-08/28/95	3	0.09	0.14	0.3	0.03	0.02	0.142	**	**	**	**
00666 PHOSPHORUS, DISSOLVED (MG/L AS P)	07/13/94-08/28/95	3	0.01	0.009	0.012	0.005	0	0.004	**	**	**	**
00671 PHOSPHORUS, DISSOLVED ORTHOPHOSPHATE (MG/L AS P)	05/11/95-08/28/95	2 ##	0.008	0.008	0.01	0.005	0	0.004	**	**	**	**
00915 CALCIUM, DISSOLVED (MG/L AS CA)	09/25/92-08/28/95	4	60.5	60.25	62	58	2.917	1.708	**	**	**	**
00925 MAGNESIUM, DISSOLVED (MG/L AS MG)	09/25/92-08/28/95	4	8.15	8.25	9.1	7.6	0.523	0.723	**	**	**	**
00930 SODIUM, DISSOLVED (MG/L AS NA)	09/25/92-08/28/95	4	36.5	38	49	30	67.333	8.206	**	**	**	**
00935 POTASSIUM, DISSOLVED (MG/L AS K)	09/25/92-08/28/95	4	9.9	10.775	15	8.3	9.309	3.051	**	**	**	**
00940 CHLORIDE, TOTAL IN WATER MG/L	09/25/92-08/28/95	4	3.5	5.75	14	2	30.917	5.56	**	**	**	**
00945 SULFATE, TOTAL (MG/L AS SO4)	09/25/92-08/28/95	4	7.5	7.25	8	6	0.917	0.957	**	**	**	**
00950 FLUORIDE, DISSOLVED (MG/L AS F)	09/25/92-08/28/95	4	0.35	0.35	0.4	0.3	0.003	0.058	**	**	**	**
00955 SILICA, DISSOLVED (MG/L AS SIO2)	09/25/92-08/28/95	4	47.5	45.25	51	35	50.917	7.136	**	**	**	**
01000 ARSENIC, DISSOLVED (UG/L AS AS)	09/25/92-08/28/95	4	8.5	9.5	13	8	5.667	2.38	**	**	**	**
01005 BARIUM, DISSOLVED (UG/L AS BA)	09/25/92-08/28/95	4	260	267.5	310	240	958.333	30.957	**	**	**	**
01010 BERYLLIUM, DISSOLVED (UG/L AS BE)	09/25/92-08/28/95	4 ##	0.25	0.25	0.25	0.25	0	0	**	**	**	**
01020 BORON, DISSOLVED (UG/L AS B)	09/25/92-08/28/95	4	75	80	100	70	200	14.142	**	**	**	**
01025 CADMIUM, DISSOLVED (UG/L AS CD)	09/25/92-08/28/95	4	0.5	0.875	2	0.5	0.563	0.75	**	**	**	**
01027 CADMIUM, TOTAL (UG/L AS CD)	09/25/92-08/28/95	4 ##	0.5	0.5	0.5	0.5	0	0	**	**	**	**
01030 CHROMIUM, DISSOLVED (UG/L AS CR)	09/25/92-08/28/95	4 ##	2.5	2.5	2.5	2.5	0	0	**	**	**	**
01035 COBALT, DISSOLVED (UG/L AS CO)	09/25/92-08/28/95	4 ##	1.5	1.5	1.5	1.5	0	0	**	**	**	**

** - Less than 9 observations   ## - Computed with 50% or more of the total observations as values that were half the detection limit   p - Has a corresponding time series plot

Parameter Inventory for Station: BADL0015

Parameter		Period of Record	Obs	Median	Mean	Maximum	Minimum	Variance	Std Dev	10th	25th	75th	90th
01040	COPPER, DISSOLVED (UG/L AS CU)	09/25/92-08/28/95	4##	5	5	5	5			**	**	**	**
01046	IRON, DISSOLVED (UG/L AS FE)	09/25/92-08/28/95	4	104.5	104	160	47	2288.667	47.84	**	**	**	**
01049	LEAD, DISSOLVED (UG/L AS PB)	09/25/92-08/28/95	4##	5	11.25	30	5	156.25	12.5	**	**	**	**
01056	MANGANESE, DISSOLVED (UG/L AS MN)	09/25/92-08/28/95	4	21.5	23	39	10	163.333	12.78	**	**	**	**
01057	THALLIUM, DISSOLVED (UG/L AS TL)	05/11/95-08/28/95	2##	0.25	0.25	0.25	0.25			**	**	**	**
01060	MOLYBDENUM, DISSOLVED (UG/L AS MO)	09/25/92-08/28/95	4##	5	6.25	10	5	6.25	2.5	**	**	**	**
01065	NICKEL, DISSOLVED (UG/L AS NI)	09/25/92-08/28/95	4##	5	5	5	5			**	**	**	**
01075	SILVER, DISSOLVED (UG/L AS AG)	09/25/92-08/28/95	4##	0.5	0.625	1	0.5	0.063	0.25	**	**	**	**
01080	STRONTIUM, DISSOLVED (UG/L AS SR)	09/25/92-08/28/95	4	445	445	470	420	433.333	20.817	**	**	**	**
01085	VANADIUM, DISSOLVED (UG/L AS V)	09/25/92-08/28/95	4##	3	2.125	3	0.5	1.563	1.25	**	**	**	**
01090	ZINC, DISSOLVED (UG/L AS ZN)	09/25/92-08/28/95	4	22	22	26	18	16.667	4.082	**	**	**	**
01130	LITHIUM, DISSOLVED (UG/L AS LI)	09/25/92-08/28/95	4##	11	11	11	11	0.063	0.25	**	**	**	**
01145	SELENIUM, DISSOLVED (UG/L AS SE)	09/25/92-08/28/95	4##	0.5	0.625	0.5	0.5	0	0	**	**	**	**
03515	BETA, DISSOLVED GROSS, AS CS-137, PC/L	07/13/94-07/13/94	1	0.05	0.05	0.05	0.05	0	0	**	**	**	**
04126	RADIUM 226, DISSOLVED, WATER (AS TH-230) PCI/L	07/13/94-07/13/94	1##	3	3	3	3	0	0	**	**	**	**
09510	ALPHA, DISSOLVED, WATER (AS TH-230) PCI/L	07/13/94-07/13/94	1	3.3	3.3	3.3	3.3	0	0	**	**	**	**
22703	URANIUM, NATURAL, DISSOLVED	07/13/94-07/13/94	1	3.3	3.3	3.3	3.3	0	0	**	**	**	**
31501	COLIFORM,TOT.MEMBRANE FILTER,IMMED M-ENDO MED,35C	07/13/94-05/11/95	2	136.5	136.5	140	133	24.5	4.95	**	**	**	**
31501	LOG COLIFORM,TOT.MEMBRANE FILTER,IMMED M-ENDO MED,3	07/13/94-05/11/95	2	2.135	2.135	2.146	2.124	0	0.016	**	**	**	**
31501	GM COLIFORM,TOT.MEMBRANE FILTER,IMMED M-ENDO MED,3	GEOMETRIC MEAN =			136.455								
31504	COLIFORM,TOT.MEMBR FILTER,IMMED,LES ENDO AGAR,35C	08/28/95-08/28/95	1	453	453	453	453	0	0	**	**	**	**
31504	LOG COLIFORM,TOT.MEMBR FILTER,IMMED,LES ENDO AGAR,	08/28/95-08/28/95	1	2.656	2.656	2.656	2.656	0	0	**	**	**	**
31504	GM COLIFORM,TOT.MEMBR FILTER,IMMED,LES ENDO AGAR,3	GEOMETRIC MEAN =			453								
31625	FECAL COLIFORM, MF,M-FC, 0.7 UM	07/13/94-08/28/95	3	360	373	547	212	28183	167.878	**	**	**	**
31625	LOG FECAL COLIFORM, MF,M-FC, 0.7 UM	07/13/94-08/28/95	3	2.556	2.54	2.738	2.326	0.043	0.206	**	**	**	**
31625	GM FECAL COLIFORM, MF,M-FC, 0.7 UM	GEOMETRIC MEAN =			346.903								
31633	E.COLI,THERMOTOL.MF,M-TEC,IN SITU UREASE #/100ML	07/13/94-08/28/95	3	105	568.667	1600	1	800440.333	894.673	**	**	**	**
31633	LOG E.COLI,THERMOTOL.MF,M-TEC,IN SITU UREASE #/100	07/13/94-08/28/95	3	2.021	1.742	3.204	0	2.625	1.62	**	**	**	**
31633	GM E.COLI,THERMOTOL.MF,M-TEC,IN SITU UREASE #/100M	GEOMETRIC MEAN =			55.178								
31673	FECAL STREPTOCOCCI,MBR FILT,KF AGAR,35C,48HR	07/13/94-08/28/95	3	392	461	966	25	224941	474.279	**	**	**	**
31673	LOG FECAL STREPTOCOCCI,MBR FILT,KF AGAR,35C,48HR	07/13/94-08/28/95	3	2.593	2.325	2.985	1.398	0.683	0.827	**	**	**	**
31673	GM FECAL STREPTOCOCCI,MBR FILT,KF AGAR,35C,48HR	GEOMETRIC MEAN =			211.544								
70300	RESIDUE,TOTAL FILTRABLE (DRIED AT 180C),MG/L	09/25/92-08/28/95	4	325	336	384	310	1224	34.986	**	**	**	**
71832	HYDROXIDE,INCREMENTAL TITRATION,(OH) FIELD MG/L	08/28/95-08/28/95	4##	0	0	0	0	0	0	**	**	**	**
71890	MERCURY, DISSOLVED (UG/L AS HG)	09/25/92-08/28/95	4##	0.05	0.05	0.05	0.05	0	0	**	**	**	**
75986	ALPHA GROSS,1 SIGMA PRC EST AS NAT U,DISS,WTR UG/L	07/13/94-07/13/94	1	2.74	2.74	2.74	2.74	0	0	**	**	**	**
75987	ALPHA GROSS,DISS,1 SIGMA PRC EST AS TH230,WTR PC/L	07/13/94-07/13/94	1	1.97	1.97	1.97	1.97	0	0	**	**	**	**
75988	BETA GROSS,DISS,1 SIGMA PRC EST AS SR90/Y90 PC/L	07/13/94-07/13/94	1	1.49	1.49	1.49	1.49	0	0	**	**	**	**
75989	BETA GROSS,1 SIGMA PRC EST AS CS-137,DISS,WTR PC/L	07/13/94-07/13/94	1	1.98	1.98	1.98	1.98	0	0	**	**	**	**
75990	URANIUM,NATURAL,1 SIGMA PRC EST,DISS,WATER UG/L	07/13/94-07/13/94	1	11	11	11	11	0	0	**	**	**	**
76001	RADIUM 226,1 SIGMA PRC EST,DISSOLVED,WATER PC/L	07/13/94-07/13/94	1	0.115	0.115	0.115	0.115	0	0	**	**	**	**
80030	ALPHA,DISSOLVED GROSS,AS URANIUM-NATURAL,UG/L	07/13/94-07/13/94	1	4.2	4.2	4.2	4.2	0	0	**	**	**	**
80050	BETA,DISSOLVED GROSS,AS SR-Y-90, PC/L	07/13/94-07/13/94	1	8.4	8.4	8.4	8.4	0	0	**	**	**	**

** - Less than 9 observations   ## - Computed with 50% or more of the total observations as values that were half the detection limit   p - Has a corresponding time series plot

EPA Water Quality Criteria Analysis for Station: BADL0015

Parameter		Std Type	Std Value	Total Obs	Exceed Standard	Prop Exceeding	10/01-1/31 Obs	10/01-1/31 Exceed	10/01-1/31 Prop	2/01-4/14 Obs	2/01-4/14 Exceed	2/01-4/14 Prop	4/15-6/30 Obs	4/15-6/30 Exceed	4/15-6/30 Prop	7/01-9/30 Obs	7/01-9/30 Exceed	7/01-9/30 Prop
00300	OXYGEN, DISSOLVED	Other-Lo Lim	4	1	0	0.00										1	0	0.00
00400	PH	Other-Hi Lim	9	4	0	0.00							1	0	0.00	3	0	0.00
00403	PH, LAB	Other-Lo Lim	6.5	4	0	0.00							1	0	0.00	3	0	0.00
		Other-Hi Lim	9	4	0	0.00							1	0	0.00	3	0	0.00
00613	NITRITE NITROGEN, DISSOLVED AS N	Drinking Water	1	2	0	0.00							1	0	0.00	1	0	0.00
00631	NITRITE PLUS NITRATE, DISS 1 DET	Drinking Water	10	3	0	0.00										3	0	0.00
00940	CHLORIDE, TOTAL IN WATER	Fresh Acute	860	4	0	0.00							1	0	0.00	3	0	0.00
		Drinking Water	250	4	0	0.00							1	0	0.00	3	0	0.00
00945	SULFATE, TOTAL (AS SO4)	Drinking Water	250	4	0	0.00							1	0	0.00	3	0	0.00

&- Below detection limit observations, for which half the detection limit exceeded the criterion, were excluded from the criterion comparison for this parameter

EPA Water Quality Criteria Analysis for Station: BADL0015

Parameter	Std Type	Std Value	Total Obs	Exceed Standard	Prop Exceeding	10/01-1/31 Obs	Exceed	Prop	2/01-4/14 Obs	Exceed	Prop	4/15-6/30 Obs	Exceed	Prop	7/01-9/30 Obs	Exceed	Prop
00950 FLUORIDE, DISSOLVED AS F	Drinking Water	4	4	0	0.00							1	0	0.00	3	0	0.00
01000 ARSENIC, DISSOLVED	Fresh Acute	360	4	0	0.00							1	0	0.00	3	0	0.00
	Drinking Water	50	4	0	0.00							1	0	0.00	3	0	0.00
01005 BARIUM, DISSOLVED	Drinking Water	2000	4	0	0.00							1	0	0.00	3	0	0.00
01010 BERYLLIUM, DISSOLVED	Drinking Water	4	4	0	0.00							1	0	0.00	3	0	0.00
01025 CADMIUM, DISSOLVED	Fresh Acute	3.9	4	0	0.00							1	0	0.00	3	0	0.00
	Drinking Water	5	4	0	0.00							1	0	0.00	3	0	0.00
01027 CADMIUM, TOTAL	Fresh Acute	3.9	3	0	0.00							1	0	0.00	2	0	0.00
	Drinking Water	5	3	0	0.00							1	0	0.00	2	0	0.00
01030 CHROMIUM, DISSOLVED	Drinking Water	100	4	0	0.00							1	0	0.00	3	0	0.00
01040 COPPER, DISSOLVED	Fresh Acute	18	4	0	0.00							1	0	0.00	3	0	0.00
	Drinking Water	1300	4	0	0.00							1	0	0.00	3	0	0.00
01049 LEAD, DISSOLVED	Fresh Acute	82	4	1	0.25							1	1	1.00	3	0	0.00
	Drinking Water	15	4	0	0.00							1	0	0.00	3	0	0.00
01057 THALLIUM, DISSOLVED	Fresh Acute	1400	2	0	0.00							1	0	0.00	1	0	0.00
	Drinking Water	2	2	0	0.00							1	0	0.00	1	0	0.00
01065 NICKEL, DISSOLVED	Fresh Acute	1400	4	0	0.00							1	0	0.00	3	0	0.00
	Drinking Water	100	4	0	0.00							1	0	0.00	3	0	0.00
01075 SILVER, DISSOLVED	Fresh Acute	4.1	4	0	0.00							1	0	0.00	3	0	0.00
	Drinking Water	100	4	0	0.00							1	0	0.00	3	0	0.00
01090 ZINC, DISSOLVED	Fresh Acute	120	4	0	0.00							1	0	0.00	3	0	0.00
	Drinking Water	5000	4	0	0.00							1	0	0.00	3	0	0.00
01145 SELENIUM, DISSOLVED	Fresh Acute	20	4	0	0.00							1	0	0.00	3	0	0.00
	Drinking Water	50	4	0	0.00							1	0	0.00	3	0	0.00
22703 URANIUM, NATURAL DISSOLVED	Drinking Water	20	1	0	0.00							1	0	0.00			
31501 COLIFORM, TOTAL, MEMBRANE FILTER, IMMED	Other-Hi Lim	1000	2	0	0.00										1	0	0.00
31504 COLIFORM,TOT,MEMBRANE FILTR,IMMED LES-E	Other-Hi Lim	1000	1	0	0.00										1	0	0.00
31625 FECAL COLIFORM, MF	Other-Hi Lim	200	3	3	1.00							1	1	1.00	2	2	1.00
71890 MERCURY, DISSOLVED	Fresh Acute	2.4	4	0	0.00							1	0	0.00	3	0	0.00
	Drinking Water	2	4	0	0.00							1	0	0.00	3	0	0.00

& - Below detection limit observations, for which half the detection limit exceeded the criterion, were excluded from the criterion comparison for this parameter

Station Inventory for Station: BADL0016

NPS Station ID: BADL0016
Location: 41N45W27BBAA
Station Type: /TYPA/AMBNT/SPRING
RMI-Indexes:
RMI-Miles:
HUC: 10140201
Major Basin:
Minor Basin:
RF1 Index: 10140201
RF3 Index: 10140210288200 00
Description:

LAT/LON: 43 506115/-102 669449

Agency: 112WRD
FIPS State/County: 46113 SOUTH DAKOTA/SHANNON
STORET Station ID(s): 433022102401001
Within Park Boundary: No

Depth of Water: 0
Elevation: 0

Aquifer:
Water Body Id:
ECO Region:
Distance from RF1: 2 90
Distance from RF3: 0 06

RF1 Mile Point: 0 000
RF3 Mile Point: 0 00

On/Off RF1:
On/Off RF3:

Date Created: 05/01/93

Parameter Inventory for Station: BADL0016

Parameter	Period of Record	Obs	Median	Mean	Maximum	Minimum	Variance	Std Dev	10th	25th	75th	90th
00010 TEMPERATURE, WATER (DEGREES CENTIGRADE)	09/30/92-07/10/95	4	13 25	13 175	14 6	11 6	1 549	1 245	**	**	**	**
00020 TEMPERATURE, AIR (DEGREES CENTIGRADE)	09/30/92-07/10/95	4	17 5	19 5	28	15	37 667	6 137	**	**	**	**
00025 BAROMETRIC PRESSURE (MM OF HG)	09/30/92-09/30/92	1	686	686	686	686	0	0	**	**	**	**
00061 FLOW, STREAM, INSTANTANEOUS CFS	09/30/92-07/10/95	4	0 02	0 018	0 02	0 01	0	0 005	**	**	**	**
00065 STAGE, STREAM (FEET)	07/27/94-07/27/94	1	0 13	0 13	0 13	0 13	0	0	**	**	**	**
00095 SPECIFIC CONDUCTANCE (UMHOS/CM @ 25C)	09/30/92-07/10/95	4	520 5	535 5	582	519	961 667	31 011	**	**	**	**
00300 OXYGEN, DISSOLVED MG/L	07/27/94-07/10/95	2	8 2	8 2	9	7 4	1 28	1 131	**	**	**	**
00400 PH (STANDARD UNITS)	09/30/92-07/10/95	4	7 775	7 8	7 9	7 75	0 005	0 069	**	**	**	**
00400 CONVERTED PH (STANDARD UNITS)	09/30/92-07/10/95	4	7 775	7 796	7 9	7 75	0 005	0 069	**	**	**	**
00400 MICRO EQUIVALENTS/LITER OF H+ COMPUTED FROM PH	09/30/92-07/10/95	4	0 017	0 016	0 018	0 013	0 005	0 002	**	**	**	**
00403 PH, LAB, STANDARD UNITS SU	09/30/92-07/10/95	4	7 85	7 825	8	7 6	0 029	0 171	**	**	**	**
00403 CONVERTED PH, LAB, STANDARD UNITS	09/30/92-07/10/95	4	7 847	7 799	8	7 6	0 03	0 173	**	**	**	**
00403 MICRO EQUIVALENTS/LITER OF H+ COMPUTED FROM PH	09/30/92-07/10/95	4	0 014	0 016	0 025	0 01	0	0 007	**	**	**	**
00410 ALKALINITY, TOTAL (MG/L AS CACO3)	07/27/94-07/27/94	1	204	204	204	204	0	0	**	**	**	**
00440 BICARBONATE ION (MG/L AS HCO3)	07/27/94-07/27/94	1	249	249	249	249	0	0	**	**	**	**
00453 BICARBONATE,WATER,DISS,INCR TIT,FIELD,AS HCO3,MG/L	07/10/95-07/10/95	1	256	256	256	256	0	0	**	**	**	**
00608 NITROGEN, AMMONIA, DISSOLVED (MG/L AS N)	07/27/94-07/10/95	3 ##	0 02	0 016	0 02	0 008	0	0 007	**	**	**	**
00613 NITRITE NITROGEN, DISSOLVED (MG/L AS N)	04/14/95-07/10/95	2 ##	0 008	0 008	0 01	0 005	0	0 004	**	**	**	**
00623 NITROGEN, KJELDAHL, DISSOLVED (MG/L AS N)	07/27/94-07/10/95	3 ##	0 1	0 1	0 1	0 1	0	0	**	**	**	**
00631 NITRITE PLUS NITRATE, DISS 1 DET (MG/L AS N)	07/27/94-07/10/95	3	2 4	2 333	2 5	2 1	0 043	0 208	**	**	**	**
00666 PHOSPHORUS, DISSOLVED (MG/L AS P)	07/27/94-07/10/95	3	0 03	0 033	0 04	0 029	0	0 006	**	**	**	**
00671 PHOSPHORUS, DISSOLVED ORTHOPHOSPHATE (MG/L AS P)	04/14/95-07/10/95	2	0 045	0 045	0 05	0 04	0	0 007	**	**	**	**
00915 CALCIUM, DISSOLVED (MG/L AS CA)	09/30/92-07/10/95	4	34 5	34 5	36	33	1 667	1 291	**	**	**	**
00925 MAGNESIUM, DISSOLVED (MG/L AS MG)	09/30/92-07/10/95	4	5 9	5 9	6	5 8	0 007	0 082	**	**	**	**
00930 SODIUM, DISSOLVED (MG/L AS NA)	09/30/92-07/10/95	4	64 5	64 25	66	62	2 917	1 708	**	**	**	**
00935 POTASSIUM, DISSOLVED (MG/L AS K)	09/30/92-07/10/95	4	11	11	11	11	0	0	**	**	**	**
00940 CHLORIDE,TOTAL IN WATER MG/L	09/30/92-07/10/95	4	9 5	9 75	11	9	0 917	0 957	**	**	**	**
00945 SULFATE, TOTAL (MG/L AS SO4)	09/30/92-07/10/95	4	31 5	31 75	34	30	4 25	2 062	**	**	**	**
00950 FLUORIDE, DISSOLVED (MG/L AS F)	09/30/92-07/10/95	4	0 3	0 3	0 3	0 3	0	0	**	**	**	**
00955 SILICA, DISSOLVED (MG/L AS SI02)	09/30/92-07/10/95	4	51	50 75	51	50	0 25	0 5	**	**	**	**
01000 ARSENIC, DISSOLVED (UG/L AS AS)	09/30/92-07/10/95	4	16	14 75	16	11	6 25	2 5	**	**	**	**
01005 BARIUM, DISSOLVED (UG/L AS BA)	09/30/92-07/10/95	4	70 5	70 75	73	69	2 917	1 708	**	**	**	**
01010 BERYLLIUM, DISSOLVED (UG/L AS BE)	09/30/92-07/10/95	4 ##	0 25	0 25	0 25	0 25	0	0	**	**	**	**
01020 BORON, DISSOLVED (UG/L AS B)	09/30/92-07/10/95	4	145	145	150	140	33 333	5 774	**	**	**	**
01025 CADMIUM, DISSOLVED (UG/L AS CD)	09/30/92-07/10/95	4 ##	0 5	0 5	0 5	0 5	0	0	**	**	**	**
01027 CADMIUM, TOTAL (UG/L AS CD)	09/30/92-07/10/95	3 ##	0 5	0 5	0 5	0 5	0	0	**	**	**	**
01030 CHROMIUM, DISSOLVED (UG/L AS CR)	09/30/92-07/10/95	4 ##	2 5	2 5	2 5	2 5	0	0	**	**	**	**
01035 COBALT, DISSOLVED (UG/L AS CO)	09/30/92-07/10/95	4 ##	1 5	1 5	1 5	1 5	0	0	**	**	**	**

** - Less than 9 observations   ## - Computed with 50% or more of the total observations as values that were half the detection limit   p - Has a corresponding time series plot

135

Parameter Inventory for Station: BADL0016

Parameter		Period of Record	Obs	Median	Mean	Maximum	Minimum	Variance	Std Dev	10th	25th	75th	90th
01040	COPPER, DISSOLVED (UG/L AS CU)	09/30/92-07/10/95	4 ##	5	5	5	5	0	0	**	**	**	**
01046	IRON, DISSOLVED (UG/L AS FE)	09/30/92-07/10/95	4 ##	1.5	3.125	8	1.5	10.563	3.25	**	**	**	**
01049	LEAD, DISSOLVED (UG/L AS PB)	09/30/92-07/10/95	4 ##	5	5	5	5	0	0	**	**	**	**
01056	MANGANESE, DISSOLVED (UG/L AS MN)	09/30/92-07/10/95	4 ##	0.5	0.5	0.5	0.5	0	0	**	**	**	**
01057	THALLIUM, DISSOLVED (UG/L AS TL)	04/14/95-07/10/95	2 ##	0.25	0.25	0.25	0.25	0	0	**	**	**	**
01060	MOLYBDENUM, DISSOLVED (UG/L AS MO)	09/30/92-07/10/95	4 ##	5	5	5	5	0	0	**	**	**	**
01065	NICKEL, DISSOLVED (UG/L AS NI)	09/30/92-07/10/95	4 ##	5	5	5	5	0	0	**	**	**	**
01075	SILVER, DISSOLVED (UG/L AS AG)	09/30/92-07/10/95	4 ##	0.5	0.5	0.5	0.5	0	0	**	**	**	**
01080	STRONTIUM, DISSOLVED (UG/L AS SR)	09/30/92-07/10/95	4	315	315	320	310	33.333	5.774	**	**	**	**
01085	VANADIUM, DISSOLVED (UG/L AS V)	09/30/92-07/10/95	4	11	10.75	12	9	1.583	1.258	**	**	**	**
01090	ZINC, DISSOLVED (UG/L AS ZN)	09/30/92-07/10/95	4 ##	3.25	4.75	11	1.5	20.083	4.481	**	**	**	**
01130	LITHIUM, DISSOLVED (UG/L AS LI)	09/30/92-07/10/95	4 ##	25.5	25.5	27	24	1.667	1.291	**	**	**	**
01145	SELENIUM, DISSOLVED (UG/L AS SE)	04/14/95-07/10/95	2	7	7	10	4	18	4.243	**	**	**	**
03515	BETA, DISSOLVED GROSS, AS CS-137, PC/L	07/27/94-07/27/94	1	14	14	14	14	0	0	**	**	**	**
04126	ALPHA, DISSOLVED, WATER (AS TH-230) PCI/L	07/27/94-07/27/94	1	8.2	8.2	8.2	8.2	0	0	**	**	**	**
09510	RADIUM 226, DISSOLVED, PLANCHET COUNT	07/27/94-07/27/94	1	0.1	0.1	0.1	0.1	0	0	**	**	**	**
22703	URANIUM, NATURAL, DISSOLVED	07/27/94-07/27/94	1	11	11	11	11	0	0	**	**	**	**
31501	COLIFORM,TOT,MEMBRANE FILTER,IMMED M-ENDO MED,35C	07/27/94-04/14/95	1	88	88	88	88	0	0	**	**	**	**
31501	LOG COLIFORM,TOT,MEMBRANE FILTER,IMMED M-ENDO MED,3	07/27/94-04/14/95	1	1.944	1.944	1.944	1.944	0	0	**	**	**	**
31501	GM COLIFORM,TOT,MEMBRANE FILTER,IMMED M-ENDO MED,3	GEOMETRIC MEAN =			88								
31503	COLIFORM,TOT,MEMBR FILTER,DELAYED,M-ENDO MED,35 C	07/10/95-07/10/95	1	45	45	45	45	0	0	**	**	**	**
31503	LOG COLIFORM,TOT,MEMBR FILTER,DELAYED,M-ENDO MED,3	07/10/95-07/10/95	1	1.653	1.653	1.653	1.653	0	0	**	**	**	**
31503	GM COLIFORM,TOT,MEMBR FILTER,DELAYED,M-ENDO MED,35	GEOMETRIC MEAN =			45								
31625	FECAL COLIFORM, MF,M-FC, 0.7 UM	07/27/94-04/14/95	1	111	111	111	111	0	0	**	**	**	**
31625	LOG FECAL COLIFORM, MF,M-FC, 0.7 UM	07/27/94-04/14/95	1	2.045	2.045	2.045	2.045	0	0	**	**	**	**
31625	GM FECAL COLIFORM, MF,M-FC, 0.7 UM	GEOMETRIC MEAN =			111								
31633	E COLI,THERMOTOL,MF,M-TEC,IN SITU UREASE #/100ML	07/27/94-04/14/95	2 ##	1.5	1.5	2	1	0.5	0.707	**	**	**	**
31633	LOG E COLI,THERMOTOL,MF M-TEC,IN SITU UREASE #/100	07/27/94-04/14/95	2 ##	0.151	0.151	0.301	0	0.045	0.213	**	**	**	**
31633	GM E COLI,THERMOTOL,MF M-TEC,IN SITU UREASE #/100M	GEOMETRIC MEAN =			1.414								
31673	FECAL STREPTOCOCCI, MBR FILT, KF AGAR 35C,48HR	07/27/94-04/14/95	1	4200	4200	4200	4200	0	0	**	**	**	**
31673	LOG FECAL STREPTOCOCCI, MBR FILT,KF AGAR,35C,48HR	07/27/94-04/14/95	1	3.623	3.623	3.623	3.623	0	0	**	**	**	**
31673	GM FECAL STREPTOCOCCI, MBR FILT,KF AGAR,35C,48HR	GEOMETRIC MEAN =			4200								
39086	ALKALINITY,WATER,DISS,INCR TIT FIELD,AS CACO3,MG/L	07/10/95-07/10/95	1	210	210	210	210	0	0	**	**	**	**
70300	RESIDUE,TOTAL FILTRABLE (DRIED AT 180C),MG/L	09/30/92-07/10/95	4 ##	335.5	335	348	321	126	11.225	**	**	**	**
71890	MERCURY, DISSOLVED (UG/L AS HG)	09/30/92-07/10/95	4 ##	0.05	0.05	0.05	0.05	0	0	**	**	**	**
75986	ALPHA GROSS,1 SIGMA PRC EST AS NAT U,DISS,WTR UG/L	07/27/94-07/27/94	1	5.2	5.2	5.2	5.2	0	0	**	**	**	**
75987	ALPHA GROSS,DISS,1 SIGMA PRC EST AS TH230,WTR PC/L	07/27/94-07/27/94	1	3.84	3.84	3.84	3.84	0	0	**	**	**	**
75988	BETA GROSS,DISS,1 SIGMA PRC EST AS SR90/Y90 PC/L	07/27/94-07/27/94	1	1.81	1.81	1.81	1.81	0	0	**	**	**	**
75989	BETA GROSS,1 SIGMA PRC EST AS CS-137 DISS,WTR PC/L	07/27/94-07/27/94	1	2.4	2.4	2.4	2.4	0	0	**	**	**	**
75990	URANIUM,NATURAL,1 SIGMA PRC EST,DISS,WATER UG/L	07/27/94-07/27/94	1	3.2	3.2	3.2	3.2	0	0	**	**	**	**
76001	RADIUM 226,1 SIGMA PRC EST DISSOLVED,WATER PC/L	07/27/94-07/27/94	1	0.125	0.125	0.125	0.125	0	0	**	**	**	**
80030	ALPHA,DISSOLVED GROSS,AS URANIUM-NATURAL,UG/L	07/27/94-07/27/94	1	11	11	11	11	0	0	**	**	**	**
80050	BETA,DISSOLVED GROSS,AS SR-Y-90, PC/L	07/27/94-07/27/94	1	11	11	11	11	0	0	**	**	**	**

** - Less than 9 observations   ## - Computed with 50% or more of the total observations as values that were half the detection limit   p - Has a corresponding time series plot

EPA Water Quality Criteria Analysis for Station: BADL0016

Parameter		Std Type	Std Value	Total Obs	Exceed Standard	Prop Exceeding	10/01-11/31 Obs	Exceed	Prop	2/01-4/14 Obs	Exceed	Prop	4/15-6/30 Obs	Exceed	Prop	7/01-9/30 Obs	Exceed	Prop
00300	OXYGEN, DISSOLVED	Other-Lo Lim	4	2	0	0.00										2	0	0.00
00400	PH	Other-Hi Lim	9	4	0	0.00	1	0	0.00	1	0	0.00				3	0	0.00
00400		Other-Lo Lim	6.5	4	0	0.00	1	0	0.00	1	0	0.00				3	0	0.00
00403	PH, LAB	Other-Hi Lim	9	4	0	0.00	1	0	0.00	1	0	0.00				3	0	0.00
00403		Other-Lo Lim	6.5	4	0	0.00	1	0	0.00	1	0	0.00				3	0	0.00
00613	NITRITE NITROGEN, DISSOLVED AS N	Drinking Water	1	1	0	0.00										1	0	0.00
00631	NITRITE PLUS NITRATE, DISS 1 DET	Drinking Water	10	2	0	0.00										2	0	0.00
00940	CHLORIDE,TOTAL IN WATER	Fresh Acute	860	4	0	0.00	1	0	0.00							3	0	0.00
00940		Drinking Water	250	4	0	0.00	1	0	0.00							3	0	0.00
00945	SULFATE, TOTAL (AS SO4)	Drinking Water	250	4	0	0.00	1	0	0.00							3	0	0.00

& - Below detection limit observations, for which half the detection limit exceeded the criterion, were excluded from the criterion comparison for this parameter

136

EPA Water Quality Criteria Analysis for Station: BADL0016

Parameter	Std Type	Std Value	Total Obs	Exceed Standard	Prop Exceeding	10/01-1/31 Obs	Exceed	Prop	2/01-4/14 Obs	Exceed	Prop	4/15-6/30 Obs	Exceed	Prop	7/01-9/30 Obs	Exceed	Prop
00950 FLUORIDE, DISSOLVED AS F	Drinking Water	4	4	0	0.00				1	0	0.000				3	0	0.00
01000 ARSENIC, DISSOLVED	Fresh Acute	360	4	0	0.00				1	0	0.000				3	0	0.00
	Drinking Water	50	4	0	0.00				1	0	0.000				3	0	0.00
01005 BARIUM, DISSOLVED	Drinking Water	2000	4	0	0.00				1	0	0.000				3	0	0.00
01010 BERYLLIUM, DISSOLVED	Fresh Acute	130	4	0	0.00				1	0	0.000				3	0	0.00
	Drinking Water	4	4	0	0.00				1	0	0.000				3	0	0.00
01025 CADMIUM, DISSOLVED	Fresh Acute	3.9	4	0	0.00				1	0	0.000				3	0	0.00
01027 CADMIUM, TOTAL	Drinking Water	5	3	0	0.00				1	0	0.000				2	0	0.00
	Fresh Acute	3.9	3	0	0.00				1	0	0.000				3	0	0.00
01030 CHROMIUM, DISSOLVED	Drinking Water	100	4	0	0.00				1	0	0.000				3	0	0.00
01040 COPPER, DISSOLVED	Fresh Acute	18	4	0	0.00				1	0	0.000				3	0	0.00
	Drinking Water	1300	4	0	0.00				1	0	0.000				3	0	0.00
01049 LEAD, DISSOLVED	Fresh Acute	82	4	0	0.00				1	0	0.000				3	0	0.00
	Drinking Water	15	4	0	0.00				1	0	0.000				3	0	0.00
01057 THALLIUM, DISSOLVED	Fresh Acute	1400	2	0	0.00				1	0	0.000				1	0	0.00
	Drinking Water	2	2	0	0.00				1	0	0.000				3	0	0.00
01065 NICKEL, DISSOLVED	Fresh Acute	1400	4	0	0.00				1	0	0.000				3	0	0.00
	Drinking Water	100	4	0	0.00				1	0	0.000				3	0	0.00
01075 SILVER, DISSOLVED	Fresh Acute	4.1	4	0	0.00				1	0	0.000				3	0	0.00
	Drinking Water	100	4	0	0.00				1	0	0.000				3	0	0.00
01090 ZINC, DISSOLVED	Fresh Acute	120	4	0	0.00				1	0	0.000				3	0	0.00
	Drinking Water	5000	4	0	0.00				1	0	0.000				3	0	0.00
01145 SELENIUM, DISSOLVED	Fresh Acute	20	2	0	0.00				1	0	0.000				1	0	0.00
	Drinking Water	50	2	0	0.00				1	0	0.000				1	0	0.00
22703 URANIUM, NATURAL DISSOLVED	Drinking Water	20	1	0	0.00				1	0	0.000						
31501 COLIFORM, TOTAL, MEMBRANE FILTER, IMMED	Other-Hi Lim	1000	1	0	0.00				1	0	0.000						
31503 COLIFORM,TOT,MEMBRANE FILTR,DELAY M-END	Other-Hi Lim	1000	1	0	0.00				1	0	0.000						
31625 FECAL COLIFORM, MF	Other-Hi Lim	200	1	0	0.00				1	0	0.00						
71890 MERCURY, DISSOLVED	Fresh Acute	2.4	4	0	0.00				1	0	0.00				3	0	0.00
	Drinking Water	2	4	0	0.00				1	0	0.00				3	0	0.00

& - Below detection limit observations, for which half the detection limit exceeded the criterion, were excluded from the criterion comparison for this parameter

137

Station Inventory for Station: BADL0017

NPS Station ID: BADL0017
Location: 41N45W 6DDCC
Station Type: /TYPA/AMBNT/SPRING
RMI-Indexes:
HUC: 10140201
Major Basin:
Minor Basin:
RF1 Index: 101402010
RF3 Index: 101402010288 00
Description:

LAT/LON: 43 550281/-102 718060

Agency: 112WRD
FIPS State/County: 46113 SOUTH DAKOTA/SHANNON
STORET Station ID(s): 43330110243050l
Within Park Boundary: Yes

Date Created: 05/01/93

Depth of Water: 0
Elevation: 0

Aquifer:
Water Body Id:
ECO Region:
Distance from RF1: 2 90
Distance from RF3: 0 06

RF1 Mile Point: 0 000
RF3 Mile Point: 0 00

On/Off RF1:
On/Off RF3:

Parameter Inventory for Station: BADL0017

Parameter		Period of Record	Obs	Median	Mean	Maximum	Minimum	Variance	Std Dev	10th	25th	75th	90th
00010	TEMPERATURE, WATER (DEGREES CENTIGRADE)	09/21/92-09/21/92	1	19	19	19	19	0	0	**	**	**	**
00020	TEMPERATURE, AIR (DEGREES CENTIGRADE)	09/21/92-09/21/92	1	24	24	24	24	0	0	**	**	**	**
00025	BAROMETRIC PRESSURE (MM OF HG)	09/21/92-09/21/92	1	680	680	680	680	0	0	**	**	**	**
00061	FLOW, STREAM, INSTANTANEOUS CFS	09/21/92-09/21/92	1	0 003	0 003	0 003	0 003	0	0	**	**	**	**
00095	SPECIFIC CONDUCTANCE (UMHOS/CM @ 25C)	09/21/92-09/21/92	1	678	678	678	678	0	0	**	**	**	**
00400	PH (STANDARD UNITS)	09/21/92-09/21/92	1	8 45	8 45	8 45	8 45	0	0	**	**	**	**
00400	CONVERTED PH (STANDARD UNITS)	09/21/92-09/21/92	1	8 45	8 45	8 45	8 45	0	0	**	**	**	**
00400	MICRO EQUIVALENTS/LITER OF H+ COMPUTED FROM PH	09/21/92-09/21/92	1	0 004	0 004	0 004	0 004	0	0	**	**	**	**
00403	PH, LAB, STANDARD UNITS SU	09/21/92-09/21/92	1	8 1	8 1	8 1	8 1	0	0	**	**	**	**
00403	CONVERTED PH, LAB, STANDARD UNITS	09/21/92-09/21/92	1	8 1	8 1	8 1	8 1	0	0	**	**	**	**
00403	MICRO EQUIVALENTS/LITER OF H+ COMPUTED FROM PH	09/21/92-09/21/92	1	0 008	0 008	0 008	0 008	0	0	**	**	**	**
00915	CALCIUM, DISSOLVED (MG/L AS CA)	09/21/92-09/21/92	1	31	31	31	31	0	0	**	**	**	**
00925	MAGNESIUM, DISSOLVED (MG/L AS MG)	09/21/92-09/21/92	1	4 2	4 2	4 2	4 2	0	0	**	**	**	**
00930	SODIUM, DISSOLVED (MG/L AS NA)	09/21/92-09/21/92	1	120	120	120	120	0	0	**	**	**	**
00935	POTASSIUM, DISSOLVED (MG/L AS K)	09/21/92-09/21/92	1	15	15	15	15	0	0	**	**	**	**
00940	CHLORIDE, TOTAL IN WATER MG/L	09/21/92-09/21/92	1	25	25	25	25	0	0	**	**	**	**
00945	SULFATE, TOTAL (MG/L AS SO4)	09/21/92-09/21/92	1	49	49	49	49	0	0	**	**	**	**
00950	FLUORIDE, DISSOLVED (MG/L AS F)	09/21/92-09/21/92	1	0 2	0 2	0 2	0 2	0	0	**	**	**	**
00955	SILICA, DISSOLVED (MG/L AS SI02)	09/21/92-09/21/92	1	55	55	55	55	0	0	**	**	**	**
01005	ARSENIC, DISSOLVED (UG/L AS AS)	09/21/92-09/21/92	1	15	15	15	15	0	0	**	**	**	**
01010	BARIUM, DISSOLVED (UG/L AS BA)	09/21/92-09/21/92	1	52	52	52	52	0	0	**	**	**	**
01020	BERYLLIUM, DISSOLVED (UG/L AS BE)	09/21/92-09/21/92	##	0 25	0 25	0 25	0 25	0	0	**	**	**	**
01025	BORON, DISSOLVED (UG/L AS B)	09/21/92-09/21/92	1	240	240	240	240	0	0	**	**	**	**
01027	CADMIUM, DISSOLVED (UG/L AS CD)	09/21/92-09/21/92	##	0 5	0 5	0 5	0 5	0	0	**	**	**	**
01030	CADMIUM, TOTAL (UG/L AS CD)	09/21/92-09/21/92	##	0 5	0 5	0 5	0 5	0	0	**	**	**	**
01035	CHROMIUM, DISSOLVED (UG/L AS CR)	09/21/92-09/21/92	##	2 5	2 5	2 5	2 5	0	0	**	**	**	**
01040	COBALT, DISSOLVED (UG/L AS CO)	09/21/92-09/21/92	##	1 5	1 5	1 5	1 5	0	0	**	**	**	**
01046	COPPER, DISSOLVED (UG/L AS CU)	09/21/92-09/21/92	##	5	5	5	5	0	0	**	**	**	**
01049	IRON, DISSOLVED (UG/L AS FE)	09/21/92-09/21/92	1	270	270	270	270	0	0	**	**	**	**
01056	LEAD, DISSOLVED (UG/L AS PB)	09/21/92-09/21/92	##	5	5	5	5	0	0	**	**	**	**
01060	MANGANESE, DISSOLVED (UG/L AS MN)	09/21/92-09/21/92	1	30	30	30	30	0	0	**	**	**	**
01065	MOLYBDENUM, DISSOLVED (UG/L AS MO)	09/21/92-09/21/92	##	5	5	5	5	0	0	**	**	**	**
01075	NICKEL, DISSOLVED (UG/L AS NI)	09/21/92-09/21/92	##	5	5	5	5	0	0	**	**	**	**
01080	SILVER, DISSOLVED (UG/L AS AG)	09/21/92-09/21/92	##	0 5	0 5	0 5	0 5	0	0	**	**	**	**
01085	STRONTIUM, DISSOLVED (UG/L AS SR)	09/21/92-09/21/92	1	270	270	270	270	0	0	**	**	**	**
01090	VANADIUM, DISSOLVED (UG/L AS V)	09/21/92-09/21/92	1	11	11	11	11	0	0	**	**	**	**
01130	ZINC, DISSOLVED (UG/L AS ZN)	09/21/92-09/21/92	1	6	6	6	6	0	0	**	**	**	**
	LITHIUM, DISSOLVED (UG/L AS LI)	09/21/92-09/21/92	1	57	57	57	57	0	0	**	**	**	**

** - Less than 9 observations   ## - Computed with 50% or more of the total observations as values that were half the detection limit   p - Has a corresponding time series plot

138

Parameter Inventory for Station: BADL0017

Parameter		Period of Record	Obs	Median	Mean	Minimum	Maximum	Variance	Std Dev	10th	25th	75th	90th
70300	RESIDUE,TOTAL FILTRABLE (DRIED AT 180C),MG/L	09/21/92-09/21/92	1 ##	457	457	457	457	0	0	**	**	**	**
71890	MERCURY, DISSOLVED (UG/L AS HG)	09/21/92-09/21/92	1 ##	0 05	0 05	0 05	0 05	0	0	**	**	**	**

** - Less than 9 observations   ## - Computed with 50% or more of the total observations as values that were half the detection limit   p - Has a corresponding time series plot

EPA Water Quality Criteria Analysis for Station: BADL0017

Parameter		Std Type	Std Value	Total Obs	Exceed Standard	Prop Exceeding	10/01-1/31 Obs	10/01-1/31 Exceed	10/01-1/31 Prop	2/01-4/14 Obs	2/01-4/14 Exceed	2/01-4/14 Prop	4/15-6/30 Obs	4/15-6/30 Exceed	4/15-6/30 Prop	7/01-9/30 Obs	7/01-9/30 Exceed	7/01-9/30 Prop
00400	PH	Other-Hi Lim	9	1	0	0.00										1	0	0.00
		Other-Lo Lim	6 5	1	0	0.00										1	0	0.00
00403	PH, LAB	Other-Hi Lim	9	1	0	0.00										1	0	0.00
		Other-Lo Lim	6 5	1	0	0.00										1	0	0.00
00940	CHLORIDE,TOTAL IN WATER	Fresh Acute	860	1	0	0.00										1	0	0.00
00945	SULFATE, TOTAL (AS SO4)	Drinking Water	250	1	0	0.00										1	0	0.00
00950	FLUORIDE, DISSOLVED AS F	Drinking Water	250	1	0	0.00										1	0	0.00
01000	ARSENIC, DISSOLVED	Fresh Acute	4	1	0	0.00										1	0	0.00
		Drinking Water	360	1	0	0.00										1	0	0.00
01005	BARIUM, DISSOLVED	Drinking Water	50	1	0	0.00										1	0	0.00
01010	BERYLLIUM, DISSOLVED	Fresh Acute	2000	1	0	0.00										1	0	0.00
		Drinking Water	130	1	0	0.00										1	0	0.00
01025	CADMIUM, DISSOLVED	Fresh Acute	4	1	0	0.00										1	0	0.00
		Drinking Water	3 9	1	0	0.00										1	0	0.00
01027	CADMIUM, TOTAL	Fresh Acute	5	1	0	0.00										1	0	0.00
01030	CHROMIUM, DISSOLVED	Fresh Acute	3 9	1	0	0.00										1	0	0.00
		Drinking Water	5	1	0	0.00										1	0	0.00
01040	COPPER, DISSOLVED	Fresh Acute	100	1	0	0.00										1	0	0.00
		Drinking Water	18	1	0	0.00										1	0	0.00
01049	LEAD, DISSOLVED	Fresh Acute	1300	1	0	0.00										1	0	0.00
		Drinking Water	82	1	0	0.00										1	0	0.00
01065	NICKEL, DISSOLVED	Fresh Acute	15	1	0	0.00										1	0	0.00
		Drinking Water	1400	1	0	0.00										1	0	0.00
01075	SILVER, DISSOLVED	Fresh Acute	100	1	0	0.00										1	0	0.00
01090	ZINC, DISSOLVED	Drinking Water	4 1	1	0	0.00										1	0	0.00
		Drinking Water	100	1	0	0.00										1	0	0.00
71890	MERCURY, DISSOLVED	Fresh Acute	120	1	0	0.00										1	0	0.00
		Drinking Water	5000	1	0	0.00										1	0	0.00
		Fresh Acute	2 4	1	0	0.00										1	0	0.00
		Drinking Water	2	1	0	0.00										1	0	0.00

& - Below detection limit observations, for which half the detection limit exceeded the criterion, were excluded from the criterion comparison for this parameter

139

Station Inventory for Station: BADL0018

NPS Station ID: BADL0018
Location: 41N46W12DACD
Station Type: /TYPA/AMBNT/SPRING
RMI-Indexes:
RMI-Miles:
HUC: 10140201
Major Basin:
Minor Basin:
RF1 Index: 10140201
RF3 Index: 10140201028200 00
Description:

LAT/LON: 43 540559/-102 737781

Agency: 112WRD
FIPS State/County: 46113 SOUTH DAKOTA/SHANNON
STORET Station ID(s) 433226102441601
Within Park Boundary: Yes

Depth of Water: 0
Elevation: 0

Aquifer:
Water Body Id:
ECO Region:
Distance from RF1: 2 90
Distance from RF3: 0 06

RF1 Mile Point: 0 000
RF3 Mile Point: 0 00

On/Off RF1:
On/Off RF3:

Date Created: 05/01/93

Parameter Inventory for Station: BADL0018

Parameter		Period of Record	Obs	Median	Mean	Maximum	Minimum	Variance	Std Dev	10th	25th	75th	90th
00010	TEMPERATURE, WATER (DEGREES CENTIGRADE)	09/21/92-06/21/95	4	13 7	13 275	14 7	11	2 549	1 597	**	**	**	**
00020	TEMPERATURE, AIR (DEGREES CENTIGRADE)	09/21/92-06/21/95	2	23 5	23 25	26	20	10 25	3 202	**	**	**	**
00025	BAROMETRIC PRESSURE (MM OF HG)	09/21/92-07/14/94	2	672	672	680	664	128	11 314	**	**	**	**
00061	FLOW, STREAM, INSTANTANEOUS CFS	09/21/92-06/21/95	4	0 02	0 023	0 03	0 02	0	0 005	**	**	**	**
00095	SPECIFIC CONDUCTANCE (UMHOS/CM @ 25C)	09/21/92-06/21/95	4	621 5	621	626	615	22	4 69	**	**	**	**
00300	OXYGEN, DISSOLVED MG/L	06/21/95-06/21/95	1	8 2	8 2	8 2	8 2			**	**	**	**
00400	PH (STANDARD UNITS)	09/21/92-06/21/95	4	7 625	7 67	7 89	7 54	0 027	0 164	**	**	**	**
00400	CONVERTED PH (STANDARD UNITS)	09/21/92-06/21/95	4	7 619	7 648	7 89	7 54	0 027	0 166	**	**	**	**
00400	MICRO EQUIVALENTS/LITER OF H+ COMPUTED FROM PH	09/21/92-06/21/95	4	0 024	0 022	0 029	0 013	0	0 008	**	**	**	**
00403	PH, LAB, STANDARD UNITS	09/21/92-06/21/95	4	7 8	7 85	8 1	7 7	0 03	0 173	**	**	**	**
00403	CONVERTED PH, LAB, STANDARD UNITS	09/21/92-06/21/95	4	7 8	7 827	8 1	7 7	0 031	0 175	**	**	**	**
00403	MICRO EQUIVALENTS/LITER OF H+ COMPUTED FROM PH	09/21/92-06/21/95	4	0 016	0 015	0 02	0 008	0	0 005	**	**	**	**
00452	CARBONATE,WATER,DISS,INCR TIT, FIELD, AS CO3, MG/L	06/21/95-06/21/95	1		0	0	0			**	**	**	**
00453	BICARBONATE,WATER,DISS,INCR TIT,FIELD,AS HCO3,MG/L	06/21/95-06/21/95	1	242	242	242	242			**	**	**	**
00608	NITROGEN, AMMONIA, DISSOLVED (MG/L AS N)	07/14/94-06/21/95	3 ##	0 008	0 011	0 02	0 005	0	0 008	**	**	**	**
00613	NITRITE NITROGEN, DISSOLVED (MG/L AS N)	04/27/95-06/21/95	2 ##	0 005	0 005	0 005	0 005	0	0	**	**	**	**
00623	NITROGEN, KJELDAHL, DISSOLVED (MG/L AS N)	07/14/94-06/21/95	3 ##	0 1	0 1	0 1	0 1	0	0	**	**	**	**
00631	NITRITE PLUS NITRATE, DISS 1 DET (MG/L AS N)	07/14/94-06/21/95	3	4	2 701	4 1	0 003	5 463	2 337	**	**	**	**
00666	PHOSPHORUS, DISSOLVED (MG/L AS P)	04/27/95-06/21/95	2	0 011	0 012	0 02	0 005	0	0 008	**	**	**	**
00671	PHOSPHORUS, DISSOLVED ORTHOPHOSPHATE (MG/L AS P)	04/27/95-06/21/95	2	0 01	0 01	0 01	0 01	0	0	**	**	**	**
00915	CALCIUM, DISSOLVED (MG/L AS CA)	09/21/92-06/21/95	4	28	27 75	28	27	0 25	0 5	**	**	**	**
00925	MAGNESIUM, DISSOLVED (MG/L AS MG)	09/21/92-06/21/95	4	4	4	4	4	0	0	**	**	**	**
00930	SODIUM, DISSOLVED (MG/L AS NA)	09/21/92-06/21/95	4	96 5	96 75	98	96	0 917	0 957	**	**	**	**
00935	POTASSIUM, DISSOLVED (MG/L AS K)	09/21/92-06/21/95	4	11	10 75	11	10	0 25	0 5	**	**	**	**
00940	CHLORIDE,TOTAL IN WATER MG/L	09/21/92-06/21/95	4	24	25	28	24	4	2	**	**	**	**
00945	SULFATE, TOTAL (MG/L AS SO4)	09/21/92-06/21/95	4	66	65 5	69	61	11 667	3 416	**	**	**	**
00950	FLUORIDE, DISSOLVED (MG/L AS F)	09/21/92-06/21/95	4	0 25	0 25	0 3	0 2	0 003	0 058	**	**	**	**
00955	SILICA, DISSOLVED (MG/L AS SIO2)	09/21/92-06/21/95	4	59	59 5	61	59	1	1	**	**	**	**
01000	ARSENIC, DISSOLVED (UG/L AS AS)	09/21/92-06/21/95	4	17	17 5	19	17	1	1	**	**	**	**
01005	BARIUM, DISSOLVED (UG/L AS BA)	09/21/92-06/21/95	4 ##	84 5	84 25	86	82	2 917	1 708	**	**	**	**
01010	BERYLLIUM, DISSOLVED (UG/L AS BE)	09/21/92-06/21/95	4 ##	0 25	0 25	0 25	0 25			**	**	**	**
01020	BORON, DISSOLVED (UG/L AS B)	09/21/92-06/21/95	4	220	220	230	210	66 667	8 165	**	**	**	**
01025	CADMIUM, DISSOLVED (UG/L AS CD)	09/21/92-06/21/95	4 ##	0 5	0 625	1	0 5	0 063	0 25	**	**	**	**
01027	CADMIUM, TOTAL (UG/L AS CD)	09/21/92-06/21/95	3 ##	0 5	0 5	0 5	0 5	0	0	**	**	**	**
01030	CHROMIUM, DISSOLVED (UG/L AS CR)	09/21/92-06/21/95	4 ##	2 5	2 5	2 5	2 5	0	0	**	**	**	**
01035	COBALT, DISSOLVED (UG/L AS CO)	09/21/92-06/21/95	4 ##	1 5	1 5	1 5	1 5	0	0	**	**	**	**
01040	COPPER, DISSOLVED (UG/L AS CU)	09/21/92-06/21/95	4 ##	5	5	5	5	0	0	**	**	**	**
01046	IRON, DISSOLVED (UG/L AS FE)	09/21/92-06/21/95	4 ##	1 5	2 375	5	1 5	3 063	1 75	**	**	**	**

** - Less than 9 observations   ## - Computed with 50% or more of the total observations as values that were half the detection limit   p - Has a corresponding time series plot

Parameter Inventory for Station: BADL0018

Parameter	Period of Record	Obs	Median	Mean	Maximum	Minimum	Variance	Std Dev	10th	25th	75th	90th
01049 LEAD, DISSOLVED (UG/L AS PB)	09/21/92-06/21/95	4##	5	5	5	5	0	0	**	**	**	**
01056 MANGANESE, DISSOLVED (UG/L AS MN)	09/21/92-06/21/95	4##	0.5	0.625	1	0.5	0.063	0.25	**	**	**	**
01057 THALLIUM, DISSOLVED (UG/L AS TL)	04/27/95-06/21/95	2##	0.25	0.25	0.25	0.25	0	0	**	**	**	**
01060 MOLYBDENUM, DISSOLVED (UG/L AS MO)	09/21/92-06/21/95	4	15	13.75	20	5	56.25	7.5	**	**	**	**
01065 NICKEL, DISSOLVED (UG/L AS NI)	09/21/92-06/21/95	4##	5	5	5	5	0	0	**	**	**	**
01075 SILVER, DISSOLVED (UG/L AS AG)	09/21/92-06/21/95	4##	0.5	0.5	0.5	0.5	0	0	**	**	**	**
01080 STRONTIUM, DISSOLVED (UG/L AS SR)	09/21/92-06/21/95	4	310	307.5	320	290	225	15	**	**	**	**
01085 VANADIUM, DISSOLVED (UG/L AS V)	09/21/92-06/21/95	4##	7	7.25	9	6	1.583	1.258	**	**	**	**
01090 ZINC, DISSOLVED (UG/L AS ZN)	09/21/92-06/21/95	4##	6.25	7.75	17	1	58.083	7.621	**	**	**	**
01130 LITHIUM, DISSOLVED (UG/L AS LI)	09/21/92-06/21/95	4	48	46	51	37	38.667	6.218	**	**	**	**
03515 BETA, DISSOLVED GROSS, AS CS-137, PC/L	07/14/94-07/14/94	1	17	17	17	17	0	0	**	**	**	**
04126 ALPHA, DISSOLVED, WATER (AS TH-230) PCI/L	07/14/94-07/14/94	1	18	18	18	18	0	0	**	**	**	**
09510 RADIUM 226, DISSOLVED, PLANCHET COUNT	07/14/94-07/14/94	1	0.1	0.1	0.1	0.1	0	0	**	**	**	**
22703 URANIUM, NATURAL, DISSOLVED	07/14/94-07/14/94	1##	13	13	13	13	0	0	**	**	**	**
31501 COLIFORM,TOT,MEMBRANE FILTER,IMMED M-ENDO MED,35C	07/14/94-04/27/95	1##	2	2	2	2	0	0	**	**	**	**
31501 GM COLIFORM,TOT,MEMBRANE FILTER,IMMED M-ENDO MED,3	GEOMETRIC MEAN = 2											
31503 COLIFORM,TOT,MEMBR FILTER,DELAYED,M-ENDO MED,35 C	06/21/95-06/21/95	1	220	220	220	220	0	0	**	**	**	**
31503 LOG COLIFORM,TOT,MEMBR FILTER,DELAYED,M-ENDO MED,3	06/21/95-06/21/95	1	2.342	2.342	2.342	2.342	0	0	**	**	**	**
31503 GM COLIFORM,TOT,MEMBR FILTER,DELAYED,M-ENDO MED,35	GEOMETRIC MEAN = 220											
31625 FECAL COLIFORM, MF-FC, 0.7 UM	07/14/94-06/21/95	2##	36	36	70	2	2312	48.083	**	**	**	**
31625 LOG FECAL COLIFORM, MF-FC, 0.7 UM	07/14/94-06/21/95	2##	1.073	1.073	1.845	0.301	1.192	1.092	**	**	**	**
31625 GM FECAL COLIFORM, MF,M-FC, 0.7 UM	GEOMETRIC MEAN = 11.832											
31633 E.COLI,THERMOTOL,MF,M-TEC,IN SITU UREASE #/100ML	07/14/94-06/21/95	2	161	161	230	92	9522	97.581	**	**	**	**
31633 LOG E COLI,THERMOTOL,MF,M-TEC,IN SITU UREASE #/100	07/14/94-06/21/95	2	2.163	2.163	2.362	1.964	0.079	0.281	**	**	**	**
31633 GM E COLI,THERMOTOL,MF,M-TEC,IN SITU UREASE #/100M	GEOMETRIC MEAN = 145.465											
31673 FECAL STREPTOCOCCI,MBR FILT,KF AGAR,35C,48HR	07/14/94-06/21/95	3##	5	25.667	70	2	1476.333	38.423	**	**	**	**
31673 LOG FECAL STREPTOCOCCI,MBR FILT,KF AGAR,35C,48HR	07/14/94-06/21/95	3##	0.699	0.948	1.845	0.301	0.643	0.802	**	**	**	**
31673 GM FECAL STREPTOCOCCI, MBR FILT,KF AGAR,35C,48HR	GEOMETRIC MEAN = 8.879											
39086 ALKALINITY,WATER,DISS,INCR TIT FIELD,AS CACO3,MG/L	06/21/95-06/21/95	1	198	198	198	198	0	0	**	**	**	**
70300 RESIDUE,TOTAL FILTRABLE (DRIED AT 180C),MG/L	09/21/92-06/21/95	4##	423	423	434	412	113.333	10.646	**	**	**	**
71890 MERCURY, DISSOLVED (UG/L AS HG)	09/21/92-06/21/95	4##	0.05	0.05	0.05	0.05	0	0	**	**	**	**
75986 ALPHA GROSS,1 SIGMA PRC EST AS NAT U,DISS,WTR UG/L	07/14/94-07/14/94	1	7.97	7.97	7.97	7.97	0	0	**	**	**	**
75987 ALPHA GROSS,DISS,1 SIGMA PRC EST AS TH230,WTR PC/L	07/14/94-07/14/94	1	6.1	6.1	6.1	6.1	0	0	**	**	**	**
75988 BETA GROSS,DISS,1 SIGMA PRC EST AS SR90/Y90 PC/L	07/14/94-07/14/94	1	2.05	2.05	2.05	2.05	0	0	**	**	**	**
75989 BETA GROSS,1 SIGMA PRC EST AS CS-137,DISS,WTR PC/L	07/14/94-07/14/94	1	2.75	2.75	2.75	2.75	0	0	**	**	**	**
75990 URANIUM,NATURAL,1 SIGMA PRC EST,DISS WATER UG/L	07/14/94-07/14/94	1	3.7	3.7	3.7	3.7	0	0	**	**	**	**
76001 RADIUM 226,1 SIGMA PRC EST,DISSOLVED,WATER PC/L	07/14/94-07/14/94	1	0.125	0.125	0.125	0.125	0	0	**	**	**	**
80030 ALPHA,DISSOLVED GROSS,AS URANIUM-NATURAL,UG/L	07/14/94-07/14/94	1	23	23	23	23	0	0	**	**	**	**
80050 BETA,DISSOLVED GROSS,AS SR-Y-90, PC/L	07/14/94-07/14/94	1	13	13	13	13	0	0	**	**	**	**

** - Less than 9 observations   ## - Computed with 50% or more of the total observations as values that were half the detection limit   p - Has a corresponding time series plot

EPA Water Quality Criteria Analysis for Station: BADL0018

Parameter	Std Type	Std Value	Total Obs	Exceed Standard	Prop Exceeding	10/01-1/31 Obs	Exceed	Prop	2/01-4/14 Obs	Exceed	Prop	4/15-6/30 Obs	Exceed	Prop	7/01-9/30 Obs	Exceed	Prop
00300 OXYGEN, DISSOLVED	Other-Lo Lim	4	4	0	0.00		0	0.00		0	0.00	2	0	0.00	2	0	0.00
00400 PH	Other-Hi Lim	9	4	0	0.00		0	0.00		0	0.00	2	0	0.00	2	0	0.00
00400 PH	Other-Lo Lim	6.5	4	0	0.00		0	0.00		0	0.00	2	0	0.00	2	0	0.00
00403 PH,LAB	Other-Hi Lim	9	4	0	0.00		0	0.00		0	0.00	2	0	0.00	2	0	0.00
00403 PH,LAB	Other-Lo Lim	6.5	4	0	0.00		0	0.00		0	0.00	2	0	0.00	2	0	0.00
00613 NITRITE NITROGEN, DISSOLVED AS N	Drinking Water	1	2	0	0.00		0	0.00		0	0.00	1	0	0.00	1	0	0.00
00631 NITRITE PLUS NITRATE, DISS 1 DET	Drinking Water	10	3	0	0.00		0	0.00		0	0.00	2	0	0.00	2	0	0.00
00940 CHLORIDE, TOTAL IN WATER	Fresh Acute	860	4	0	0.00		0	0.00		0	0.00	2	0	0.00	2	0	0.00
00945 SULFATE, TOTAL (AS SO4)	Drinking Water	250	4	0	0.00		0	0.00		0	0.00	2	0	0.00	2	0	0.00
00950 FLUORIDE, DISSOLVED AS F	Drinking Water	4	4	0	0.00		0	0.00		0	0.00	2	0	0.00	2	0	0.00
01000 ARSENIC, DISSOLVED	Fresh Acute	360	4	0	0.00		0	0.00		0	0.00	2	0	0.00	2	0	0.00
01000 ARSENIC, DISSOLVED	Drinking Water	50	4	0	0.00		0	0.00		0	0.00	2	0	0.00	2	0	0.00

& - Below detection limit observations, for which half the detection limit exceeded the criterion, were excluded from the criterion comparison for this parameter

EPA Water Quality Criteria Analysis for Station: BADL0018

Parameter		Std Type	Std Value	Total Obs	Exceed Standard	Prop Exceeding	10/01-1/31 Obs	Exceed	Prop	2/01-4/14 Obs	Exceed	Prop	4/15-6/30 Obs	Exceed	Prop	7/01-9/30 Obs	Exceed	Prop
01005	BARIUM, DISSOLVED	Drinking Water	2000	4	0	0.00							2	0	0.00	2	0	0.00
01010	BERYLLIUM, DISSOLVED	Fresh Acute	130	4	0	0.00							2	0	0.00	2	0	0.00
01025	CADMIUM, DISSOLVED	Drinking Water	4	4	0	0.00							2	0	0.00	2	0	0.00
		Fresh Acute	3 9	4	0	0.00							2	0	0.00	2	0	0.00
01027	CADMIUM, TOTAL	Drinking Water	5	3	0	0.00							2	0	0.00	1	0	0.00
		Fresh Acute	3 9	3	0	0.00							2	0	0.00	1	0	0.00
01030	CHROMIUM, DISSOLVED	Drinking Water	100	4	0	0.00							2	0	0.00	2	0	0.00
01040	COPPER, DISSOLVED	Fresh Acute	18	4	0	0.00							2	0	0.00	2	0	0.00
		Drinking Water	1300	4	0	0.00							2	0	0.00	2	0	0.00
01049	LEAD, DISSOLVED	Fresh Acute	82	4	0	0.00							2	0	0.00	2	0	0.00
		Drinking Water	15	4	0	0.00							2	0	0.00	2	0	0.00
01057	THALLIUM, DISSOLVED	Fresh Acute	1400	2	0	0.00							2	0	0.00			
		Drinking Water	2	2	0	0.00							2	0	0.00			
01065	NICKEL, DISSOLVED	Fresh Acute	1400	4	0	0.00							2	0	0.00	2	0	0.00
		Drinking Water	100	4	0	0.00							2	0	0.00	2	0	0.00
01075	SILVER, DISSOLVED	Fresh Acute	4 1	4	0	0.00							2	0	0.00	2	0	0.00
		Drinking Water	100	4	0	0.00							2	0	0.00	2	0	0.00
01090	ZINC, DISSOLVED	Fresh Acute	120	4	0	0.00							2	0	0.00	2	0	0.00
		Drinking Water	5000	4	0	0.00							2	0	0.00	2	0	0.00
22703	URANIUM, NATURAL DISSOLVED	Drinking Water	20	1	0	0.00							1	0	0.00	1	0	0.00
31501	COLIFORM, TOTAL, MEMBRANE FILTER, IMMED	Other-Hi Lim	1000	1	0	0.00							1	0	0.00			
31503	COLIFORM,TOT.MEMBRANE FILTR,DELAY M-END	Other-Hi Lim	1000	1	0	0.00							1	0	0.00			
31625	FECAL COLIFORM, MF	Other-Hi Lim	200	2	0	0.00							2	0	0.00			
71890	MERCURY, DISSOLVED	Fresh Acute	2 4	4	0	0.00							2	0	0.00	2	0	0.00
		Drinking Water	2	4	0	0.00							2	0	0.00	2	0	0.00

& - Below detection limit observations, for which half the detection limit exceeded the criterion, were excluded from the criterion comparison for this parameter

142

Station Inventory for Station: BADL0019

NPS Station ID: BADL0019
Location: 4 IN46W24DCBA
Station Type: /TYPA/AMBNT/SPRING
RMI-Indexes:
HUC: 10140201
Major Basin:
Minor Basin:
RF1 Index: 10140201
RF3 Index: 10140201028200.00
Description:

LAT/LON: 43.509726/-102.738060

Agency: 112WRD
FIPS State/County: 46113 SOUTH DAKOTA/SHANNON
STORET Station ID(s): 433035102441701
Within Park Boundary: No

Depth of Water: 0
Elevation: 0

Aquifer:
Water Body Id:
ECO Region:
Distance from RF1: 2.90
Distance from RF3: 0.06

RF1 Mile Point: 0.000
RF3 Mile Point: 0.00

Date Created: 05/01/93

On/Off RF1:
On/Off RF3:

Parameter Inventory for Station: BADL0019

Parameter	Period of Record	Obs	Median	Mean	Maximum	Minimum	Variance	Std Dev	10th	25th	75th	90th
00010 TEMPERATURE, WATER (DEGREES CENTIGRADE)	10/08/92-08/29/95	3	15	14.3	17.8	10.1	15.19	3.897	**	**	**	**
00020 TEMPERATURE, AIR (DEGREES CENTIGRADE)	10/08/92-08/29/95	3	23	17.667	24	6	102.333	10.116	**	**	**	**
00025 BAROMETRIC PRESSURE (MM OF HG)	10/08/92-10/08/92	1	681	681	681	681	0	0	**	**	**	**
00061 FLOW, STREAM, INSTANTANEOUS CFS	10/08/92-08/29/95		0.02	0.02	0.03	0.01	0	0.008	**	**	**	**
00095 SPECIFIC CONDUCTANCE (UMHOS/CM @ 25C)	10/08/92-08/29/95	4	511.5	518.25	550	500	477.583	21.854	**	**	**	**
00300 OXYGEN, DISSOLVED MG/L	07/27/94-08/29/95	2	9.25	9.25	9.7	8.8	0.405	0.636	**	**	**	**
00400 PH (STANDARD UNITS)	10/08/92-08/29/95	4	8.32	8.323	8.41	8.24	0.005	0.071	**	**	**	**
00400 CONVERTED PH (STANDARD UNITS)	10/08/92-08/29/95	4	8.32	8.318	8.41	8.24	0.005	0.072	**	**	**	**
00400 MICRO EQUIVALENTS/LITER OF H+ COMPUTED FROM PH	10/08/92-08/29/95	4	0.005	0.005	0.006	0.004	0	0.001	**	**	**	**
00403 PH, LAB, STANDARD UNITS SU	10/08/92-08/29/95	4	8.15	8.15	8.3	8	0.017	0.129	**	**	**	**
00403 CONVERTED PH, LAB, STANDARD UNITS	10/08/92-08/29/95	4	8.147	8.136	8.3	8	0.017	0.13	**	**	**	**
00403 MICRO EQUIVALENTS/LITER OF H+ COMPUTED FROM PH	10/08/92-08/29/95	4	0.007	0.007	0.01	0.005	0	0.002	**	**	**	**
00410 ALKALINITY, TOTAL (MG/L AS CACO3)	07/27/94-07/27/94	1	204	204	204	204	0	0	**	**	**	**
00419 ALKALINITY, CARBONATE, INCREMENTAL TITR FIELD MG/L	08/29/95-08/29/95	1	212	212	212	212	0	0	**	**	**	**
00440 BICARBONATE ION (MG/L AS HCO3)	07/27/94-07/27/94	1	249	249	249	249	0	0	**	**	**	**
00447 CARBONATE, INCREMENTAL TITRATION, (CO3) FIELD MG/L	08/29/95-08/29/95	1	2	2	2	2	0	0	**	**	**	**
00450 BICARBONATE, INCREMENTAL TITRATION, (HCO3) FIELD MG/L	08/29/95-08/29/95	1	259	259	259	259	0	0	**	**	**	**
00608 NITROGEN, AMMONIA, DISSOLVED (MG/L AS N)	07/27/94-08/29/95	3	0.03	0.023	0.03	0.008	0	0.013	**	**	**	**
00613 NITRITE NITROGEN, DISSOLVED (MG/L AS N)	05/12/95-08/29/95	2 ##	0.008	0.008	0.01	0.005	0	0.004	**	**	**	**
00623 NITROGEN, KJELDAHL, DISSOLVED (MG/L AS N)	07/27/94-08/29/95	3	0.2	0.2	0.3	0.1	0.01	0.1	**	**	**	**
00631 NITRITE PLUS NITRATE, DISS 1 DET (MG/L AS N)	07/27/94-08/29/95	3 ##	4	3.8	4.2	3.2	0.28	0.529	**	**	**	**
00666 PHOSPHORUS, DISSOLVED (MG/L AS P)	07/27/94-08/29/95	3 ##	0.005	0.006	0.007	0.005	0	0.001	**	**	**	**
00671 PHOSPHORUS, DISSOLVED ORTHOPHOSPHATE (MG/L AS P)	05/12/95-08/29/95	2 ##	0.008	0.008	0.01	0.005	0	0.004	**	**	**	**
00915 CALCIUM, DISSOLVED (MG/L AS CA)	10/08/92-08/29/95	4	27.5	27.75	29	27	0.917	0.957	**	**	**	**
00925 MAGNESIUM, DISSOLVED (MG/L AS MG)	10/08/92-08/29/95	4	3.05	3.05	3.2	2.9	0.017	0.129	**	**	**	**
00930 SODIUM, DISSOLVED (MG/L AS NA)	10/08/92-08/29/95	4	73.5	73.75	75	73	0.917	0.957	**	**	**	**
00935 POTASSIUM, DISSOLVED (MG/L AS K)	10/08/92-08/29/95	4	13	13.25	14	13	0.25	0.5	**	**	**	**
00940 CHLORIDE, TOTAL IN WATER MG/L	10/08/92-08/29/95	4	11.5	11.25	13	9	2.917	1.708	**	**	**	**
00945 SULFATE, TOTAL (MG/L AS SO4)	10/08/92-08/29/95	4	35.5	34.75	39	29	18.917	4.349	**	**	**	**
00950 FLUORIDE, DISSOLVED (MG/L AS F)	10/08/92-08/29/95	4	0.3	0.275	0.3	0.2	0.002	0.05	**	**	**	**
00955 SILICA, DISSOLVED (MG/L AS SI02)	10/08/92-08/29/95	4	57	56.75	57	56	0.25	0.5	**	**	**	**
01000 ARSENIC, DISSOLVED (UG/L AS AS)	10/08/92-08/29/95	4	9.5	9.75	12	8	2.917	1.708	**	**	**	**
01005 BARIUM, DISSOLVED (UG/L AS BA)	10/08/92-08/29/95	4	46	47	52	44	12.667	3.559	**	**	**	**
01010 BERYLLIUM, DISSOLVED (UG/L AS BE)	10/08/92-08/29/95	4 ##	0.25	0.25	0.25	0.25	0	0	**	**	**	**
01020 BORON, DISSOLVED (UG/L AS B)	10/08/92-08/29/95	4	175	175	180	170	33.333	5.774	**	**	**	**
01025 CADMIUM, DISSOLVED (UG/L AS CD)	10/08/92-08/29/95	4 ##	0.5	0.5	0.5	0.5	0	0	**	**	**	**
01027 CADMIUM, TOTAL (UG/L AS CD)	10/08/92-08/29/95	3 ##	0.5	0.875	2.5	0.5	0.563	0.75	**	**	**	**
01030 CHROMIUM, DISSOLVED (UG/L AS CR)	10/08/92-08/29/95	4 ##	2.5	2.5	2.5	2.5	0	0	**	**	**	**

** - Less than 9 observations   ## - Computed with 50% or more of the total observations as values that were half the detection limit   p - Has a corresponding time series plot

143

Parameter Inventory for Station: BADL0019

Parameter		Period of Record	Obs	Median	Mean	Maximum	Minimum	Variance	Std Dev	10th	25th	75th	90th
01035	COBALT, DISSOLVED (UG/L AS CO)	10/08/92-08/29/95	4 ##	5	5	5	5	0	0	**	**	**	**
01040	COPPER, DISSOLVED (UG/L AS CU)	10/08/92-08/29/95	4 ##	5	5	5	5	0	0	**	**	**	**
01046	IRON, DISSOLVED (UG/L AS FE)	10/08/92-08/29/95	4 ##	2.25	2.25	3	1.5	0.75	0.866	**	**	**	**
01049	LEAD, DISSOLVED (UG/L AS PB)	10/08/92-08/29/95	4 ##	5	11.25	30	5	156.25	12.5	**	**	**	**
01056	MANGANESE, DISSOLVED (UG/L AS MN)	10/08/92-08/29/95	4 ##	0.5	0.875	2	0.5	0.563	0.75	**	**	**	**
01057	THALLIUM, DISSOLVED (UG/L AS TL)	05/12/95-08/29/95	2 ##	0.25	0.25	0.25	0.25	0	0	**	**	**	**
01060	MOLYBDENUM DISSOLVED (UG/L AS MO)	10/08/92-08/29/95	4 ##	5	5	5	5	0	0	**	**	**	**
01065	NICKEL, DISSOLVED (UG/L AS NI)	10/08/92-08/29/95	4 ##	5	5	5	5	0	0	**	**	**	**
01075	SILVER, DISSOLVED (UG/L AS AG)	10/08/92-08/29/95	4 ##	0.5	0.5	0.5	0.5	0	0	**	**	**	**
01080	STRONTIUM, DISSOLVED (UG/L AS SR)	10/08/92-08/29/95	4	360	360	370	350	66.667	8.165	**	**	**	**
01085	VANADIUM, DISSOLVED (UG/L AS V)	10/08/92-08/29/95	4	9	8.5	9	7	1	1	**	**	**	**
01090	ZINC, DISSOLVED (UG/L AS ZN)	10/08/92-08/29/95	4 ##	1.5	2.375	9	1.5	3.063	1.75	**	**	**	**
01130	LITHIUM, DISSOLVED (UG/L AS LI)	10/08/92-08/29/95	4 ##	36.5	35.25	38	30	12.917	3.594	**	**	**	**
01145	SELENIUM, DISSOLVED (UG/L AS SE)	10/08/92-08/29/95	4	10	9	10	6	4	2	**	**	**	**
03515	BETA, DISSOLVED GROSS, AS CS-137, PC/L	07/27/94-07/27/94	1	18	18	18	18	0	0	**	**	**	**
04126	ALPHA, DISSOLVED, WATER (AS TH-230) PCI/L	07/27/94-07/27/94	1	11	11	11	11	0	0	**	**	**	**
09510	RADIUM 226, DISSOLVED, PLANCHET COUNT	07/27/94-07/27/94	1 ##	0.05	0.05	0.05	0.05	0	0	**	**	**	**
22703	URANIUM, NATURAL, DISSOLVED	07/27/94-07/27/94	1	9.6	9.6	9.6	9.6	0	0	**	**	**	**
31501	COLIFORM,TOT,MEMBRANE FILTER,IMMED M-ENDO MED,35C	07/27/94-05/12/95	2	60	60	60	60	0	0	**	**	**	**
31501	GM COLIFORM,TOT,MEMBRANE FILTER,IMMED M-ENDO MED,	07/27/94-05/12/95	2	1.778	1.778	1.778	1.778	0	0	**	**	**	**
	GEOMETRIC MEAN = 60												
31504	COLIFORM,TOT,MEMBR FILTER,IMMED,LES ENDO AGAR,35C	08/29/95-08/29/95	1	26	26	26	26	0	0	**	**	**	**
31504	GM COLIFORM,TOT,MEMBR FILTER,IMMED,LES ENDO AGAR,3	08/29/95-08/29/95	1	1.415	1.415	1.415	1.415	0	0	**	**	**	**
	GEOMETRIC MEAN = 26												
31625	FECAL COLIFORM, MF,M-FC, 0.7 UM	07/27/94-08/29/95	3	44	104.333	225	44	10920.333	104.5	**	**	**	**
31625	GM FECAL COLIFORM, MF,M-FC, 0.7 UM	07/27/94-08/29/95	3	1.643	1.88	2.352	1.643	0.167	0.409	**	**	**	**
	GEOMETRIC MEAN = 75.805												
31633	E COLI,THERMOTOL,MF,M-TEC,IN SITU UREASE #/100ML	07/27/94-08/29/95	3	2	2901	8700	1	25221301	5022.081	**	**	**	**
31633	GM E COLI,THERMOTOL,MF,M-TEC,IN SITU UREASE #/100M	07/27/94-08/29/95	3	0.301	1.414	3.94	0	4.808	2.193	**	**	**	**
	GEOMETRIC MEAN = 25.913												
31673	FECAL STREPTOCOCCI, MBR FILT,KF AGAR,35C,48HR	07/27/94-08/29/95	3	420	295.333	420	46	46625.333	215.929	**	**	**	**
31673	GM FECAL STREPTOCOCCI, MBR FILT,KF AGAR,35C,48HR	07/27/94-08/29/95	3	2.623	2.303	2.623	1.663	0.308	0.555	**	**	**	**
	GEOMETRIC MEAN = 200.949												
70300	RESIDUE,TOTAL FILTRABLE (DRIED AT 180C),MG/L	10/08/92-08/29/95	4	355	356.75	370	347	124.917	11.177	**	**	**	**
71890	MERCURY, DISSOLVED (UG/L AS HG)	10/08/92-08/29/95	4 ##	0.05	0.05	0.05	0.05	0	0	**	**	**	**
75986	ALPHA GROSS,1 SIGMA PRC EST AS NAT U,DISS,WTR UG/L	07/27/94-07/27/94	1	6.83	6.83	6.83	6.83	0	0	**	**	**	**
75987	ALPHA GROSS,DISS,1 SIGMA PRC EST AS TH230,WTR PC/L	07/27/94-07/27/94	1	4.43	4.43	4.43	4.43	0	0	**	**	**	**
75988	BETA GROSS,DISS,1 SIGMA PRC EST AS SR90/Y90 PC/L	07/27/94-07/27/94	1	2.05	2.05	2.05	2.05	0	0	**	**	**	**
75989	BETA GROSS, 1 SIGMA PRC EST AS CS-137,DISS,WTR PC/L	07/27/94-07/27/94	1	2.72	2.72	2.72	2.72	0	0	**	**	**	**
75990	URANIUM,NATURAL,1 SIGMA PRC EST,DISS,WATER UG/L	07/27/94-07/27/94	1	2.8	2.8	2.8	2.8	0	0	**	**	**	**
76001	RADIUM 226,1 SIGMA PRC EST DISSOLVED,WATER PC/L	07/27/94-07/27/94	1	0.13	0.13	0.13	0.13	0	0	**	**	**	**
80030	ALPHA,DISSOLVED GROSS,AS URANIUM-NATURAL,UG/L	07/27/94-07/27/94	1	18	18	18	18	0	0	**	**	**	**
80050	BETA,DISSOLVED GROSS,AS SR-Y-90, PC/L	07/27/94-07/27/94	1	14	14	14	14	0	0	**	**	**	**

** - Less than 9 observations   ## - Computed with 50% or more of the total observations as values that were half the detection limit   p - Has a corresponding time series plot

EPA Water Quality Criteria Analysis for Station: BADL0019

Parameter		Std Type	Std Value	Total Obs	Exceed Standard	Prop Exceeding	10/01-1/31 Obs	Exceed	Prop	2/01-4/14 Obs	Exceed	Prop	4/15-6/30 Obs	Exceed	Prop	7/01-9/30 Obs	Exceed	Prop
00300	OXYGEN, DISSOLVED	Other-Lo Lim	4	2	0	0.00	—			1	0	0.00	0	0	0.00	2	0	0.00
00400	PH	Other-Hi Lim	9	4	0	0.00	1	0	0.00	1	0	0.00	0	0	0.00	2	0	0.00
00400	PH	Other-Lo Lim	6.5	4	0	0.00	1	0	0.00	1	0	0.00	0	0	0.00	2	0	0.00
00403	PH, LAB	Other-Hi Lim	9	4	0	0.00	1	0	0.00	1	0	0.00	0	0	0.00	2	0	0.00
00403	PH, LAB	Other-Lo Lim	6.5	4	0	0.00	1	0	0.00	1	0	0.00	0	0	0.00	2	0	0.00
00613	NITRITE NITROGEN, DISSOLVED AS N	Drinking Water	1	2	0	0.00	—			1	0	0.00	0	0	0.00	1	0	0.00
00631	NITRITE PLUS NITRATE, DISS 1 DET	Drinking Water	10	3	0	0.00	1	0	0.00	1	0	0.00	0	0	0.00	2	0	0.00
00940	CHLORIDE, TOTAL IN WATER	Fresh Acute	860	4	0	0.00	1	0	0.00	1	0	0.00	0	0	0.00	2	0	0.00
00940	CHLORIDE, TOTAL IN WATER	Drinking Water	250	4	0	0.00	1	0	0.00	1	0	0.00	0	0	0.00	2	0	0.00
00945	SULFATE, TOTAL (AS SO4)	Drinking Water	250	4	0	0.00	1	0	0.00	1	0	0.00	0	0	0.00	2	0	0.00

& - Below detection limit observations, for which half the detection limit exceeded the criterion, were excluded from the criterion comparison for this parameter

EPA Water Quality Criteria Analysis for Station: BADL0019

Parameter		Std Type	Std Value	Total Obs	Exceed Standard	Prop Exceeding	10/01-1/31 Obs	Exceed	Prop	2/01-4/14 Obs	Exceed	Prop	4/15-6/30 Obs	Exceed	Prop	7/01-9/30 Obs	Exceed	Prop
00950	FLUORIDE, DISSOLVED AS F	Drinking Water	4	4	0	0.00	1	0	0.00				1	0	0.00	2	0	0.00
01000	ARSENIC, DISSOLVED	Fresh Acute	360	4	0	0.00	1	0	0.00				1	0	0.00	2	0	0.00
01005	BARIUM, DISSOLVED	Drinking Water	50	4	0	0.00	1	0	0.00				1	0	0.00	2	0	0.00
01010	BERYLLIUM, DISSOLVED	Drinking Water	2000	4	0	0.00	1	0	0.00				1	0	0.00	2	0	0.00
01025	CADMIUM, DISSOLVED	Fresh Acute	130	4	0	0.00	1	0	0.00				1	0	0.00	2	0	0.00
		Drinking Water	4	4	0	0.00	1	0	0.00				1	0	0.00	2	0	0.00
01027	CADMIUM, TOTAL	Drinking Water	3.9	3	0	0.00	1	0	0.00				1	0	0.00	1	0	0.00
		Fresh Acute	5	3	0	0.00	1	0	0.00				1	0	0.00	1	0	0.00
01030	CHROMIUM, DISSOLVED	Drinking Water	3.9	4	0	0.00	1	0	0.00				1	0	0.00	2	0	0.00
01040	COPPER, DISSOLVED	Drinking Water	5	4	0	0.00	1	0	0.00				1	0	0.00	2	0	0.00
		Fresh Acute	100	4	0	0.00	1	0	0.00				1	0	0.00	2	0	0.00
01049	LEAD, DISSOLVED	Drinking Water	18	4	0	0.00	1	0	0.00				1	0	0.00	2	0	0.00
		Fresh Acute	1300	4	0	0.00	1	0	0.00				1	0	0.00	2	0	0.00
01057	THALLIUM, DISSOLVED	Fresh Acute	82	4	1	0.25	1	0	0.00				1	1	1.00	2	0	0.00
		Drinking Water	15	2	0	0.00	1	0	0.00				1	0	0.00	1	0	0.00
01065	NICKEL, DISSOLVED	Fresh Acute	1400	2	0	0.00	1	0	0.00							1	0	0.00
		Drinking Water	2	4	0	0.00	1	0	0.00				1	0	0.00	2	0	0.00
01075	SILVER, DISSOLVED	Fresh Acute	1400	4	0	0.00	1	0	0.00				1	0	0.00	2	0	0.00
		Drinking Water	100	4	0	0.00	1	0	0.00				1	0	0.00	2	0	0.00
01090	ZINC, DISSOLVED	Fresh Acute	4.1	4	0	0.00	1	0	0.00				1	0	0.00	2	0	0.00
		Drinking Water	100	4	0	0.00	1	0	0.00				1	0	0.00	2	0	0.00
01145	SELENIUM, DISSOLVED	Fresh Acute	120	4	0	0.00	1	0	0.00				1	0	0.00	2	0	0.00
		Drinking Water	5000	4	0	0.00	1	0	0.00				1	0	0.00	2	0	0.00
22703	URANIUM, NATURAL DISSOLVED	Fresh Acute	20	4	0	0.00	1	0	0.00				1	0	0.00	2	0	0.00
		Drinking Water	50	4	0	0.00							1	0	0.00	2	0	0.00
31501	COLIFORM, TOTAL, MEMBRANE FILTER, IMMED	Drinking Water	20	1	0	0.00										1	0	0.00
31504	COLIFORM,TOT,MEMBRANE FILTR,IMMED LES-E	Other-Hi Lim	1000	2	0	0.00							1	0	0.00	1	0	0.00
31625	FECAL COLIFORM, MF	Other-Hi Lim	1000	1	0	0.00							1	0	0.00			
		Other-Hi Lim	200	3	1	0.33							1	0	0.00	2	1	0.50
71890	MERCURY, DISSOLVED	Fresh Acute	2.4	4	0	0.00	1	0	0.00				1	0	0.00	2	0	0.00
		Drinking Water	2	4	0	0.00	1	0	0.00				1	0	0.00	2	0	0.00

& - Below detection limit observations, for which half the detection limit exceeded the criterion, were excluded from the criterion comparison for this parameter

145

Station Inventory for Station: BADL0020

NPS Station ID: BADL0020
Location: 41N46W25BCA W DAGMAN
Station Type: /TYPA/AMBNT/SPRING
RMI-Indexes:
HUC: 10140201
Major Basin:
Minor Basin:
RF1 Index: 101402010
RF3 Index: 10140201027400 00
Description:

LAT/LON: 43 501115/-102 751670

Depth of Water: 0
Elevation: 0

RF1 Mile Point: 0 000
RF3 Mile Point: 0 20

Agency: 112WRD
FIPS State/County: 46113 SOUTH DAKOTA/SHANNON
STORET Station IDs): 433004102450601
Within Park Boundary: No

Aquifer:
Water Body Id:
ECO Region:
Distance from RF1: 3 50
Distance from RF3: 0 19

Date Created: 01/16/79

On/Off RF1:
On/Off RF3:

Parameter Inventory for Station: BADL0020

Parameter		Period of Record	Obs	Median	Mean	Maximum	Minimum	Variance	Std Dev	10th	25th	75th	90th
00095	SPECIFIC CONDUCTANCE (UMHOS/CM @ 25C)	11/29/62-11/29/62	1	436	436	436	436	0	0	**	**	**	**
00400	PH (STANDARD UNITS)	11/29/62-11/29/62	1	7 5	7 5	7 5	7 5	0	0	**	**	**	**
00400	CONVERTED PH (STANDARD UNITS)	11/29/62-11/29/62	1	7 5	7 5	7 5	7 5	0	0	**	**	**	**
00400	MICRO EQUIVALENTS/LITER OF H+ COMPUTED FROM PH	11/29/62-11/29/62	1	0 032	0 032	0 032	0 032	0	0	**	**	**	**
00405	CARBON DIOXIDE (MG/L AS CO2)	11/29/62-11/29/62	1	12	12	12	12	0	0	**	**	**	**
00410	ALKALINITY, TOTAL (MG/L AS CACO3)	11/29/62-11/29/62	1	189	189	189	189	0	0	**	**	**	**
00440	BICARBONATE ION (MG/L AS HCO3)	11/29/62-11/29/62	1	230	230	230	230	0	0	**	**	**	**
00445	CARBONATE ION (MG/L AS CO3)	11/29/62-11/29/62	1	0	0	0	0	0	0	**	**	**	**
00618	NITRATE NITROGEN, DISSOLVED (MG/L AS N)	11/29/62-11/29/62	1	1 6	1 6	1 6	1 6	0	0	**	**	**	**
00900	HARDNESS, TOTAL (MG/L AS CACO3)	11/29/62-11/29/62	1	121	121	121	121	0	0	**	**	**	**
00902	HARDNESS, NON-CARBONATE (MG/L AS CACO3)	11/29/62-11/29/62	1	0	0	0	0	0	0	**	**	**	**
00915	CALCIUM, DISSOLVED (MG/L AS CA)	11/29/62-11/29/62	1	26	26	26	26	0	0	**	**	**	**
00925	MAGNESIUM, DISSOLVED (MG/L AS MG)	11/29/62-11/29/62	1	14	14	14	14	0	0	**	**	**	**
00930	SODIUM, DISSOLVED (MG/L AS NA)	11/29/62-11/29/62	1	43	43	43	43	0	0	**	**	**	**
00931	SODIUM ADSORPTION RATIO	11/29/62-11/29/62	1	1 7	1 7	1 7	1 7	0	0	**	**	**	**
00932	SODIUM, PERCENT	11/29/62-11/29/62	1	41	41	41	41	0	0	**	**	**	**
00935	POTASSIUM, DISSOLVED (MG/L AS K)	11/29/62-11/29/62	1	11	11	11	11	0	0	**	**	**	**
00940	CHLORIDE, TOTAL IN WATER MG/L	11/29/62-11/29/62	1	13	13	13	13	0	0	**	**	**	**
00945	SULFATE, TOTAL (MG/L AS SO4)	11/29/62-11/29/62	1	20	20	20	20	0	0	**	**	**	**
00950	FLUORIDE, DISSOLVED (MG/L AS F)	11/29/62-11/29/62	1	0 3	0 3	0 3	0 3	0	0	**	**	**	**
00955	SILICA, DISSOLVED (MG/L AS SI02)	11/29/62-11/29/62	1	58	58	58	58	0	0	**	**	**	**
01020	BORON, DISSOLVED (UG/L AS B)	11/29/62-11/29/62	1	150	150	150	150	0	0	**	**	**	**
70300	RESIDUE, TOTAL FILTRABLE (DRIED AT 180C) MG/L	11/29/62-11/29/62	1	320	320	320	320	0	0	**	**	**	**
70301	SOLIDS, DISSOLVED-SUM OF CONSTITUENTS (MG/L)	11/29/62-11/29/62	1	306	306	306	306	0	0	**	**	**	**
71851	NITRATE NITROGEN, DISSOLVED (MG/L AS NO3)	11/29/62-11/29/62	1	6 9	6 9	6 9	6 9	0	0	**	**	**	**
71883	MANGANESE, TOTAL ELEMENTAL (UG/L AS MN)	11/29/62-11/29/62	1	0	0	0	0	0	0	**	**	**	**
71885	IRON (UG/L AS FE)	11/29/62-11/29/62	1	10	10	10	10	0	0	**	**	**	**

** - Less than 9 observations   ## - Computed with 50% or more of the total observations as values that were half the detection limit   p - Has a corresponding time series plot

EPA Water Quality Criteria Analysis for Station: BADL0020

Parameter	Std Type	Std Value	Total Obs	Exceed Standard	Prop Exceeding	10/01-1/31 Obs	Exceed	Prop	2/01-4/14 Obs	Exceed	Prop	4/15-6/30 Obs	Exceed	Prop	7/01-9/30 Obs	Exceed	Prop
00400 PH	Other-Hi Lim	9	1	0	0 00	1	0	0 00									
	Other-Lo Lim	6 5	1	0	0 00	1	0	0 00									

& - Below detection limit observations, for which half the detection limit exceeded the criterion, were excluded from the criterion comparison for this parameter

EPA Water Quality Criteria Analysis for Station: BADL0020

Parameter		Std Type	Std Value	Total Obs	Exceed Standard	Prop Exceeding	10/01-1/31			2/01-4/14			4/15-6/30			7/01-9/30		
							Obs	Exceed	Prop	Obs	Exceed	Prop	Obs	Exceed	Prop	Obs	Exceed	Prop
00618	NITRATE NITROGEN, DISSOLVED AS N	Drinking Water	10	1	0	0.00	1	0	0.00									
00940	CHLORIDE,TOTAL IN WATER	Fresh Acute	860	1	0	0.00	1	0	0.00									
00945	SULFATE, TOTAL (AS SO4)	Drinking Water	250	1	0	0.00	1	0	0.00									
00950	FLUORIDE, DISSOLVED AS F	Drinking Water	4	1	0	0.00	1	0	0.00									
71851	NITRATE NITROGEN, DISSOLVED (AS NO3)	Drinking Water	44	1	0	0.00	1	0	0.00									

& - Below detection limit observations, for which half the detection limit exceeded the criterion, were excluded from the criterion comparison for this parameter

147

Station Inventory for Station: BADL0021

NPS Station ID: BADL0021
Location: 41N46W 9BBCA
Station Type: /TYPA/AMBNT/SPRING
RMI-Indexes:
RMI-Miles:
HUC: 10140201
Major Basin:
Minor Basin:
RF1 Index: 101402010
RF3 Index: 10140201028200 00
Description:

LAT/LON: 43 547782/-102 811671

Depth of Water: 0
Elevation: 0

RF1 Mile Point: 0 000
RF3 Mile Point: 0 00

Agency: 112WRD
FIPS State/County: 46113 SOUTH DAKOTA/SHANNON
STORET Station ID(s) 43325210248420I
Within Park Boundary: Yes

Aquifer:
Water Body Id:
ECO Region:
Distance from RF1: 2 90
Distance from RF3: 0 06

On/Off RF1:
On/Off RF3:

Date Created: 05/01/93

Parameter Inventory for Station: BADL0021

Parameter		Period of Record	Obs	Median	Mean	Maximum	Minimum	Variance	Std Dev	10th	25th	75th	90th
00010	TEMPERATURE, WATER (DEGREES CENTIGRADE)	10/01/92-07/13/95	4	20	19 75	20	19	0 25	0 5	**	**	**	**
00020	TEMPERATURE, AIR (DEGREES CENTIGRADE)	10/01/92-07/13/95	3	30	29 667	31	28	2 333	1 528	**	**	**	**
00025	BAROMETRIC PRESSURE (MM OF HG)	10/01/92-10/01/92	1	679	679	679	679	0	0	**	**	**	**
00061	FLOW, STREAM, INSTANTANEOUS CFS	10/01/92-07/13/95	4	0 09	0 09	0 1	0 08	0	0 008	**	**	**	**
00065	STAGE, STREAM (FEET)	07/13/95-07/13/95	1	0 28	0 28	0 28	0 28	0	0	**	**	**	**
00095	SPECIFIC CONDUCTANCE (UMHOS/CM @ 25C)	10/01/92-07/13/95	5	542	549	592	510	892	29 866	**	**	**	**
00300	OXYGEN, DISSOLVED MG/L	07/13/95-07/13/95	1	7	7	7	7			**	**	**	**
00400	PH (STANDARD UNITS)	10/01/92-07/13/95	5	8 15	8 106	8 3	7 94	0 026	0 161	**	**	**	**
00400	CONVERTED PH (STANDARD UNITS)	10/01/92-07/13/95	5	8 15	8 082	8 3	7 94	0 027	0 163	**	**	**	**
00400	MICRO EQUIVALENTS/LITER OF H+ COMPUTED FROM PH	10/01/92-07/13/95	5	0 007	0 008	0 011	0 005		0 003	**	**	**	**
00403	PH, LAB, STANDARD UNITS SU	10/01/92-07/13/95	5	7 8	7 66	8	7 2	0 118	0 344	**	**	**	**
00403	CONVERTED PH, LAB, STANDARD UNITS	10/01/92-07/13/95	5	7 8	7 549	8	7 2	0 133	0 365	**	**	**	**
00403	MICRO EQUIVALENTS/LITER OF H+ COMPUTED FROM PH	10/01/92-07/13/95	5	0 016	0 028	0 063	0 01	0 001	0 023	**	**	**	**
00453	BICARBONATE,WATER,DISS,INCR TIT FIELD,AS HCO3,MG/L	07/13/95-07/13/95	1	329	329	329	329	0	0	**	**	**	**
00608	NITROGEN, AMMONIA, DISSOLVED (MG/L AS N)	07/28/94-07/13/95	4	0 15	0 138	0 23	0 02	0 009	0 096	**	**	**	**
00613	NITRITE NITROGEN, DISSOLVED (MG/L AS N)	05/16/95-07/13/95	2	0 02	0 02	0 03	0 01		0 014	**	**	**	**
00623	NITROGEN, KJELDAHL, DISSOLVED (MG/L AS N)	07/28/94-07/13/95	4	0 75	0 725	1 3	0 1	0 283	0 532	**	**	**	**
00631	NITRITE PLUS NITRATE, DISS 1 DET (MG/L AS N)	07/28/94-07/13/95	4	2 9	2 675	3	1 9	0 276	0 525	**	**	**	**
00666	PHOSPHORUS, DISSOLVED (MG/L AS P)	07/28/94-07/13/95	4	0 07	0 062	0 09	0 02	0 001	0 031	**	**	**	**
00671	PHOSPHORUS, DISSOLVED ORTHOPHOSPHATE (MG/L AS P)	05/16/95-07/13/95	2	0 045	0 045	0 07	0 02	0 001	0 035	**	**	**	**
00915	CALCIUM, DISSOLVED (MG/L AS CA)	10/01/92-07/13/95	5	58	58 4	63	55	10 3	3 209	**	**	**	**
00925	MAGNESIUM, DISSOLVED (MG/L AS MG)	10/01/92-07/13/95	5	7 7	7 54	7 9	7 1	0 133	0 365	**	**	**	**
00930	SODIUM, DISSOLVED (MG/L AS NA)	10/01/92-07/13/95	5	48	47 6	53	44	14 3	3 782	**	**	**	**
00935	POTASSIUM, DISSOLVED (MG/L AS K)	10/01/92-07/13/95	4	13	13 75	16	13	2 25	1 5	**	**	**	**
00940	CHLORIDE,TOTAL IN WATER MG/L	10/01/92-07/13/95	5	8	8 8	11	7	2 7	1 643	**	**	**	**
00945	SULFATE, TOTAL (MG/L AS SO4)	10/01/92-07/13/95	5	23	22 8	24	21	1 2	1 095	**	**	**	**
00950	FLUORIDE, DISSOLVED (MG/L AS F)	10/01/92-07/13/95	5	0 3	0 3	0 3	0 3	0	0	**	**	**	**
00955	SILICA, DISSOLVED (MG/L AS SI02)	10/01/92-07/13/95	5	54	53 4	56	50	4 8	2 191	**	**	**	**
01000	ARSENIC, DISSOLVED (UG/L AS AS)	10/01/92-07/13/95	5	14	13 8	15	11	2 7	1 643	**	**	**	**
01005	BARIUM, DISSOLVED (UG/L AS BA)	10/01/92-07/13/95	5	210	232	270	210	920	30 332	**	**	**	**
01010	BERYLLIUM, DISSOLVED (UG/L AS BE)	10/01/92-07/13/95	5 ##	0 25	0 48	1 4	0 25	0 265	0 514	**	**	**	**
01020	BORON, DISSOLVED (UG/L AS B)	10/01/92-07/13/95	5 ##	130	134	150	130	80	8 944	**	**	**	**
01025	CADMIUM, DISSOLVED (UG/L AS CD)	10/01/92-07/13/95	5 ##	0 5	0 5	0 5	0 5	0	0	**	**	**	**
01027	CADMIUM, TOTAL (UG/L AS CD)	10/01/92-07/13/95	4 ##	0 5	0 5	0 5	0 5	0	0	**	**	**	**
01030	CHROMIUM, DISSOLVED (UG/L AS CR)	10/01/92-07/13/95	5 ##	2 5	2 5	2 5	2 5	0	0	**	**	**	**
01035	COBALT, DISSOLVED (UG/L AS CO)	10/01/92-07/13/95	5 ##	1 5	2	4	1 5	1 25	1 118	**	**	**	**
01040	COPPER, DISSOLVED (UG/L AS CU)	10/01/92-07/13/95	5 ##	5	5	5	5	0	0	**	**	**	**
01046	IRON, DISSOLVED (UG/L AS FE)	10/01/92-07/13/95	5	23	25 2	37	11	104 2	10 208	**	**	**	**

** - Less than 9 observations   ## - Computed with 50% or more of the total observations as values that were half the detection limit   p - Has a corresponding time series plot

Parameter Inventory for Station: BADL0021

Parameter		Period of Record	Obs	Median	Mean	Maximum	Minimum	Variance	Std Dev	10th	25th	75th	90th
01049	LEAD, DISSOLVED (UG/L AS PB)	10/01/92-07/13/95	5 ##	5	7	10	5	7 5	2 739	**	**	**	**
01056	MANGANESE, DISSOLVED (UG/L AS MN)	10/01/92-07/13/95	5	67	56	78	16	677 5	26 029	**	**	**	**
01057	THALLIUM, DISSOLVED (UG/L AS TL)	05/16/95-07/13/95	2 ##	0 25	0 25	0 25	0 25	0	0	**	**	**	**
01060	MOLYBDENUM, DISSOLVED (UG/L AS MO)	10/01/92-07/13/95	5 ##	5	6	10	5	5	2 236	**	**	**	**
01065	NICKEL, DISSOLVED (UG/L AS NI)	10/01/92-07/13/95	5 ##	5	5	5	5	0	0	**	**	**	**
01075	SILVER, DISSOLVED (UG/L AS AG)	10/01/92-07/13/95	5 ##	0 5	0 5	0 5	0 5	0	0	**	**	**	**
01080	STRONTIUM, DISSOLVED (UG/L AS SR)	10/01/92-07/13/95	5	450	444	480	410	830	28 81	**	**	**	**
01085	VANADIUM, DISSOLVED (UG/L AS V)	10/01/92-07/13/95	5	8	8 2	12	3	11 2	3 347	**	**	**	**
01090	ZINC, DISSOLVED (UG/L AS ZN)	10/01/92-07/13/95	5	4	4 2	5	3	9 575	3 094	**	**	**	**
01130	LITHIUM, DISSOLVED (UG/L AS LI)	10/01/92-07/13/95	5	27	26 4	36	18	42 3	6 504	**	**	**	**
01145	SELENIUM, DISSOLVED (UG/L AS SE)	10/01/92-07/13/95	5	4	3 8	5	3	0 7	0 837	**	**	**	**
03515	BETA, DISSOLVED GROSS, AS CS-137 PC/L	07/28/94-07/28/94	2	22 5	22 5	24	21	4 5	2 121	**	**	**	**
04126	ALPHA, DISSOLVED, WATER (AS TH-230) PCI/L	07/28/94-07/28/94	2	9 65	9 65	11	8 3	3 645	1 909	**	**	**	**
09510	RADIUM 226, DISSOLVED, PLANCHET COUNT	07/28/94-07/28/94	2	0 15	0 15	0 2	0 1	0 005	0 071	**	**	**	**
22703	URANIUM, NATURAL, DISSOLVED	07/28/94-07/28/94	2	11	11	12	10	2	1 414	**	**	**	**
31501	COLIFORM,TOT,MEMBRANE FILTER,IMMED M-ENDO MED,35C	07/28/94-05/16/95	2	4008 5	4008 5	8000	17	31864144 5	5644 833	**	**	**	**
31501	GM COLIFORM,TOT,MEMBRANE FILTER,IMMED M-ENDO MED,3	07/28/94-05/16/95	2	2 567	2 567	3 903	1 23	3 572	1 89	**	**	**	**
31501	GEOMETRIC MEAN =			368 782									
31503	COLIFORM,TOT,MEMBR FILTER,DELAYED,M-ENDO MED,35 C	07/13/95-07/13/95	1	11800	11800	11800	11800	0	0	**	**	**	**
31503	LOG COLIFORM,TOT,MEMBR FILTER,DELAYED,M-ENDO MED,3	07/13/95-07/13/95	1	4 072	4 072	4 072	4 072	0	0	**	**	**	**
31503	GEOMETRIC MEAN =			11800									
31625	FECAL COLIFORM, MF-FC, 0 7 UM	07/28/94-07/13/95	2	7400	7400	8800	6000	3920000	1979 899	**	**	**	**
31625	LOG FECAL COLIFORM, MF-FC, 0 7 UM	07/28/94-07/13/95	2	3 861	3 861	3 944	3 778	0 014	0 118	**	**	**	**
31625	GM FECAL COLIFORM, MF-FC, 0 7 UM	GEOMETRIC MEAN =			2 81									
31633	E COLI,THERMOTOL,MF,M-TEC,IN SITU UREASE #/100ML	07/28/94-07/13/95	3	2100	3372	8000	16	17149552	4141 202	**	**	**	**
31633	LOG E COLI,THERMOTOL,MF-TEC,IN SITU UREASE #/100	07/28/94-07/13/95	3	3 322	2 81	3 903	1 204	2 018	1 421	**	**	**	**
31633	GM E COLI,THERMOTOL,MF-M-TEC,IN SITU UREASE #/100M	GEOMETRIC MEAN =			645 371									
31673	FECAL STREPTOCOCCI, MBR FILT,KF AGAR,35C,48HR	07/28/94-07/13/95	3	4600	4924 333	10000	173	24221376 333	4921 522	**	**	**	**
31673	LOG FECAL STREPTOCOCCI, MBR FILT,KF AGAR,35C,48HR	07/28/94-07/13/95	3	3 663	3 3	4	2 238	0 875	0 935	**	**	**	**
31673	GM FECAL STREPTOCOCCI, MBR FILT,KF AGAR,35C,48HR	GEOMETRIC MEAN =			1996 494									
39086	ALKALINITY,WATER,DISS,INCR TIT FIELD,AS CACO3,MG/L	07/13/95-07/13/95	1	270	270	270	270	0	0	**	**	**	**
70300	RESIDUE,TOTAL FILTRABLE (DRIED AT 180C),MG/L	10/01/92-07/13/95	5	378	379 2	394	371	78 7	8 871	**	**	**	**
71890	MERCURY, DISSOLVED (UG/L AS HG)	07/28/94-07/28/94	5 ##	0 05	0 05	0 05	0 05	0	0	**	**	**	**
75986	ALPHA GROSS,1 SIGMA PRC EST AS NAT U,DISS,WTR UGL	07/28/94-07/28/94	2	5 79	5 79	6 28	5 3	0 48	0 693	**	**	**	**
75987	ALPHA GROSS,DISS,1 SIGMA PRC EST AS TH230,WTR PC/L	07/28/94-07/28/94	2	4 045	4 045	4 39	3 7	0 238	0 488	**	**	**	**
75988	BETA GROSS,DISS,1 SIGMA PRC EST AS SR90/Y90 PC/L	07/28/94-07/28/94	2	2 495	2 495	2 57	2 42	0 011	0 106	**	**	**	**
75989	BETA GROSS,1 SIGMA PRC EST AS CS-137,DISS,WTR PC/L	07/28/94-07/28/94	2	3 315	3 315	3 45	3 18	0 036	0 191	**	**	**	**
75990	URANIUM,NATURAL,1 SIGMA PRC EST,DISS,WATER UG/L	07/28/94-07/28/94	2	3 3	3 3	3 6	3	0 18	0 424	**	**	**	**
76001	RADIUM 226,1 SIGMA PRC EST,DISSOLVED,WATER PC/L	07/28/94-07/28/94	2	0 136	0 136	0 142	0 129	0 009	0 009	**	**	**	**
80030	ALPHA,DISSOLVED GROSS,AS URANIUM-NATURAL,UG/L	07/28/94-07/28/94	2	13 5	13 5	15	12	4 5	2 121	**	**	**	**
80050	BETA,DISSOLVED GROSS,AS SR-Y-90, PC/L	07/28/94-07/28/94	2	17	17	18	16	2	1 414	**	**	**	**

** - Less than 9 observations   ## - Computed with 50% or more of the total observations as values that were half the detection limit   p - Has a corresponding time series plot

EPA Water Quality Criteria Analysis for Station: BADL0021

Parameter		Std Type	Std Value	Total Obs	Exceed Standard	Prop Exceeding	10/01-1/31 Obs	Exceed	Prop	2/01-4/14 Obs	Exceed	Prop	4/15-6/30 Obs	Exceed	Prop	7/01-9/30 Obs	Exceed	Prop
00300	OXYGEN, DISSOLVED	Other-Lo Lim	4	1	0	0.00	—			—			0	0	0.00	1	0	0.00
00400	PH	Other-Hi Lim	9	5	0	0.00	—			—			0	0	0.00	3	0	0.00
		Other-Lo Lim	6 5	5	0	0.00	—			—			0	0	0.00	3	0	0.00
00403	PH, LAB	Other-Hi Lim	9	5	0	0.00	—			—			0	0	0.00	3	0	0.00
		Other-Lo Lim	6 5	5	0	0.00	—			—			0	0	0.00	3	0	0.00
00613	NITRITE NITROGEN, DISSOLVED AS N	Drinking Water	1	4	0	0.00	—			—			0	0	0.00	1	0	0.00
00631	NITRITE PLUS NITRATE, DISS 1 DET	Drinking Water	10	5	0	0.00	—			—			0	0	0.00	3	0	0.00
00940	CHLORIDE,TOTAL IN WATER	Fresh Acute	860	5	0	0.00	—			—			0	0	0.00	3	0	0.00
00945	SULFATE, TOTAL (AS SO4)	Drinking Water	250	5	0	0.00	—			—			0	0	0.00	3	0	0.00
00950	FLUORIDE, DISSOLVED AS F	Drinking Water	4	5	0	0.00	—			—			0	0	0.00	3	0	0.00
01000	ARSENIC, DISSOLVED	Fresh Acute	360	5	0	0.00	—			—			0	0	0.00	3	0	0.00
		Drinking Water	50	5	0	0.00	—			—			0	0	0.00	3	0	0.00

& - Below detection limit observations, for which half the detection limit exceeded the criterion, were excluded from the criterion comparison for this parameter

EPA Water Quality Criteria Analysis for Station: BADL0021

Parameter		Std Type	Std Value	Total Obs	Exceed Standard	Prop Exceeding	10/01-1/31 Obs	10/01-1/31 Exceed	10/01-1/31 Prop	2/01-4/14 Obs	2/01-4/14 Exceed	2/01-4/14 Prop	4/15-6/30 Obs	4/15-6/30 Exceed	4/15-6/30 Prop	7/01-9/30 Obs	7/01-9/30 Exceed	7/01-9/30 Prop
01005	BARIUM, DISSOLVED	Drinking Water	2000	5	0	0.00	1	0	0.00				1	0	0.00	3	0	0.00
01010	BERYLLIUM, DISSOLVED	Fresh Acute	130	5	0	0.00	1	0	0.00				1	0	0.00	3	0	0.00
01025	CADMIUM, DISSOLVED	Drinking Water	4	5	0	0.00	1	0	0.00				1	0	0.00	3	0	0.00
		Fresh Acute	3.9	5	0	0.00	1	0	0.00				1	0	0.00	3	0	0.00
01027	CADMIUM, TOTAL	Drinking Water	5	4	0	0.00	1	0	0.00				1	0	0.00	2	0	0.00
		Fresh Acute	3.9	4	0	0.00	1	0	0.00				1	0	0.00	2	0	0.00
01030	CHROMIUM, DISSOLVED	Drinking Water	100	5	0	0.00	1	0	0.00				1	0	0.00	3	0	0.00
01040	COPPER, DISSOLVED	Fresh Acute	18	5	0	0.00	1	0	0.00				1	0	0.00	3	0	0.00
		Drinking Water	1300	5	0	0.00	1	0	0.00				1	0	0.00	3	0	0.00
01049	LEAD, DISSOLVED	Fresh Acute	82	5	0	0.00	1	0	0.00				1	0	0.00	3	0	0.00
01057	THALLIUM, DISSOLVED	Drinking Water	15	5	0	0.00	1	0	0.00				1	0	0.00	3	0	0.00
		Fresh Acute	1400	2	0	0.00							1	0	0.00	1	0	0.00
01065	NICKEL, DISSOLVED	Drinking Water	2	2	0	0.00							1	0	0.00	1	0	0.00
		Fresh Acute	1400	5	0	0.00	1	0	0.00				1	0	0.00	3	0	0.00
01075	SILVER, DISSOLVED	Drinking Water	100	5	0	0.00	1	0	0.00				1	0	0.00	3	0	0.00
		Fresh Acute	4.1	5	0	0.00	1	0	0.00				1	0	0.00	3	0	0.00
01090	ZINC, DISSOLVED	Drinking Water	100	5	0	0.00	1	0	0.00				1	0	0.00	3	0	0.00
		Fresh Acute	120	5	0	0.00	1	0	0.00				1	0	0.00	3	0	0.00
01145	SELENIUM, DISSOLVED	Drinking Water	5000	5	0	0.00	1	0	0.00				1	0	0.00	3	0	0.00
		Fresh Acute	20	5	0	0.00	1	0	0.00				1	0	0.00	3	0	0.00
22703	URANIUM, NATURAL, DISSOLVED	Drinking Water	50	5	0	0.00	1	0	0.00				1	0	0.00	3	0	0.00
		Drinking Water	20	2	0	0.00							1	0	0.00	2	0	0.00
31501	COLIFORM, TOTAL, MEMBRANE FILTER, IMMED	Other-Hi Lim	1000	2	1	0.50							1	0	0.00	1	1	1.00
31503	COLIFORM,TOT,MEMBRANE FILTR,DELAY M-END	Other-Hi Lim	1000	1	1	1.00										1	1	1.00
31625	FECAL COLIFORM, MF	Other-Hi Lim	200	2	2	1.00										2	2	1.00
71890	MERCURY, DISSOLVED	Fresh Acute	2.4	5	0	0.00	1	0	0.00				1	0	0.00	3	0	0.00
		Drinking Water	2	5	0	0.00	1	0	0.00				1	0	0.00	3	0	0.00

& - Below detection limit observations, for which half the detection limit exceeded the criterion, were excluded from the criterion comparison for this parameter

150

Station Inventory for Station: BADL0022

NPS Station ID: BADL0022
Location: 40N46W18BBAD
Station Type: /TYPA/AMBNT/SPRING
RMI-Miles:
HUC: 10140201
Major Basin:
Minor Basin:
RF1 Index: 10140201
RF3 Index: 1014020102820000
Description:

LAT/LON: 43.463892/-102.813060

Depth of Water: 0
Elevation: 0

RF1 Mile Point: 0.000
RF3 Mile Point: 0.00

Agency: 112WRD
FIPS State/County: 46113 SOUTH DAKOTA/SHANNON
STORET Station ID(s): 4327501024847O1
Within Park Boundary: No

Aquifer:
Water Body Id:
ECO Region:
Distance from RF1: 2.90
Distance from RF3: 0.06

On/Off RF1:
On/Off RF3:

Date Created: 05/01/93

Parameter Inventory for Station: BADL0022

Parameter		Period of Record	Obs	Median	Mean	Maximum	Minimum	Variance	Std Dev	10th	25th	75th	90th
00010	TEMPERATURE, WATER (DEGREES CENTIGRADE)	09/29/92-09/29/92		17.9	17.9	17.9	17.9	0	0	**	**	**	**
00020	TEMPERATURE, AIR (DEGREES CENTIGRADE)	09/29/92-09/29/92		19	19	19	19	0	0	**	**	**	**
00025	BAROMETRIC PRESSURE (MM OF HG)	09/29/92-09/29/92		688	688	688	688	0	0	**	**	**	**
00095	SPECIFIC CONDUCTANCE (UMHOS/CM @ 25C)	09/29/92-09/29/92		819	819	819	819	0	0	**	**	**	**
00400	PH (STANDARD UNITS)	09/29/92-09/29/92		8.1	8.1	8.1	8.1	0	0	**	**	**	**
00400	CONVERTED PH (STANDARD UNITS)	09/29/92-09/29/92		8.1	8.1	8.1	8.1	0	0	**	**	**	**
00400	MICRO EQUIVALENTS/LITER OF H+ COMPUTED FROM PH	09/29/92-09/29/92		0.008	0.008	0.008	0.008	0	0	**	**	**	**
00403	PH, LAB, STANDARD UNITS SU	09/29/92-09/29/92		8.1	8.1	8.1	8.1	0	0	**	**	**	**
00403	CONVERTED PH, LAB, STANDARD UNITS	09/29/92-09/29/92		8.1	8.1	8.1	8.1	0	0	**	**	**	**
00403	MICRO EQUIVALENTS/LITER OF H+ COMPUTED FROM PH	09/29/92-09/29/92		0.008	0.008	0.008	0.008	0	0	**	**	**	**
00915	CALCIUM, DISSOLVED (MG/L AS CA)	09/29/92-09/29/92		6.4	6.4	6.4	6.4	0	0	**	**	**	**
00925	MAGNESIUM, DISSOLVED (MG/L AS MG)	09/29/92-09/29/92		0.4	0.4	0.4	0.4	0	0	**	**	**	**
00930	SODIUM, DISSOLVED (MG/L AS NA)	09/29/92-09/29/92		190	190	190	190	0	0	**	**	**	**
00935	POTASSIUM, DISSOLVED (MG/L AS K)	09/29/92-09/29/92		14	14	14	14	0	0	**	**	**	**
00940	CHLORIDE, TOTAL IN WATER MG/L	09/29/92-09/29/92		10	10	10	10	0	0	**	**	**	**
00945	SULFATE, TOTAL (MG/L AS SO4)	09/29/92-09/29/92		27	27	27	27	0	0	**	**	**	**
00950	FLUORIDE, DISSOLVED (MG/L AS F)	09/29/92-09/29/92		0.5	0.5	0.5	0.5	0	0	**	**	**	**
00955	SILICA, DISSOLVED (MG/L AS SI02)	09/29/92-09/29/92		68	68	68	68	0	0	**	**	**	**
01000	ARSENIC, DISSOLVED (UG/L AS AS)	09/29/92-09/29/92		13	13	13	13	0	0	**	**	**	**
01005	BARIUM, DISSOLVED (UG/L AS BA)	09/29/92-09/29/92		17	17	17	17	0	0	**	**	**	**
01010	BERYLLIUM, DISSOLVED (UG/L AS BE)	09/29/92-09/29/92	##	0.25	0.25	0.25	0.25	0	0	**	**	**	**
01020	BORON, DISSOLVED (UG/L AS B)	09/29/92-09/29/92		240	240	240	240	0	0	**	**	**	**
01025	CADMIUM, DISSOLVED (UG/L AS CD)	09/29/92-09/29/92	##	0.5	0.5	0.5	0.5	0	0	**	**	**	**
01027	CADMIUM, TOTAL (UG/L AS CD)	09/29/92-09/29/92	##	0.5	0.5	0.5	0.5	0	0	**	**	**	**
01030	CHROMIUM, DISSOLVED (UG/L AS CR)	09/29/92-09/29/92	##	2.5	2.5	2.5	2.5	0	0	**	**	**	**
01035	COBALT, DISSOLVED (UG/L AS CO)	09/29/92-09/29/92	##	1.5	1.5	1.5	1.5	0	0	**	**	**	**
01040	COPPER, DISSOLVED (UG/L AS CU)	09/29/92-09/29/92	##	5	5	5	5	0	0	**	**	**	**
01046	IRON, DISSOLVED (UG/L AS FE)	09/29/92-09/29/92		23	23	23	23	0	0	**	**	**	**
01049	LEAD, DISSOLVED (UG/L AS PB)	09/29/92-09/29/92	##	5	5	5	5	0	0	**	**	**	**
01056	MANGANESE, DISSOLVED (UG/L AS MN)	09/29/92-09/29/92	##	0.5	0.5	0.5	0.5	0	0	**	**	**	**
01060	MOLYBDENUM, DISSOLVED (UG/L AS MO)	09/29/92-09/29/92	##	5	5	5	5	0	0	**	**	**	**
01065	NICKEL, DISSOLVED (UG/L AS NI)	09/29/92-09/29/92	##	5	5	5	5	0	0	**	**	**	**
01075	SILVER, DISSOLVED (UG/L AS AG)	09/29/92-09/29/92	##	0.5	0.5	0.5	0.5	0	0	**	**	**	**
01080	STRONTIUM, DISSOLVED (UG/L AS SR)	09/29/92-09/29/92		130	130	130	130	0	0	**	**	**	**
01085	VANADIUM, DISSOLVED (UG/L AS V)	09/29/92-09/29/92		14	14	14	14	0	0	**	**	**	**
01090	ZINC, DISSOLVED (UG/L AS ZN)	09/29/92-09/29/92	##	1.5	1.5	1.5	1.5	0	0	**	**	**	**
01130	LITHIUM, DISSOLVED (UG/L AS LI)	09/29/92-09/29/92		64	64	64	64	0	0	**	**	**	**
01145	SELENIUM, DISSOLVED (UG/L AS SE)	09/29/92-09/29/92		4	4	4	4	0	0	**	**	**	**

** - Less than 9 observations   ## - Computed with 50% or more of the total observations as values that were half the detection limit   p - Has a corresponding time series plot

Parameter Inventory for Station: BADL0022

Parameter		Period of Record	Obs	Median	Mean	Maximum	Minimum	Variance	Std Dev	10th	25th	75th	90th
70300	RESIDUE,TOTAL FILTRABLE (DRIED AT 180C),MG/L	09/29/92-09/29/92	1	574	574	574	574	0	0	**	**	**	**
71890	MERCURY, DISSOLVED (UG/L AS HG)	09/29/92-09/29/92	1 ##	0 05	0 05	0 05	0 05	0	0	**	**	**	**

** - Less than 9 observations   ## - Computed with 50% or more of the total observations as values that were half the detection limit   p - Has a corresponding time series plot

EPA Water Quality Criteria Analysis for Station: BADL0022

Parameter		Std Type	Std Value	Total Obs	Exceed Standard	Prop Exceeding	10/01-1/31 Obs	Exceed	Prop	2/01-4/14 Obs	Exceed	Prop	4/15-6/30 Obs	Exceed	Prop	7/01-9/30 Obs	Exceed	Prop
00400	PH	Other-Hi Lim	9	1	0	0.00										1	0	0.00
		Other-Lo Lim	6 5	1	0	0.00										1	0	0.00
00403	PH, LAB	Other-Hi Lim	9	1	0	0.00										1	0	0.00
		Other-Lo Lim	6 5	1	0	0.00										1	0	0.00
00940	CHLORIDE, TOTAL IN WATER	Fresh Acute	860	1	0	0.00										1	0	0.00
00945	SULFATE, TOTAL (AS SO4)	Drinking Water	250	1	0	0.00										1	0	0.00
00950	FLUORIDE, DISSOLVED AS F	Drinking Water	4	1	0	0.00										1	0	0.00
01000	ARSENIC, DISSOLVED	Fresh Acute	360	1	0	0.00										1	0	0.00
		Drinking Water	50	1	0	0.00										1	0	0.00
01005	BARIUM, DISSOLVED	Drinking Water	2000	1	0	0.00										1	0	0.00
01010	BERYLLIUM, DISSOLVED	Fresh Acute	130	1	0	0.00										1	0	0.00
01025	CADMIUM, DISSOLVED	Fresh Acute	3 9	1	0	0.00										1	0	0.00
		Drinking Water	5	1	0	0.00										1	0	0.00
01027	CADMIUM, TOTAL	Fresh Acute	3 9	1	0	0.00										1	0	0.00
		Drinking Water	5	1	0	0.00										1	0	0.00
01030	CHROMIUM, DISSOLVED	Drinking Water	100	1	0	0.00										1	0	0.00
01040	COPPER, DISSOLVED	Fresh Acute	18	1	0	0.00										1	0	0.00
		Drinking Water	1300	1	0	0.00										1	0	0.00
01049	LEAD, DISSOLVED	Fresh Acute	82	1	0	0.00										1	0	0.00
		Drinking Water	15	1	0	0.00										1	0	0.00
01065	NICKEL, DISSOLVED	Fresh Acute	1400	1	0	0.00										1	0	0.00
		Drinking Water	100	1	0	0.00										1	0	0.00
01075	SILVER, DISSOLVED	Fresh Acute	4 1	1	0	0.00										1	0	0.00
		Drinking Water	100	1	0	0.00										1	0	0.00
01090	ZINC, DISSOLVED	Fresh Acute	120	1	0	0.00										1	0	0.00
		Drinking Water	5000	1	0	0.00										1	0	0.00
01145	SELENIUM, DISSOLVED	Fresh Acute	20	1	0	0.00										1	0	0.00
		Drinking Water	50	1	0	0.00										1	0	0.00
71890	MERCURY, DISSOLVED	Fresh Acute	2 4	1	0	0.00										1	0	0.00
		Drinking Water	2	1	0	0.00										1	0	0.00

& - Below detection limit observations, for which half the detection limit exceeded the criterion, were excluded from the criterion comparison for this parameter

Station Inventory for Station: BADL0023

NPS Station ID: BADL0023
Location: BATTLE CR ABV CONF WITH CHEYENNE
Station Type: /TYPA/AMBNT/STREAM
RMI-Indexes:
HUC: 10120109
Major Basin: MISSOURI RIVER BASIN
Minor Basin: CENTRAL MISSOURI BASIN
RF1 Index: 10120109010
RF3 Index: 10120109056700 16
Description:
CLASS B: MERCURY (W&F) STATION; S18,T4S,R11E

LAT/LON: 43.700838/-102.852505

Agency: 31BLHICD
FIPS State/County: 46000 SOUTH DAKOTA/
STORET Station ID(s): BT-01
Within Park Boundary: No

Depth of Water: 0
Elevation: 0

Aquifer:
Water Body Id:
ECO Region:
Distance from RF1: 0.00
Distance from RF3: 0.01

RF1 Mile Point: 2.860
RF3 Mile Point: 0.15

Date Created: / /

On/Off RF1: OFF
On/Off RF3:

Parameter Inventory for Station: BADL0023

Parameter		Period of Record	Obs	Median	Mean	Maximum	Minimum	Variance	Std Dev	10th	25th	75th	90th
00010	TEMPERATURE, WATER (DEGREES CENTIGRADE)	06/01/71-08/21/72	3	19	18	21	14	13	3.606	**	**	**	**
00020	TEMPERATURE, AIR (DEGREES CENTIGRADE)	06/01/71-08/21/72	3	23	21.667	23	19	5.333	2.309	**	**	**	**
00060	FLOW, STREAM, MEAN DAILY CFS	06/01/71-08/21/72	3	20	30	50	20	300	17.321	**	**	**	**
00070	TURBIDITY, (JACKSON CANDLE UNITS)	06/01/71-08/21/72	3	54	386.667	1055	51	335004.333	578.796	**	**	**	**
00095	SPECIFIC CONDUCTANCE (UMHOS/CM @ 25C)	06/01/71-08/21/72	3	1900	2463.333	4300	1190	2656033.333	1629.734	**	**	**	**
00300	OXYGEN, DISSOLVED MG/L	06/01/71-08/21/72	3	8.5	8.7	9.5	8.1	0.52	0.721	**	**	**	**
00400	PH (STANDARD UNITS)	06/01/71-08/21/72	3	8	7.9	8.2	7.5	0.13	0.361	**	**	**	**
00400	CONVERTED PH (STANDARD UNITS)	06/01/71-08/21/72	3	8	7.796	8.2	7.5	0.146	0.382	**	**	**	**
00400	MICRO EQUIVALENTS/LITER OF H+ COMPUTED FROM PH	06/01/71-08/21/72	3	0.01	0.016	0.032	0.006	0	0.014	**	**	**	**
00410	ALKALINITY, TOTAL (MG/L AS CACO3)	06/01/71-08/21/72	3	144	146.667	216	80	4629.333	68.039	**	**	**	**
00430	ALKALINITY, CARBONATE (MG/L AS CACO3)	06/01/71-08/21/72	3	0	0	0	0	0	0	**	**	**	**
00500	RESIDUE, TOTAL (MG/L)	06/01/71-08/21/72	3	504	964.667	1918	472	681889.333	825.766	**	**	**	**
00505	RESIDUE, TOTAL VOLATILE (MG/L)	06/01/71-08/21/72	3	124	147.333	240	78	6969.333	83.483	**	**	**	**
00515	RESIDUE, TOTAL FILTRABLE (DRIED AT 105C),MG/L	06/01/71-08/21/72	3	14	502.667	1494	0	737105.333	858.548	**	**	**	**
00520	RESIDUE, VOLATILE FILTRABLE (MG/L)	06/01/71-08/21/72	3	14	65.333	182	0	10257.333	101.278	**	**	**	**
00630	NITRITE PLUS NITRATE, TOTAL, 1 DET (MG/L AS N)	06/01/71-08/21/72	3	0.13	0.267	0.6	0.07	0.084	0.29	**	**	**	**
00650	PHOSPHATE, TOTAL (MG/L AS PO4)	06/01/71-08/21/72	3	8	6.133	9	1.4	17.053	4.13	**	**	**	**
00660	PHOSPHATE, ORTHO (MG/L AS PO4)	06/01/71-08/21/72	3	0.1	0.2	0.5	0	0.07	0.265	**	**	**	**
00900	HARDNESS, TOTAL (MG/L AS CACO3)	06/01/71-08/21/72	3	331	258	343	100	18759	136.963	**	**	**	**
00901	HARDNESS, CARBONATE (MG/L AS CACO3)	06/01/71-08/21/72	3	144	146.667	216	80	4629.333	68.039	**	**	**	**
00902	HARDNESS, NON-CARBONATE (MG/L AS CACO3)	06/01/71-08/21/72	3	127	111.333	187	20	7156.333	84.595	**	**	**	**
00915	CALCIUM, DISSOLVED (MG/L AS CA)	06/01/71-08/21/72	3	81	67.333	87	34	842.333	29.023	**	**	**	**
00925	MAGNESIUM, DISSOLVED (MG/L AS MG)	06/01/71-08/21/72	3	28	22	34	4	252	15.875	**	**	**	**
00930	SODIUM, DISSOLVED (MG/L AS NA)	06/01/71-08/21/72	3	18	13.333	18	4	65.333	8.083	**	**	**	**
00935	POTASSIUM, DISSOLVED (MG/L AS K)	06/01/71-08/21/72	3	6	5.333	7	3	4.333	2.082	**	**	**	**
00940	CHLORIDE,TOTAL IN WATER MG/L	06/01/71-08/21/72	3	3	4.667	10	1	22.333	4.726	**	**	**	**
00945	SULFATE, TOTAL (MG/L AS SO4)	06/01/71-08/21/72	3	155	132.667	215	28	9116.333	95.479	**	**	**	**
00955	SILICA, DISSOLVED (MG/L AS SI02)	06/01/71-08/21/72	3	15	15.333	17	14	2.333	1.528	**	**	**	**
01045	IRON, TOTAL (UG/L AS FE)	06/01/71-08/21/72	3	175	408.333	1000	50	266458.333	516.196	**	**	**	**
01055	MANGANESE, TOTAL (UG/L AS MN)	06/01/71-08/21/72	3	90	796.667	2300	0	1697033.333	1302.702	**	**	**	**
01145	SELENIUM, DISSOLVED (UG/L AS SE)	06/01/71-08/21/72	3	0	3.333	10	0	33.333	5.774	**	**	**	**

** - Less than 9 observations   ## - Computed with 50% or more of the total observations as values that were half the detection limit   p - Has a corresponding time series plot

EPA Water Quality Criteria Analysis for Station: BADL0023

Parameter		Std Type	Std Value	Total Obs	Exceed Standard	Prop Exceeding	10/01-1/31			2/01-4/14			4/15-6/30			7/01-9/30		
							Obs	Exceed	Prop	Obs	Exceed	Prop	Obs	Exceed	Prop	Obs	Exceed	Prop
00070	TURBIDITY, JACKSON CANDLE UNITS	Other-Hi Lim	50	3	3	1.00							1	1	1.00	2	2	1.00
00300	OXYGEN, DISSOLVED	Other-Lo Lim	4	3	0	0.00							1	0	0.00	2	0	0.00
00400	PH	Other-Hi Lim	9	3	0	0.00							1	0	0.00	2	0	0.00
		Other-Lo Lim	6 5	3	0	0.00							1	0	0.00	2	0	0.00
00630	NITRITE PLUS NITRATE, TOTAL 1 DET	Drinking Water	10	3	0	0.00							1	0	0.00	2	0	0.00
00940	CHLORIDE, TOTAL IN WATER	Fresh Acute	860	3	0	0.00							1	0	0.00	2	0	0.00
		Drinking Water	250	3	0	0.00							1	0	0.00	2	0	0.00
00945	SULFATE, TOTAL (AS SO4)	Drinking Water	250	3	0	0.00							1	0	0.00	2	0	0.00
01145	SELENIUM, DISSOLVED	Fresh Acute	20	3	0	0.00							1	0	0.00	2	0	0.00
		Drinking Water	50	3	0	0.00							1	0	0.00	2	0	0.00

& - Below detection limit observations, for which half the detection limit exceeded the criterion, were excluded from the criterion comparison for this parameter

154

Station Inventory for Station: BADL0024

NPS Station ID: BADL0024
Location: 42N47W 2ACAA
Station Type: /TYPA/AMBNT/SPRING
RMI-Indexes:
RMI-Miles:
HUC: 10120109
Major Basin:
Minor Basin:
RF1 Index: 10120109
RF3 Index: 10140201028200 00
Description:

LAT/LON: 43 64778l/-102 879170

Agency: 112WRD
FIPS State/County: 46113 SOUTH DAKOTA/SHANNON
STORET Station ID(s): 433852102524501
Within Park Boundary: Yes

Depth of Water: 0
Elevation: 0

Aquifer:
Water Body Id:
ECO Region:
Distance from RF1: 2 90
Distance from RF3: 0 06

RF1 Mile Point: 0 000
RF3 Mile Point: 0 00

On/Off RF1:
On/Off RF3:

Date Created: 05/01/93

Parameter Inventory for Station: BADL0024

Parameter	Period of Record	Obs	Median	Mean	Maximum	Minimum	Variance	Std Dev	10th	25th	75th	90th
00010 TEMPERATURE, WATER (DEGREES CENTIGRADE)	09/30/92-09/30/92	1	22 2	22 2	22 2	22 2	0	0	**	**	**	**
00020 TEMPERATURE, AIR (DEGREES CENTIGRADE)	09/30/92-09/30/92	1	28	28	28	28	0	0	**	**	**	**
00025 BAROMETRIC PRESSURE (MM OF HG)	09/30/92-09/30/92	1	685	685	685	685	0	0	**	**	**	**
00095 SPECIFIC CONDUCTANCE (UMHOS/CM @ 25C)	09/30/92-09/30/92	1	3440	3440	3440	3440	0	0	**	**	**	**
00400 PH (STANDARD UNITS)	09/30/92-09/30/92	1	8 56	8 56	8 56	8 56	0	0	**	**	**	**
00400 CONVERTED PH (STANDARD UNITS)	09/30/92-09/30/92	1	8 56	8 56	8 56	8 56	0	0	**	**	**	**
00400 MICRO EQUIVALENTS/LITER OF H+ COMPUTED FROM PH	09/30/92-09/30/92	1	0 003	0 003	0 003	0 003	0	0	**	**	**	**
00403 PH, LAB, STANDARD UNITS SU	09/30/92-09/30/92	1	8 2	8 2	8 2	8 2	0	0	**	**	**	**
00403 CONVERTED PH, LAB, STANDARD UNITS	09/30/92-09/30/92	1	8 2	8 2	8 2	8 2	0	0	**	**	**	**
00403 MICRO EQUIVALENTS/LITER OF H+ COMPUTED FROM PH	09/30/92-09/30/92	1	0 006	0 006	0 006	0 006	0	0	**	**	**	**
00915 CALCIUM, DISSOLVED (MG/L AS CA)	09/30/92-09/30/92	1	76	76	76	76	0	0	**	**	**	**
00925 MAGNESIUM, DISSOLVED (MG/L AS MG)	09/30/92-09/30/92	1	23	23	23	23	0	0	**	**	**	**
00930 SODIUM, DISSOLVED (MG/L AS NA)	09/30/92-09/30/92	1	750	750	750	750	0	0	**	**	**	**
00935 POTASSIUM, DISSOLVED (MG/L AS K)	09/30/92-09/30/92	1	21	21	21	21	0	0	**	**	**	**
00940 CHLORIDE,TOTAL IN WATER MG/L	09/30/92-09/30/92	1	110	110	110	110	0	0	**	**	**	**
00945 SULFATE, TOTAL (MG/L AS SO4)	09/30/92-09/30/92	1	1200	1200	1200	1200	0	0	**	**	**	**
00950 FLUORIDE, DISSOLVED (MG/L AS F)	09/30/92-09/30/92	1	1 4	1 4	1 4	1 4	0	0	**	**	**	**
00955 SILICA, DISSOLVED (MG/L AS SI02)	09/30/92-09/30/92	1	8 2	8 2	8 2	8 2	0	0	**	**	**	**
01000 ARSENIC, DISSOLVED (UG/L AS AS)	09/30/92-09/30/92	1	1	1	1	1	0	0	**	**	**	**
01005 BARIUM, DISSOLVED (UG/L AS BA)	09/30/92-09/30/92	1	18	18	18	18	0	0	**	**	**	**
01010 BERYLLIUM, DISSOLVED (UG/L AS BE)	09/30/92-09/30/92	##	0 75	0 75	0 75	0 75	0	0	**	**	**	**
01025 CADMIUM, DISSOLVED (UG/L AS CD)	09/30/92-09/30/92	##	1 5	1 5	1 5	1 5	0	0	**	**	**	**
01027 CADMIUM, TOTAL (UG/L AS CD)	09/30/92-09/30/92	##	0 5	0 5	0 5	0 5	0	0	**	**	**	**
01030 CHROMIUM, DISSOLVED (UG/L AS CR)	09/30/92-09/30/92	##	7 5	7 5	7 5	7 5	0	0	**	**	**	**
01035 COBALT, DISSOLVED (UG/L AS CO)	09/30/92-09/30/92	##	4 5	4 5	4 5	4 5	0	0	**	**	**	**
01040 COPPER, DISSOLVED (UG/L AS CU)	09/30/92-09/30/92	##	15	15	15	15	0	0	**	**	**	**
01046 IRON, DISSOLVED (UG/L AS FE)	09/30/92-09/30/92	##	110	110	110	110	0	0	**	**	**	**
01049 LEAD, DISSOLVED (UG/L AS PB)	09/30/92-09/30/92	##	15	15	15	15	0	0	**	**	**	**
01056 MANGANESE, DISSOLVED (UG/L AS MN)	09/30/92-09/30/92	##	3	3	3	3	0	0	**	**	**	**
01060 MOLYBDENUM, DISSOLVED (UG/L AS MO)	09/30/92-09/30/92	##	15	15	15	15	0	0	**	**	**	**
01065 NICKEL, DISSOLVED (UG/L AS NI)	09/30/92-09/30/92	##	15	15	15	15	0	0	**	**	**	**
01075 SILVER, DISSOLVED (UG/L AS AG)	09/30/92-09/30/92	##	1 5	1 5	1 5	1 5	0	0	**	**	**	**
01080 STRONTIUM, DISSOLVED (UG/L AS SR)	09/30/92-09/30/92	##	1300	1300	1300	1300	0	0	**	**	**	**
01085 VANADIUM, DISSOLVED (UG/L AS V)	09/30/92-09/30/92	##	9	9	9	9	0	0	**	**	**	**
01090 ZINC, DISSOLVED (UG/L AS ZN)	09/30/92-09/30/92	##	4 5	4 5	4 5	4 5	0	0	**	**	**	**
01130 LITHIUM, DISSOLVED (UG/L AS LI)	09/30/92-09/30/92	##	250	250	250	250	0	0	**	**	**	**
01145 SELENIUM, DISSOLVED (UG/L AS SE)	09/30/92-09/30/92	##	0 5	0 5	0 5	0 5	0	0	**	**	**	**
70300 RESIDUE,TOTAL FILTRABLE (DRIED AT 180C),MG/L	09/30/92-09/30/92	##	2600	2600	2600	2600	0	0	**	**	**	**

** - Less than 9 observations   ## - Computed with 50% or more of the total observations as values that were half the detection limit   p - Has a corresponding time series plot

Parameter Inventory for Station: BADL0024

Parameter	Period of Record	Obs	Median	Mean	Maximum	Minimum	Variance	Std Dev	10th	25th	75th	90th
71890 MERCURY, DISSOLVED (UG/L AS HG)	09/30/92-09/30/92	1	0.1	0.1	0.1	0.1	0	0	**	**	**	**

** - Less than 9 observations   ## - Computed with 50% or more of the total observations as values that were half the detection limit   p - Has a corresponding time series plot

EPA Water Quality Criteria Analysis for Station: BADL0024

Parameter	Std Type	Std Value	Total Obs	Exceed Standard	Prop Exceeding	10/01-1/31 Obs	Exceed	Prop	2/01-4/14 Obs	Exceed	Prop	4/15-6/30 Obs	Exceed	Prop	7/01-9/30 Obs	Exceed	Prop
00400 PH	Other-Hi Lim	9	1	0	0.00										1	0	0.00
	Other-Lo Lim	6.5	1	0	0.00										1	0	0.00
00403 PH, LAB	Other-Hi Lim	9	1	0	0.00										1	0	0.00
	Other-Lo Lim	6.5	1	0	0.00										1	0	0.00
00940 CHLORIDE,TOTAL IN WATER	Fresh Acute	860	1	0	0.00										1	0	0.00
00945 SULFATE, TOTAL (AS SO4)	Drinking Water	250	1	1	1.00										1	1	1.00
00950 FLUORIDE, DISSOLVED AS F	Drinking Water	4	1	0	0.00										1	0	0.00
01000 ARSENIC, DISSOLVED	Fresh Acute	360	1	0	0.00										1	0	0.00
	Drinking Water	50	1	0	0.00										1	0	0.00
01005 BARIUM, DISSOLVED	Drinking Water	2000	1	0	0.00										1	0	0.00
01010 BERYLLIUM, DISSOLVED	Fresh Acute	130	1	0	0.00										1	0	0.00
	Drinking Water	4	1	0	0.00										1	0	0.00
01025 CADMIUM, DISSOLVED	Fresh Acute	3.9	1	0	0.00										1	0	0.00
	Drinking Water	5	1	0	0.00										1	0	0.00
01027 CADMIUM, TOTAL	Fresh Acute	3.9	1	0	0.00										1	0	0.00
	Drinking Water	5	1	0	0.00										1	0	0.00
01030 CHROMIUM, DISSOLVED	Drinking Water	100	1	0	0.00										1	0	0.00
01040 COPPER, DISSOLVED	Fresh Acute	18	1	0	0.00										1	0	0.00
	Drinking Water	1300	1	0	0.00										1	0	0.00
01049 LEAD, DISSOLVED	Fresh Acute	82	1	0	0.00										1	0	0.00
	Drinking Water	15	0 &	0	0.00										1	0	0.00
01065 NICKEL, DISSOLVED	Fresh Acute	1400	1	0	0.00										1	0	0.00
	Fresh Acute	100	1	0	0.00										1	0	0.00
01075 SILVER, DISSOLVED	Fresh Acute	4.1	1	0	0.00										1	0	0.00
01090 ZINC, DISSOLVED	Drinking Water	100	1	0	0.00										1	0	0.00
	Fresh Acute	120	1	0	0.00										1	0	0.00
	Drinking Water	5000	1	0	0.00										1	0	0.00
01145 SELENIUM, DISSOLVED	Fresh Acute	20	1	0	0.00										1	0	0.00
	Drinking Water	50	1	0	0.00										1	0	0.00
71890 MERCURY, DISSOLVED	Fresh Acute	2.4	1	0	0.00										1	0	0.00
	Drinking Water	2	1	0	0.00										1	0	0.00

& - Below detection limit observations, for which half the detection limit exceeded the criterion, were excluded from the criterion comparison for this parameter

156

Station Inventory for Station: BADL0025

NPS Station ID: BADL0025
Location: 41N47W23BAAC
Station Type: /TYPA/AMBNT/SPRING
RMI-Indexes:
HUC: 10120109
Major Basin:
Minor Basin:
RF1 Index: 10120109
RF3 Index: 10140201028200 00
Description:

LAT/LON: 43 520281/-102 884726

Agency: 112WRD
FIPS State/County: 46113 SOUTH DAKOTA/SHANNON
STORET Station ID(s): 43311310253050 1
Within Park Boundary: Yes

Date Created: 05/01/93

Depth of Water: 0
Elevation: 0

Aquifer:
Water Body Id:
ECO Region:
Distance from RF1: 2 90
Distance from RF3: 0 06

RF1 Mile Point: 0 000
RF3 Mile Point: 0 00

On/Off RF1:
On/Off RF3:

Parameter Inventory for Station: BADL0025

Parameter		Period of Record	Obs	Median	Mean	Maximum	Minimum	Variance	Std Dev	10th	25th	75th	90th
00010	TEMPERATURE, WATER (DEGREES CENTIGRADE)	09/18/92-09/18/92	1	14 6	14 6	14 6	14 6	0	0	**	**	**	**
00020	TEMPERATURE, AIR (DEGREES CENTIGRADE)	09/18/92-09/18/92	1	18	18	18	18	0	0	**	**	**	**
00025	BAROMETRIC PRESSURE (MM OF HG)	09/18/92-09/18/92	1	685	685	685	685	0	0	**	**	**	**
00095	SPECIFIC CONDUCTANCE (UMHOS/CM @ 25C)	09/18/92-09/18/92	1	518	518	518	518	0	0	**	**	**	**
00400	PH (STANDARD UNITS)	09/18/92-09/18/92	1	7 28	7 28	7 28	7 28	0	0	**	**	**	**
00400	CONVERTED PH (STANDARD UNITS)	09/18/92-09/18/92	1	7 28	7 28	7 28	7 28	0	0	**	**	**	**
00400	MICRO EQUIVALENTS/LITER OF H+ COMPUTED FROM PH	09/18/92-09/18/92	1	0 052	0 052	0 052	0 052	0	0	**	**	**	**
00403	PH, LAB, STANDARD UNITS SU	09/18/92-09/18/92	1	7 6	7 6	7 6	7 6	0	0	**	**	**	**
00403	CONVERTED PH, LAB, STANDARD UNITS	09/18/92-09/18/92	1	7 6	7 6	7 6	7 6	0	0	**	**	**	**
00403	MICRO EQUIVALENTS/LITER OF H+ COMPUTED FROM PH	09/18/92-09/18/92	1	0 025	0 025	0 025	0 025	0	0	**	**	**	**
00915	CALCIUM, DISSOLVED (MG/L AS CA)	09/18/92-09/18/92	1	42	42	42	42	0	0	**	**	**	**
00925	MAGNESIUM, DISSOLVED (MG/L AS MG)	09/18/92-09/18/92	1	4 7	4 7	4 7	4 7	0	0	**	**	**	**
00930	SODIUM, DISSOLVED (MG/L AS NA)	09/18/92-09/18/92	1	63	63	63	63	0	0	**	**	**	**
00935	POTASSIUM, DISSOLVED (MG/L AS K)	09/18/92-09/18/92	1	9 7	9 7	9 7	9 7	0	0	**	**	**	**
00940	CHLORIDE,TOTAL IN WATER MG/L	09/18/92-09/18/92	1	19	19	19	19	0	0	**	**	**	**
00945	SULFATE, TOTAL (MG/L AS SO4)	09/18/92-09/18/92	1	23	23	23	23	0	0	**	**	**	**
00950	FLUORIDE, DISSOLVED (MG/L AS F)	09/18/92-09/18/92	1	0 2	0 2	0 2	0 2	0	0	**	**	**	**
00955	SILICA, DISSOLVED (MG/L AS SI02)	09/18/92-09/18/92	1	58	58	58	58	0	0	**	**	**	**
01000	ARSENIC, DISSOLVED (UG/L AS AS)	09/18/92-09/18/92	1	5	5	5	5	0	0	**	**	**	**
01005	BARIUM, DISSOLVED (UG/L AS BA)	09/18/92-09/18/92	1	210	210	210	210	0	0	**	**	**	**
01010	BERYLLIUM, DISSOLVED (UG/L AS BE)	09/18/92-09/18/92	##	0 25	0 25	0 25	0 25	0	0	**	**	**	**
01020	BORON, DISSOLVED (UG/L AS B)	09/18/92-09/18/92	##	170	170	170	170	0	0	**	**	**	**
01025	CADMIUM, DISSOLVED (UG/L AS CD)	09/18/92-09/18/92	##	0 5	0 5	0 5	0 5	0	0	**	**	**	**
01027	CADMIUM, TOTAL (UG/L AS CD)	09/18/92-09/18/92	##	0 5	0 5	0 5	0 5	0	0	**	**	**	**
01030	CHROMIUM, DISSOLVED (UG/L AS CR)	09/18/92-09/18/92	##	2 5	2 5	2 5	2 5	0	0	**	**	**	**
01035	COBALT, DISSOLVED (UG/L AS CO)	09/18/92-09/18/92	##	1 5	1 5	1 5	1 5	0	0	**	**	**	**
01040	COPPER, DISSOLVED (UG/L AS CU)	09/18/92-09/18/92	##	5	5	5	5	0	0	**	**	**	**
01046	IRON, DISSOLVED (UG/L AS FE)	09/18/92-09/18/92	##	310	310	310	310	0	0	**	**	**	**
01049	LEAD, DISSOLVED (UG/L AS PB)	09/18/92-09/18/92	##	5	5	5	5	0	0	**	**	**	**
01056	MANGANESE, DISSOLVED (UG/L AS MN)	09/18/92-09/18/92	1	190	190	190	190	0	0	**	**	**	**
01060	MOLYBDENUM, DISSOLVED (UG/L AS MO)	09/18/92-09/18/92	##	5	5	5	5	0	0	**	**	**	**
01065	NICKEL, DISSOLVED (UG/L AS NI)	09/18/92-09/18/92	##	5	5	5	5	0	0	**	**	**	**
01075	SILVER, DISSOLVED (UG/L AS AG)	09/18/92-09/18/92	##	0 5	0 5	0 5	0 5	0	0	**	**	**	**
01080	STRONTIUM, DISSOLVED (UG/L AS SR)	09/18/92-09/18/92	##	260	260	260	260	0	0	**	**	**	**
01085	VANADIUM, DISSOLVED (UG/L AS V)	09/18/92-09/18/92	1	3	3	3	3	0	0	**	**	**	**
01090	ZINC, DISSOLVED (UG/L AS ZN)	09/18/92-09/18/92	1	48	48	48	48	0	0	**	**	**	**
01130	LITHIUM, DISSOLVED (UG/L AS LI)	09/18/92-09/18/92	1	28	28	28	28	0	0	**	**	**	**
01145	SELENIUM, DISSOLVED (UG/L AS SE)	09/18/92-09/18/92	1	5	5	5	5	0	0	**	**	**	**

** - Less than 9 observations   ## - Computed with 50% or more of the total observations as values that were half the detection limit   p - Has a corresponding time series plot

157

Parameter Inventory for Station: BADL0025

Parameter		Period of Record	Obs	Median	Mean	Minimum	Maximum	Variance	Std Dev	10th	25th	75th	90th
70300	RESIDUE,TOTAL FILTRABLE (DRIED AT 180C),MG/L	09/18/92-09/18/92	1	353	353	353	353	0	0	**	**	**	**
71890	MERCURY, DISSOLVED (UG/L AS HG)	09/18/92-09/18/92	1 ##	0 05	0 05	0 05	0 05	0	0	**	**	**	**

** - Less than 9 observations   ## - Computed with 50% or more of the total observations as values that were half the detection limit   p - Has a corresponding time series plot

EPA Water Quality Criteria Analysis for Station: BADL0025

Parameter		Std Type	Std Value	Total Obs	Exceed Standard	Prop Exceeding	10/01-1/31 Obs	Exceed	Prop	2/01-4/14 Obs	Exceed	Prop	4/15-6/30 Obs	Exceed	Prop	7/01-9/30 Obs	Exceed	Prop
00400	PH	Other-Hi Lim	9	1	0	0.00										1	0	0.00
		Other-Lo Lim	6 5	1	0	0.00										1	0	0.00
00403	PH, LAB	Other-Hi Lim	9	1	0	0.00										1	0	0.00
		Other-Lo Lim	6 5	1	0	0.00										1	0	0.00
00940	CHLORIDE,TOTAL IN WATER	Fresh Acute	860	1	0	0.00										1	0	0.00
00945	SULFATE, TOTAL (AS SO4)	Drinking Water	250	1	0	0.00										1	0	0.00
00950	FLUORIDE, DISSOLVED AS F	Drinking Water	250	1	0	0.00										1	0	0.00
01000	ARSENIC, DISSOLVED	Drinking Water	4	1	0	0.00										1	0	0.00
		Fresh Acute	360	1	0	0.00										1	0	0.00
		Drinking Water	50	1	0	0.00										1	0	0.00
01005	BARIUM, DISSOLVED	Drinking Water	2000	1	0	0.00										1	0	0.00
01010	BERYLLIUM, DISSOLVED	Fresh Acute	130	1	0	0.00										1	0	0.00
		Drinking Water	4	1	0	0.00										1	0	0.00
01025	CADMIUM, DISSOLVED	Fresh Acute	3 9	1	0	0.00										1	0	0.00
01027	CADMIUM, TOTAL	Drinking Water	5	1	0	0.00										1	0	0.00
		Fresh Acute	3 9	1	0	0.00										1	0	0.00
01030	CHROMIUM, DISSOLVED	Drinking Water	100	1	0	0.00										1	0	0.00
01040	COPPER, DISSOLVED	Fresh Acute	18	1	0	0.00										1	0	0.00
		Drinking Water	1300	1	0	0.00										1	0	0.00
01049	LEAD, DISSOLVED	Fresh Acute	82	1	0	0.00										1	0	0.00
		Drinking Water	15	1	0	0.00										1	0	0.00
01065	NICKEL, DISSOLVED	Fresh Acute	1400	1	0	0.00										1	0	0.00
		Drinking Water	100	1	0	0.00										1	0	0.00
01075	SILVER, DISSOLVED	Fresh Acute	4 1	1	0	0.00										1	0	0.00
01090	ZINC, DISSOLVED	Fresh Acute	100	1	0	0.00										1	0	0.00
		Drinking Water	120	1	0	0.00										1	0	0.00
01145	SELENIUM, DISSOLVED	Fresh Acute	5000	1	0	0.00										1	0	0.00
		Drinking Water	20	1	0	0.00										1	0	0.00
71890	MERCURY, DISSOLVED	Fresh Acute	50	1	0	0.00										1	0	0.00
		Drinking Water	2 4	1	0	0.00										1	0	0.00
		Drinking Water	2	1	0	0.00										1	0	0.00

& - Below detection limit observations, for which half the detection limit exceeded the criterion, were excluded from the criterion comparison for this parameter

158

Station Inventory for Station: BADL0026

Date Created: 12/30/89

NPS Station ID: BADL0026
Location: BATTLE CR BELOW HERMOSA SD
Station Type: /TYPA/AMBNT/STREAM
RMI-Miles:
HUC: 10120109
Major Basin:
Minor Basin:
RF1 Index: 10120109
RF3 Index: 101201090047000 00
Description:

LAT/LON: 43.725004/-102.904170

Agency: 112WRD
FIPS State/County: 46033 SOUTH DAKOTA/CUSTER
STORET Station ID(s): 06406500
Within Park Boundary: No

Depth of Water: 0
Elevation: 0

Aquifer:
Water Body Id:
ECO Region:
Distance from RF1: 15.40
Distance from RF3: 0.02

RF1 Mile Point: 0.000
RF3 Mile Point: 3.75

On/Off RF1:
On/Off RF3:

Parameter Inventory for Station: BADL0026

Parameter	Parameter	Period of Record	Obs	Median	Mean	Maximum	Minimum	Variance	Std Dev	10th	25th	75th	90th
00010	TEMPERATURE, WATER (DEGREES CENTIGRADE)	03/09/89-09/10/96	71	14	12.123	28	0	66.565	8.159	0	4	19	22
00020	TEMPERATURE, AIR (DEGREES CENTIGRADE)	03/09/89-09/10/96	68	18.5	15.944	32	-20	116.874	10.811	4.45	-2	24.875	29.1
00025	BAROMETRIC PRESSURE (MM OF HG)	09/07/93-08/30/94	2	687	687	694	680	98	9.899	**	**	**	**
00061	FLOW, STREAM, INSTANTANEOUS CFS	03/09/89-09/10/96	73	12	69.718	1355	0.2	34510.957	185.771	1	4	39.5	169.8
00065	STAGE, STREAM (FEET)	10/01/92-08/30/94	12	0	0.418	2.94	0	0.985	0.993	**	0	0	**
00076	TURBIDITY,HACH TURBIDIMETER (FORMAZIN TURB UNIT)	09/07/93-08/30/94	2	4.9	4.9	6.7	3.1	6.48	2.546	**	**	**	2.679
00095p	SPECIFIC CONDUCTANCE (UMHOS/CM @ 25C)	03/09/89-09/10/96	69	650	683.826	1860	148	74500.146	272.947	339	534	840	1067
00300	OXYGEN, DISSOLVED MG/L	09/07/93-08/30/94	2	9.5	9.5	9.9	9.1	0.32	0.566	**	**	**	**
00400	PH (STANDARD UNITS)	09/07/93-08/30/94	2	8.175	8.175	8.4	7.95	0.101	0.318	**	**	**	**
00400	CONVERTED PH (STANDARD UNITS)	09/07/93-08/30/94	2	8.119	8.119	8.4	7.95	0.107	0.328	**	**	**	**
00400	MICRO EQUIVALENTS/LITER OF H+ COMPUTED FROM PH	09/07/93-08/30/94	2	0.008	0.008	0.011	0.004	0	0.005	**	**	**	**
00403	PH, LAB, STANDARD UNITS SU	09/07/93-08/30/94	2	8.05	8.05	8.1	8	0.005	0.071	**	**	**	**
00403	CONVERTED PH, LAB, STANDARD UNITS	09/07/93-08/30/94	2	8.047	8.047	8.1	8	0.005	0.071	**	**	**	**
00403	MICRO EQUIVALENTS/LITER OF H+ COMPUTED FROM PH	09/07/93-08/30/94	2	0.009	0.009	0.01	0.008	0	0.001	**	**	**	**
00530	RESIDUE, TOTAL NONFILTRABLE (MG/L)	09/07/93-08/30/94	2	27	27	38	16	242	15.556	**	**	**	**
00608	NITROGEN, AMMONIA, DISSOLVED (MG/L AS N)	09/07/93-08/30/94	2##	0.03	0.03	0.04	0.02	0	0.014	**	**	**	**
00613	NITRITE NITROGEN, DISSOLVED (MG/L AS N)	09/07/93-08/30/94	2##	0.005	0.005	0.005	0.005	0	0	**	**	**	**
00623	NITROGEN, KJELDAHL, DISSOLVED (MG/L AS N)	09/07/93-08/30/94	2##	0.1	0.1	0.1	0.1	0	0	**	**	**	**
00631	NITRITE PLUS NITRATE, DISS 1 DET (MG/L AS N)	09/07/93-08/30/94	2##	0.025	0.025	0.025	0.025	0	0	**	**	**	**
00666	PHOSPHORUS, DISSOLVED (MG/L AS P)	09/07/93-08/30/94	2##	0.007	0.007	0.008	0.006	0	0.001	**	**	**	**
00671	PHOSPHORUS, DISSOLVED ORTHOPHOSPHATE (MG/L AS P)	09/07/93-09/07/93	1##	0.005	0.005	0.005	0.005			**	**	**	**
00723	CYANIDE, DISSOLVED STD METHOD (UG/L)	09/07/93-08/30/94	2	0.005	0.005	0.005	0.005	0	0	**	**	**	**
00915	CALCIUM, DISSOLVED (MG/L AS CA)	09/07/93-08/30/94	2	84.5	84.5	89	80	40.5	6.364	**	**	**	**
00925	MAGNESIUM, DISSOLVED (MG/L AS MG)	09/07/93-08/30/94	2	25	25	28	22	18	4.243	**	**	**	**
00930	SODIUM, DISSOLVED (MG/L AS NA)	09/07/93-08/30/94	2	19	19	24	14	50	7.071	**	**	**	**
00935	POTASSIUM, DISSOLVED (MG/L AS K)	09/07/93-08/30/94	2	5.5	5.5	5.7	5.3	0.08	0.283	**	**	**	**
00940	CHLORIDE,TOTAL IN WATER MG/L	09/07/93-08/30/94	2	8.5	8.5	9	8	0.5	0.707	**	**	**	**
00945	SULFATE, TOTAL (MG/L AS SO4)	09/07/93-08/30/94	2	165	165	190	140	1250	35.355	**	**	**	**
00950	FLUORIDE, DISSOLVED (MG/L AS F)	09/07/93-08/30/94	2	0.3	0.3	0.3	0.3	0	0	**	**	**	**
00955	SILICA, DISSOLVED (MG/L AS SI02)	09/07/93-08/30/94	2	11.45	11.45	13	9.9	4.805	2.192	**	**	**	**
01000	ARSENIC, DISSOLVED (UG/L AS AS)	09/07/93-08/30/94	2	7	7	8	6	2	1.414	**	**	**	**
01002	ARSENIC, TOTAL (UG/L AS AS)	09/07/93-08/30/94	2	8.5	8.5	10	7	4.5	2.121	**	**	**	**
01005	BARIUM, DISSOLVED (UG/L AS BA)	09/07/93-08/30/94	2	72.5	72.5	74	71	4.5	2.121	**	**	**	**
01010	BERYLLIUM, DISSOLVED (UG/L AS BE)	09/07/93-08/30/94	2	0.25	0.25	0.25	0.25	0	0	**	**	**	**
01020	BORON, DISSOLVED (UG/L AS B)	09/07/93-08/30/94	2	50	50	60	40	200	14.142	**	**	**	**
01025	CADMIUM, DISSOLVED (UG/L AS CD)	09/07/93-08/30/94	2##	3.5	3.5	5	2.5	4.5	2.121	**	**	**	**
01030	CHROMIUM, DISSOLVED (UG/L AS CR)	09/07/93-08/30/94	2##	3.25	3.25	4	2.5	1.125	1.061	**	**	**	**
01035	COBALT, DISSOLVED (UG/L AS CO)	08/30/94-08/30/94	1##	1.5	1.5	1.5	1.5	0	0	**	**	**	**

** - Less than 9 observations   ## - Computed with 50% or more of the total observations as values that were half the detection limit   p - Has a corresponding time series plot

Parameter Inventory for Station: BADL0026

Parameter		Period of Record	Obs	Median	Mean	Maximum	Minimum	Variance	Std Dev	10th	25th	75th	90th
01040	COPPER, DISSOLVED (UG/L AS CU)	09/07/93-08/30/94	2 ##	3	3	5	1	8	2.828	**	**	**	**
01045	IRON, TOTAL (UG/L AS FE)	09/07/93-08/30/94	2	430	430	570	290	39200	197.99	**	**	**	**
01046	IRON, DISSOLVED (UG/L AS FE)	09/07/93-08/30/94	2	5	5	5	5	0	0	**	**	**	**
01049	LEAD, DISSOLVED (UG/L AS PB)	09/07/93-08/30/94	2 ##	2.75	2.75	5	0.5	10.125	3.182	**	**	**	**
01056	MANGANESE, DISSOLVED (UG/L AS MN)	09/07/93-08/30/94	2 ##	3	3	4	2	2	1.414	**	**	**	**
01060	MOLYBDENUM, DISSOLVED (UG/L AS MO)	08/30/94-08/30/94	1 ##	20	20	20	20	0	0	**	**	**	**
01065	NICKEL, DISSOLVED (UG/L AS NI)	08/30/94-08/30/94	1 ##	5	5	5	5	0	0	**	**	**	**
01075	SILVER, DISSOLVED (UG/L AS AG)	08/30/94-08/30/94	1 ##	0.5	0.5	0.5	0.5	0	0	**	**	**	**
01080	STRONTIUM, DISSOLVED (UG/L AS SR)	08/30/94-08/30/94	1 ##	750	750	750	750	0	0	**	**	**	**
01085	VANADIUM, DISSOLVED (UG/L AS V)	08/30/94-08/30/94	1 ##	3	3	4	3	0	0	**	**	**	**
01090	ZINC, DISSOLVED (UG/L AS ZN)	09/07/93-08/30/94	2 ##	2.75	2.75	3	1.5	3.125	1.768	**	**	**	**
01092	ZINC, TOTAL (UG/L AS ZN)	09/07/93-08/30/94	2 ##	5	5	5	5	0	0	**	**	**	**
01095	ANTIMONY, DISSOLVED (UG/L AS SB)	09/07/93-08/30/94	2 ##	1	1	1	1	0	0	**	**	**	**
01130	LITHIUM, DISSOLVED (UG/L AS LI)	09/07/93-08/30/94	2 ##	32.5	32.5	39	26	84.5	9.192	**	**	**	**
01145	SELENIUM, DISSOLVED (UG/L AS SE)	09/07/93-08/30/94	2 ##	0.5	0.5	0.5	0.5	0	0	**	**	**	**
01147	SELENIUM, TOTAL (UG/L AS SE)	09/07/93-08/30/94	2 ##	0.5	0.5	0.5	0.5	0	0	**	**	**	**
03515	BETA, DISSOLVED GROSS, AS CS-137, PC/L	09/07/93-08/30/94	2	9.1	9.1	11	7.2	7.22	2.687	**	**	**	**
04126	ALPHA, DISSOLVED, WATER (AS TH-230) PCI/L	09/07/93-08/30/94	2	7.4	7.4	9.9	4.9	12.5	3.536	**	**	**	**
22703	URANIUM, NATURAL, DISSOLVED	09/07/93-08/30/94	2	5.1	5.1	6	4.2	1.62	1.273	**	**	**	**
31625	FECAL COLIFORM, MF,M-FC, 0.7 UM	09/07/93-08/30/94	2	500	500	920	80	352800	593.97	**	**	**	**
31625	LOG FECAL COLIFORM, MF,M-FC, 0.7 UM	09/07/93-08/30/94	2	2.433	2.433	2.964	1.903	0.563	0.75	**	**	**	**
	GEOMETRIC MEAN =				271.293								
31673	FECAL STREPTOCOCCI, MBR FILT,KF AGAR,35C,48HR	09/07/93-08/30/94	2	424.5	424.5	760	89	225120.5	474.469	**	**	**	**
31673	LOG FECAL STREPTOCOCCI, MBR FILT,KF AGAR,35C,48HR	09/07/93-08/30/94	2	2.415	2.415	2.881	1.949	0.434	0.659	**	**	**	**
	GEOMETRIC MEAN =				260.077								
32730	PHENOLICS, TOTAL, RECOVERABLE (UG/L)	09/07/93-08/30/94	2 ##	0.5	0.5	0.5	0.5	0	0	**	**	**	**
70300	RESIDUE,TOTAL FILTRABLE (DRIED AT 180C),MG/L	09/07/93-08/30/94	2 ##	438	438	491	385	5618	74.953	**	**	**	**
71890	MERCURY, DISSOLVED (UG/L AS HG)	09/07/93-08/30/94	2 ##	0.05	0.05	0.05	0.05	0	0	**	**	**	**
75986	ALPHA GROSS,1 SIGMA PRC EST AS NAT U,DISS,WTR UG/L	09/07/93-08/30/94	2	4.365	4.365	5.2	3.53	1.394	1.181	**	**	**	**
75987	ALPHA GROSS,DISS,1 SIGMA PRC EST AS TH230,WTR PC/L	09/07/93-08/30/94	2	3.115	3.115	3.72	2.51	0.732	0.856	**	**	**	**
75988	BETA GROSS,DISS,1 SIGMA PRC EST AS SR90,Y90 PC/L	09/07/93-08/30/94	2	1.345	1.345	1.45	1.24	0.022	0.148	**	**	**	**
75989	BETA GROSS,1 SIGMA PRC EST AS CS-137,DISS,WTR PC/L	09/07/93-08/30/94	2	1.79	1.79	1.95	1.63	0.051	0.226	**	**	**	**
75990	URANIUM,NATURAL,1 SIGMA PRC EST,DISS,WATER UG/L	09/07/93-08/30/94	2	0.75	0.75	0.9	0.6	0.045	0.212	**	**	**	**
80030	ALPHA,DISSOLVED GROSS,AS URANIUM-NATURAL,UG/L	09/07/93-08/30/94	2	10.45	10.45	14	6.9	25.205	5.02	**	**	**	**
80050	BETA,DISSOLVED GROSS,AS SR-Y-90, PC/L	09/07/93-08/30/94	2	6.85	6.85	8.2	5.5	3.645	1.909	**	**	**	**
80154	SUSP. SEDIMENT CONCENTRATION-EVAP. AT 110C (MG/L)	08/30/94-08/30/94	1	116	116	116	116	0	0	**	**	**	**

** - Less than 9 observations   ## - Computed with 50% or more of the total observations as values that were half the detection limit   p - Has a corresponding time series plot

EPA Water Quality Criteria Analysis for Station: BADL0026

Parameter		Std Type	Std Value	Total Obs	Exceed Standard	Prop Exceeding	10/01-1/31 Obs	Exceed	Prop	2/01-4/14 Obs	Exceed	Prop	4/15-6/30 Obs	Exceed	Prop	7/01-9/30 Obs	Exceed	Prop
00076	TURBIDITY, HACH TURBIDIMETER	Other-Hi Lim	50	2	0	0.00										2	0	0.00
00300	OXYGEN, DISSOLVED	Other-Lo Lim	4	2	0	0.00										2	0	0.00
00400	PH	Other-Hi Lim	9	2	0	0.00										2	0	0.00
		Other-Lo Lim	6.5	2	0	0.00										2	0	0.00
00403	PH, LAB	Other-Hi Lim	9	2	0	0.00										2	0	0.00
		Other-Lo Lim	6.5	2	0	0.00										2	0	0.00
00613	NITRITE NITROGEN, DISSOLVED AS N	Drinking Water	1	2	0	0.00										2	0	0.00
00631	NITRITE PLUS NITRATE, DISS 1 DET	Drinking Water	10	2	0	0.00										2	0	0.00
00723	CYANIDE, DISSOLVED STD METHOD	Fresh Acute	22	1	0	0.00										1	0	0.00
00940	CHLORIDE, TOTAL IN WATER	Drinking Water	200	1	0	0.00										1	0	0.00
		Fresh Acute	860	2	0	0.00										2	0	0.00
00945	SULFATE, TOTAL (AS SO4)	Drinking Water	250	2	0	0.00										2	0	0.00
00950	FLUORIDE, DISSOLVED AS F	Drinking Water	250	2	0	0.00										2	0	0.00
		Drinking Water	4	2	0	0.00										2	0	0.00
01000	ARSENIC, DISSOLVED	Fresh Acute	360	2	0	0.00										2	0	0.00
		Drinking Water	50	2	0	0.00										2	0	0.00
01002	ARSENIC, TOTAL	Fresh Acute	360	2	0	0.00										2	0	0.00
		Drinking Water	50	2	0	0.00										2	0	0.00

& - Below detection limit observations, for which half the detection limit exceeded the criterion, were excluded from the criterion comparison for this parameter

160

EPA Water Quality Criteria Analysis for Station: BADL0026

Parameter		Std Type	Std Value	Total Obs	Exceed Standard	Prop Exceeding	10/01-1/31			2/01-4/14			4/15-6/30			7/01-9/30		
							Obs	Exceed	Prop	Obs	Exceed	Prop	Obs	Exceed	Prop	Obs	Exceed	Prop
01005	BARIUM DISSOLVED	Drinking Water	2000	2	0	0.00										2	0	0.00
01010	BERYLLIUM, DISSOLVED	Fresh Acute	130	2	0	0.00										2	0	0.00
		Drinking Water	4	2	0	0.00										2	0	0.00
01025	CADMIUM, DISSOLVED	Fresh Acute	3 9	1 &	0	0.00										1	0	0.00
		Drinking Water	5	1 &	0	0.00										1	0	0.00
01030	CHROMIUM, DISSOLVED	Drinking Water	100	2	0	0.00										2	0	0.00
01040	COPPER, DISSOLVED	Fresh Acute	18	2	0	0.00										2	0	0.00
		Drinking Water	1300	2	0	0.00										2	0	0.00
01049	LEAD, DISSOLVED	Fresh Acute	82	2	0	0.00										2	0	0.00
		Drinking Water	15	2	0	0.00										2	0	0.00
01065	NICKEL, DISSOLVED	Fresh Acute	1400	1	0	0.00										1	0	0.00
		Drinking Water	100	1	0	0.00										1	0	0.00
01075	SILVER, DISSOLVED	Fresh Acute	4 1	1	0	0.00										1	0	0.00
		Drinking Water	100	1	0	0.00										1	0	0.00
01090	ZINC, DISSOLVED	Fresh Acute	120	2	0	0.00										2	0	0.00
		Drinking Water	5000	2	0	0.00										2	0	0.00
01092	ZINC, TOTAL	Fresh Acute	120	2	0	0.00										2	0	0.00
		Drinking Water	5000	2	0	0.00										2	0	0.00
01095	ANTIMONY, DISSOLVED	Fresh Acute	88	2	0	0.00										2	0	0.00
		Drinking Water	6	2	0	0.00										2	0	0.00
01145	SELENIUM, DISSOLVED	Fresh Acute	20	2	0	0.00										2	0	0.00
		Drinking Water	50	2	0	0.00										2	0	0.00
01147	SELENIUM, TOTAL	Fresh Acute	20	2	0	0.00										2	0	0.00
		Drinking Water	50	2	0	0.00										2	0	0.00
22703	URANIUM, NATURAL DISSOLVED	Drinking Water	20	2	0	0.00										2	0	0.00
31625	FECAL COLIFORM, MF	Other-Hi Lim	200	2	1	0.50										2	1	0.50
71890	MERCURY, DISSOLVED	Fresh Acute	2 4	2	0	0.00										2	0	0.00
		Drinking Water		2	0	0.00										2	0	0.00

& - Below detection limit observations, for which half the detection limit exceeded the criterion, were excluded from the criterion comparison for this parameter

161

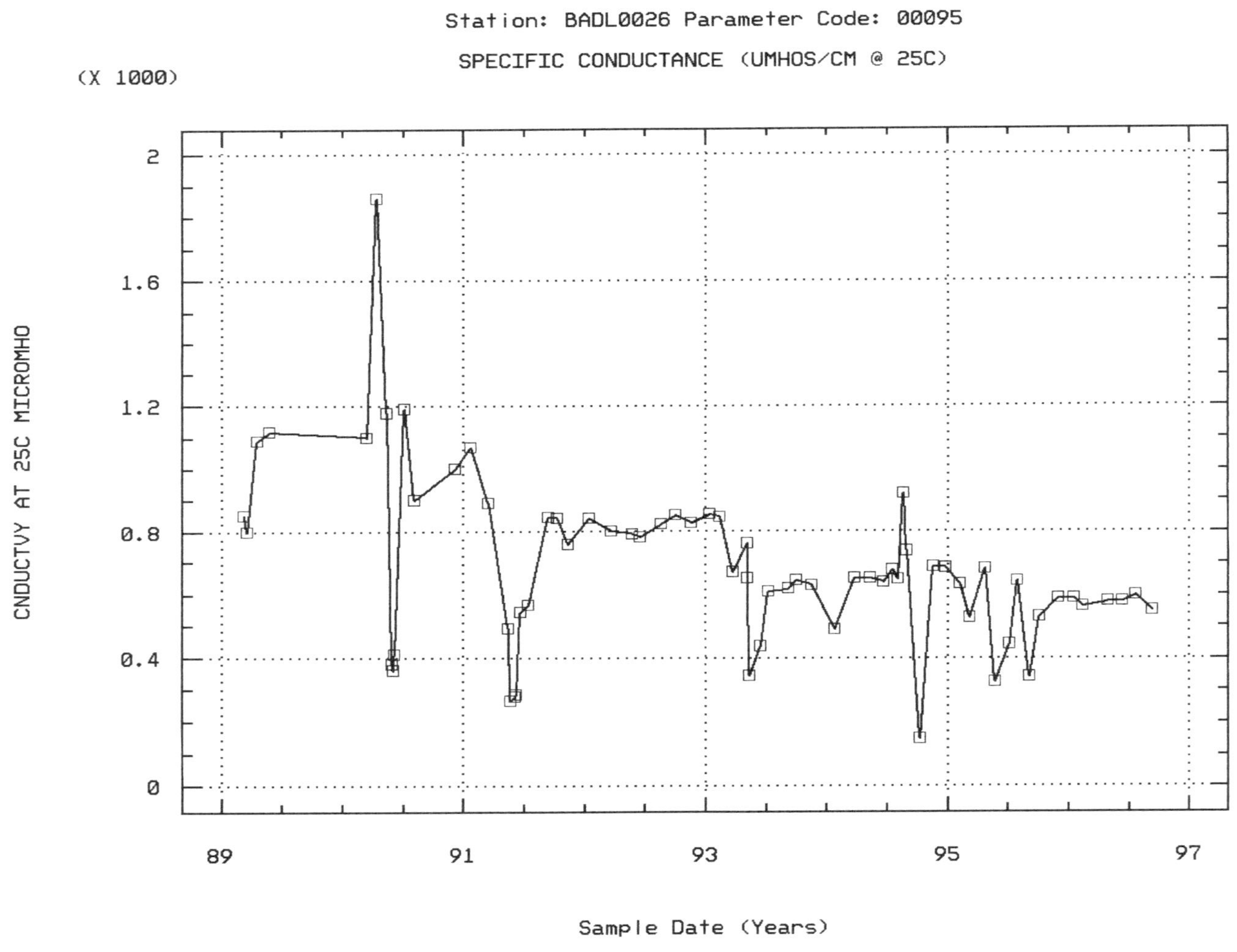

Station: BADL0026 Parameter Code: 00095
SPECIFIC CONDUCTANCE (UMHOS/CM @ 25C)
(X 1000)
CNDUCTVY AT 25C MICROMHO
2
1.6
1.2
0.8
0.4
0
89
91
93
95
97
Sample Date (Years)

BATTLE CR BELOW HERMOSA SD

Annual Analysis for 1989 - Station BADL0026

Parameter		Period of Record	Obs	Median	Mean	Maximum	Minimum	Variance	Std Dev	10th	25th	75th	90th
00010	TEMPERATURE, WATER (DEGREES CENTIGRADE)	03/09/89-09/10/96	4	7.25	7.125	13.5	0.5	35.229	5.935	**	**	**	**
00020	TEMPERATURE, AIR (DEGREES CENTIGRADE)	03/09/89-09/10/96	4	12	12.25	17	8	13.583	3.686	**	**	**	**
00061	FLOW, STREAM, INSTANTANEOUS CFS	03/09/89-09/10/96	4	2	2.55	6	0.2	6.677	2.584	**	**	**	**
00095p	SPECIFIC CONDUCTANCE (UMHOS/CM @ 25C)	03/09/89-09/10/96	4	970	965	1120	800	26700	163.401	**	**	**	**

** - Less than 9 observations   ## - Computed with 50% or more of the total observations as values that were half the detection limit   p - Has a corresponding box-and-whisker plot

Annual Analysis for 1990 - Station BADL0026

Parameter		Period of Record	Obs	Median	Mean	Maximum	Minimum	Variance	Std Dev	10th	25th	75th	90th
00010	TEMPERATURE, WATER (DEGREES CENTIGRADE)	03/09/89-09/10/96	9	17	14.556	26	0	70.965	8.424	0	7.25	21.5	26
00020	TEMPERATURE, AIR (DEGREES CENTIGRADE)	03/09/89-09/10/96	8	18.5	16.313	31	-6.5	151.21	12.297	**	**	**	**
00061	FLOW, STREAM, INSTANTANEOUS CFS	03/09/89-09/10/96	10	2	77.44	351	0.5	16242.28	127.445	0.54	0.975	129.25	342.7
00095p	SPECIFIC CONDUCTANCE (UMHOS/CM @ 25C)	03/09/89-09/10/96	9	1000	931.111	1860	360	240936.111	490.852	360	395	1185	1860

** - Less than 9 observations   ## - Computed with 50% or more of the total observations as values that were half the detection limit   p - Has a corresponding box-and-whisker plot

Annual Analysis for 1991 - Station BADL0026

Parameter		Period of Record	Obs	Median	Mean	Maximum	Minimum	Variance	Std Dev	10th	25th	75th	90th
00010	TEMPERATURE, WATER (DEGREES CENTIGRADE)	03/09/89-09/10/96	13	16.5	13.769	23	0	55.942	7.479	1.6	5.5	19.75	22.2
00020	TEMPERATURE, AIR (DEGREES CENTIGRADE)	03/09/89-09/10/96	12	19.5	17.667	32	4	104.742	10.234	4.15	5.75	27	31.1
00061	FLOW, STREAM, INSTANTANEOUS CFS	03/09/89-09/10/96	14	28	122.986	525	0.8	36918.372	192.142	1.9	4.75	185.75	489
00095p	SPECIFIC CONDUCTANCE (UMHOS/CM @ 25C)	03/09/89-09/10/96	12	553.5	592.5	1067	267	79703	282.317	270.9	281.25	844.5	1014.5

** - Less than 9 observations   ## - Computed with 50% or more of the total observations as values that were half the detection limit   p - Has a corresponding box-and-whisker plot

Annual Analysis for 1992 - Station BADL0026

Parameter		Period of Record	Obs	Median	Mean	Maximum	Minimum	Variance	Std Dev	10th	25th	75th	90th
00010	TEMPERATURE, WATER (DEGREES CENTIGRADE)	03/09/89-09/10/96	7	11.5	10.714	19.5	0	57.821	7.604	**	**	**	**
00020	TEMPERATURE, AIR (DEGREES CENTIGRADE)	03/09/89-09/10/96	7	16	13.714	29	0.5	110.488	10.511	**	**	**	**
00061	FLOW, STREAM, INSTANTANEOUS CFS	03/09/89-09/10/96	7	4	4	6	1	3.667	1.915	***	**	**	**
00095p	SPECIFIC CONDUCTANCE (UMHOS/CM @ 25C)	03/09/89-09/10/96	7	823	815.286	850	778	708.905	26.625	**	**	**	**

** - Less than 9 observations   ## - Computed with 50% or more of the total observations as values that were half the detection limit   p - Has a corresponding box-and-whisker plot

Annual Analysis for 1993 - Station BADL0026

Parameter		Period of Record	Obs	Median	Mean	Maximum	Minimum	Variance	Std Dev	10th	25th	75th	90th
00010	TEMPERATURE, WATER (DEGREES CENTIGRADE)	03/09/89-09/10/96	11	15	10.682	19.5	0	55.264	7.434	0.1	0	16.5	18.9
00020	TEMPERATURE, AIR (DEGREES CENTIGRADE)	03/09/89-09/10/96	10	20	12.35	24.5	-20	187.558	13.695	0.1	11.5	23.375	-15.55
00061	FLOW, STREAM, INSTANTANEOUS CFS	03/09/89-09/10/96	11	19	42.182	141	4	2376.164	48.746	4.2	8	102	133.4
00095p	SPECIFIC CONDUCTANCE (UMHOS/CM @ 25C)	03/09/89-09/10/96	11	644	642.091	853	343	23493.491	153.276	361.4	609	761	851.4

** - Less than 9 observations   ## - Computed with 50% or more of the total observations as values that were half the detection limit   p - Has a corresponding box-and-whisker plot

163

Annual Analysis for 1994 - Station BADL0026

Parameter	Period of Record	Obs	Median	Mean	Maximum	Minimum	Variance	Std Dev	10th	25th	75th	90th
00010 TEMPERATURE, WATER (DEGREES CENTIGRADE)	03/09/89-09/10/96	11	18	13 064	28	0	102 815	10 14	0 1	1	19 5	26 8
00020 TEMPERATURE, AIR (DEGREES CENTIGRADE)	03/09/89-09/10/96	11	21 7	16 882	30	-5	172 194	13 122	1 2	6	28	29 8
00061 FLOW, STREAM, INSTANTANEOUS CFS	03/09/89-09/10/96	11	6	7 818	16	1	30 964	5 564	1	3	13	15 8
00095p SPECIFIC CONDUCTANCE (UMHOS/CM @ 25C)	03/09/89-09/10/96	11	651	629 818	920	148	35761 164	189 106	216	637	688	883 2

** - Less than 9 observations   ## - Computed with 50% or more of the total observations as values that were half the detection limit   p - Has a corresponding box-and-whisker plot

Annual Analysis for 1995 - Station BADL0026

Parameter	Period of Record	Obs	Median	Mean	Maximum	Minimum	Variance	Std Dev	10th	25th	75th	90th
00010 TEMPERATURE, WATER (DEGREES CENTIGRADE)	03/09/89-09/10/96	10	10 25	11 05	24	1	61 025	7 812	1 1	4 25	19 25	23 6
00020 TEMPERATURE, AIR (DEGREES CENTIGRADE)	03/09/89-09/10/96	10	15 5	16 3	31	3	83 733	9 151	3 4	8 5	22 875	30 75
00061 FLOW, STREAM, INSTANTANEOUS CFS	03/09/89-09/10/96	10	24 5	179 4	1355	9	173804 711	416 899	9	15 75	123 75	1238 4
00095p SPECIFIC CONDUCTANCE (UMHOS/CM @ 25C)	03/09/89-09/10/96	9	526	521 889	681	324	16865 611	129 868	324	391	637 5	681

** - Less than 9 observations   ## - Computed with 50% or more of the total observations as values that were half the detection limit   p - Has a corresponding box-and-whisker plot

Annual Analysis for 1996 - Station BADL0026

Parameter	Period of Record	Obs	Median	Mean	Maximum	Minimum	Variance	Std Dev	10th	25th	75th	90th
00010 TEMPERATURE, WATER (DEGREES CENTIGRADE)	03/09/89-09/10/96	6	14 75	12 583	24	0	118 042	10 865	**	**	**	**
00020 TEMPERATURE, AIR (DEGREES CENTIGRADE)	03/09/89-09/10/96	6	21	20 75	30	9 5	69 575	8 341	**	**	**	**
00061 FLOW, STREAM, INSTANTANEOUS CFS	03/09/89-09/10/96	6	25 5	35 167	91	21	752 567	27 433	**	**	**	**
00095p SPECIFIC CONDUCTANCE (UMHOS/CM @ 25C)	03/09/89-09/10/96	6	575 5	573 167	595	546	318 167	17 837	**	**	**	**

** - Less than 9 observations   ## - Computed with 50% or more of the total observations as values that were half the detection limit   p - Has a corresponding box-and-whisker plot

Station: BADL0026 Parameter Code: 00095

SPECIFIC CONDUCTANCE (UMHOS/CM @ 25C)

BATTLE CR BELOW HERMOSA SD

165

Seasonal Analysis for Season #1: 10/01 to 1/31 - Station BADL0026

Parameter		Period of Record	Obs	Median	Mean	Maximum	Minimum	Variance	Std Dev	10th	25th	75th	90th
00010	TEMPERATURE, WATER (DEGREES CENTIGRADE)	03/09/89-09/10/96	17	1	4.765	15	0	35.785	5.982	0		11.25	15
00020	TEMPERATURE, AIR (DEGREES CENTIGRADE)	03/09/89-09/10/96	17	-6.5	8	32	-20	186.938	13.673	0.9	3.25	23.5	21.6
00061	FLOW, STREAM, INSTANTANEOUS CFS	03/09/89-09/10/96	17	6	10.635	25	0.8	74.611	8.638	1.76	4	18.5	25
00095p	SPECIFIC CONDUCTANCE (UMHOS/CM @ 25C)	03/09/89-09/10/96	17	688	706.412	1067	148	46423.882	215.462	420	586	845	1013.4

** - Less than 9 observations   ## - Computed with 50% or more of the total observations as values that were half the detection limit   p - Has a corresponding box-and-whisker plot

Seasonal Analysis for Season #2: 2/01 to 4/14 - Station BADL0026

Parameter		Period of Record	Obs	Median	Mean	Maximum	Minimum	Variance	Std Dev	10th	25th	75th	90th
00010	TEMPERATURE, WATER (DEGREES CENTIGRADE)	03/09/89-09/10/96	12	3.25	3.625	8	0	9.097	3.016	0	0.625	6.75	7.85
00020	TEMPERATURE, AIR (DEGREES CENTIGRADE)	03/09/89-09/10/96	12	4	10.75	23	-1	49.477	7.034	3.9	8	15.75	22.4
00061	FLOW, STREAM, INSTANTANEOUS CFS	03/09/89-09/10/96	12	6	8.533	26	0.5	62.568	7.91	0.62	3	13.5	24.2
00095p	SPECIFIC CONDUCTANCE (UMHOS/CM @ 25C)	03/09/89-09/10/96	12	800	848.833	1860	524	127138.333	356.565	534.8	637.25	881.5	1632

** - Less than 9 observations   ## - Computed with 50% or more of the total observations as values that were half the detection limit   p - Has a corresponding box-and-whisker plot

Seasonal Analysis for Season #3: 4/15 to 6/30 - Station BADL0026

Parameter		Period of Record	Obs	Median	Mean	Maximum	Minimum	Variance	Std Dev	10th	25th	75th	90th
00010	TEMPERATURE, WATER (DEGREES CENTIGRADE)	03/09/89-09/10/96	26	15.75	15.288	24	6	19.883	4.459	9.1	11.875	18.125	21.3
00020	TEMPERATURE, AIR (DEGREES CENTIGRADE)	03/09/89-09/10/96	23	19	17.696	30	4	48.108	6.936	7.8	12	22	28
00061	FLOW, STREAM, INSTANTANEOUS CFS	03/09/89-09/10/96	28	68	160.864	1355	0.2	77579.819	278.532	1	9.5	177	460.2
00095p	SPECIFIC CONDUCTANCE (UMHOS/CM @ 25C)	03/09/89-09/10/96	24	558.5	578.667	1180	267	73310.928	270.76	280	347.25	741	1105

** - Less than 9 observations   ## - Computed with 50% or more of the total observations as values that were half the detection limit   p - Has a corresponding box-and-whisker plot

Seasonal Analysis for Season #4: 7/01 to 9/30 - Station BADL0026

Parameter		Period of Record	Obs	Median	Mean	Maximum	Minimum	Variance	Std Dev	10th	25th	75th	90th
00010	TEMPERATURE, WATER (DEGREES CENTIGRADE)	03/09/89-09/10/96	16	19.75	21.169	28	16.5	8.869	2.978	18.25	19.275	23	26.6
00020	TEMPERATURE, AIR (DEGREES CENTIGRADE)	03/09/89-09/10/96	16	26.5	25.763	31	19	15.156	3.893	20.05	22.025	28.875	31
00061	FLOW, STREAM, INSTANTANEOUS CFS	03/09/89-09/10/96	16	11.5	18.875	102	1	667.45	25.835	1	1.25	25	60.7
00095p	SPECIFIC CONDUCTANCE (UMHOS/CM @ 25C)	03/09/89-09/10/96	16	645	693.813	1190	339	42410.296	205.938	411.8	572.5	840.25	1001

** - Less than 9 observations   ## - Computed with 50% or more of the total observations as values that were half the detection limit   p - Has a corresponding box-and-whisker plot

Station: BADL0026 Parameter Code: 00095

SPECIFIC CONDUCTANCE (UMHOS/CM @ 25C)

(X 1000)

Season

CNDUCTVY AT 25C MICROMHO

BATTLE CR BELOW HERMOSA SD

167

Station Inventory for Station: BADL0027

Date Created 09/03/88

NPS Station ID: BADL0027
Location: CHEYENNE RIVER NR FAIRBURN, SD
Station Type: /TYPA/AMBNT/STREAM
RMI-Indexes:
HUC:
Major Basin:
Minor Basin:
RF1 Index:
RF3 Index: 10120109043700 00
Description:

LAT/LON: 43 700004/-102 909727

Agency: 112WRD
FIPS State/County: 46033 SOUTH DAKOTA/CUSTER
STORET Station ID(s): 06403700
Within Park Boundary: No

Aquifer:
Water Body Id:
ECO Region:
Distance from RF1: 2 60
Distance from RF3: 0 01

Depth of Water: 0
Elevation: 0

RF1 Mile Point: 0 000
RF3 Mile Point: 0 04

On/Off RF1:
On/Off RF3:

Parameter Inventory for Station: BADL0027

Parameter		Period of Record	Obs	Median	Mean	Maximum	Minimum	Variance	Std Dev	10th	25th	75th	90th
00010	TEMPERATURE, WATER (DEGREES CENTIGRADE)	05/06/88-09/08/94	6	18 35	19 033	31	9 5	49 247	7 018	**	**	**	**
00020	TEMPERATURE, AIR (DEGREES CENTIGRADE)	05/06/88-09/08/94	6	21	23 867	40	13 7	81 647	9 036	**	**	**	**
00025	BAROMETRIC PRESSURE (MM OF HG)	04/20/94-09/08/94	2	686 5	686 5	690	683	24 5	4 95	**	**	**	**
00061	FLOW, STREAM, INSTANTANEOUS CFS	05/06/88-09/08/94	6	68 5	71 667	117	38	679 467	26 067	**	**	**	**
00095	SPECIFIC CONDUCTANCE (UMHOS/CM @ 25C)	05/06/88-09/08/94	6	2640	2600	2750	2320	22040	148 459	**	**	**	**
00300	OXYGEN, DISSOLVED MG/L	05/06/88-09/08/94	6	9 15	9 067	11 2	6 6	2 239	1 496	**	**	**	**
00400	PH (STANDARD UNITS)	05/06/88-09/08/94	6	8 225	8 247	8 33	8 18	0 004	0 065	**	**	**	**
00400	CONVERTED PH (STANDARD UNITS)	05/06/88-09/08/94	6	8 224	8 243	8 33	8 18	0 004	0 065	**	**	**	**
00400	MICRO EQUIVALENTS/LITER OF H+ COMPUTED FROM PH	05/06/88-09/08/94	6	0 006	0 006	0 007	0 005	0	0 001	**	**	**	**
00403	PH, LAB, STANDARD UNITS SU	05/06/88-09/08/94	6	8 05	7 967	8 1	7 7	0 031	0 175	**	**	**	**
00403	CONVERTED PH, LAB, STANDARD UNITS	05/06/88-09/08/94	6	8 047	7 935	8 1	7 7	0 032	0 178	**	**	**	**
00403	MICRO EQUIVALENTS/LITER OF H+ COMPUTED FROM PH	05/06/88-09/08/94	6	0 009	0 012	0 02	0 008	0 005	0 005	**	**	**	**
00410	ALKALINITY, TOTAL (MG/L AS CACO3)	06/20/88-10/31/88	2	169	169	203	135	2312	48 083	**	**	**	**
00419	ALKALINITY, CARBONATE, INCREMENTAL TITR FIELD MG/L	06/20/88-06/20/88	1	136	136	136	136	0	0	**	**	**	**
00440	BICARBONATE ION (MG/L AS HCO3)	06/20/88-06/20/88	1	166	166	166	166	0	0	**	**	**	**
00445	CARBONATE ION (MG/L AS CO3)	06/20/88-06/20/88	1	0	0	0	0	0	0	**	**	**	**
00452	CARBONATE, WATER, DISS, INCR TIT, FIELD, AS CO3, MG/L	10/31/88-04/20/94	2	0	0	0	0	0	0	**	**	**	**
00453	BICARBONATE, WATER, DISS, INCR TIT, FIELD, AS HCO3, MG/L	10/31/88-04/20/94	2	226 5	226 5	247	206	840 5	28 991	**	**	**	**
00631	NITRITE PLUS NITRATE, DISS 1 DET (MG/L AS N)	05/06/88-10/31/88	4	0 55	0 65	1 3	0 2	0 217	0 465	**	**	**	**
00915	CALCIUM, DISSOLVED (MG/L AS CA)	05/06/88-09/08/94	6	250	250	270	230	360 4	18 974	**	**	**	**
00925	MAGNESIUM, DISSOLVED (MG/L AS MG)	05/06/88-09/08/94	6	81	83	92	74	50 4	7 099	**	**	**	**
00930	SODIUM, DISSOLVED (MG/L AS NA)	05/06/88-09/08/94	6	255	246 667	260	210	386 667	19 664	**	**	**	**
00935	POTASSIUM, DISSOLVED (MG/L AS K)	05/06/88-09/08/94	6	11	12 167	17	10	8 167	2 858	**	**	**	**
00940	CHLORIDE, TOTAL IN WATER MG/L	05/06/88-09/08/94	6	135	135	170	110	430	20 736	**	**	**	**
00945	SULFATE, TOTAL (MG/L AS SO4)	05/06/88-09/08/94	6 ##	1200	1161 667	1300	970	12816 667	113 211	**	**	**	**
01000	ARSENIC, DISSOLVED (UG/L AS AS)	05/06/88-09/08/94	6 ##	0 75	0 75	1	0 5	0 075	0 274	**	**	**	**
01020	BORON, DISSOLVED (UG/L AS B)	05/06/88-09/08/94	6	320	316 667	370	240	1866 667	43 205	**	**	**	**
01025	CADMIUM, DISSOLVED (UG/L AS CD)	05/06/88-09/08/94	6 ##	0 5	0 583	1	0 5	0 042	0 204	**	**	**	**
01030	CHROMIUM, DISSOLVED (UG/L AS CR)	05/06/88-09/08/94	6 ##	1 25	1 25	2	0 5	0 675	0 822	**	**	**	**
01040	COPPER, DISSOLVED (UG/L AS CU)	05/06/88-09/08/94	6 ##	0 75	0 917	2	0 5	0 342	0 585	**	**	**	**
01049	LEAD, DISSOLVED (UG/L AS PB)	05/06/88-09/08/94	6 ##	2 5	2 417	8	0 5	4 042	2 01	**	**	**	**
01060	MOLYBDENUM, DISSOLVED (UG/L AS MO)	05/06/88-09/08/94	6	7	7	8	6	0 8	0 894	**	**	**	**
01085	VANADIUM, DISSOLVED (UG/L AS V)	05/06/88-09/08/94	6	2	2 083	4	0 5	1 642	1 281	**	**	**	**
01090	ZINC, DISSOLVED (UG/L AS ZN)	05/06/88-09/08/94	6 ##	12 5	17 5	50	5	307 5	17 536	**	**	**	**
01106	ALUMINUM, DISSOLVED (UG/L AS AL)	04/20/94-09/08/94	2 ##	12 5	12 5	20	5	112 5	10 607	**	**	**	**
01145	SELENIUM, DISSOLVED (UG/L AS SE)	05/06/88-09/08/94	6	2 5	2 417	15	0 5	1 442	1 201	**	**	**	**
22703	URANIUM, NATURAL, DISSOLVED	05/06/88-09/08/94	6	13 5	12 833	15	10	3 767	1 941	**	**	**	**
39024	PROPAZINE, COULSON CONDUCTIVITY, WATER SAMPL(UG/L)	05/06/88-10/31/88	4 ##	0 05	0 05	0 05	0 05	0	0	**	**	**	**

** - Less than 9 observations   ## - Computed with 50% or more of the total observations as values that were half the detection limit   p - Has a corresponding time series plot

Parameter Inventory for Station: BADL0027

Parameter		Period of Record	Obs	Median	Mean	Maximum	Minimum	Variance	Std Dev	10th	25th	75th	90th
39030	TREFLAN, MICROCOULOMETRIC, WATER SAMPLE (UG/L)	05/06/88-10/31/88	4 ##	0.05	0.05	0.05	0.05	0	0	**	**	**	**
39051	METHOMYL IN WHOLE WATER (UG/L)	05/06/88-10/31/88	4 ##	0.25	0.25	0.25	0.25	0	0	**	**	**	**
39052	PROPHAM IN WHOLE WATER (UG/L)	05/06/88-10/31/88	4 ##	0.25	0.25	0.25	0.25	0	0	**	**	**	**
39054	SIMETRYNE IN WHOLE WATER (UG/L)	05/06/88-10/31/88	4 ##	0.05	0.05	0.05	0.05	0	0	**	**	**	**
39055	SIMAZINE IN WHOLE WATER (UG/L)	05/06/88-10/31/88	4 ##	0.05	0.063	0.1	0.05	0.001	0.025	**	**	**	**
39056	PROMETONE IN WHOLE WATER (UG/L)	05/06/88-10/31/88	4 ##	0.05	0.088	0.2	0.05	0.006	0.075	**	**	**	**
39057	PROMETRYNE IN WHOLE WATER (UG/L)	05/06/88-10/31/88	4 ##	0.05	0.05	0.05	0.05	0	0	**	**	**	**
39086	ALKALINITY,WATER,DISS,INCR TIT,FIELD,AS CACO3,MG/L	10/31/88-04/20/94	2	186	186	203	169	578	24.042	**	**	**	**
39630	ATRAZINE(AATREX) IN WHOLE WATER SAMPLE (UG/L)	05/06/88-10/31/88	4	0.1	0.125	0.2	0.1	0.003	0.05	**	**	**	**
39750	SEVIN IN WHOLE WATER SAMPLE (UG/L)	05/06/88-10/31/88	4 ##	0.25	0.25	0.25	0.25	0	0	**	**	**	**
70300	RESIDUE,TOTAL FILTRABLE (DRIED AT 180C),MG/L	05/06/88-09/08/94	6	2140	2091.667	2240	1830	23416.667	153.025	**	**	**	**
71890	MERCURY, DISSOLVED (UG/L AS HG)	05/06/88-09/08/94	6 ##	0.05	0.075	0.2	0.05	0.004	0.061	**	**	**	**
75990	URANIUM,NATURAL,1 SIGMA PRC EST,DISS,WATER UG/L	04/20/94-09/08/94	2	1.8	1.8	1.9	1.7	0.02	0.141	**	**	**	**
77825	ALACHLOR WHOLE WATER,UG/L	05/06/88-10/31/88	4 ##	0.05	0.05	0.05	0.05	0	0	**	**	**	**
81757	CYANAZINE IN THE WHOLE WATER SAMPLE UG/L	05/06/88-10/31/88	4 ##	0.05	0.05	0.05	0.05	0	0	**	**	**	**
82184	AMETRYNE (GESAPAX OR EVIK) TOTAL UG/L	05/06/88-10/31/88	4 ##	0.05	0.05	0.05	0.05	0	0	**	**	**	**
82611	METRIBUZIN, WHOLE WATER, TOTAL RECOVERABLE UG/L	05/06/88-10/31/88	4 ##	0.05	0.05	0.05	0.05	0	0	**	**	**	**
82612	METOLACHLOR, WHOLE WATER, TOTAL RECOVERABLE UG/L	05/06/88-10/31/88	4 ##	0.05	0.05	0.05	0.05	0	0	**	**	**	**

** - Less than 9 observations   ## - Computed with 50% or more of the total observations as values that were half the detection limit   p - Has a corresponding time series plot

EPA Water Quality Criteria Analysis for Station: BADL0027

Parameter		Std Type	Std Value	Total Obs	Exceed Standard	Prop Exceeding	10/01-1/31 Obs	Exceed	Prop	2/01-4/14 Obs	Exceed	Prop	4/15-6/30 Obs	Exceed	Prop	7/01-9/30 Obs	Exceed	Prop
00300	OXYGEN, DISSOLVED	Other-Lo Lim	4	6	0	0.00	—	0	0.00			0.00	3	0	0.00	2	0	0.00
00400	PH	Other-Hi Lim	9	6	0	0.00	—	0	0.00			0.00	3	0	0.00	2	0	0.00
00403	PH, LAB	Other-Lo Lim	6.5	6	0	0.00	—	0	0.00			0.00	3	0	0.00	2	0	0.00
		Other-Hi Lim	9	6	0	0.00	—	0	0.00			0.00	3	0	0.00	2	0	0.00
		Other-Lo Lim	6.5	6	0	0.00	—	0	0.00			0.00	3	0	0.00	2	0	0.00
00631	NITRITE PLUS NITRATE, DISS. 1 DET	Drinking Water	10	4	0	0.00	—	0	0.00			0.00	2	0	0.00	1	0	0.00
00940	CHLORIDE, TOTAL IN WATER	Fresh Acute	860	6	0	0.00	—	0	0.00			0.00	3	0	0.00	2	0	0.00
00945	SULFATE, TOTAL (AS SO4)	Drinking Water	250	6	6	1.00	—	1	1.00			1.00	3	3	1.00	2	2	1.00
01000	ARSENIC, DISSOLVED	Fresh Acute	360	6	0	0.00	—	0	0.00			0.00	3	0	0.00	2	0	0.00
01025	CADMIUM, DISSOLVED	Drinking Water	50	6	0	0.00	—	0	0.00			0.00	3	0	0.00	2	0	0.00
01030	CHROMIUM, DISSOLVED	Fresh Acute	3.9	6	0	0.00	—	0	0.00			0.00	3	0	0.00	2	0	0.00
01040	COPPER, DISSOLVED	Drinking Water	5	6	0	0.00	—	0	0.00			0.00	3	0	0.00	2	0	0.00
01049	LEAD, DISSOLVED	Drinking Water	100	6	0	0.00	—	0	0.00			0.00	3	0	0.00	2	0	0.00
01090	ZINC, DISSOLVED	Fresh Acute	18	6	0	0.00	—	0	0.00			0.00	3	0	0.00	2	0	0.00
01145	SELENIUM, DISSOLVED	Drinking Water	1300	6	0	0.00	—	0	0.00			0.00	3	0	0.00	2	0	0.00
		Fresh Acute	82	6	0	0.00	—	0	0.00			0.00	3	0	0.00	2	0	0.00
		Drinking Water	15	6	0	0.00	—	0	0.00			0.00	3	0	0.00	2	0	0.00
		Fresh Acute	120	6	0	0.00	—	0	0.00			0.00	3	0	0.00	2	0	0.00
		Drinking Water	5000	6	0	0.00	—	0	0.00			0.00	3	0	0.00	2	0	0.00
		Fresh Acute	20	6	0	0.00	—	0	0.00			0.00	3	0	0.00	2	0	0.00
		Drinking Water	50	6	0	0.00	—	0	0.00			0.00	3	0	0.00	2	0	0.00
22703	URANIUM, NATURAL DISSOLVED	Drinking Water	20	6	0	0.00	—	0	0.00			0.00	3	0	0.00	2	0	0.00
39055	SIMAZINE IN WHOLE WATER	Drinking Water	4	4	0	0.00	—	0	0.00			0.00	2	0	0.00	1	0	0.00
39630	ATRAZINE(AATREX) IN WHOLE WATER SAMPLE	Drinking Water	3	4	0	0.00	—	0	0.00			0.00	2	0	0.00	1	0	0.00
71890	MERCURY, DISSOLVED	Fresh Acute	2.4	6	0	0.00	—	0	0.00			0.00	3	0	0.00	2	0	0.00
		Drinking Water	2	6	0	0.00	—	0	0.00			0.00	3	0	0.00	2	0	0.00

& - Below detection limit observations, for which half the detection limit exceeded the criterion, were excluded from the criterion comparison for this parameter

169

Station Inventory for Station: BADL0028

NPS Station ID: BADL0028
Location: FRENCH CREEK ABV CONF W CHEYENNE
Station Type: /TYPA/AMBNT/STREAM
RMI-Indexes:
RMI-Miles:
HUC: 10120109
Major Basin: MISSOURI RIVER BASIN
Minor Basin: CENTRAL MISSOURI BASIN
RF1 Index: 10120109022
RF3 Index: 10120111038600 00
Description:
CLASS C, S4,T5S,R10E

LAT/LON: 43 645838/-102 930838

Agency: 31BLHICD
FIPS State/County: 46000 SOUTH DAKOTA/
STORET Station ID(s): FN-01
Within Park Boundary: No

Depth of Water: 0
Elevation: 0

Aquifer:
Water Body Id:
ECO Region:
Distance from RF1: 0 00
Distance from RF3: 0 27

RF1 Mile Point: 0 640
RF3 Mile Point: 2 54

On/Off RF1: OFF
On/Off RF3:

Date Created: / /

Parameter Inventory for Station: BADL0028

Parameter	Period of Record	Obs	Median	Mean	Maximum	Minimum	Variance	Std Dev	10th	25th	75th	90th
00010 TEMPERATURE, WATER (DEGREES CENTIGRADE)	06/03/71 -06/03/71	1	20	20	20	20	0	0	**	**	**	**
00020 TEMPERATURE, AIR (DEGREES CENTIGRADE)	06/03/71 -06/03/71	1	29	29	29	29	0	0	**	**	**	**
00060 FLOW, STREAM, MEAN DAILY CFS	06/03/71 -06/03/71	1	15	15	15	15	0	0	**	**	**	**
00070 TURBIDITY, (JACKSON CANDLE UNITS)	06/03/71 -06/03/71	1	585	585	585	585	0	0	**	**	**	**
00095 SPECIFIC CONDUCTANCE (UMHOS/CM @ 25C)	06/03/71 -06/03/71	1	2180	2180	2180	2180	0	0	**	**	**	**
00300 OXYGEN, DISSOLVED MG/L	06/03/71 -06/03/71	1	7 8	7 8	7 8	7 8	0	0	**	**	**	**
00400 PH (STANDARD UNITS)	06/03/71 -06/03/71	1	8	8	8	8	0	0	**	**	**	**
00400 CONVERTED PH (STANDARD UNITS)	06/03/71 -06/03/71	1	8	8	8	8	0	0	**	**	**	**
00400 MICRO EQUIVALENTS/LITER OF H+ COMPUTED FROM PH	06/03/71 -06/03/71	1	0 01	0 01	0 01	0 01	0	0	**	**	**	**
00410 ALKALINITY, TOTAL (MG/L AS CACO3)	06/03/71 -06/03/71	1	128	128	128	128	0	0	**	**	**	**
00430 ALKALINITY, CARBONATE (MG/L AS CACO3)	06/03/71 -06/03/71	1	0	0	0	0	0	0	**	**	**	**
00500 RESIDUE, TOTAL (MG/L)	06/03/71 -06/03/71	1	578	578	578	578	0	0	**	**	**	**
00505 RESIDUE, TOTAL VOLATILE (MG/L)	06/03/71 -06/03/71	1	120	120	120	120	0	0	**	**	**	**
00515 RESIDUE, TOTAL FILTRABLE (DRIED AT 105C),MG/L	06/03/71 -06/03/71	1	282	282	282	282	0	0	**	**	**	**
00520 RESIDUE, VOLATILE FILTRABLE (MG/L)	06/03/71 -06/03/71	1	40	40	40	40	0	0	**	**	**	**
00630 NITRITE PLUS NITRATE, TOTAL 1 DET (MG/L AS N)	06/03/71 -06/03/71	1	1 5	1 5	1 5	1 5	0	0	**	**	**	**
00650 PHOSPHATE, TOTAL (MG/L AS PO4)	06/03/71 -06/03/71	1	20	20	20	20	0	0	**	**	**	**
00660 PHOSPHATE, ORTHO (MG/L AS PO4)	06/03/71 -06/03/71	1	0 9	0 9	0 9	0 9	0	0	**	**	**	**
00900 HARDNESS, TOTAL (MG/L AS CACO3)	06/03/71 -06/03/71	1	146	146	146	146	0	0	**	**	**	**
00901 HARDNESS, CARBONATE (MG/L AS CACO3)	06/03/71 -06/03/71	1	128	128	128	128	0	0	**	**	**	**
00902 HARDNESS, NON-CARBONATE (MG/L AS CACO3)	06/03/71 -06/03/71	1	18	18	18	18	0	0	**	**	**	**
00915 CALCIUM, DISSOLVED (MG/L AS CA)	06/03/71 -06/03/71	1	41	41	41	41	0	0	**	**	**	**
00925 MAGNESIUM, DISSOLVED (MG/L AS MG)	06/03/71 -06/03/71	1	9	9	9	9	0	0	**	**	**	**
00930 SODIUM, DISSOLVED (MG/L AS NA)	06/03/71 -06/03/71	1	7	7	7	7	0	0	**	**	**	**
00935 POTASSIUM, DISSOLVED (MG/L AS K)	06/03/71 -06/03/71	1	6	6	6	6	0	0	**	**	**	**
00940 CHLORIDE,TOTAL IN WATER MG/L	06/03/71 -06/03/71	1	13	13	13	13	0	0	**	**	**	**
00945 SULFATE, TOTAL (MG/L AS SO4)	06/03/71 -06/03/71	1	52	52	52	52	0	0	**	**	**	**
00955 SILICA, DISSOLVED (MG/L AS SI02)	06/03/71 -06/03/71	1	14	14	14	14	0	0	**	**	**	**
01045 IRON, TOTAL (UG/L AS FE)	06/03/71 -06/03/71	1	500	500	500	500	0	0	**	**	**	**
01055 MANGANESE, TOTAL (UG/L AS MN)	06/03/71 -06/03/71	1	1800	1800	1800	1800	0	0	**	**	**	**
01145 SELENIUM, DISSOLVED (UG/L AS SE)	06/03/71 -06/03/71	1	0	0	0	0	0	0	**	**	**	**

** - Less than 9 observations   ## - Computed with 50% or more of the total observations as values that were half the detection limit   p - Has a corresponding time series plot

EPA Water Quality Criteria Analysis for Station: BADL0028

Parameter		Std Type	Std Value	Total Obs	Exceed Standard	Prop Exceeding	10/01-1/31 Obs	Exceed	Prop	2/01-4/14 Obs	Exceed	Prop	4/15-6/30 Obs	Exceed	Prop	7/01-9/30 Obs	Exceed	Prop
00070	TURBIDITY, JACKSON CANDLE UNITS	Other-Hi Lim	50	1	1	1.00							1	1	1.00			
00300	OXYGEN, DISSOLVED	Other-Lo Lim	4	1	0	0.00							1	0	0.00			
00400	PH	Other-Hi Lim	9	1	0	0.00							1	0	0.00			
		Other-Lo Lim	6 5	1	0	0.00							1	0	0.00			
00630	NITRITE PLUS NITRATE, TOTAL 1 DET	Drinking Water	10	1	0	0.00							1	0	0.00			
00940	CHLORIDE,TOTAL IN WATER	Fresh Acute	860	1	0	0.00							1	0	0.00			
		Drinking Water	250	1	0	0.00							1	0	0.00			
00945	SULFATE, TOTAL (AS SO4)	Drinking Water	250	1	0	0.00							1	0	0.00			
01145	SELENIUM, DISSOLVED	Fresh Acute	20	1	0	0.00							1	0	0.00			
		Drinking Water	50	1	0	0.00							1	0	0.00			

& - Below detection limit observations, for which half the detection limit exceeded the criterion, were excluded from the criterion comparison for this parameter

171

Station Inventory for Station: BADL0029

NPS Station ID: BADL0029
Location: FRENCH CREEK NEAR RED SHIRT, SD
Station Type: /TYPA/AMBNT/STREAM
RMI-Indexes:
RMI-Miles:
HUC: 10120109
Major Basin:
Minor Basin:
RF1 Index: 10120109
RF3 Index: 10140201028200 00
Description:

LAT/LON: 43 642226/-102 947226

Agency: 112WRD
FIPS State/County: 46033 SOUTH DAKOTA/CUSTER
STORET Station ID(s) 43383210256500
Within Park Boundary: No

Depth of Water: 0
Elevation: 0

Aquifer:
Water Body Id:
ECO Region:
Distance from RF1: 2 90
Distance from RF3: 0 06

RF1 Mile Point: 0 000
RF3 Mile Point: 0 00

On/Off RF1:
On/Off RF3:

Parameter Inventory for Station: BADL0029

Parameter		Period of Record	Obs	Median	Mean	Maximum	Minimum	Variance	Std Dev	10th	25th	75th	90th
00010	TEMPERATURE, WATER (DEGREES CENTIGRADE)	06/30/95-11/21/95	4	20 75	17 125	24	3	98 063	9 903	**	**	**	**
00020	TEMPERATURE, AIR (DEGREES CENTIGRADE)	06/30/95-07/23/96	4	25	24 25	34	13	90 917	9 535	**	**	**	**
00061	FLOW, STREAM, INSTANTANEOUS CFS	06/30/95-07/23/96	5	5	21 18	66	0 9	807 212	28 411	**	**	**	**
00095	SPECIFIC CONDUCTANCE (UMHOS/CM @ 25C)	06/30/95-11/21/95	4	863 5	1165	2470	463	83404 4 667	913 26	**	**	**	**

** - Less than 9 observations   ## - Computed with 50% or more of the total observations as values that were half the detection limit   p - Has a corresponding time series plot

********** No EPA Water Quality Criteria exist to compare against the data at this station **********

Station Inventory for Station: BADL0030

NPS Station ID: BADL0030
Location: 41N47W 7DCBA
Station Type: /TYPA/AMBNT/SPRING
RMI-Indexes:
HUC: 10120109
Major Basin:
Minor Basin:
RF1 Index: 10120109
RF3 Index: 10140201028200 00
Description:

LAT/LON: 43 538337/-102 960560

Depth of Water: 0
Elevation: 0

RF1 Mile Point 0 000
RF3 Mile Point 0 00

Agency: 112WRD
FIPS State/County: 46113 SOUTH DAKOTA/SHANNON
STORET Station ID(s) 4332181025573801
Within Park Boundary: No

Aquifer:
Water Body Id:
ECO Region:
Distance from RF1: 2 90
Distance from RF3: 0 06

Date Created: 05/01/93

On/Off RF1:
On/Off RF3:

Parameter Inventory for Station: BADL0030

Parameter		Period of Record	Obs	Median	Mean	Maximum	Minimum	Variance	Std Dev	10th	25th	75th	90th
00025	BAROMETRIC PRESSURE (MM OF HG)	11/04/92-11/04/92	1	687	687	687	687	0	0	**	**	**	**
00095	SPECIFIC CONDUCTANCE (UMHOS/CM @ 25C)	11/04/92-11/04/92	1	2680	2680	2680	2680	0	0	**	**	**	**
00400	PH (STANDARD UNITS)	11/04/92-11/04/92	1	6 8	6 8	6 8	6 8	0	0	**	**	**	**
00400	CONVERTED PH (STANDARD UNITS)	11/04/92-11/04/92	1	6 8	6 8	6 8	6 8	0	0	**	**	**	**
00400	MICRO EQUIVALENTS/LITER OF H+ COMPUTED FROM PH	11/04/92-11/04/92	1	0 158	0 158	0 158	0 158	0	0	**	**	**	**
00403	PH, LAB, STANDARD UNITS SU	11/04/92-11/04/92	1	7	7	7	7	0	0	**	**	**	**
00403	CONVERTED PH, LAB, STANDARD UNITS	11/04/92-11/04/92	1	7	7	7	7	0	0	**	**	**	**
00403	MICRO EQUIVALENTS/LITER OF H+ COMPUTED FROM PH	11/04/92-11/04/92	1	0 1	0 1	0 1	0 1	0	0	**	**	**	**
00915	CALCIUM, DISSOLVED (MG/L AS CA)	11/04/92-11/04/92	1	280	280	280	280	0	0	**	**	**	**
00925	MAGNESIUM, DISSOLVED (MG/L AS MG)	11/04/92-11/04/92	1	72	72	72	72	0	0	**	**	**	**
00930	SODIUM, DISSOLVED (MG/L AS NA)	11/04/92-11/04/92	1	300	300	300	300	0	0	**	**	**	**
00935	POTASSIUM, DISSOLVED (MG/L AS K)	11/04/92-11/04/92	1	22	22	22	22	0	0	**	**	**	**
00940	CHLORIDE, TOTAL IN WATER MG/L	11/04/92-11/04/92	1	10	10	10	10	0	0	**	**	**	**
00945	SULFATE, TOTAL (MG/L AS SO4)	11/04/92-11/04/92	1	1000	1000	1000	1000	0	0	**	**	**	**
00955	SILICA, DISSOLVED (MG/L AS SI02)	11/04/92-11/04/92	1	13	13	13	13	0	0	**	**	**	**
01000	ARSENIC, DISSOLVED (UG/L AS AS)	11/04/92-11/04/92	##	0 5	0 5	0 5	0 5	0	0	**	**	**	**
01005	BARIUM, DISSOLVED (UG/L AS BA)	11/04/92-11/04/92	##	22	22	22	22	0	0	**	**	**	**
01010	BERYLLIUM, DISSOLVED (UG/L AS BE)	11/04/92-11/04/92	##	0 75	0 75	0 75	0 75	0	0	**	**	**	**
01020	BORON, DISSOLVED (UG/L AS B)	11/04/92-11/04/92	1	470	470	470	470	0	0	**	**	**	**
01025	CADMIUM, DISSOLVED (UG/L AS CD)	11/04/92-11/04/92	##	1 5	1 5	1 5	1 5	0	0	**	**	**	**
01027	CADMIUM, TOTAL (UG/L AS CD)	11/04/92-11/04/92	##	0 5	0 5	0 5	0 5	0	0	**	**	**	**
01030	CHROMIUM, DISSOLVED (UG/L AS CR)	11/04/92-11/04/92	##	7 5	7 5	7 5	7 5	0	0	**	**	**	**
01035	COBALT, DISSOLVED (UG/L AS CO)	11/04/92-11/04/92	##	4 5	4 5	4 5	4 5	0	0	**	**	**	**
01040	COPPER, DISSOLVED (UG/L AS CU)	11/04/92-11/04/92	##	15	15	15	15	0	0	**	**	**	**
01046	IRON, DISSOLVED (UG/L AS FE)	11/04/92-11/04/92	1	390	390	390	390	0	0	**	**	**	**
01049	LEAD, DISSOLVED (UG/L AS PB)	11/04/92-11/04/92	##	15	15	15	15	0	0	**	**	**	**
01060	MOLYBDENUM, DISSOLVED (UG/L AS MO)	11/04/92-11/04/92	##	15	15	15	15	0	0	**	**	**	**
01065	NICKEL, DISSOLVED (UG/L AS NI)	11/04/92-11/04/92	##	15	15	15	15	0	0	**	**	**	**
01075	SILVER, DISSOLVED (UG/L AS AG)	11/04/92-11/04/92	##	1 5	1 5	1 5	1 5	0	0	**	**	**	**
01080	STRONTIUM, DISSOLVED (UG/L AS SR)	11/04/92-11/04/92	1	3700	3700	3700	3700	0	0	**	**	**	**
01085	VANADIUM, DISSOLVED (UG/L AS V)	11/04/92-11/04/92	1	9	9	9	9	0	0	**	**	**	**
01090	ZINC, DISSOLVED (UG/L AS ZN)	11/04/92-11/04/92	1	16	16	16	16	0	0	**	**	**	**
01130	LITHIUM, DISSOLVED (UG/L AS LI)	11/04/92-11/04/92	1	370	370	370	370	0	0	**	**	**	**
01145	SELENIUM, DISSOLVED (UG/L AS SE)	11/04/92-11/04/92	##	0 5	0 5	0 5	0 5	0	0	**	**	**	**
70300	RESIDUE, TOTAL FILTRABLE (DRIED AT 180C),MG/L	11/04/92-11/04/92	1	2210	2210	2210	2210	0	0	**	**	**	**
71890	MERCURY, DISSOLVED (UG/L AS HG)	11/04/92-11/04/92	##	0 05	0 05	0 05	0 05	0	0	**	**	**	**

** - Less than 9 observations   ## - Computed with 50% or more of the total observations as values that were half the detection limit   p - Has a corresponding time series plot

EPA Water Quality Criteria Analysis for Station: BADL0030

Parameter	Std Type	Std Value	Total Obs	Exceed Standard	Prop Exceeding	10/01-1/31 Obs	10/01-1/31 Exceed	10/01-1/31 Prop	2/01-4/14 Obs	2/01-4/14 Exceed	2/01-4/14 Prop	4/15-6/30 Obs	4/15-6/30 Exceed	4/15-6/30 Prop	7/01-9/30 Obs	7/01-9/30 Exceed	7/01-9/30 Prop
00400 PH	Other-Hi Lim	9	1	0	0.00	1	0	0.00									
	Other-Lo Lim	6.5	1	0	0.00	1	0	0.00									
00403 PH, LAB	Other-Hi Lim	9	1	0	0.00	1	0	0.00									
	Other-Lo Lim	6.5	1	0	0.00	1	0	0.00									
00940 CHLORIDE,TOTAL IN WATER	Fresh Acute	860	1	0	0.00	1	0	0.00									
	Drinking Water	250	1	0	0.00	1	0	0.00									
00945 SULFATE, TOTAL (AS SO4)	Drinking Water	250	1	1	1.00	1	1	1.00									
01000 ARSENIC, DISSOLVED	Fresh Acute	360	1	0	0.00	1	0	0.00									
	Drinking Water	50	1	0	0.00	1	0	0.00									
01005 BARIUM, DISSOLVED	Drinking Water	2000	1	0	0.00	1	0	0.00									
01010 BERYLLIUM, DISSOLVED	Fresh Acute	130	1	0	0.00	1	0	0.00									
	Drinking Water	4	1	0	0.00	1	0	0.00									
01025 CADMIUM, DISSOLVED	Fresh Acute	3.9	1	0	0.00	1	0	0.00									
	Drinking Water	5	1	0	0.00	1	0	0.00									
01027 CADMIUM, TOTAL	Fresh Acute	3.9	1	0	0.00	1	0	0.00									
	Drinking Water	5	1	0	0.00	1	0	0.00									
01030 CHROMIUM, DISSOLVED	Drinking Water	100	1	0	0.00	1	0	0.00									
01040 COPPER, DISSOLVED	Fresh Acute	18	1	0	0.00	1	0	0.00									
	Drinking Water	1300	1	0	0.00	1	0	0.00									
01049 LEAD, DISSOLVED	Fresh Acute	82	0 &	0	0.00	1	0	0.00									
	Drinking Water	15	1	0	0.00	1	0	0.00									
01065 NICKEL, DISSOLVED	Fresh Acute	1400	1	0	0.00	1	0	0.00									
	Drinking Water	100	1	0	0.00	1	0	0.00									
01075 SILVER, DISSOLVED	Fresh Acute	4.1	1	0	0.00	1	0	0.00									
	Drinking Water	100	1	0	0.00	1	0	0.00									
01090 ZINC, DISSOLVED	Fresh Acute	120	1	0	0.00	1	0	0.00									
	Drinking Water	5000	1	0	0.00	1	0	0.00									
01145 SELENIUM, DISSOLVED	Fresh Acute	20	1	0	0.00	1	0	0.00									
	Drinking Water	50	1	0	0.00	1	0	0.00									
71890 MERCURY, DISSOLVED	Fresh Acute	2.4	1	0	0.00	1	0	0.00									
	Drinking Water	2	1	0	0.00	1	0	0.00									

& - Below detection limit observations, for which half the detection limit exceeded the criterion, were excluded from the criterion comparison for this parameter

Station Inventory for Station: BADL0031

NPS Station ID: BADL0031
Location: 42N48W25ABCC
Station Type: /TYPA/AMBNT/SPRING
RMI-Indexes:
HUC: 10120109
Major Basin:
Minor Basin:
RF1 Index: 10120109
RF3 Index: 10140201028200 00
Description:

LAT/LON: 43 590005/-102 981670

Agency: 112WRD
FIPS State/County: 46113 SOUTH DAKOTA/SHANNON
STORET Station ID(s): 43352410258540 1
Within Park Boundary: No

Depth of Water: 0
Elevation: 0

Aquifer:
Water Body Id:
ECO Region:
Distance from RF1: 2 90
Distance from RF3: 0 06

RF1 Mile Point: 0 000
RF3 Mile Point: 0 00

On/Off RF1:
On/Off RF3:

Date Created: 05/01/93

Parameter Inventory for Station: BADL0031

Parameter	Period of Record	Obs	Median	Mean	Maximum	Minimum	Variance	Std Dev	10th	25th	75th	90th
00010 TEMPERATURE, WATER (DEGREES CENTIGRADE)	11/05/92-07/12/95	3	19	15 533	19 7	7 9	43 823	6 62	**	**	**	**
00020 TEMPERATURE, AIR (DEGREES CENTIGRADE)	11/05/92-07/12/95	3	27	20 5	31	3 5	220 75	14 858	**	**	**	**
00025 BAROMETRIC PRESSURE (MM OF HG)	11/05/92-08/03/94	2	681 5	681 5	685	678	24 5	4 95	**	**	**	**
00061 FLOW, STREAM, INSTANTANEOUS CFS	11/05/92-07/12/95	4	0 01	0 013	0 02	0 01	0	0 005	**	**	**	**
00095 SPECIFIC CONDUCTANCE (UMHOS/CM @ 25C)	11/05/92-07/12/95	4	534 5	513 75	537	449	1864 917	43 185	**	**	**	**
00300 OXYGEN, DISSOLVED MG/L	08/03/94-07/12/95	2	7 6	7 6	7 7	7 5	0 02	0 141	**	**	**	**
00400 PH (STANDARD UNITS)	11/05/92-07/12/95	4	8 03	8 108	8 74	7 63	0 217	0 465	**	**	**	**
00400 CONVERTED PH (STANDARD UNITS)	11/05/92-07/12/95	4	8 024	7 957	8 74	7 63	0 247	0 497	**	**	**	**
00400 MICRO EQUIVALENTS/LITER OF H+ COMPUTED FROM PH	11/05/92-07/12/95	4	0 009	0 011	0 023	0 002	0	0 009	**	**	**	**
00403 PH, LAB, STANDARD UNITS	11/05/92-07/12/95	4	7 75	7 875	8 3	7 7	0 083	0 287	**	**	**	**
00403 CONVERTED PH, LAB, STANDARD UNITS	11/05/92-07/12/95	4	7 747	7 818	8 3	7 7	0 087	0 295	**	**	**	**
00403 MICRO EQUIVALENTS/LITER OF H+ COMPUTED FROM PH	11/05/92-07/12/95	4	0 018	0 015	0 02	0 005	0	0 007	**	**	**	**
00410 ALKALINITY, TOTAL (MG/L AS CACO3)	08/03/94-08/03/94	1	224	224	224	224	0	0	**	**	**	**
00440 BICARBONATE ION (MG/L AS HCO3)	08/03/94-08/03/94	1	273	273	273	273	0	0	**	**	**	**
00453 BICARBONATE, WATER, DISS, INCR TIT, FIELD, AS HCO3, MG/L	07/12/95-07/12/95	1	298	298	298	298	0	0	**	**	**	**
00608 NITROGEN, AMMONIA, DISSOLVED (MG/L AS N)	08/03/94-07/12/95	3 ##	0 008	0 015	0 03	0 008	0	0 013	**	**	**	**
00613 NITRITE NITROGEN, DISSOLVED (MG/L AS N)	05/16/95-07/12/95	2 ##	0 005	0 005	0 005	0 005	0	0	**	**	**	**
00623 NITROGEN, KJELDAHL, DISSOLVED (MG/L AS N)	08/03/94-07/12/95	3 ##	0 2	0 2	0 3	0 1	0 01	0 1	**	**	**	**
00631 NITRITE PLUS NITRATE, DISS 1 DET (MG/L AS N)	08/03/94-07/12/95	3 ##	0 025	0 02	0 025	0 01	0	0 009	**	**	**	**
00666 PHOSPHORUS, DISSOLVED (MG/L AS P)	08/03/94-07/12/95	3	0 015	0 017	0 03	0 005	0	0 013	**	**	**	**
00671 PHOSPHORUS, DISSOLVED ORTHOPHOSPHATE (MG/L AS P)	05/16/95-07/12/95	2 ##	0 005	0 005	0 005	0 005	0	0	**	**	**	**
00915 CALCIUM, DISSOLVED (MG/L AS CA)	11/05/92-07/12/95	4	58 5	54 5	59	42	69 667	8 347	**	**	**	**
00925 MAGNESIUM, DISSOLVED (MG/L AS MG)	11/05/92-07/12/95	4	11 5	11 5	12	11	0 333	0 577	**	**	**	**
00930 SODIUM, DISSOLVED (MG/L AS NA)	11/05/92-07/12/95	4	37 5	37 75	39	37	0 917	0 957	**	**	**	**
00935 POTASSIUM, DISSOLVED (MG/L AS K)	11/05/92-07/12/95	4	6 05	5 975	6 3	5 5	0 116	0 34	**	**	**	**
00940 CHLORIDE, TOTAL IN WATER MG/L	11/05/92-07/12/95	4	5 5	5 5	6	5	0 333	0 577	**	**	**	**
00945 SULFATE, TOTAL (MG/L AS SO4)	11/05/92-07/12/95	4	45	44 5	45	43	1	1	**	**	**	**
00950 FLUORIDE, DISSOLVED (MG/L AS F)	11/05/92-07/12/95	4	0 4	0 4	0 8	0 4	0 04	0 2	**	**	**	**
00955 SILICA, DISSOLVED (MG/L AS SIO2)	11/05/92-07/12/95	4	18	17 25	19	14	5 583	2 363	**	**	**	**
01000 ARSENIC, DISSOLVED (UG/L AS AS)	11/05/92-07/12/95	4 ##	0 5	0 625	1	0 5	0 063	0 25	**	**	**	**
01005 BARIUM, DISSOLVED (UG/L AS BA)	11/05/92-07/12/95	4 ##	150	140	160	100	733 333	27 08	**	**	**	**
01010 BERYLLIUM, DISSOLVED (UG/L AS BE)	11/05/92-07/12/95	4	0 25	0 25	0 25	0 25	0	0	**	**	**	**
01020 BORON, DISSOLVED (UG/L AS B)	11/05/92-07/12/95	4	60	62 5	80	50	158 333	12 583	**	**	**	**
01025 CADMIUM, DISSOLVED (UG/L AS CD)	11/05/92-07/12/95	4 ##	0 5	0 5	0 5	0 5	0	0	**	**	**	**
01027 CADMIUM, TOTAL (UG/L AS CD)	11/05/92-07/12/95	4 ##	0 5	0 5	0 5	0 5	0	0	**	**	**	**
01030 CHROMIUM, DISSOLVED (UG/L AS CR)	11/05/92-07/12/95	4 ##	2 5	2 5	2 5	2 5	0	0	**	**	**	**
01035 COBALT, DISSOLVED (UG/L AS CO)	11/05/92-07/12/95	4 ##	1 5	1 5	1 5	1 5	0	0	**	**	**	**
01040 COPPER, DISSOLVED (UG/L AS CU)	11/05/92-07/12/95	4 ##	5	5	5	5	0	0	**	**	**	**

** - Less than 9 observations   ## - Computed with 50% or more of the total observations as values that were half the detection limit   p - Has a corresponding time series plot

Parameter Inventory for Station: BADL0031

Parameter		Period of Record	Obs	Median	Mean	Maximum	Minimum	Variance	Std Dev	10th	25th	75th	90th
01046	IRON, DISSOLVED (UG/L AS FE)	11/05/92-07/12/95	4	22 5	26	52	7	475 333	21 802	**	**	**	**
01049	LEAD, DISSOLVED (UG/L AS PB)	11/05/92-07/12/95	4 ##	7 5	10	20	5	50	7 071	**	**	**	**
01056	MANGANESE, DISSOLVED (UG/L AS MN)	11/05/92-07/12/95	4	3 5	21 5	78	1	1420 333	37 687	**	**	**	**
01057	THALLIUM, DISSOLVED (UG/L AS TL)	05/16/95-07/12/95	2 ##	0 25	0 25	0 25	0 25	0	0	**	**	**	**
01060	MOLYBDENUM, DISSOLVED (UG/L AS MO)	11/05/92-07/12/95	4 ##	5	5	5	5	0	0	**	**	**	**
01065	NICKEL, DISSOLVED (UG/L AS NI)	11/05/92-07/12/95	4 ##	5	5	5	5	0	0	**	**	**	**
01075	SILVER, DISSOLVED (UG/L AS AG)	11/05/92-07/12/95	4 ##	0 5	0 5	0 5	0 5	0	0	**	**	**	**
01080	STRONTIUM, DISSOLVED (UG/L AS SR)	11/05/92-07/12/95	4 ##	480	472 5	500	430	891 667	29 861	**	**	**	**
01085	VANADIUM, DISSOLVED (UG/L AS V)	11/05/92-07/12/95	4 ##	3	3	3	3	0	0	**	**	**	**
01090	ZINC, DISSOLVED (UG/L AS ZN)	11/05/92-07/12/95	4 ##	2 75	3 25	6	1 5	4 75	2 179	**	**	**	**
01130	LITHIUM, DISSOLVED (UG/L AS LI)	11/05/92-07/12/95	4	28	27	28	24	4	2	**	**	**	**
01145	SELENIUM, DISSOLVED (UG/L AS SE)	11/05/92-07/12/95	4	3 5	3 5	5	2	1 667	1 291	**	**	**	**
03515	BETA, DISSOLVED GROSS, AS CS-137, PC/L	08/03/94-08/03/94	1	10	10	10	10	0	0	**	**	**	**
04126	ALPHA, DISSOLVED, WATER (AS TH-230) PCI/L	08/03/94-08/03/94	1	6 5	6 5	6 5	6 5	0	0	**	**	**	**
09510	RADIUM 226, DISSOLVED, PLANCHET COUNT	08/03/94-08/03/94	1	0 2	0 2	0 2	0 2	0	0	**	**	**	**
22703	URANIUM, NATURAL, DISSOLVED	08/03/94-08/03/94	1	7	7	7	7	0	0	**	**	**	**
31501	COLIFORM,TOT,MEMBRANE FILTER,IMMED M-ENDO MED,35C	08/03/94-05/16/95	2	378	378	440	316	7688	87 681	**	**	**	**
31501	LOG COLIFORM,TOT,MEMBRANE FILTER,IMMED M-ENDO MED,	08/03/94-05/16/95	2	2 572	2 572	2 643	2 5	0 01	0 102	**	**	**	**
31501	GM COLIFORM,TOT,MEMBRANE FILTER,IMMED M-ENDO MED,3	GEOMETRIC MEAN =			372 881								
31503	COLIFORM,TOT,MEMBR FILTER,DELAYED,M-ENDO MED,35 C	07/12/95-07/12/95	1	1740	1740	1740	1740	0	0	**	**	**	**
31503	LOG COLIFORM,TOT,MEMBR FILTER,DELAYED,M-ENDO MED,3	07/12/95-07/12/95	1	3 241	3 241	3 241	3 241	0	0	**	**	**	**
31503	GM COLIFORM,TOT,MEMBR FILTER,DELAYED,M-ENDO MED,35	GEOMETRIC MEAN =			1740								
31625	FECAL COLIFORM, MF,M-FC, 0 7 UM	08/03/94-07/12/95	3	102	142 333	300	25	20126 333	141 867	**	**	**	**
31625	LOG FECAL COLIFORM, MF,M-FC, 0 7 UM	08/03/94-07/12/95	3	2 009	1 961	2 477	1 398	0 293	0 541	**	**	**	**
31625	GM FECAL COLIFORM, MF,M-FC, 0 7 UM	GEOMETRIC MEAN =			91 458								
31633	E COLI,THERMOTOL,MF,M-TEC,IN SITU UREASE #/100ML	08/03/94-07/12/95	3	90	113 333	200	50	6033 333	77 675	**	**	**	**
31633	LOG E COLI,THERMOTOL,MF M-TEC,IN SITU UREASE #/100	08/03/94-07/12/95	3	1 954	1 985	2 301	1 699	0 091	0 302	**	**	**	**
31633	GM E COLI,THERMOTOL,MF,M-TEC,IN SITU UREASE #/100M	GEOMETRIC MEAN =			96 549								
31673	FECAL STREPTOCOCCI, MBR FILT,KF AGAR,35C,48HR	08/03/94-07/12/95	3	613	474 667	687	124	93594 333	305 932	**	**	**	**
31673	LOG FECAL STREPTOCOCCI, MBR FILT,KF AGAR,35C,48HR	08/03/94-07/12/95	3	2 787	2 573	2 837	2 093	0 173	0 416	**	**	**	**
31673	GM FECAL STREPTOCOCCI, MBR FILT,KF AGAR,35C,48HR	GEOMETRIC MEAN =			373 777								
39086	ALKALINITY,WATER,DISS,INCR TTT,FIELD,AS CACO3,MG/L	07/12/95-07/12/95	1	244	244	244	244	0	0	**	**	**	**
70300	RESIDUE,TOTAL FILTRABLE (DRIED AT 180C),MG/L	11/05/92-07/12/95	4 ##	324 5	310 75	332	262	1070 25	32 715	**	**	**	**
71890	MERCURY, DISSOLVED (UG/L AS HG)	11/05/92-07/12/95	4 ##	0 05	0 05	0 05	0 05	0	0	**	**	**	**
75986	ALPHA GROSS,1 SIGMA PRC EST AS NAT U,DISS,WTR UG/L	08/03/94-08/03/94	1	4 14	4 14	4 14	4 14	0	0	**	**	**	**
75987	ALPHA GROSS,DISS,1 SIGMA PRC EST AS TH230,WTR PC/L	08/03/94-08/03/94	1	3 18	3 18	3 18	3 18	0	0	**	**	**	**
75988	BETA GROSS,DISS,1 SIGMA PRC EST AS SR90/Y90 PC/L	08/03/94-08/03/94	1	1 44	1 44	1 44	1 44	0	0	**	**	**	**
75989	BETA GROSS,1 SIGMA PRC EST AS CS-137,DISS,WTR PC/L	08/03/94-08/03/94	1	1 93	1 93	1 93	1 93	0	0	**	**	**	**
75990	URANIUM,NATURAL,1 SIGMA PRC EST,DISS,WATER UG/L	08/03/94-08/03/94	1	2 1	2 1	2 1	2 1	0	0	**	**	**	**
76001	RADIUM 226,1 SIGMA PRC EST,DISS,WATER PC/L	08/03/94-08/03/94	1	0 167	0 167	0 167	0 167	0	0	**	**	**	**
80030	ALPHA,DISSOLVED GROSS,AS URANIUM-NATURAL,UG/L	08/03/94-08/03/94	1	8 4	8 4	8 4	8 4	0	0	**	**	**	**
80050	BETA,DISSOLVED GROSS,AS SR-Y-90, PC/L	08/03/94-08/03/94	1	7 7	7 7	7 7	7 7	0	0	**	**	**	**

** - Less than 9 observations   ## - Computed with 50% or more of the total observations as values that were half the detection limit   p - Has a corresponding time series plot

EPA Water Quality Criteria Analysis for Station: BADL0031

Parameter		Std Type	Std Value	Total Obs	Exceed Standard	Prop Exceeding	10/01-1/31 Obs	Exceed	Prop	2/01-4/14 Obs	Exceed	Prop	4/15-6/30 Obs	Exceed	Prop	7/01-9/30 Obs	Exceed	Prop
00300	OXYGEN, DISSOLVED	Other-Lo Lim	4	2	0	0.00										2	0	0.00
00400	PH	Other-Hi Lim	9	4	0	0.00	1	0	0.00	1	0	0.00	1	0	0.00	2	0	0.00
00400	PH	Other-Lo Lim	6 5	4	0	0.00	1	0	0.00	1	0	0.00	1	0	0.00	2	0	0.00
00403	PH, LAB	Other-Hi Lim	9	4	0	0.00							1	0	0.00	1	0	0.00
00403	PH, LAB	Other-Lo Lim	6 5	2	0	0.00							1	0	0.00	2	0	0.00
00613	NITRITE NITROGEN, DISSOLVED AS N	Drinking Water	1	2	0	0.00	1	0	0.00							2	0	0.00
00631	NITRITE PLUS NITRATE, DISS 1 DET	Drinking Water	10	3	0	0.00	1	0	0.00							2	0	0.00
00940	CHLORIDE, TOTAL IN WATER	Fresh Acute	860	4	0	0.00				1	0	0.00				2	0	0.00
00945	SULFATE, TOTAL (AS SO4)	Drinking Water	250	4	0	0.00				1	0	0.00				2	0	0.00
00950	FLUORIDE, DISSOLVED AS F	Drinking Water	4	4	0	0.00				1	0	0.00				2	0	0.00

176

& - Below detection limit observations, for which half the detection limit exceeded the criterion, were excluded from the criterion comparison for this parameter

EPA Water Quality Criteria Analysis for Station: BADL0031

Parameter		Std Type	Std Value	Total Obs	Exceed Standard	Prop Exceeding	10/01-1/31 Obs	Exceed	Prop	2/01-4/14 Obs	Exceed	Prop	4/15-6/30 Obs	Exceed	Prop	7/01-9/30 Obs	Exceed	Prop
01000	ARSENIC, DISSOLVED	Fresh Acute	360	4	0	0.00	1	0	0.00				1	0	0.00	2	0	0.00
		Drinking Water	50	4	0	0.00	1	0	0.00				1	0	0.00	2	0	0.00
01005	BARIUM, DISSOLVED	Drinking Water	2000	4	0	0.00	1	0	0.00				1	0	0.00	2	0	0.00
01010	BERYLLIUM, DISSOLVED	Fresh Acute	130	4	0	0.00	1	0	0.00				1	0	0.00	2	0	0.00
		Drinking Water	4	4	0	0.00	1	0	0.00				1	0	0.00	2	0	0.00
01025	CADMIUM, DISSOLVED	Fresh Acute	3.9	4	0	0.00	1	0	0.00				1	0	0.00	2	0	0.00
		Drinking Water	5	4	0	0.00	1	0	0.00				1	0	0.00	2	0	0.00
01027	CADMIUM, TOTAL	Fresh Acute	3.9	4	0	0.00	1	0	0.00				1	0	0.00	2	0	0.00
		Drinking Water	5	4	0	0.00	1	0	0.00				1	0	0.00	2	0	0.00
01030	CHROMIUM, DISSOLVED	Drinking Water	100	4	0	0.00	1	0	0.00				1	0	0.00	2	0	0.00
01040	COPPER, DISSOLVED	Fresh Acute	18	4	0	0.00	1	0	0.00				1	0	0.00	2	0	0.00
		Drinking Water	1300	4	0	0.00	1	0	0.00				1	0	0.00	2	0	0.00
01049	LEAD, DISSOLVED	Fresh Acute	82	4	1	0.25	1	0	0.00				1	1	1.00	2	0	0.00
		Drinking Water	15	4	0	0.00	1	0	0.00				1	0	0.00	2	0	0.00
01057	THALLIUM, DISSOLVED	Fresh Acute	1400	2	0	0.00	1	0	0.00							1	0	0.00
		Drinking Water	2	2	0	0.00	1	0	0.00							1	0	0.00
01065	NICKEL, DISSOLVED	Fresh Acute	1400	4	0	0.00	1	0	0.00				1	0	0.00	2	0	0.00
		Drinking Water	100	4	0	0.00	1	0	0.00				1	0	0.00	2	0	0.00
01075	SILVER, DISSOLVED	Fresh Acute	4.1	4	0	0.00	1	0	0.00				1	0	0.00	2	0	0.00
		Drinking Water	100	4	0	0.00	1	0	0.00				1	0	0.00	2	0	0.00
01090	ZINC, DISSOLVED	Fresh Acute	120	4	0	0.00	1	0	0.00				1	0	0.00	2	0	0.00
		Drinking Water	5000	4	0	0.00	1	0	0.00				1	0	0.00	2	0	0.00
01145	SELENIUM, DISSOLVED	Fresh Acute	20	4	0	0.00	1	0	0.00				1	0	0.00	2	0	0.00
		Drinking Water	50	4	0	0.00	1	0	0.00				1	0	0.00	2	0	0.00
22703	URANIUM, NATURAL DISSOLVED	Drinking Water	20	2	0	0.00							1	0	0.00	1	0	0.00
31501	COLIFORM, TOTAL, MEMBRANE FILTER, IMMED	Other-Hi Lim	1000	2	0	0.00							1	0	0.00	1	0	0.00
31503	COLIFORM,TOT.MEMBRANE FILTR,DELAY M-END	Other-Hi Lim	1000	1	1	1.00							1	1	1.00			
31625	FECAL COLIFORM, MF	Other-Hi Lim	200	3	1	0.33	1	0	0.00				1	1	1.00	1	0	0.00
71890	MERCURY, DISSOLVED	Fresh Acute	2.4	4	0	0.00	1	0	0.00				1	0	0.00	2	0	0.00
		Drinking Water	2	4	0	0.00	1	0	0.00				1	0	0.00	2	0	0.00

& - Below detection limit observations, for which half the detection limit exceeded the criterion, were excluded from the criterion comparison for this parameter

177

EPA Water Quality Criteria Analysis for Entire BADL Study Area

Parameter		Std Type	Std Value	Total Obs	Exceed Standard	Prop Exceeding	10/01-1/31 Obs	Exceed	Prop	2/01-4/14 Obs	Exceed	Prop	4/15-6/30 Obs	Exceed	Prop	7/01-9/30 Obs	Exceed	Prop
00070	TURBIDITY, JACKSON CANDLE UNITS	Other-Hi Lim	50	4	4	1.00							2	2	1.00	2	2	1.00
00076	TURBIDITY, HACH TURBIDIMETER	Other-Hi Lim	50	2	0	0.00										2	0	0.00
00300	OXYGEN, DISSOLVED	Other-Lo Lim	4	42	0	0.00	5	0	0.00				11	0	0.00	26	0	0.00
00400	PH	Other-Hi Lim	9	107	0	0.00	20	0	0.00	8	0	0.00	30	0	0.00	49	0	0.00
00400	PH	Other-Lo Lim	6.5	107	0	0.00	20	0	0.00	8	0	0.00	30	0	0.00	49	1	0.00
00403	PH, LAB	Other-Hi Lim	9	63	3	0.05	7	0	0.00	2	0	0.00	17	2	0.12	37	1	0.03
00403	PH, LAB	Other-Lo Lim	6.5	63	0	0.00	7	0	0.00	2	0	0.00	17	0	0.00	37	0	0.00
00613	NITRITE NITROGEN, DISSOLVED AS N	Drinking Water	1	28	0	0.00	1	0	0.00	2	0	0.00	13	0	0.00	12	0	0.00
00618	NITRATE NITROGEN, DISSOLVED AS N	Drinking Water	10	4	0	0.00	3	0	0.00				1	0	0.00			
00630	NITRITE PLUS NITRATE, TOTAL 1 DET	Drinking Water	10	16	0	0.00	1	0	0.00	1	0	0.00	5	0	0.00	9	0	0.00
00631	NITRITE PLUS NITRATE, DISS 1 DET	Drinking Water	10	34	0	0.00	1	0	0.00	1	0	0.00	12	0	0.00	20	0	0.00
00723	CYANIDE, DISSOLVED STD METHOD	Fresh Acute	22	1	0	0.00				1	0	0.00				1	0	0.00
00723		Drinking Water	200	1	0	0.00				1	0	0.00				1	0	0.00
00940	CHLORIDE, TOTAL IN WATER	Fresh Acute	860	101	0	0.00	20	0	0.00	8	0	0.00	28	0	0.00	45	0	0.00
00940		Drinking Water	250	101	0	0.00	20	0	0.00	8	0	0.00	28	0	0.00	45	0	0.00
00941	CHLORIDE, DISSOLVED IN WATER	Fresh Acute	860	1	0	0.00				1	0	0.00						
00941		Drinking Water	250	1	0	0.00				1	0	0.00						
00945	SULFATE, TOTAL (AS SO4)	Drinking Water	250	101	13	0.13	19	2	0.11	8	0	0.00	29	5	0.17	45	6	0.13
00946	SULFATE, DISSOLVED (AS SO4)	Drinking Water	250	3	0	0.00				3	0	0.00						
00950	FLUORIDE, DISSOLVED AS F	Drinking Water	4	82	0	0.00	17	0	0.00	7	0	0.00	20	0	0.00	38	0	0.00
01000	ARSENIC, DISSOLVED	Fresh Acute	360	50	0	0.00	6	0	0.00	1	0	0.00	13	0	0.00	30	0	0.00
01000		Drinking Water	50	50	0	0.00	6	0	0.00	1	0	0.00	13	0	0.00	30	0	0.00
01002	ARSENIC, TOTAL	Fresh Acute	360	4	0	0.00	1	0	0.00	1	0	0.00				2	0	0.00
01002		Drinking Water	50	4	0	0.00	1	0	0.00	1	0	0.00				2	0	0.00
01005	BARIUM, DISSOLVED	Drinking Water	2000	44	0	0.00	5	0	0.00	1	0	0.00	10	0	0.00	28	0	0.00
01007	BARIUM, TOTAL	Drinking Water	2000	2	0	0.00	1	0	0.00	1	0	0.00						
01010	BERYLLIUM, DISSOLVED	Fresh Acute	130	44	0	0.00	5	0	0.00	1	0	0.00	10	0	0.00	28	0	0.00
01010		Drinking Water	4	44	0	0.00	5	0	0.00	1	0	0.00	10	0	0.00	28	0	0.00
01012	BERYLLIUM, TOTAL	Fresh Acute	130	2	0	0.00	1	0	0.00	1	0	0.00						
01012		Drinking Water	4	2	0	0.00	1	0	0.00	1	0	0.00						
01025	CADMIUM, DISSOLVED	Fresh Acute	3.9	49 &	0	0.00	6	0	0.00	1	0	0.00	13	0	0.00	29	0	0.00
01025		Drinking Water	5	49 &	0	0.00	6	0	0.00	1	0	0.00	13	0	0.00	29	0	0.00
01027	CADMIUM, TOTAL	Fresh Acute	3.9	35	0	0.00	2	0	0.00	1	0	0.00	13	0	0.00	19	0	0.00
01030	CHROMIUM, DISSOLVED	Drinking Water	100	50	0	0.00	6	0	0.00	1	0	0.00	13	0	0.00	30	0	0.00
01034	CHROMIUM, TOTAL	Drinking Water	100	3	0	0.00	1	0	0.00	1	0	0.00				2	0	0.00
01040	COPPER, DISSOLVED	Fresh Acute	18	51	1	0.02	7	1	0.14	1	0	0.00	13	0	0.00	30	0	0.00
01040		Drinking Water	1300	51	0	0.00	7	0	0.00	1	0	0.00	13	0	0.00	30	0	0.00
01042	COPPER, TOTAL	Fresh Acute	18	2	0	0.00	1	0	0.00	1	0	0.00						
01042		Drinking Water	1300	2	0	0.00	1	0	0.00	1	0	0.00						
01049	LEAD, DISSOLVED	Fresh Acute	82	50	0	0.00	6	0	0.00	1	0	0.00	13	0	0.00	30	0	0.00
01049		Drinking Water	15	48 &	4	0.08	5	0	0.00	1	0	0.00	13	4	0.31	29	0	0.00
01057	THALLIUM, DISSOLVED	Fresh Acute	1400	18	0	0.00				1	0	0.00	10	0	0.00	7	0	0.00
01057		Drinking Water	2	18	0	0.00				1	0	0.00	10	0	0.00	7	0	0.00
01065	NICKEL, DISSOLVED	Fresh Acute	1400	43	0	0.00	5	0	0.00	1	0	0.00	10	0	0.00	27	0	0.00
01065		Drinking Water	100	43	0	0.00	5	0	0.00	1	0	0.00	10	0	0.00	27	0	0.00
01067	NICKEL, TOTAL	Fresh Acute	1400	2	0	0.00	1	0	0.00	1	0	0.00						
01067		Drinking Water	100	2	0	0.00	1	0	0.00	1	0	0.00						
01075	SILVER, DISSOLVED	Fresh Acute	4.1	43	0	0.00	5	0	0.00	1	0	0.00	10	0	0.00	27	0	0.00
01075		Drinking Water	100	43	0	0.00	5	0	0.00	1	0	0.00	10	0	0.00	27	0	0.00
01077	SILVER, TOTAL	Fresh Acute	4.1	2	0	0.00	1	0	0.00	1	0	0.00						
01077		Drinking Water	100	2	0	0.00	1	0	0.00	1	0	0.00						
01090	ZINC, DISSOLVED	Fresh Acute	120	51	0	0.00	7	0	0.00	1	0	0.00	13	0	0.00	30	0	0.00
01090		Drinking Water	5000	51	0	0.00	7	0	0.00	1	0	0.00	13	0	0.00	30	0	0.00
01092	ZINC, TOTAL	Fresh Acute	120	4	0	0.00	1	0	0.00	1	0	0.00				2	0	0.00
01092		Drinking Water	5000	2	0	0.00	1	0	0.00	1	0	0.00						
01095	ANTIMONY, DISSOLVED	Drinking Water	6	2	0	0.00				1	0	0.00				2	0	0.00
01145	SELENIUM, DISSOLVED	Fresh Acute	20	47	0	0.00	6	0	0.00	1	0	0.00	13	0	0.00	27	0	0.00
01145		Drinking Water	50	47	0	0.00	6	0	0.00	1	0	0.00	13	0	0.00	27	0	0.00
01147	SELENIUM, TOTAL	Fresh Acute	20	4	0	0.00	1	0	0.00	1	0	0.00				2	0	0.00
01147		Drinking Water	50	4	0	0.00	1	0	0.00	1	0	0.00				2	0	0.00

& - Below detection limit observations, for which half the detection limit exceeded the criterion, were excluded from the criterion comparison for this parameter

178

EPA Water Quality Criteria Analysis for Entire BADL Study Area

Parameter		Std Type	Std Value	Total Obs	Exceed Standard	Prop Exceeding	10/01-1/31			2/01-4/14			4/15-6/30			7/01-9/30		
							Obs	Exceed	Prop	Obs	Exceed	Prop	Obs	Exceed	Prop	Obs	Exceed	Prop
22703	URANIUM, NATURAL DISSOLVED	Drinking Water	20	18	0	0.00	1	0	0.00				3	0	0.00	14	0	0.00
31501	COLIFORM, TOTAL, MEMBRANE FILTER, IMMED	Other-Hi Lim	1000	13	2	0.15							5	0	0.00	8	2	0.25
31503	COLIFORM,TOT,MEMBRANE FILTR,DELAY M-END	Other-Hi Lim	1000	6	3	0.50							2	0	0.00	4	3	0.75
31504	COLIFORM,TOT,MEMBRANE FILTR,IMMED LES-E	Other-Hi Lim	1000	3	1	0.33										3	1	0.33
31616	FECAL COLIFORM, MEMBRANE FILTER, BROTH	Other-Hi Lim	200	7	0	0.00				1	0	0.00	3	0	0.00	3	0	0.00
31625	FECAL COLIFORM, MF	Other-Hi Lim	200	22	11	0.50							7	2	0.29	15	9	0.60
39055	SIMAZINE IN WHOLE WATER	Drinking Water	4	4	0	0.00	1	0	0.00				2	0	0.00	1	0	0.00
39630	ATRAZINE(AATREX) IN WHOLE WATER SAMPLE	Drinking Water	3	4	0	0.00	1	0	0.00				2	0	0.00	1	0	0.00
71851	NITRATE NITROGEN, DISSOLVED (AS NO3)	Drinking Water	44	39	0	0.00	13	0	0.00	6	0	0.00	10	0	0.00	10	0	0.00
71856	NITRITE NITROGEN, DISSOLVED (AS NO2)	Fresh Acute	3.3	14	5	0.36	4	2	0.50	1	0	0.00	2	0	0.00	7	3	0.43
71890	MERCURY, DISSOLVED	Drinking Water	2	50	0	0.00	6	0	0.00	1	0	0.00	13	0	0.00	30	0	0.00

& - Below detection limit observations, for which half the detection limit exceeded the criterion, were excluded from the criterion comparison for this parameter

NPS Servicewide Inventory and Monitoring Program Level I

Water Quality Parameter Inventory Data Evaluation and Analysis:

Missing Level I Groups

No STORET Data Within the BADL Study Area Exist for These Groups:

Chlorophyll*

---

*Not A Priority Parameter

NPS Servicewide Inventory and Monitoring Program Level I

Water Quality Parameter Inventory Data Evaluation and Analysis:

Present Level I Groups

STORET Data Within the BADL Study Area Exist for These Groups:

Alkalinity		Total Obs	01/01/85 to 09/10/96	01/01/75 to 12/31/84	Before 01/01/75	Total Stations
00410	ALKALINITY, TOTAL (MG/L AS CACO3)	113	12	12	89	19
00415	ALKALINITY, PHENOLPHTHALEIN (MG/L)	4	4	0	0	2
00430	ALKALINITY, CARBONATE (MG/L AS CACO3)	4	0	0	4	2
00440	BICARBONATE ION (MG/L AS HCO3)	110	6	0	104	10
00445	CARBONATE ION (MG/L AS CO3)	105	1	0	104	5
		336	23	12	301	38 (19)[1]

pH		Total Obs	01/01/85 to 09/10/96	01/01/75 to 12/31/84	Before 01/01/75	Total Stations
00400	PH (STANDARD UNITS)	172	54	10	108	29
00403	PH, LAB (STANDARD UNITS)	63	54	9	0	22
		235	108	19	108	51 (30)[1]

Conductivity		Total Obs	01/01/85 to 09/10/96	01/01/75 to 12/31/84	Before 01/01/75	Total Stations
00095	SPECIFIC CONDUCTANCE (UMHOS/CM @ 25C)	238	122	8	108	28
		238	122	8	108	28 (28)[1]

Dissolved Oxygen		Total Obs	01/01/85 to 09/10/96	01/01/75 to 12/31/84	Before 01/01/75	Total Stations
00300	OXYGEN, DISSOLVED (MG/L)	43	28	8	7	19
		43	28	8	7	19 (19)[1]

Water Temperature		Total Obs	01/01/85 to 09/10/96	01/01/75 to 12/31/84	Before 01/01/75	Total Stations
00010	TEMPERATURE, WATER (DEGREES CENTIGRADE)	247	123	7	117	26
00011	TEMPERATURE, WATER (DEGREES FAHRENHEIT)	8	0	8	0	3
		255	123	15	117	29 (26)[1]

Flow		Total Obs	01/01/85 to 09/10/96	01/01/75 to 12/31/84	Before 01/01/75	Total Stations
00060	FLOW, STREAM, MEAN DAILY CFS	193	0	0	193	3
00061	FLOW, STREAM, INSTANTANEOUS CFS	121	121	0	0	13
00065	STAGE, STREAM (FEET)	20	20	0	0	6
		334	141	0	193	22 (16)[1]

Clarity/Turbidity		Total Obs	01/01/85 to 09/10/96	01/01/75 to 12/31/84	Before 01/01/75	Total Stations
00070	TURBIDITY, (JACKSON CANDLE UNITS)	4	0	0	4	2
00076	TURBIDITY, HACH TURBIDIMETER (FORMAZIN TURB UNIT)	2	2	0	0	1
00078	TRANSPARENCY, SECCHI DISC (METERS)	2	2	0	0	1
00530	RESIDUE, TOTAL NONFILTRABLE (MG/L)	14	6	8	0	6
		22	10	8	4	10 (8)[1]

[1] Since a station can have data for more than one of the parameters in the parameter group, the number in the parenthesis is the number of unique stations having data for this parameter group

Nitrate/Nitrogen		Total Obs	01/01/85 to 09/10/96	01/01/75 to 12/31/84	Before 01/01/75	Total Stations
00608	NITROGEN, AMMONIA, DISSOLVED (MG/L AS N)	32	30	2	0	11
00610	NITROGEN, AMMONIA, TOTAL (MG/L AS N)	10	4	6	0	5
00618	NITRATE NITROGEN, DISSOLVED (MG/L AS N)	4	0	1	3	4
00623	NITROGEN, KJELDAHL, DISSOLVED (MG/L AS N)	30	30	0	0	10
00625	NITROGEN, KJELDAHL, TOTAL (MG/L AS N)	12	4	8	0	5
00630	NITRITE PLUS NITRATE, TOTAL 1 DET (MG/L AS N)	16	4	8	4	7
00631	NITRITE PLUS NITRATE, DISS 1 DET (MG/L AS N)	34	34	0	0	11
71846	NITROGEN, AMMONIA, DISSOLVED (MG/L AS NH4)	18	0	0	18	1
71851	NITRATE NITROGEN, DISSOLVED (MG/L AS NO3)	104	0	0	104	4
71856	NITRITE NITROGEN, DISSOLVED (MG/L AS NO2)	14	0	0	14	1
		274	106	25	143	59 (23)[1]

Phosphate/Phosphorus		Total Obs	01/01/85 to 09/10/96	01/01/75 to 12/31/84	Before 01/01/75	Total Stations
00650	PHOSPHATE, TOTAL (MG/L AS PO4)	16	0	1	15	4
00660	PHOSPHATE, ORTHO (MG/L AS PO4)	5	0	1	4	3
00665	PHOSPHORUS, TOTAL (MG/L AS P)	6	4	2	0	4
00666	PHOSPHORUS, DISSOLVED (MG/L AS P)	30	30	0	0	10
00671	PHOSPHORUS, DISSOLVED ORTHOPHOSPHATE (MG/L AS P)	29	24	5	0	15
70505	PHOSPHORUS, TOTAL, COLORIMETRIC METHOD (MG/L AS P)	8	0	8	0	3
70507	PHOSPHORUS, IN TOTAL ORTHOPHOSPHATE (MG/L AS P)	3	0	3	0	3
		97	58	20	19	42 (21)[1]

Sulfates/Total Dissolved Solids/Hardness		Total Obs	01/01/85 to 09/10/96	01/01/75 to 12/31/84	Before 01/01/75	Total Stations
00900	HARDNESS, TOTAL (MG/L AS CACO3)	109	0	1	108	7
00945	SULFATE, TOTAL (MG/L AS SO4)	166	50	8	108	27
00946	SULFATE, DISSOLVED (MG/L AS SO4)	3	0	3	0	2
70300	RESIDUE, TOTAL FILTRABLE (DRIED AT 180C), (MG/L)	158	54	0	104	22
		436	104	12	320	58 (30)[1]

Bacteria		Total Obs	01/01/85 to 09/10/96	01/01/75 to 12/31/84	Before 01/01/75	Total Stations
31501	COLIFORM, TOT, MEMBRANE FILTER,IMMED M-ENDOMED,35C	17	17	0	0	9
31503	COLIFORM, TOT, MEMBRANE FILTER,DELAY,M-ENDOMED,35C	6	6	0	0	6
31504	COLIFORM, TOT, MEMBRANE FILTER,L-ENDAGAR,35C	3	3	0	0	3
31616	FECAL COLIFORM, MEMBR FILTER, M-FC BROTH, 44 5C	7	0	7	0	3
31625	FECAL COLIFORM, MF, M-FC, 0 7 UM	27	27	0	0	10
31673	FECAL STREPTOCOCCI, MBR FILT, KF AGAR, 35C, 48HR	27	27	0	0	10
		87	80	7	0	41 (13)[1]

[1] Since a station can have data for more than one of the parameters in the parameter group, the number in the parenthesis is the number of unique stations having data for this parameter group

Toxic Elements		Total Obs	01/01/85 to 09/10/96	01/01/75 to 12/31/84	Before 01/01/75	Total Stations
01095	ANTIMONY, DISSOLVED (UG/L AS SB)	2	2	0	0	1
01000	ARSENIC, DISSOLVED (UG/L AS AS)	50	50	0	0	16
01002	ARSENIC, TOTAL (UG/L AS AS)	4	2	2	0	3
01010	BERYLLIUM, DISSOLVED (UG/L AS BE)	44	44	0	0	15
01012	BERYLLIUM, TOTAL (UG/L AS BE)	2	0	2	0	2
01025	CADMIUM, DISSOLVED (UG/L AS CD)	50	50	0	0	16
01027	CADMIUM, TOTAL (UG/L AS CD)	35	35	0	0	14
01030	CHROMIUM, DISSOLVED (UG/L AS CR)	50	50	0	0	16
01034	CHROMIUM, TOTAL (UG/L AS CR)	3	0	2	1	3
01040	COPPER, DISSOLVED (UG/L AS CU)	51	50	0	1	17
01042	COPPER, TOTAL (UG/L AS CU)	2	0	2	0	2
01049	LEAD, DISSOLVED (UG/L AS PB)	50	50	0	0	16
71890	MERCURY, DISSOLVED (UG/L AS HG)	50	50	0	0	16
01065	NICKEL, DISSOLVED (UG/L AS NI)	43	43	0	0	15
01067	NICKEL, TOTAL (UG/L AS NI)	2	0	2	0	2
01145	SELENIUM, DISSOLVED (UG/L AS SE)	47	43	0	4	16
01147	SELENIUM, TOTAL (UG/L AS SE)	4	2	2	0	3
01075	SILVER, DISSOLVED (UG/L AS AG)	43	43	0	0	15
01077	SILVER, TOTAL (UG/L AS AG)	2	0	2	0	2
01057	THALLIUM, DISSOLVED (UG/L AS TL)	18	18	0	0	9
01090	ZINC, DISSOLVED (UG/L AS ZN)	51	50	0	1	17
01092	ZINC, TOTAL (UG/L AS ZN)	4	2	2	0	3
00723	CYANIDE, DISSOLVED STD METHOD (UG/L)	1	1	0	0	1
		608	585	16	7	220 (21)[1]

[1]Since a station can have data for more than one of the parameters in the parameter group, the number in the parenthesis is the number of unique stations having data for this parameter group

NPS Servicewide Inventory and Monitoring Program Level I

Water Quality Parameter Inventory Data Evaluation and Analysis:

Park Summary: Level I Group Currentness and Distribution

Parameter Group	Total Obs.	Obs. Since 1985	% Obs. Since 1985	Stations Measuring This Group	% of Total Stations Measuring This Group	Obs. Per Station Measuring This Group	Period of Record For This Group	Observations Per Year of Period of Record
Alkalinity	336	23	6.8	19	61.3	17.7	11/29/62-08/03/94	10.6
pH	235	108	46.0	30	96.8	7.8	11/29/62-08/29/95	7.2
Conductivity	238	122	51.3	28	90.3	8.5	11/29/62-09/10/96	7.0
Dissolved Oxygen	43	28	65.1	19	61.3	2.3	01/31/67-08/29/95	1.5
Water Temperature	255	123	48.2	26	83.9	9.8	08/13/64-09/10/96	7.9
Flow	334	141	42.2	16	51.6	20.9	08/13/64-09/10/96	10.4
Clarity/Turbidity	22	10	45.5	8	25.8	2.8	06/01/71-08/30/94	0.9
Nitrate/Nitrogen	274	106	38.7	23	74.2	11.9	11/29/62-08/29/95	8.4
Phosphate/Phosphorus	97	58	59.8	21	67.7	4.6	11/04/64-08/29/95	3.1
Chlorophyll	0	0	0.0	0	0.0	0.0	No Data For Group	0.0
Sulfates/Total Dissolved Solids/Hardness	436	104	23.9	30	96.8	14.5	11/29/62-08/29/95	13.3
Bacteria	87	80	92.0	13	41.9	6.7	03/15/78-08/29/95	5.0
Toxic Elements	608	585	96.2	21	67.7	29.0	01/31/67-08/29/95	21.3

Water Quality Observations

Outside STORET Edit Criteria for BADL

(Disposition: X = Discarded, Blank = Retained)

NPS Station ID	Parameter		Date	Time	Parameter Value	Agency	STORET Station ID	Disposition
BADL0011	32730	PHENOLICS, TOTAL, RECOVERABLE (UG/L)	651108		8300.0000000	112WRD	06446200	
BADL0011	32730	PHENOLICS, TOTAL, RECOVERABLE (UG/L)	651111		7800.0000000	112WRD	06446200	
BADL0011	32730	PHENOLICS, TOTAL, RECOVERABLE (UG/L)	660309		8000.0000000	112WRD	06446200	
BADL0011	32730	PHENOLICS, TOTAL, RECOVERABLE (UG/L)	660314		8200.0000000	112WRD	06446200	
BADL0011	32730	PHENOLICS, TOTAL, RECOVERABLE (UG/L)	660401		7800.0000000	112WRD	06446200	
BADL0011	32730	PHENOLICS, TOTAL, RECOVERABLE (UG/L)	660525		7800.0000000	112WRD	06446200	
BADL0011	32730	PHENOLICS, TOTAL, RECOVERABLE (UG/L)	660628		4000.0000000	112WRD	06446200	
BADL0011	32730	PHENOLICS, TOTAL, RECOVERABLE (UG/L)	660914		8000.0000000	112WRD	06446200	

185

APPENDICES

Appendix A

Computer Files Transmitted With

Park Baseline Water Quality Data Inventory and Analysis

Computer disk(s) accompanying this report include up to seven (depending on the presence or absence of certain data elements) compressed (ZIP) files containing digital copies of nearly all the tables, figures, and other materials used to produce this report. To decompress these files, you must use the commonly available shareware program PKUNZIP. The command to type at the DOS prompt is:

PKUNZIP -E *COMPRESS.ZIP FILENAME.EXT*

where COMPRESS.ZIP is the name of one of the seven compressed (ZIP) files listed below and FILENAME.EXT is the name of the file you wish to extract. If you want to decompress all of the files in COMPRESS.ZIP, simply omit the FILENAME.EXT. To obtain a listing of all the files compressed into a particular ZIP file, type the following:

PKUNZIP -V *COMPRESS.ZIP* |MORE

where COMPRESS.ZIP is the name of one of the seven compressed ZIP files listed below. If a ZIP file spans multiple disks, use the last disk of the series (span) when obtaining a listing of all the files compressed into a particular ZIP file. Once you see the file you wish to obtain, substitute this file name for FILENAME.EXT in the first command line above to extract and decompress this particular file.

Included on one of the disk(s) accompanying this report is a program named PRINTZIP. This program will decompress ZIP files which don't span multiple disks and print certain files to a Hewlett-Packard (or compatible) Laser Printer. To use PRINTZIP, however, you must still have a copy of PKUNZIP in a directory listed in your path or in the same directory as the PRINTZIP program. PRINTZIP provides an easy, menu-driven interface for using PKUNZIP to decompress files and then send them to the printer. PRINTZIP allows you to send individual files, groups of files, or all files to the printer. PRINTZIP will not work with ZIP files that span multiple disks.

The following compressed (ZIP) files are included on the disk(s) accompanying this report:

(1)   BADLTABS.ZIP

      This compressed file contains all the tables presented in the report. The files compressed into this file include:

      (a) BADLSITE.DOC   -   Descriptive listing of select fields from the industrial facilities discharges, drinking water intakes, and EPA-USGS stream gages databases.

      (b) BADLAGNC.DOC   -   Contacts for agencies whose data were retrieved within the study area.

      (c) BADLAGNQ.DOC   -   Number of stations, observations, and parameters retrieved by agency code within the study area and park.

(d) BADLOV0.DOC   -   Overview of park and retrieved data.

(e) BADLOV1.DOC   -   Station period of record table.

(f) BADLOV2.DOC   -   Parameter period of record table.

(g) BADLOV3.DOC   -   Station/parameter period of record table.

(h) BADLINV.DOC   -   Station by station descriptive statistics over the entire period of record and comparison against EPA Water Quality Criteria for each station.

(i) BADLSEAN.DOC   -   Seasonal and annual water quality descriptive statistics at stations with water quality data meeting the default seasonal and annual criteria.

(j) BADLEPAS.DOC   -   EPA Water Quality Criteria comparison for data at all stations combined within the study area.

(k) BADLIDEA.DOC   -   Comparison of downloaded STORET data with NPS Servicewide Inventory and Monitoring Program "Level I" water quality parameters.

(l) BADLBAD.DOC   -   Water quality observation values that were outside the range of one of 190 STORET edit criteria and were either discarded or retained.

All these compressed document files are in ASCII format and contain printer codes appropriate to Hewlett-Packard (or compatible) Laser Printers. While at the DOS prompt, any of these document files may be printed directly to a Hewlett-Packard (or compatible) Laser Printer by using the PRINT command. For example, if the document BADLOV1.DOC is in the subdirectory C:\WATER, you could type: PRINT C:\WATER\BADLOV1.DOC. This will print the file to your local or networked Hewlett-Packard (or compatible) Laser Printer attached to parallel port one (LPT1:). Alternatively, you can use the PRINTZIP program to decompress and print any of these files provided the ZIP file doesn't span multiple disks. These ASCII files can also be imported into word-processed documents, but the printer codes will then have to be removed.

(2)   BADLFIGS.ZIP

This compressed file contains graphics files for all the statistical figures (time series plots; annual box and whiskers plots; seasonal box and whiskers plots) in the report in two different formats: Computer Graphic Metafile (CGM) and Hewlett-Packard Printer Control Language (PCL). The files are named with the last three digits of the Station Name followed by the five digit STORET code. The file name extension begins with either a 1 (time series), 2 (annual), or 3 (seasonal) and then either GM for CGM or CL for PCL. For example, 00100300.2GM would denote the file contains an annual box and whiskers plot in CGM format for parameter 00300 (dissolved oxygen) at station BADL0001. While at the DOS prompt, any PCL file can be printed directly to a Hewlett-Packard (or compatible) Laser Printer by using the COPY command. For example, if the graphic 00100300.2CL (an annual box and whiskers plot of parameter 00300, dissolved oxygen, at station BADL0001) is in the subirectory C:\WATER, you would type: COPY C:\WATER\00100300.2CL LPT1: /B. This will print the file to your local or networked Hewlett-Packard (or compatible) Laser Printer attached to parallel port one (LPT1:). The /B is necessary because the PCL file is in a binary format. Alternatively, you can use the PRINTZIP program to decompress and print any of the PCL files provided the ZIP file doesn't span multiple disks. The CGM files can be imported and/or edited in most graphics packages, including WordPerfect.

(3)   BADLPARM.ZIP

This file compresses BADLPARM.DBF which contains all the actual values (raw data) of all the water quality data downloaded from STORET and summarized in the report. The detailed database structure for this file is contained in Appendix B.

(4)   BADLSITE.ZIP

This compressed file contains up to five geo-referenced, DBASE III+ compatible site (point location) files documenting the location in the study area of water quality monitoring stations, industrial facilities discharges, drinking water intakes, water gages, and water impoundments. These files include:

(a) BADLWQ.DBF   -   All water quality monitoring station locations within the project's study area downloaded from STORET.

(b) BADLIFD.DBF   -   All municipal and industrial facility discharges within the project's study area downloaded from the IFD database.

(c) BADLDRIN.DBF   -   All drinking water intakes within the project's study area downloaded from the DRINKS database.

(d) BADLGAGE.DBF   -   All water gages within the project's study area downloaded from the GAGES database.

(e) BADLDAMS.DBF   -   All water impoundments within the project's study area downloaded from the DAMS database.

The absence of any of these files indicates that none of the particular sites were found within the study area. Detailed database structures for each of these files are contained in Appendix B.

(5)   BADLMISC.ZIP

This compressed file contains a variety of graphic and document files that are contained in the report. They are grouped into this miscellaneous compressed (ZIP) file because they don't fit neatly into any of the other compressed files. The files contained in this compressed file include:

(a) BADLEXEC.DOC   -   WordPerfect Ver. 5.1 copy of the Executive Summary in the report.

(b) BADLTOC.DOC   -   WordPerfect Ver. 5.1 copy of the report's Table of Contents.

(c) INTRO.DOC   -   WordPerfect Ver. 5.1 copy of all the text in the report from the Introduction through the Interpretive Guide to Water Quality Results.

(d) APPENDIX.DOC   -   WordPerfect Ver. 5.1 copy of all the Appendices in the report.

(e) BADLREGI   -   PCL and CLP (Windows Clipboard) copies of map displaying the regional location of the park and study area.

(f) BADLWQ   -   PCL and CLP (Windows Clipboard) copies of park maps displaying water quality station locations within the park's study area. If, due to scaling and aesthetic concerns, multiple maps were needed, these files will have alphabetically ordered suffixes (BADLWQA, BADLWQB, BADLWQC, etc.) and the index map name will end with an ampersand (&).

(g) BADLIDG	-	PCL and CLP (Windows Clipboard) copies of park maps displaying locations of industrial facilities discharges, drinking water intakes, and stream gages within the park's study area. If, due to scaling and aesthetic concerns, multiple maps were needed, these files will have alphabetically ordered suffixes (BADLIDGA, BADLIDGB, BADLIDGC, etc.) and the index map name will end with an ampersand (&). If no industrial facilities discharges, drinking water intakes, water gages, or water impoundments exist within the park's study area, these files will not be in the compressed (ZIP) file.
(h) BADLSEHY	-	PCL and CLP (Windows Clipboard) copies of the hydrographs or other materials used by WRD staff as the basis for a first attempt at a seasonal analysis of the park's water quality data.

Other materials may also be included in this miscellaneous compressed (ZIP) file as warranted by conditions at the park. As with BADLFIGS.ZIP and BADLTABS.ZIP, you can use the PRINTZIP program to print any of the PCL files in BADLMISC.ZIP provided the ZIP file doesn't span multiple disks. You should not, however, use PRINTZIP to print the WordPerfect document files. The CLP (Windows Clipboard) files can be imported (pasted) and/or edited in most Windows-based word processors and graphics packages.

(6)   BADLRF3.ZIP

This compressed file contains the Environmental Protection Agency's River Reach File Ver. 3.0 provisional data for the USGS catalog unit(s) encompassing the study area. The attribute data exist in both ASCII and DBASE III+ format, while the geographic traces exist in ASCII format. This compressed file contains four files for each catalog unit that touches the study area. Catalog units are identified by unique 8-character numeric names which identify the region, subregion, accounting unit, and catalog unit. Examples (your 8-character numeric names will be different) of the file types included in this compressed file are:

(a) 12345678.RF3	-	ASCII formatted attribute file from the River Reach File for all hydrographic traces within the catalog unit.
(b) 12345678.DBF	-	DBASE III+ formatted attribute file from the River Reach File for all hydrographic traces within the catalog unit.
(c) 12345678.TRC	-	ASCII formatted geographic file from the River Reach File containing digital, geo-referenced descriptions of all hydrographic traces within the catalog unit at a scale of 1:100,000 suitable for import into a geographic information system.
(d) 12345678.CUB	-	ASCII formatted geographic file from the River Reach File containing a digital, geo-referenced description of the catalog unit boundary suitable for import into a geographic information system.

Detailed database structures for RF3-related files are contained in Appendix B.

(7)   <u>BADLWQMW.ZIP</u>

Between 2000 and 2002, all Baseline Water Quality Data Inventory and Analysis Reports were compiled or re-compiled in Microsoft Word 2000 (Ver. 9.0) format. This complete, digital version of the report will be made available through various means, including the Internet. Although the reports can be opened in Microsoft Word 1997 (Ver. 8.0), the time series and annual and seasonal box-plots may not be centered appropriately on a page due to discrepancies with how Word 2000 formats pictures and how Word 1997 formatted pictures. Consequently, Word 2000 is the recommended software for viewing the report. Prior to printing the report from Word, be sure to enable "Print Text as Graphics" or "Print True Type Font as Graphics" in the Printer Properties. This ensures a more faithful reproduction of the maps included in the Word document.

The Microsoft Word version of the Baseline Water Quality Data Inventory and Analysis Report may differ slightly from the original analog version. Reports issued during 1994-1996 didn't have as many "bells-and-whistles" as subsequent reports. In compiling digital Microsoft Word versions of these earlier reports, attempts were made to bring these 1994-1996 reports up to the current standard wherever feasible and practicable. Unfortunately, some changes were not feasible or practicable. For example, water quality criteria screens were added or modified over time when newer criteria became available. The digital Microsoft Word version of Appendix F presents the latest criteria screening parameters and values. Some of these parameters and/or values may not have been screened against in the EPA water quality criteria analyses for each station and the entire study area in the 1994-1996 analog versions of the report. Similarly, the Introduction, Methodology, and Interpretive Guide to Water Quality Results may mention certain features that aren't included in the 1994-1996 reports. Additionally, to prepare a Microsoft Word version of this report, data were processed through different versions of software than used originally. Consequently, some results presented in the Overview and Executive Summary may differ slightly from those presented in the analog report (eg. # of In Park and Longer Term Stations).

Appendix B

Water Quality Database File Structures

The following table provides the DBASE III+ database field structure for all the water quality parameter data downloaded from STORET. This data will allow parks or other interested parties to replicate the statistical analyses and graphics contained in this report; perform more sophisticated analyses; or to establish a baseline park water quality database.

Parameter Data File: BADLPARM.DBF in BADLPARM.ZIP				
Field Name	**Start**	**Stop**	**Length**	**Field Description**
NPSSTATID	1	8	8	NPS Station ID (NPS park code + 4 digit sequence number)
BEGDATE	9	14	6	Measurement Start Date [yymmdd]
BEGTIME	15	18	4	Measurement Start Time [hhmm]
PARMCODE	19	23	5	STORET Parameter Code
PARMVALU	24	39	16.7	Parameter Value
REMARK	40	40	1	Parameter Remark Value
				A=Value is Mean of 2 or More Determinations
				B=Results Based Upon Colony Counts Outside Acceptable Range
				C=Value Calculated
				D=Field Measurement
				E=Extra Sample Taken in Compositing Process
				F=Female Species
				G=Maximum of 2 or More Determinations
				H=Based on Field Kit Determination
				I=Value is Less Than Practical Quantitation Limit and Greater Than or Equal to the Method Detection Limit
				J=Estimated. Not the Result of Analytic Measurement
				K=Off-scale Low. Actual Value Not Known. But Known to be Less Than Value Shown
				L=Off-scale High. Actual Value Not Known. But Known to be Greater Than Value Shown

Parameter Data File: BADLPARM.DBF in BADLPARM.ZIP				
Field Name	**Start**	**Stop**	**Length**	**Field Description**
				M=Presence Verified, But Not Quantified, Below Quantification Limit; For Species, Male; For Oxygen Reduction Potential, Indicates a Negative Value
				N=Presumptive Evidence of Presence
				O=Analysis Lost
				P=Too Numerous to Count
				Q=Exceeded Normal Holding Time
				R=Significant Rain in Last 48 Hours
				S=Laboratory test
				T=Less Than Detection Criteria
				U=Analyzed For But Not Detected, Value is Detection Limit For Process Used; If Species, Undetermined
				V=Analyte was Detected in Sample and Method Blank
				W=Less Than Lowest Value Reportable Under Remark "T"
				X=Quasi Vertically-Integrated Sample
				Y=Analysis of Unpreserved Sample
				Z=Too Many Colonies Were Present to Count (TNTC), Value Represents Filtration Value
				$=Calculated By Retrieval Software
MEDIA	41	46	6	Sample Media
DEPTH	47	55	9.3	Depth of Sample [in feet]
ENDDATE	56	61	6	Measurement End Date [yymmdd] [all composite samples]
ENDTIME	62	65	4	Measurement End Time [hhmm] [all composite samples]
SAMPTYPE	66	69	4	Type of Sample ["sophisticated" composite samples]
				C=Continuous Collection
				G=Collection of Individual Grab Samples
				GNxx=xx is the Number of Individual Grab Samples
				B=N/A

Parameter Data File: BADLPARM.DBF in BADLPARM.ZIP				
Field Name	Start	Stop	Length	Field Description
COMPTYPE	70	70	1	Composite Value Type ["sophisticated" composite samples]
				A=Average
				H=Maximum
				L=Minimum
				N=Number of Observations
				#=Number of Observations
				S=Standard Deviation
				U=Sum of Squares
				V=Variance
				C=Coefficient of Error
				X=Coefficient of Variance
				E=Skewness
				F=Kurtosis
				Z=Number of Observations That Exceed an Established Limit
				%=Precision
				$=Accuracy
				B=N/A
				D=Indicates Replicate Sample
COMPST	71	71	1	Composite Space/Time Indicator
				S=Space
				T=Time
				B=Space and Time
				F=Flow Proportional
				1-9=Replicate Number

Note: DBASE III+ record lengths will be one greater than the last stop column displayed (71 here) because DBASE III+ reserves the first space/column of every record for a deletion flag. Hence, DBASE III+ will display a record length of 72 for this database.

The following table provides the DBASE III+ database field structure for all the water quality station locations downloaded from STORET. As this file is geo-referenced, it should import easily into the park's Geographic Information System.

Water Quality Station Data File: BADLWQ.DBF in BADLSITE.ZIP				
Field Name	Start	Stop	Length	Field Description
NPSSTATID	1	8	8	NPS Station ID (NPS park code + 4 digit sequence number)
AGENCY	9	16	8	Agency Code of Station Owner
STORIDP	17	31	15	STORET Primary Station Code
STORIDS1	32	43	12	STORET First Secondary Station Code
STORIDS2	44	55	12	STORET Second Secondary Station Code
STORIDS3	56	65	10	STORET Third Secondary Station Code
LATITUDE	66	73	8	Station Latitude [degrees:minutes:seconds]
LONGITUDE	74	82	9	Station Longitude [degrees:minutes:seconds]
LAT	83	93	11.6	Station Latitude [decimal degrees. (-) below equator]
LON	94	104	11.6	Station Longitude [decimal degrees. (-) western hemisphere]
LLPREC	105	105	1	Latitude/Longitude Precision Code
RMI	106	329	224	River Mile Index
STATLOC	330	377	48	Station Location Description
CNTYCODE	378	382	5	FIPS State/County Code
STNAME	383	398	16	State Name
CNTYNAME	399	418	20	County Name
HYDUNIT	419	426	8	Hydrologic Unit Code (MAJ/MIN/SUB = Catalog Unit)
MAJBASN	427	450	24	Major Basin Name
MINBASN	451	490	40	Minor Basin Name
STATTYPE	491	550	60	Station Type
STORDATE	551	556	6	Date Station was Stored in STORET
RF1INDEX	557	567	11	RF1 Reach Number Location [2]
RF1MILE	568	575	8.3	Mile Point on RF1 Reach [2]
RF1LOC	576	578	3	Indicates the Location as ON or OFF RF1 Reach [2]
RF1DIST	579	584	6.2	Distance From RF1 Reach

				Water Quality Station Data File: BADLWQ.DBF in BADLSITE.ZIP	
Field Name	**Start**	**Stop**	**Length**	**Field Description**
RF3INDEX	585	601	17	RF3 Reach Number Location [3]
RF3MILE	602	607	6.2	Mile point on RF3 Reach [3]
RF3LOC	608	610	3	Indicates the Location as ON or OFF RF3 Reach [2]
RF3DIST	611	616	6.2	Distance From RF3 Reach
DEPH2O	617	620	4	Depth of Water at Station Location [in feet]
ELEV	621	625	5	Station Elevation
ECOREG	626	628	3	ECO Region
H2OBODY	629	678	50	Waterbody ID
AQUIFERS	679	718	40	Aquifer Description
STATDESC1	719	790	72	Station Sentence Description
STATDESC2	791	862	72	Station Sentence Description
STATDESC3	863	934	72	Station Sentence Description
STATDESC4	935	1006	72	Station Sentence Description
STATDESC5	1007	1078	72	Station Sentence Description
STATDESC6	1079	1150	72	Station Sentence Description
STATDESC7	1151	1222	72	Station Sentence Description
STATDESC8	1223	1294	72	Station Sentence Description
STATDESC9	1295	1366	72	Station Sentence Description
STATDESC10	1367	1438	72	Station Sentence Description
STATDESC11	1439	1510	72	Station Sentence Description
STATDESC12	1511	1582	72	Station Sentence Description
STATDESC13	1583	1654	72	Station Sentence Description
STATDESC14	1655	1726	72	Station Sentence Description
STATDESC15	1727	1798	72	Station Sentence Description
STATLOCKED	1799	1799	1	Station Locked (Logical) True/False

The following table provides the DBASE III+ database field structures for the EPA Industrial Facilities Discharge database. As this file is geo-referenced, it should import easily into the park's Geographic Information System.

Industrial Facilities Discharges File: BADLIFD.DBF in BADLSITE.ZIP				
Field Name	**Start**	**Stop**	**Length**	**Field Description**
SITEID	1	9	9	Site Identifier (NPDES Number)
LATITUDE	10	17	8	Facility Latitude (Degrees:Minutes:Seconds)
LONGITUDE	18	26	9	Facility Longitude (Degrees:Minutes:Seconds)
LAT	27	37	11.6	Facility Latitude (decimal degrees. (-) below equator)
LON	38	48	11.6	Facility Longitude (decimal degrees. (-) west. hem.)
RF1INDEX	49	59	11	RF1 Reach Number Location
RF1MILE	60	65	6.2	Mile Point on RF1 Reach
RF1DIST	66	71	6.2	Distance From RF1 Reach
RF3INDEX	72	88	17	RF3 Reach Number Location
RF3MILE	89	94	6.2	Mile Point on RF3 Reach
RF3DIST	95	100	6.2	Distance From RF3 Reach
ADR	101	125	25	Address
BFL	126	132	7.2	Total Direct Combined C&P Flow (1000 GPD)
CCFLG	133	133	1	Coastal County Flag "Y"/"N"/"E"=Estuary
CC1	134	138	5	City Code #1 (EPA Code)
CFL	139	145	7.2	Total Direct Cooling Flow (1000 GPD)
CNC	146	148	3	County Code (FIPS)
CTY	149	168	20	City Name
CZIP	169	177	9	Canadian Zip Code
DNB	178	186	9	Dunn & Bradstreet Number
DNBFLG	187	187	1	Dunn & Bradstreet PCS Source Flag
EGF	188	202	15.4	Flow From Effluent Guidelines (1000 GPD)
EGS	203	208	6	Effluent Guidelines Subcategory
EXPDT	209	216	8	Expiration Date (mm/dd/yy)
E308SN	217	220	4	Effluent Guidelines Survey Number
FAC	221	229	9	SCS Facility Identifier (Cross-Reference)
FDS	230	232	3	Facility Data Source

Industrial Facilities Discharges File: BADLIFD.DBF in BADLSITE.ZIP				
Field Name	Start	Stop	Length	Field Description
FFL	233	239	7.2	Total Facility Flow (1000 GPD)
FHF	240	240	1	Fac. Hit Flag (Reach File) V=Versar Assumed
FLOTYP	241	243	3	I=Blow Down. R=Bottom Ash. S=Fly Ash
FLR	244	250	7.2	Flow Recvd-Industrial (1000 GPD) Permit Data
FRDS	251	259	9	FRDS ID# - XREF To Water Supply
FRW	260	289	30	Facility Receiving Water Name
FS1	290	293	4	Facility SIC Code (From PCS)
FS2	294	297	4	Facility SIC Code #1
FS3	298	301	4	Facility SIC Code #2
FS4	302	305	4	Facility SIC Code #3
FS5	306	309	4	Facility SIC Code #4
FUD	310	317	8	Facility Level Last Date Updated (mm/dd/yy)
IACC	318	318	1	Inactive/Active Indicator ("I" or "A")
ICAT	319	320	2	WQAB Industrial Category
ICAT2	321	322	2	WQAB Industrial Category 2
ICAT3	323	324	2	WQAB Industrial Category 3
IFL	325	331	7	Total Indirect Flow (1000 GPD)
IFT	332	332	1	Illinois Facility Type (A thru Z)
IG1	333	334	2	Facility Industrial Group #1
IG2	335	336	2	Facility Industrial Group #2
IJCN	337	346	10	Canadian Record Identifier
INACT	347	353	7	Inactive/Rescinded P=Based on Permit;A=Actual
INDCNT	354	357	4	Computed Number of Indirect Dischargers
LATLON	358	372	15	Polygon Retrieval Lat/Long.
MAJ	373	373	1	Major-Minor Flag (From PCS)
MAPID	374	377	4	Map Identifier
MJMN	378	381	4	Major/Minor Basin (EPA-STORET)
NAM	382	441	60	Facility Name
NDC	442	444	3	Number of Discharges (Pipes)

Industrial Facilities Discharges File: BADLIFD.DBF in BADLSITE.ZIP				
Field Name	**Start**	**Stop**	**Length**	**Field Description**
NDSFLO	445	451	7.2	NEEDS Flow (1000 GPD)
NDSIFLO	452	458	7.2	NEEDS Industrial Flow (1000 GPD)
NID	459	462	4	Number of Indirect Dischargers
NPC	463	463	1	NEEDS Pre-Treatment Code "Y"=Yes. "N"=No
NPS	464	464	1	NPDES Facility Source/Status
NSN	465	473	9	NEEDS Survey Number
NTC	474	474	1	NEEDS Treatment Code
OCP	475	480	6	Organic Chemical Producers ID Number
ODESCC	481	481	1	ODES Coastal County "Y"=Yes: "N"=No
OFL	482	488	7.2	Total Non-Direct Other Flow (1000 GPD)
OWN	489	491	3	Ownership Code
PFL	492	498	7.2	Total Direct Process Flow (1000 GPD)
REG	499	500	2	EPA Region
REGKEY	501	504	4	Region Key
RSLOFLO	505	511	7.2	Receiving Stream Low Flow
RSMNFLO	512	518	7.2	Receiving Stream Mean Flow
STA	519	520	2	State Postal Abbreviation
STAID	521	535	15	State Identifier
STC	536	537	2	State Code (FIPS)
STCITY	538	544	7	State/City Code
TFLOW	545	551	7.2	Type Flow (1000 GPD)
UFL	552	558	7.2	Total Direct Undefined Flow (1000 GPD)
XEGS	559	561	3	Effluent Guidelines Subcat Index
XKEY	562	562	1	"1"."2"."3"."4"."5"."6"."7"."8"."9"
XNME	563	565	3	GLP.DIR.F2C.ENF.CET.LAG.PPB,M85,M86
ZIP	566	570	5	Zip Code

The following table provides the DBASE III+ database field structures for drinking water intakes from the EPA DRINKS database. As this file is geo-referenced, it should import easily into the park's Geographic Information System.

Drinking Water Intakes File: BADLDRIN.DBF in BADLSITE.ZIP				
Field Name	**Start**	**Stop**	**Length**	**Field Description**
SITEID	1	20	20	Site Identifier
LATITUDE	21	28	8	Facility Latitude (Degrees:Minutes:Seconds)
LONGITUDE	29	37	9	Facility Longitude (Degrees:Minutes:Seconds)
LAT	38	48	11.6	Facility Latitude (decimal degrees, (-) below equator)
LON	49	59	11.6	Facility Longitude (decimal degrees, (-) west, hem.)
RF1INDEX	60	70	11	RF1 Reach Number Location
RF1MILE	71	76	6.2	Mile Point on RF1 Reach
RF1DIST	77	82	6.2	Distance From RF1 Reach
RF3INDEX	83	99	17	RF3 Reach Number Location
RF3MILE	100	105	6.2	Mile Point on RF3 Reach
RF3DIST	106	111	6.2	Distance From RF3 Reach
AQCD	112	115	4	Aquifer Code
ASC	116	138	23	STORET Agency/Station Code
AVGD	139	142	4	Average Depth
BUY	143	143	1	Purchase Code
CC1	144	148	5	City Code #1 (EPA Code)
CNC	149	151	3	County Code (FIPS)
CNME	152	166	15	Contact Name
CNN	167	186	20	County Name
CTITLE	187	201	15	Contact Title
CTY	202	221	20	City Name
DUD	222	229	8	Date of Update
FRDS	230	238	9	FRDS ID# - Cross-Reference
GEOAG	239	258	20	Geologic Age
GEOCDE	259	261	3	Geologic Age Code
IDAT	262	269	8	Date (mm/dd/yy)

Drinking Water Intakes File: BADLDRIN.DBF in BADLSITE.ZIP				
Field Name	Start	Stop	Length	Field Description
INTAKET	270	270	1	Type Source G/S/B
INTRVWR	271	285	15	Interviewer
MAXD	286	289	4	Maximum Depth
MILES	290	296	7.2	Miles
MIND	297	300	4	Minimum Depth
NAME	301	320	20	Name
NPD	321	329	9	NPDES# XREF to IFD Database
NWLS	330	332	3	Number of Wells
OWN	333	335	3	Ownership
PAVGF	336	342	7.2	Production Avg. Daily (Gal/Day)
PCTSUP	343	345	3	%Surface / %Ground
PHONE	346	355	10	Telephone Number
PMAXF	356	362	7.2	Production Max. Daily (Gal/Day)
POPSV	363	371	9	Population Served
REG	372	373	2	EPA Region
SHLAT	374	379	6	Sitehelp Latitude (DDMMSS)
SHLNG	380	386	7	Sitehelp Longitude (DDDMMSS)
SHMILES	387	393	7.2	Sitehelp Miles
SHNME	394	403	10	Sitehelp Source Name
SHPCT	404	410	7.2	Sitehelp Percent of Reach Miles
SRC	411	413	3	Sitehelp Source Code
STA	414	415	2	State Abbreviation
STC	416	417	2	State Code (FIPS)
TUF	418	424	7.2	Total Utility Flow
TYPCDE	425	425	1	Type Code
UHF	426	426	1	Utility Hit Flag (Reach File)
VCDE	427	427	1	Versar Code='V'=>25K: '*'=<25K POPSVD
WFPC	428	428	1	Wellfield Precision Code
WFTYP	429	429	1	Well Type (Cassing.Artesian.Infiltration.etc.)

Drinking Water Intakes File: BADLDRIN.DBF in BADLSITE.ZIP				
Field Name	Start	Stop	Length	Field Description
WUN	430	449	20	Water Utility Name

The following table provides the DBASE III+ database field structures for the Water Gage database. As this file is geo-referenced, it should import easily into the park's Geographic Information System.

Water Gage File: BADLGAGE.DBF in BADLSITE.ZIP				
Field Name	**Start**	**Stop**	**Length**	**Field Description**
SITEID	1	20	20	Site Identifier
LATITUDE	21	28	8	Facility Latitude (DDMMSS)
LONGITUDE	29	37	9	Facility Longitude (DDDMMSS)
LAT	38	48	11.6	Facility Latitude (decimal degrees. (-) below equator)
LON	49	59	11.6	Facility Longitude (decimal degrees. (-) west. hem.)
RF1INDEX	60	70	11	RF1 Reach Number Location
RF1MILE	71	76	6.2	Mile Point on RF1 Reach
RF1DIST	77	82	6.2	Distance From RF1 Reach
RF3INDEX	83	99	17	RF3 Reach Number Location
RF3MILE	100	105	6.2	Mile Point on RF3 Reach
RF3DIST	106	111	6.2	Distance From RF3 Reach
JAN	112	118	7.2	Monthly Flow - January
FEB	119	125	7.2	Monthly Flow - February
MAR	126	132	7.2	Monthly Flow - March
APR	133	139	7.2	Monthly Flow - April
MAY	140	146	7.2	Monthly Flow - May
JUN	147	153	7.2	Monthly Flow - June
JUL	154	160	7.2	Monthly Flow - July
AUG	161	167	7.2	Monthly Flow - August
SEP	168	174	7.2	Monthly Flow - September
OCT	175	181	7.2	Monthly Flow - October
NOV	182	188	7.2	Monthly Flow - November
DEC	189	195	7.2	Monthly Flow - December
RGN	196	197	2	Region Code
AREA	198	204	7.2	Drainage Area (SQ.MI.)
DUD	205	212	8	Date of Update

				Water Gage File: BADLGAGE.DBF in BADLSITE.ZIP
Field Name	**Start**	**Stop**	**Length**	**Field Description**
FBCF	213	213	1	Flag - Basic Characteristic File ('Y')
FDFF	214	214	1	Flag - Daily Flows File ('Y')
FQMINV	215	224	10	IHS Pt. Files Index
GHF	225	225	1	Hit Flag (Reach File)
ICDE	226	226	1	Integrity Code
LFVEL	227	233	7.2	Low Flow Velocity
METHOD	234	236	3	Calculation Method Code
MFVEL	237	243	7.2	Mean Flow Velocity
MNFLO	244	250	7.2	USGS Mean Annual Flow
NME	251	298	48	Station Name
SHLAT	299	304	6	Sitehelp Latitude (DDMMSS)
SHLNG	305	311	7	Sitehelp Longitude (DDDMMSS)
SHMILES	312	318	7.2	Sitehelp Miles
SHNME	319	328	10	Sitehelp Source Name
SHPCT	329	335	7.2	Sitehelp Percent of Reach Miles
SITE	336	337	2	Site Location
SRC	338	340	3	Sitehelp Source Code
STCTY	341	345	5	State/County Numeric Code
SVTEN	346	352	7.2	USGS 7-10 Year Flow
BEG_WYR	353	356	4	Beginning Water Year
END_WYR	357	359	4	Ending Water Year
ELEV	361	368	8.2	Elevation (Feet)
WELL_DP	369	376	8.2	Well Depth (Feet)

The following table provides the DBASE III+ database field structures for the Water Impoundment database. As this file is geo-referenced, it should import easily into the park's Geographic Information System.

Water Impoundment File: BADLDAMS.DBF in BADLSITE.ZIP				
Field Name	**Start**	**Stop**	**Length**	**Field Description**
SITEID	1	7	7	Site Identifier
SOURCE	8	10	3	Source of Data
ST1	11	12	2	Primary State Code Abbreviation
STCTY1	13	17	5	State/County Numeric Code
NAME	18	47	30	Official Name of Dam
LATITUDE	48	53	6	Facility Latitude (DDMMSS)
LONGITUDE	54	60	7	Facility Longitude (DDDMMSS)
LAT	61	70	10.6	Facility Latitude (decimal degrees, (-) below equator)
LON	71	81	11.6	Facility Longitude (decimal degrees, (-) west. hem.)
INME	82	111	30	Impoundment Name
RNME	112	139	28	River, Stream, or Tributary Name on Which Dam Built
CUSEGMI	140	149	10	Catalog Unit, Segment, and Segment Length
REGN	150	151	2	Water Resources Council Region Code
RGBSN	152	155	4	Water Resources Region/Basin Code
CU	156	163	8	Catalog Unit
SEG	164	166	3	Reach Segment of Dam
SEGL	167	171	5.2	Reach Segment Length
PURP	172	172	1	Major Purpose of Dam
				I=Irrigation
				H=Hydroelectric
				N=Navigation
				S=Water Supply
				R=Recreation
				P=Stock/Farm Pond
				D=Debris Control
				F=Flood Control

Water Impoundment File: BADLDAMS.DBF in BADLSITE.ZIP				
Field Name	Start	Stop	Length	Field Description
				O=Other
FRF3	173	189	17	RF3 Reach Number Location
FRF3MI	190	194	5	Mile Point on RF3 Reach
PURPKEY	195	195	1	Purpose Key
PUR2	196	196	1	Purpose of Dam 2 (See Above)
PUR3	197	197	1	Purpose of Dam 3 (See Above)
PUR4	198	198	1	Purpose of Dam 4 (See Above)
PUR5	199	199	1	Purpose of Dam 5 (See Above)
PUR6	200	200	1	Purpose of Dam 6 (See Above)
PUR7	201	201	1	Purpose of Dam 7 (See Above)
PUR8	202	202	1	Purpose of Dam 8 (See Above)
PUR9	203	203	1	Purpose of Dam 9 (See Above)
PUR10	204	204	1	Purpose of Dam 10 (See Above)
TYPDAM	205	206	2	Major Dam Portion Type
				RE=Earth
				VA=Vaulted Arch
				CD=Buttress
				PG=Gravity
				ER=Rockfill
				MV=Multi-Arch
				OT=Other
YRCMP	207	210	4	Year Dam Completed
SHGT	211	214	4	Structural Height (Feet)
HHGT	215	218	4	Hydraulic Height (Feet)
VNORM	219	236	8	Normal Storage of Impoundment (Acre-Feet)
VMAX	227	234	8	Maximum Storage of Impoundment (Acre-Feet)
LCRST	235	239	5	Crest Length of Dam (Feet)
TSPL	240	240	1	Spillway Type
				C=Controlled

Water Impoundment File: BADLDAMS.DBF in BADLSITE.ZIP				
Field Name	Start	Stop	Length	Field Description
				U=Uncontrolled
				N=None
				X=Unknown
WSPL	241	244	4	Dam Spillway Width (Feet)
QMAX	245	251	7	Maximum Spillway Discharge (CFS)
PINS	252	258	7.2	Quantity of Installed Power (Megawatts)
PPRO	259	265	7.2	Quantity of Proposed Power (Megawatts)
LOCK	266	266	1	Number of Navigational Locks
OWNR	267	290	24	Name of Impoundment Owner
PFOWN	291	291	1	Ownership Code
				N=Non-Federal
				G=Federal Government Agency
				C=Corps of Engineers
				X=Unknown
FEDR	292	292	1	Federally Regulated (Y=Yes, N=No, X=Unknown)
FLND	293	293	1	Private Dam on Federal Land (Y=Yes, N=No, X=Unknown)
SCSA	294	294	1	Type of Soil Conservation Service Assistance
				N=No Assistance
				T=Technical Assistance
				F=Financial Assistance
				B=Both Technical and Financial Assistance
				X=Unknown
DHAZ	295	295	1	Degree of Downstream Hazard
				1=High (More than a Few Lives Lost; Excessive Economic Loss)
				2=Significant (A Few Lives Lost; Appreciable Economic Loss)
				3=Low (No Lives Expected Lost; Minimal Economic Loss)
DCITY	296	319	24	Nearest Downstream City

Water Impoundment File: BADLDAMS.DBF in BADLSITE.ZIP				
Field Name	**Start**	**Stop**	**Length**	**Field Description**
POP	320	326	7	Population of Downstream City
DMILE	327	331	5.2	Distance of Downstream City From Dam (Miles)
RET	332	342	11.2	Retention Coefficient (Dimensionless)
MIX	343	353	11.2	Mixing Coefficient (Dimensionless)
SAREA	354	361	8	Surface Area of Impoundment (Acres)
SAFLG	362	362	1	Surface Area Flag (C=Calc., M=Measured. O=Other)
ILNTH	363	367	5	Length of Impoundment (Feet)
ILFLG	368	368	1	Impoundment Length Flag (C=Calc., M=Measured. O=Other)
UPKEY	369	374	6	Update Key (YYMMDD)

The following table provides the ASCII and DBASE III+ database field structures for the EPA River Reach File Ver. 3.0 (1:100,000 scale hydrography) attributes. The actual numeric file names will vary depending on the catalog unit(s). This information can be readily incorporated into the park's Geographic Information System.

RF3 Structure File: 12345678.RF3 and 12345678.DBF in BADLRF3.ZIP				
Field Name	**Start**	**Stop**	**Length**	**Field Description**
CATUNIT	1	8	8	Cataloging Unit (CU)
SEGM	9	12	4	Segment Number (SEG)
MI	13	17	5.2	Mile Point (MI)
UPMI	18	22	5.2	Upstream Mile Pt.
SEQNO	23	33	11.6	Hydro Sequence No.
RFLAG	34	34	1	Reach Flag (0.1)
OWFLAG	35	35	1	Open Water Flag (0.1)
TFLAG	36	36	1	Terminal Flag (0.1)
SFLAG	37	37	1	Start Flag (0.1)
RCHTYPE	38	38	1	Reach Type Code
LEV	39	40	2	Stream Level
JUNC	41	42	2	Level of Downstream Reach
DIVERGENCE	43	43	1	Divergence Code
STARTCU	44	51	8	Start CU
STRTSG	52	55	4	Start SEG
STOPCU	56	63	8	Stop CU
STOPSG	64	67	4	Stop SEG
USDIR	68	68	1	Upstream Direction
TERMID	69	73	5	Terminal Stream ID
TRMBLV	74	74	1	Terminal Base Level
PNAME	75	104	30	Primary Name
PNMCD	105	115	11	Primary Name Code
CNAME	116	145	30	Complement Name
CNMCD	146	156	11	Complement Name Code

RF3 Structure File: 12345678.RF3 and 12345678.DBF in BADLRF3.ZIP				
Field Name	**Start**	**Stop**	**Length**	**Field Description**
OWNAME	157	186	30	Open Water Name
OWNMCD	187	197	11	Open Water Name Code
DSCU	198	205	8	Downstream CU
DSSEG	206	209	4	Downstream SEG
DSMI	210	214	5.2	Downstream MI
CCU	215	222	8	Complement CU
CSEG	223	226	4	Complement SEG
CMILE	227	231	5.2	Complement MI
CDIR	232	232	1	Complement Direction
ULCU	233	240	8	Upstream Left CU
ULSEG	241	244	4	Upstream Left SEG
ULMI	245	249	5.2	Upstream Left MI
URCU	250	257	8	Upstream Right CU
URSEG	258	261	4	Upstream Right SEG
URMI	262	266	5.2	Upstream Right MI
SEGL	267	272	6.2	Reach Length (Miles)
RFORGFLAG	273	273	1	RF Orgin flag(1.2.3)
ALTPNMCD	274	281	8	Alt. Primary Name Code
ALTOWNMC	282	289	8	Alt. OW Name Code
DLAT	290	297	8.4	Downstream Latitude
DLONG	298	305	8.4	Downstream Longitude
ULAT	306	313	8.4	Upstream Latitude
ULONG	314	321	8.4	Upstream Longitude
MINLAT	322	329	8.4	Minimum Latitude
MINLONG	330	337	8.4	Minimum Longitude
MAXLAT	338	345	8.4	Maximum Latitude
MAXLONG	346	353	8.4	Maximum Longitude
NDLGREC	354	357	4	No. of DLG Records
LL1KEY1	358	367	10	Starting DLG LL Key1

Field Name	Start	Stop	Length	Field Description
LL2KEY1	368	377	10	Ending DLG LL Key1
LL1KEY2	378	387	10	Starting DLG LL Key2
LL2KEY2	388	497	10	Ending DLG LL Key2
LL1KEY3	398	407	10	Starting DLG LL Key3
LL2KEY3	408	417	10	Ending DLG LL Key3
LL1KEY4	418	427	10	Starting DLG LL Key4
LL2KEY4	428	437	10	Ending DLG LL Key4
LL1KEY5	438	447	10	Starting DLG LL Key5
LL2KEY5	448	457	10	Ending DLG LL Key5
LL1KEY6	458	467	10	Starting DLG LL Key6
LL2KEY6	468	477	10	Ending DLG LL Key6
LL1KEY7	478	487	10	Starting DLG LL Key7
LL2KEY7	488	597	10	Ending DLG LL Key7
LL1KEY8	498	507	10	Starting DLG LL Key8
LL2KEY8	508	517	10	Ending DLG LL Key8
LL1KEY9	518	527	10	Starting DLG LL Key9
LL2KEY9	528	537	10	Ending DLG LL Key9
LL1KEY10	538	547	10	Start DLG LL Key 10
LL2KEY10	548	557	10	Ending DLG LL Key10
LN1AT2	558	561	4	DLG Line Attr. 1
LN2AT2	562	565	4	DLG Line Attr. 2
AREA1	566	569	4	DLG Area ID 1
AREA2	570	573	4	DLG Area ID 2
AR1AT2	574	577	4	DLG Area Attribute
AR1AT4	578	581	4	DLG Area Attribute
AR2AT2	582	585	4	DLG Area Attribute
AR2AT4	586	589	4	DLG Area Attribute
UPDATE1	590	595	6	Update Date #1 (mmddyy)
UPDTCD1	596	603	8	Update Type Code #1

RF3 Structure File: 12345678.RF3 and 12345678.DBF in BADLRF3.ZIP

Field Name	Start	Stop	Length	Field Description
UPDTSRC1	604	611	8	Update Source #1
UPDATE2	612	617	6	Update Date #2 (mmddyy)
UPDTCD2	618	625	8	Update Type Code#2
UPDTSRC2	626	633	8	Update Source #2
UPDATE3	634	639	6	Update Date #3 (mmddyy)
UPDTCD3	640	647	8	Update Type Code #3
UPDTSRC3	648	655	8	Update Source #3
DIVCU	656	663	8	Divergent CU
DIVSEG	664	667	4	Divergent SEG
DIVMILE	668	672	5.2	Divergent MI
DLGID	673	678	6	DLG Number Special Use For Internal State Codes
FILLER	678	685	7	Filler: Future Use

RF3 Structure File: 12345678.RF3 and 12345678.DBF in BADLRF3.ZIP

Note: The structure for the .DBF file varies slightly from the RF3 structure displayed here in that the fields UPDATE1, UPDATE2, and UPDATE3 have a width of 8 and the last two fields, DLGID and FILLER, have been replaced with a field named ID of length 17. This ID field combines the CATUNIT, SEGM, and MI fields.

The following table provides the ASCII database field structures for the EPA River Reach File Ver. 3.0 (1:100,000 scale hydrography) traces. The actual numeric file names will vary depending on the catalog unit(s). This file contains the actual hydrographic network and is suitable for conversion into a variety of Geographic Information System formats.

RF3 Trace File: 12345678.TRC in BADLRF3.ZIP				
Field Name	**Start**	**Stop**	**Length**	**Field Description**
(Header Record)				
CATUNIT	1	8	8	Cataloging Unit
SEGM	9	12	4	Segment Number
MI	13	17	5.2	Mile Point
NPTS	18	21	4	Number of Lat/Lon Coordinates
(Coordinate Record)				
LATITUDE	1	8	8.4	Latitude in Decimal
LONGITUDE	9	16	8.4	Longitude in Decimal
FILLER	17	21	5	

The following table provides the ASCII database field structures for the EPA River Reach File Ver. 3.0 (1:100.000 scale hydrography) catalog unit boundary file. The actual numeric file names will vary depending on the catalog unit(s). This file contains the actual catalog unit boundary and is suitable for conversion into a variety of Geographic Information System formats.

Catalog Unit Boundary File: 12345678.CUB in BADLRF3.ZIP
First Line = Catalog Unit Number (8 Characters)
Subsequent Lines:
L=DDMMSS.L=DDDMMSS.L=DDMMSS.L=DDDMMSS.L=DDMMSS.L=DDDMMSS. ...
Example:
02070010
L=391259.L=0770809.L=391220.L=0770749.L=391147.L=0770715.L=391120.L=0770633.
L=391058.L=0770535.L=391042.L=0770520.L=391016.L=0770427.L=390948.L=0770416.
L=390526.L=0765331.L=390500.L=0765149.L=390456.L=0765139.L=390357.L=0765123.
...
L=390744.L=0771007.L=390826.L=0771022.L=390910.L=0771022.L=390950.L=0771003.
L=391107.L=0770922.
There can be as many as four latitude/longitude pairs per line.

The following table provides the DBASE III+ database field structure of the Water Resources Division's "encyclopedia" file that documents the minimum and maximum parameter values found and the park(s) where they occurred. This file is intended for Water Resources Division internal use. but will be available to anyone upon request after Baseline Water Quality Data Inventory and Analysis reports have been completed for all parks.

Encyclopedia File: WRD File For Internal Use Only				
Field Name	Start	Stop	Length	Field Description
PARM	1	5	5	STORET Parameter Code
PARMNAME	6	45	40	Parameter Name
MINVAL	46	61	16.7	Minimum Value
MINVALPARK	62	65	4	Park Unit with Minimum Value
MAXVAL	66	71	16.7	Maximum Value
MAXVALPARK	72	75	4	Park Unit with Maximum Value

Appendix C

STORET Water Quality Control/Edit Checking

The following table provides the high and low values used by STORET since November 1983 for 190 common water quality parameters to screen or error check data. Data entered into STORET prior to November 1983, however, were not subjected to this edit/bounds check. Additionally, data from the USGS WATSTORE system that is loaded into STORET is never subjected to these edit criteria and agencies entering data in STORET can override these edit criteria to enter data values that fall outside a range. As a consequence, all data downloaded from STORET for the purposes of this project were filtered through these edit criteria to document values outside the generally accepted ranges. Decisions were then made on a case-by-case basis to retain or discard obviously incorrect data. Refer to the Water Quality Observations Outside STORET Edit Criteria section of the Interpretive Guide To Water Quality Results chapter for more information on this subject.

STORET Code	STORET Parameter Description	High Value	Low Value
00010	TEMPERATURE. WATER (DEGREES CENTIGRADE)	37.0	-2.0
00011	TEMPERATURE. WATER (DEGREES FAHRENHEIT)	98.0	31.0
00020	TEMPERATURE. AIR (DEGREES CENTIGRADE)	52.0	-40.0
00021	TEMPERATURE. AIR (DEGREES FAHRENHEIT)	125.0	-40.0
00026	TOXICS-IDENTIFY DATA COLLECTION BY EPA DIRECTIVE	1990.9	1977.0
00032	CLOUD COVER (PERCENT)	101.0	0.0
00035	WIND VELOCITY (MILES PER HOUR)	85.0	0.0
00036	WIND DIRECTION IN DEGREES FROM TRUE N (CLOCKWISE)	361.0	0.0
00045	PRECIPITATION. TOTAL (INCHES PER DAY)	15.0	0.0
00070	TURBIDITY. (JACKSON CANDLE UNITS)	1500.0	0.0
00074	TURBIDITY. TRANSMISSOMETER. PERCENT TRANSMISSION	101.0	0.0
00075	TURBIDITY. HELLIGE (PPM AS SILICON DIOXIDE)	500.0	0.0
00076	TURBIDITY.HACH TURBIDIMETER (FORMAZIN TURB UNIT)	1000.0	0.0
00077	TRANSPARENCY. SECCHI DISC (INCHES)	600.0	0.0
00080	COLOR (PLATINUM-COBALT UNITS)	500.0	0.0
00081	COLOR.APPARENT(UNFILTERED SAMPLE) PLAT-COB UNITS	500.0	0.0
00085	ODOR (THRESHOLD NUMBER AT ROOM TEMPERATURE)	250.0	0.0
00094	SPECIFIC CONDUCTANCE.FIELD (UMHOS/CM @ 25C)	60000.0	1.0
00095	SPECIFIC CONDUCTANCE (UMHOS/CM @ 25C)	60000.0	1.0
00299	OXYGEN. DISSOLVED. ANALYSIS BY PROBE (MG/L)	30.0	0.0

STORET Code	STORET Parameter Description	High Value	Low Value
00300	OXYGEN, DISSOLVED (MG/L)	30.0	0.0
00301	OXYGEN, DISSOLVED, PERCENT OF SATURATION%	200.0	0.0
00310	BOD, 5 DAY, 20 DEG C (MG/L)	150.0	0.0
00335	COD, .025N K2CR2O7 (MG/L)	1000.0	0.0
00340	COD, .25N K2CR2O7 (MG/L)	1000.0	0.0
00365	CHLORINE DEMAND, 15 MINUTE (MG/L)	15.0	0.0
00400	PH (STANDARD UNITS)	12.0	0.9
00403	PH, LAB, STANDARD UNITS, (STANDARD UNITS)	12.0	0.9
00405	CARBON DIOXIDE (MG/L AS CO2)	100.0	0.0
00406	PH, FIELD (STANDARD UNITS)	12.0	0.9
00410	ALKALINITY, TOTAL (MG/L AS CACO3)	1000.0	0.0
00415	ALKALINITY, PHENOLPHTHALEIN (MG/L)	750.0	0.0
00435	ACIDITY, TOTAL (MG/L AS CACO3)	1000.0	0.0
00436	ACIDITY, MINERAL (METHYL ORANGE) (MG/L AS CACO3)	1000.0	0.0
00437	ACIDITY, CO2 (PHENOLPHTHALEIN) (MG/L AS CACO3)	750.0	0.0
00440	BICARBONATE ION (MG/L AS HCO3)	450.0	0.0
00445	CARBONATE ION (MG/L AS CO3)	100.0	0.0
00480	SALINITY - PARTS PER THOUSAND	40.0	0.0
00500	RESIDUE, TOTAL (MG/L)	15000.0	0.0
00505	RESIDUE, TOTAL VOLATILE (MG/L)	10000.0	0.0
00510	RESIDUE, TOTAL FIXED (MG/L)	10000.0	0.0
00515	RESIDUE, TOTAL FILTRABLE (DRIED AT 105C), (MG/L)	20000.0	0.0
00520	RESIDUE, VOLATILE FILTRABLE (MG/L)	10000.0	0.0
00525	RESIDUE, FIXED FILTRABLE (MG/L)	10000.0	0.0
00530	RESIDUE, TOTAL NONFILTRABLE (MG/L)	10000.0	0.0
00535	RESIDUE, VOLATILE NONFILTRABLE (MG/L)	10000.0	0.0
00540	RESIDUE, FIXED NONFILTRABLE (MG/L)	10000.0	0.0
00545	RESIDUE, SETTLEABLE (ML/L)	1000.0	0.0
00546	RESIDUE, SETTLEABLE (MG/L)	1000.0	0.0

STORET Code	STORET Parameter Description	High Value	Low Value
00550	OIL & GREASE (SOXHLET EXTRACTION) TOTAL.REC., (MG/L)	250.0	0.0
00600	NITROGEN. TOTAL (MG/L AS N)	100.0	0.0
00605	NITROGEN. ORGANIC. TOTAL (MG/L AS N)	15.0	0.0
00608	NITROGEN. AMMONIA. DISSOLVED (MG/L AS N)	25.0	0.0
00610	NITROGEN. AMMONIA. TOTAL (MG/L AS N)	20.0	0.0
00615	NITRITE NITROGEN. TOTAL (MG/L AS N)	5.0	0.0
00620	NITRATE NITROGEN. TOTAL (MG/L AS N)	50.0	0.0
00625	NITROGEN. KJELDAHL. TOTAL. (MG/L AS N)	50.0	0.0
00630	NITRITE PLUS NITRATE. TOTAL 1 DET. (MG/L AS N)	55.0	0.0
00635	NITROGEN. AMMONIA & ORG.. TOTAL 1 DET (MG/L AS N)	70.0	0.0
00650	PHOSPHATE. TOTAL (MG/L AS PO4)	30.0	0.0
00653	PHOSPHATE. TOTAL SOLUBLE (MG/L)	30.0	0.0
00655	PHOSPHATE. POLY (MG/L AS PO4)	30.0	0.0
00660	PHOSPHATE. ORTHO (MG/L AS PO4)	30.0	0.0
00665	PHOSPHORUS. TOTAL (MG/L AS P)	10.0	0.0
00666	PHOSPHORUS. DISSOLVED (MG/L AS P)	10.0	0.0
00680	CARBON. TOTAL ORGANIC (MG/L AS C)	100.0	0.0
00681	CARBON. DISSOLVED ORGANIC (MG/L AS C)	100.0	0.0
00685	CARBON. TOTAL INORGANIC (MG/L AS C)	100.0	0.0
00690	CARBON. TOTAL (MG/L AS C)	150.0	0.0
00720	CYANIDE. TOTAL (MG/L AS CN)	10.0	0.0
00745	SULFIDE. TOTAL (MG/L AS S)	1500.0	0.0
00746	SULFIDE. DISSOLVED (MG/L AS S)	1500.0	0.0
00760	SULFITE WASTE LIQUOR. PEARL BENSON INDEX (MG/L)	150.0	0.0
00900	HARDNESS. TOTAL (MG/L AS CACO3)	5000.0	0.0
00910	CALCIUM (MG/L AS CACO3)	3000.0	0.0
00915	CALCIUM. DISSOLVED (MG/L AS CA)	1000.0	0.0
00916	CALCIUM. TOTAL (MG/L AS CA)	1000.0	0.0
00920	MAGNESIUM (MG/L AS CACO3)	3000.0	0.0

STORET Code	STORET Parameter Description	High Value	Low Value
00925	MAGNESIUM. DISSOLVED (MG/L AS MG)	1000.0	0.0
00927	MAGNESIUM. TOTAL (MG/L AS MG)	1000.0	0.0
00929	SODIUM, TOTAL (MG/L AS NA)	5000.0	0.0
00930	SODIUM. DISSOLVED (MG/L AS NA)	5000.0	0.0
00931	SODIUM ADSORPTION RATIO	50.0	0.0
00935	POTASSIUM. DISSOLVED (MG/L AS K)	175.0	0.0
00937	POTASSIUM. TOTAL MG/L AS K)	175.0	0.0
00940	CHLORIDE. TOTAL IN WATER. (MG/L)	22000.0	0.0
00945	SULFATE, TOTAL (MG/L AS SO4)	2500.0	0.0
00946	SULFATE. DISSOLVED (MG/L AS SO4)	2500.0	0.0
00950	FLUORIDE, DISSOLVED (MG/L AS F)	15.0	0.0
00951	FLUORIDE. TOTAL (MG/L AS F)	15.0	0.0
00955	SILICA. DISSOLVED (MG/L AS SI02)	2000.0	0.0
00956	SILICA. TOTAL (MG/L AS SI02)	2000.0	0.0
01000	ARSENIC. DISSOLVED (UG/L AS AS)	5000.0	0.0
01002	ARSENIC. TOTAL (UG/L AS AS)	5000.0	0.0
01005	BARIUM. DISSOLVED (UG/L AS BA)	2000.0	0.0
01007	BARIUM, TOTAL (UG/L AS BA)	2000.0	0.0
01010	BERYLLIUM. DISSOLVED (UG/L AS BE)	2000.0	0.0
01012	BERYLLIUM. TOTAL (UG/L AS BE)	2000.0	0.0
01020	BORON. DISSOLVED (UG/L AS B)	5000.0	0.0
01022	BORON. TOTAL (UG/L AS B)	5000.0	0.0
01025	CADMIUM. DISSOLVED (UG/L AS CD)	500.0	0.0
01027	CADMIUM, TOTAL (UG/L AS CD)	500.0	0.0
01030	CHROMIUM. DISSOLVED (UG/L AS CR)	2000.0	0.0
01032	CHROMIUM. HEXAVALENT (UG/L AS CR)	2000.0	0.0
01033	CHROMIUM. TRI-VAL (UG/L AS CR)	2000.0	0.0
01034	CHROMIUM. TOTAL (UG/L AS CR)	2000.0	0.0
01040	COPPER. DISSOLVED (UG/L AS CU)	2000.0	0.0

STORET Code	STORET Parameter Description	High Value	Low Value
01042	COPPER. TOTAL (UG/L AS CU)	5000.0	0.0
01045	IRON. TOTAL (UG/L AS FE)	56000.0	0.0
01046	IRON. DISSOLVED (UG/L AS FE)	56000.0	0.0
01047	IRON. FERROUS (UG/L AS FE)	56000.0	0.0
01049	LEAD. DISSOLVED (UG/L AS PB)	1000.0	0.0
01051	LEAD. TOTAL (UG/L AS PB)	1000.0	0.0
01055	MANGANESE. TOTAL (UG/L AS MN)	5000.0	0.0
01056	MANGANESE. DISSOLVED (UG/L AS MN)	5000.0	0.0
01065	NICKEL. DISSOLVED (UG/L AS NI)	2000.0	0.0
01067	NICKEL. TOTAL (UG/L AS NI)	2000.0	0.0
01075	SILVER. DISSOLVED (UG/L AS AG)	5000.0	0.0
01077	SILVER. TOTAL (UG/L AS AG)	5000.0	0.0
01090	ZINC. DISSOLVED (UG/L AS ZN)	25000.0	0.0
01092	ZINC. TOTAL (UG/L AS ZN)	25000.0	0.0
01105	ALUMINUM. TOTAL (UG/L AS AL)	20000.0	0.0
01106	ALUMINUM. DISSOLVED (UG/L AS AL)	20000.0	0.0
01145	SELENIUM. DISSOLVED (UG/L AS SE)	100.0	0.0
01501	ALPHA. TOTAL	200.0	0.0
01503	ALPHA. DISSOLVED	75.0	0.0
01505	ALPHA. SUSPENDED	150.0	0.0
03501	BETA. TOTAL	3500.0	0.0
03503	BETA. DISSOLVED	3000.0	0.0
03505	BETA. SUSPENDED	1500.0	0.0
09503	RADIUM 226. DISSOLVED	500.0	0.0
13501	STRONTIUM 90. TOTAL	500.0	0.0
22703	URANIUM. NATURAL. DISSOLVED	500.0	0.0
31501	COLIFORM. TOT.MEMBRANE FILTER.IMMED.M-ENDO MED. 35C	24000000.0	0.0
31502	COLIFORM. TOTAL. 10/ML	24000000.0	0.0
31503	COLIFORM. TOT.MEMBR FILTER. DELAYED.M-ENDO MED. 35C	24000000.0	0.0

STORET Code	STORET Parameter Description	High Value	Low Value
31504	COLIFORM, TOT.MEMBR FILTER,IMMED,LES ENDO AGAR, 35C	24000000.0	0.0
31613	FECAL COLIFORM, MEMBR FILTER, M-FC AGAR,44.5C, 24HR	10000000.0	0.0
31615	FECAL COLIFORM, MPN, EC MED, 44.5C (TUBE 31614)	10000000.0	0.0
31616	FECAL COLIFORM, MEMBR FILTER,M-FC BROTH, 44.5C	10000000.0	0.0
31672	FECAL STREPTOCOCCI,PLATE COUNT M-ENTER AGAR,35C48HR	500000.0	0.0
31673	FECAL STREPTOCOCCI, MBR FILT, KF AGAR, 35C, 48HR	500000.0	0.0
31677	FECAL STREPTOCOCCI,MPN,AD-EVA, 35C (TUBE 31678)	500000.0	0.0
31679	FECAL STREPTOCOCCI, MF M-ENTEROCOCCUS AGAR,35C,48H	500000.0	0.0
31749	PLATE COUNT, TOTAL, TPC AGAR, 20C, 48 HRS	99999999.0	0.0
31751	PLATE COUNT, TOTAL, TPC AGAR, 35C, 24 HRS	99999999.0	0.0
32210	CHLOROPHYLL-A UG/L TRICHROMATIC UNCORRECTED	500.0	0.0
32211	CHLOROPHYLL-A UG/L SPECTROPHOTOMETRIC ACID. METH.	750.0	0.0
32212	CHLOROPHYLL-B UG/L TRICHROMATIC UNCORRECTED	1000.0	0.0
32214	CHLOROPHYLL-C UG/L TRICHROMATIC UNCORRECTED	200.0	0.0
32217	CHLOROPHYLL A UG/L FLUOROMETRIC UNCORRECTED	500.0	0.0
32218	PHEOPHYTIN-A UG/L SPECTROPHOTOMETRIC ACID. METH.	200.0	0.0
32219	PHEOPHYTIN RATIO(OD 663)SPECTRO,BEFORE/AFTER ACID	2.0	0.0
32221	CHLOROPHYLL A,% OF(PHEOPHYTIN A+CHL A),SPEC-ACID.	101.0	0.0
32230	CHLOROPHYLL A (MG/L)	0.5	0.0
32231	CHLOROPHYLL B (MG/L)	0.8	0.0
32232	CHLOROPHYLL C (MG/L)	0.2	0.0
32234	CHLOROPHYLL, TOTAL (A+B+C) (MG/L)	1.0	0.0
32270	CHLOROFORM EXTRACTABLES TOTAL IN MG PER LITER	5.0	0.0
32730	PHENOLICS, TOTAL, RECOVERABLE (UG/L)	1500.0	0.0
38260	METHYLENE BLUE ACTIVE SUBST. (DETERGENTS, ETC.)	10.0	0.0
39330	ALDRIN IN WHOLE WATER SAMPLE (UG/L)	20.0	0.0
39340	GAMMA-BHC(LINDANE),WHOLE WATER, (UG/L)	20.0	0.0
39350	CHLORDANE(TECH MIX & METABS),WHOLE WATER, (UG/L)	20.0	0.0
39360	DDD IN WHOLE WATER SAMPLE (UG/L)	20.0	0.0

STORET Code	STORET Parameter Description	High Value	Low Value
39365	DDE IN WHOLE WATER SAMPLE (UG/L)	20.0	0.0
39370	DDT IN WHOLE WATER SAMPLE (UG/L)	20.0	0.0
39380	DIELDRIN IN WHOLE WATER SAMPLE (UG/L)	20.0	0.0
39390	ENDRIN IN WHOLE WATER SAMPLE (UG/L)	20.0	0.0
39400	TOXAPHENE IN WHOLE WATER SAMPLE (UG/L)	20.0	0.0
39410	HEPTACHLOR IN WHOLE WATER SAMPLE (UG/L)	20.0	0.0
39420	HEPTACHLOR EPOXIDE IN WHOLE WATER SAMPLE (UG/L)	20.0	0.0
39480	METHOXYCHLOR IN WHOLE WATER SAMPLE (UG/L)	20.0	0.0
39516	PCBS IN WHOLE WATER SAMPLE (UG/L)	20.0	0.0
39530	MALATHION IN WHOLE WATER SAMPLE (UG/L)	20.0	0.0
39540	PARATHION IN WHOLE WATER SAMPLE (UG/L)	20.0	0.0
39600	METHYL PARATHION IN WHOLE WATER SAMPLE (UG/L)	20.0	0.0
39782	LINDANE IN WHOLE WATER SAMPLE (UG/L)	20.0	0.0
50060	CHLORINE. TOTAL RESIDUAL (MG/L)	5.0	0.0
60050	ALGAE. TOTAL (CELLS/ML)	700000.0	0.0
70300	RESIDUE. TOTAL FILTRABLE (DRIED AT 180C). (MG/L)	4000.0	0.0
70505	PHOSPHATE. TOTAL.COLORIMETRIC METHOD (MG/L AS P)	10.0	0.0
70507	PHOSPHORUS. IN TOTAL ORTHOPHOSPHATE (MG/L AS P)	10.0	0.0
71850	NITRATE NITROGEN. TOTAL (MG/L AS NO3)	65.0	0.0
71886	PHOSPHORUS. TOTAL. AS PO4 - (MG/L)	30.0	0.0
71890	MERCURY. DISSOLVED (UG/L AS HG)	10.0	0.0
71895	MERCURY. SUSPENDED (UG/L AS HG)	10.0	0.0
71900	MERCURY. TOTAL (UG/L AS HG)	10.0	0.0
74010	IRON. TOTAL (MG/L AS FE)	56000.0	0.0

Appendix D

STORET Administrative Parameters

STORET Code	Description of STORET Administrative Parameters
00022	LENGTH OF EXPOSURE OF SAMPLE OR TEST - DAYS
00026	TOXICS-IDENTIFY DATA COLLECTION BY EPA DIRECTIVE
00027	CODE NO FOR AGENCY COLLECTING SAMPLE
00028	CODE NO FOR AGENCY ANALYZING SAMPLE
00029	NUMBER USED IN SAMPLE ACCOUNTING PROCEDURE
00063	SAMPLING POINTS. NUMBER OF IN A CROSS SECTION
00073	SAMPLE LOC CODE DEFINED BY THERMAL STRUCT & DEPTH
00111	RATIO OF FECAL COLIFORM TO FECAL STREPTOCOCCI
00115	SAMPLE TREATMENT CODE (1=RAW.2=TREATED)
00116	INTENSIVE SURVEY IDENTIFICATION NUMBER
00145	TOTAL PRODUCTION OF PRODUCT MANUFACTURED TONS/DAY
01273	TOTAL ACID PRIORITY POLLUTANTS MG/L
01274	TOTAL BASE-NEUTRAL PRIORITY POLLUTANTS MG/L
01275	TOTAL VOLATILE PRIORITY POLLUTANTS MG/L
01365	ANALYSIS DATE (DIOXIN) (YYMMDD)
04177	SAMPLE STABILIZATION. RECOVERY TEST CODE
04178	FIELD PROTOCOL(CONFDNCE ASSIGNED FIELD SAMPLE) CODE
04179	SAMPLE STATION LOCKED CODE
04180	CONDITION OF STATION SITE CODE
04181	LABORATORY QA/QC PLAN CONFIDENCE CODE
04182	SAMPLE TYPE CODE
04183	SAMPLE REMARKS CODE
30333	BAG MESH SIZE. BEDLOAD SAMPLER. MM
34772	NPDES NUMBER. CROSS REFERENCE CODE
34785	GAGE TYPE. METHOD CODE

STORET Code	Description of STORET Administrative Parameters
45575	GC MAKE AND MODEL INFORMATION CODE
45576	GC DETECTOR TYPE CODE
45577	GC COLUMN TYPE CODE
45580	METHOD OF ANALYSIS CODE
45581	LABORATORY LOCATION CODE
46107	SAMPLE LOCATION CODE (TREATMENT PLANT OPERATION)
46390	TOXICITY CHARACTERISTIC LEACHING PROCEDURE P OR F
46396	PROCESS TO SIGNIFICANTLY REDUCE PATHOGENS YES OR NO
46397	PROCESS TO FURTHER REDUCE PATHOGENS YES OR NO
47001	PERMIT EXPIRATION DATE (JULIAN CALENDAR)
47044	OBSERVATIONS.WASTE SITE-SEVERITY OF PROBLEMS CODE
47460	SUBSAMPLE - DECIMAL FRACTION OF WHOLE NUMBER
47477	COMPOSITION AND/OR DISPOSITION OF CATCH NUM CODE
70231	CURRENT DIRECTION (DEGREES FROM DOWNSTREAM FLOW)
71999	SAMPLE PURPOSE CODE
72032	NUMBER OF SPILLWAY GATES OPEN
73672	DATE OF ANALYSIS YYMMDD
73673	DATE OF EXTRACTION YYMMDD
74031	GRANT, PROJECT COST ELIGIBLE FOR CONSTRUCTION
74032	GRANT, AMOUNT OF PL 660 GRANT FOR THIS PROJECT
74033	GRANT, FEDERAL, OTHER THAN PL 660 GRANT
74034	GRANT, FUTURE PL 660 WHICH MAY APPLY TO THIS PROJ
74035	GRANT, TOTAL FEDERAL, WHICH APPLIES TO THIS PROJ
74036	GRANT, PROJ NUMBER ASSIGNED TO THIS APPLICATION
74037	GRANT, TYPE OF PROJECT TO WHICH GRANT APPLIES
74038	GRANT, STATUS OF PROJECT TO WHICH GRANT APPLIES
74039	PCS/STORET WATER QUALITY FILE INTERFACE YR/MO/DAY
74040	SURVEY NUMBER YYMMNO
74041	STORET STORAGE TRANSACTION DATE YR/MO/DAY

STORET Code	Description of STORET Administrative Parameters
74050	RADIOACTIVITY. GENERAL (PERMIT)
74051	ALGICIDES. GENERAL (PERMIT)
74052	CHLORINATED HYDROCARBONS. GENERAL (PERMIT)
74053	PESTICIDES. GENERAL (PERMIT)
74056	COLIFORM. TOTAL. GENERAL (PERMIT)
74065	STREAM FLOW CLASS
74066	ANNUAL RUNOFF
74067	SOIL CLASSIFICATION
74068	WATER QUALITY DESIGNATED USE CLASSIFICATION (IA)
74100	PRIMARY 1972 SIC CODE
74101	SECONDARY 1972 SIC CODE
74102	SECONDARY 1972 SIC CODE
74103	SECONDARY 1972 SIC CODE
74200	SAMPLE PRESERVATION METHODS ONE OR MORE IN COMB.
74205	LAND RESOURCE AREA (IOWA)
74206	SOIL EROSION POTENTIAL (IOWA)
74209	WATER QUALITY INDEX - STATE OF ILLINOIS. EPA
74210	FOREST STREAM WATER QUALITY INDEX CALC. NUMBER
74990	FISH SPECIES NUMERIC CODE - F&W SERVICE
74995	ANATOMY CODE
75000	SPECIES CODE-REMARK=SEX (M=MALE.F=FEMALE.U=UNK.)
81028	WITHDRAWAL OF GROUNDWATER (MILLION GAL/DAY)
82258	WATER CLASSIFICATION CODE (1-9) CODE
82292	DATA RELAY GROUND STATION SOURCE NODE CODE. CODE
82309	CONTAMINATION SOURCE POSSIBLE CODES NUMERIC CODE
82310	DEPTH CONFIDENCE IN REPORTED VALUES NUMERIC CODES
82373	FREQUENCY OF SAMPLING M=MON.Q=QUAR.Y=YR.R=RNFFCODE
82519	DRILLER REGISTRATION NUMBER ALPHA-NUMERIC CODE
82562	NARRATIVE REQUIREMENT EXCEEDANCES INTEGER

STORET Code	Description of STORET Administrative Parameters
82576	DAILY EXCURSION TIME, WATER MIN
82577	MONTHLY EXCURSION TIME, WATER TOTAL MIN
82578	DAY/MAXIMUM EXCURSION TIME, WATER MIN
82579	CODE NUMBER FOR PERSON COLLECTING SAMPLE
84002	CODE, GENERAL INFORMATION - ALPHA, NUMERIC CODE
84003	WATER SHED ID NUMBER (IOWA)
84005	FISH SPECIES CODE-FISH & WILDLIFE SER
84006	OWNERSHIP CLASSIFICATION OF LAKE, ILLINOIS SYSTEM
84010	PUBLIC ACCESS TO LAKE ILLINOIS SYSTEM
84011	CONFIDENCE CODE FOR GLC CONFIRMATION CODE
84012	PATIENT PARAMETERS (AGE, SEX, WT, ETC.) CODE
84013	SAMPLE PARAMETERS D=DESIGN SPECIMEN, S=SURPLUS
84027	CODE NUMBER FOR AGENCY COLLECTING SAMPLE
84028	CODE NO FOR AGENCY ANALYZING SAMPLE
84029	NUMBER USED IN SAMPLE ACCOUNTING PROCEDURE FIELD
84033	EGD ANALYTICAL DATA COMPLETENESS Y=YES N=NO CODE
84034	EGD SMPL NO.(SMPL.IDENT) NUMERIC=SCS ALPH+4NUM=JRB
84035	EGD SAMPLE CLASSIFICATION CATEGORY ALPHA CODE
84036	EGD INDUSTRIAL CATEGORY NUMERIC CODE
84037	EGD INDUSTRIAL CATEGORY NAME ALPHA CODE
84038	EGD LABORATORY NUMERIC CODE
84039	EGD LABORATORY NAME ALPHA CODE
84040	EGD SAMPLE STATUS (1-5,9,AND BLANK) NUMERIC CODE
84041	EGD ACID STATUS (1-5,9,AND BLANK) NUMERIC CODE
84042	EGD BASE STATUS (1-5,9AND BLANK) NUMERIC CODE
84043	EGD PESTICIDE STATUS (1-5,9,AND BLANK) NUMERIC CODE
84044	EGD VOA FRACT. STATUS INDICATOR (1-5,9,BLANK) CODE
84045	EGD ACID EXTRACT DATE (YYMMDD) NUMERIC CODE
84046	EGD BASE EXTRACTION DATE (YYMMDD) NUMERIC CODE

STORET Code	Description of STORET Administrative Parameters
84047	EGD PESTICIDE EXTRACTION DATE (YYMMDD) NUMERIC CODE
84048	EGD VOA FRACTION INJECTION DATE YYMMDD NUMERIC CODE
84049	EGD ACID CONC. FACTOR (FIVE NUMERIC DIGITS) CODE
84050	EGD BASE CONC.FACTOR (FIVE NUMERIC DIGITS) CODE
84051	EGD PESTICIDE CONC.FACTOR (FIVE NUMERIC DIGITS) CODE
84052	EGD VOA FRACTION CONC. FACTOR (5 NUMERIC DIGITS) CODE
84053	SAMPLE TYPE AND FREQUENCY OF COLLECTION CODE
84054	LITHOLOGY ALPHA-NUMERIC CODE
84055	AVAILABLE LOGS ALPHA-NUMERIC CODE
84056	WATER USE CATEGORY ALPHA-NUMERIC CODE
84057	INSPECTION TYPE ALPHA-NUMERIC CODE
84058	HYDROGEOLOGIC SYSTEM ALPHA-NUMERIC CODE
84059	WELL OWNERSHIP ALPHA-NUMERIC CODE
84060	TOPOGRAPHY ALPHA-NUMERIC CODE
84061	WELL USE ALPHA-NUMERIC CODE
84062	MEASURING POINT DESCRIPTION ALPHA-NUMERIC CODE
84063	DRILLING METHOD ALPHA-NUMERIC CODE
84064	WELL DATA AVAILABILITY ALPHA-NUMERIC CODE
84065	PERMIT COMPLIANCE DATA ALPHA-NUMERIC CODE
84067	NATURE OF MONITORING ALPHA-NUMERIC CODE
84073	REPLACES EXISTING WELL ALPHA-NUMERIC CODE
84074	AQUIFER TYPE (SEE USGS HANDBOOK) ALPHA CODE
84075	WELL PERMIT NUMBER ALPHA-NUMERIC CODE
84076	TSD MONITORING WELL TYPE ALPHA CODE
84077	TSD MONITORING WELL SAMPLING METHOD ALPHA CODE
84083	POLLUTION VERIFICATION ALPHA CODE
84084	WELL SAMPLE PURPOSE ALPHA CODE
84090	SAMPLE FILE CONTROL PROJECT IDENTIFICATION A-CODE
84091	INFILTRATION DATE/BEGINNING 'YYMMDD'

STORET Code	Description of STORET Administrative Parameters
84092	INFILTRATION DATE/ENDING 'YYMMDD'
84093	ENFORCEMENT FORM #2-C.DATA IDENTIFICATION CODE
84102	SAMPLE SPECIES-SUB ID ALPHA CODE
84103	DIOXIN LABORATORY ALPHA CODE
84104	DIOXIN STUDY ALPHA CODE
84112	SOURCE OF GEOHYDROLOGIC DATA CODE
84119	SOURCE OF EVACUATION DATA CODE
84121	REGULATING AGENCY CODE
84122	SAMPLE PURPOSE CODE
84126	SOURCE OF DEPTH DATA CODE
84127	METHOD OF DEPTH MEASUREMENT CODE
84128	SOURCE OF WATER-LEVEL DATA CODE
84129	DATA QUALITY
84141	LAKE. PHYSICAL CONDITION AT SAMPLE TIME, 1-5, CODE
84142	LAKE.RECREATIONAL SUITABILITY @ SMPL TIME,1-5, CODE
84164	SAMPLER TYPE, CODE
85300	PROBLEM CODE NES SURVEY
85327	WATER LEVEL AT SAMPLE COLLECTION TIME-CODE-NES
85332	CLOUD COVER AT SAMPLE COLLECTION TIME-CODE-NES
85553	WELL COMPLETION DATE (MONTH/YEAR)
85554	WELL WORKOVER DATE, LATEST (MONTH/YEAR)

Appendix E

STORET Parameters Not Suitable for Statistical Analysis

STORET Code	Description of STORET Parameters Not Suitable for Statistical Analysis
00001	X-SEC. LOC.. HORIZ (FT. FROM R BANK LOOK UPSTR.)
00002	X-SEC. LOC.. HORIZ (% FROM R BANK LOOK UPSTR.)
00003	SAMPLING STATION LOCATION. VERTICAL (FEET)
00005	X-SEC. LOC.. VERTICAL (PERCENT OF TOTAL DEPTH)
00006	DISTANCE FROM LOCATION IN X MILES
00007	DISTANCE FROM LOCATION IN Y MILES
00008	NUMBER USED IN SAMPLE ACCOUNTING PROCEDURE
00009	X-SEC. LOC.(FT FROM LEFT BANK LOOKING DOWNSTRM)
00027	CODE NO FOR AGENCY COLLECTING SAMPLE
00028	CODE NO FOR AGENCY ANALYZING SAMPLE
00033	WEATHER CODE FOR OCEAN-OBSERV. (WMO CODE 4677)
00037	WIND FORCE (BEAUFORT UNITS)
00038	WIND DIRECTION (WMO CODES 0885 + 0887)
00041	WEATHER (WMO CODE 4501)
00042	ALTITUDE IN FEET ABOVE MEAN SEA LEVEL
00043	CLOUD TYPE (WMO CODE 0500)
00044	CLOUD AMOUNT (WMO CODE 2700)
00047	TOTAL PARTIAL PRESSURE DISSOLVED GASES (MM HG)
00048	TOTAL PARTIAL PRESSURE DISSOLVED GASES (% SAT)
00049	SURFACE AREA IN SQUARE MILES
00050	EVAPORATION. TOTAL (INCHES PER DAY)
00051	SURFACE AREA IN SQUARE FEET
00053	SURFACE AREA. ACRES
00054	RESERVOIR STORAGE - ACRE FEET
00063	SAMPLING POINTS. NUMBER OF IN A CROSS SECTION
00067	TIDE STAGE

STORET Code	Description of STORET Parameters Not Suitable for Statistical Analysis
00069	SEA WAVES(0=NONE;1=0-3";2=4-20";3=21-48";4=4-8')
00097	SAMPLING STATION LOCATION, VERTICAL (FEET)
00098	SAMPLING STATION LOCATION, VERTICAL (METERS)
00111	RATIO OF FECAL COLIFORM TO FECAL STREPTOCOCCI
00115	SAMPLE TREATMENT CODE (1=RAW,2=TREATED)
01300	OIL-GREASE (SEVERITY)
01305	DETERGENT SUDS (SEVERITY)
01310	GAS BUBBLES (SEVERITY)
01315	SLUDGE, FLOATING (SEVERITY)
01320	GARBAGE, FLOATING (SEVERITY)
01325	ALGAE, FLOATING MATS (SEVERITY)
01330	ODOR, ATMOSPHERIC (SEVERITY)
01331	TASTE (SEVERITY)
01335	SEWAGE SOLIDS, FRESH, FLOATING (SEVERITY)
01340	FISH, DEAD (SEVERITY)
01345	DEBRIS, FLOATING (SEVERITY)
01350	TURBIDITY (SEVERITY)
01351	FLOW, STRM,1DRY,2LOW,3NORM,4FLOOD,5ABOVE NORM,CODE
01355	ICE COVER, FLOATING OR SOLID (SEVERITY)
03595	BIOASSAY (96 HR), EFFLUENT, TOTAL CODE
03596	BIOASSAY (48 HR), EFFLUENT, TOTAL CODE
03597	BIOASSAY (24 HR), EFFLUENT, TOTAL CODE
03598	TOXICITY, EFFLUENT, TOTAL CODE
03599	TOXICITY, CHOICE OF SPECIES, EFFLUENT CODE
03600	TOXICITY, TROUT, EFFLUENT, TOTAL CODE
03601	TOXICITY, SAND DOLLAR, EFFLUENT CODE
03602	BIOCHEMICAL OXYGEN DEMAND, EFFLUENT, TOTAL CODE
03603	SOLIDS, TOTAL SUSPENDABLE, EFFLUENT, TOTAL CODE
03605	FLOW METER CALIBRATION, WATER CODE

STORET Code	Description of STORET Parameters Not Suitable for Statistical Analysis
03717	ONCORHYNCHUS MYKISS, WATER CODE
04117	TETHER LINE USED FOR COLLECTING SAMPLE CODE
04160	HALOCARBONS, PURGEABLE, SCAN, EFFLUENT CODE
04161	HALOCARBONS, PURGEABLE, SCAN, SLUDGE CODE
04162	AROMATIC, PURGEABLE, SCAN, EFFLUENT CODE
04163	AROMATIC, PURGEABLE, SCAN, SLUDGE CODE
04164	PHENOLIC, TOTAL, SCAN, EFFLUENT CODE
04165	PHENOLIC, TOTAL, SCAN, SLUDGE CODE
04166	PCB, TOTAL, SCAN, EFFLUENT CODE
04167	PCB, TOTAL, SCAN, SLUDGE CODE
04174	FREE LIQUIDS IN SEWAGE SLUDGE CODE
34765	AVIAN NUMERICAL SPECIES CODE (BIRDS)
34766	MAMMALIAN NUMERICAL SPECIES CODE
34771	MACROPHYTE, INSTREAM, VISUAL SIGHTING CODE
34773	ODOR, AMBIENT WATER CODE
34774	FISH, INSTREAM, VISUAL SIGHTING CODE
34775	STREAMBANK CHANNEL ALTERATIONS CODE
34776	HYDRAULIC STRUCTURES, INSTREAM CODE
34780	LAND USE, ADJACENT STREAM CODE
34781	SAMPLE POINTS, # OF LONGTDNL TRANSECTS, REACH CODE
34782	STREAM STAGE TREND CODE
34789	HABITATS, TYPES SAMPLED CODE
45613	FLOATING SOLIDS/VISIBLE FOAM, VISUAL, YES=1, NO=0, CODE
45614	SANITARY WASTE DISCHARGE ASSESSMENT, YES=1, NO=0, CODE
45615	INTERMITTENT DISCHARGE ASSESSMENT, YES=1, NO=0, CODE
46001	WATER APPEARANCE CODE (BASED ON FIELD ASSESSMENT)
46478	EQUIPMENT INSPECTION, VISUAL CODE
46486	TOXICITY,ACUTE 24HR(STATIC)CERIODAPHNIA (P/F) CODE
47454	FLOW METER REVOLUTIONS NUMBER

STORET Code	Description of STORET Parameters Not Suitable for Statistical Analysis
47455	LATITUDE. STARTING. OF A SAMPLE TOW DDMMSS
47456	LONGITUDE. STARTING. OF A SAMPLE TOW DDDMMSS
47457	LATITUDE. FINISHING. OF A SAMPLE TOW DDMMSS
47458	LONGITUDE. FINISHING. OF A SAMPLE TOW DDDMMSS
47459	LENGTH FREQUENCY NUMBER
47461	TIME THAT THE EQUIPMENT WAS SAMPLING MINUTES
47476	DIRECTION OF TOW IN RELATION TO CURRENT NUM CODE
50044	HYDROGRAPH LIMB. 1BASE. 2RISING. 3PEAK. 4FALLING. CODE
61390	DIATOMS.FIRST DOMINANT SPECIES OF UNITS - CODE
61391	DIATOMS.SECOND DOMINANT SPECIES OF UNITS - CODE
61392	DIATOMS.THIRD DOMINANT SPECIES OF UNITS - CODE
61393	DIATOMS.FOURTH DOMINANT SPECIES OF UNITS - CODE
70220	WAVE DIRECTION (WMO CODES 0885 + 0887)
70222	WAVE HEIGHT (WMO CODE 1555)
70223	WAVE PERIOD (WMO CODE 3155)
71090	BIVALVE SPECIES CODE
71500	EQUITABILITY INDEX.BENTHIC MACROINVER CODE
72000	ELEVATION OF LAND SURFACE DATUM (FT. ABOVE MSL)
72001	DEPTH. TOTAL OF HOLE (FT BELOW LAND SURFACE DATUM)
72002	DEPTH TO TOP OF WATER-BEARING ZONE SAMPLED (FT)
72003	DEPTH TO BOTTOM OF WATER-BEARING ZONE SAMPLED (FT)
72004	PUMP OR FLOW PERIOD PRIOR TO SAMPLING MINUTES
72005	SAMPLE SOURCE CODE (BM WELL DATA)
72006	SAMPLING CONDITION CODE (BM WELL DATA)
72007	FORMATION NAME CODE (BM WELL DATA)
72017	SERIES CODE (BM WELL DATA)
72018	SYSTEM CODE (BM WELL DATA)
72111	DIRECT READOUT GROUND STATN TRANSMIT EROR CODE NUM
74054	FECAL STREPTOCOCCI. GENERAL (PERMIT)

STORET Code	Description of STORET Parameters Not Suitable for Statistical Analysis
74055	FECAL COLIFORM, GENERAL (PERMIT)
80889	ACTIVATED SLUDGE PROCESS MODIFICATION CODE
81024	DRAINAGE AREA IN SQUARE MILES (SQ. MI.)
81637	SHELLFISH SPECIES NUMERIC CODE
82289	LAGOON OBSERVATION, VISUAL, Y=YES N=NO CODE
82398	SAMPLING METHOD (CODES)
82524	STORAGE COEFFICIENT NUMERICAL CODE
82923	ATMOSPHERIC DEPOSITION TYPE, WET CODE
83205	ATMOSPHERIC DEPOSITION TYPE, BULK CODE
84000	GEOLOGIC AGE CODE (SEE USGS CATALOG)
84001	AQUIFER NAME CODE (SEE USGS CATALOG)
84004	LAKE TYPE ILLINOIS CLASSIFICATION SYSTEM
84007	ANATOMY ALPHA CODE
84008	LIFE STYLE/HABITAT OF THE INDIVIDUALS IN THE SAMPLE
84009	SHELLFISH SPECIES ALPHANUMERIC CODE
84014	SPECIES SEX CODE
84030	CLOUD AMOUNT ALPHA WEATHER CODES
84031	PHYSICAL WEATHER ALPHA WEATHER CODES
84032	STREAM CONDITION ALPHA WEATHER CODES
84066	OIL AND GREASE, VISUAL, ALPHA-NUMERIC CODE
84068	SERIES CODE ALPHA-NUMERIC CODE
84069	FORMATION CODE ALPHA-NUMERIC CODE
84070	METHOD OF TESTING WELL YIELD ALPHA-NUMERIC CODE
84071	WATER LEVEL MEASUREMENT CONDITIONS ALPHA-NUM CODE
84072	WATER LEVEL MEASUREMENT METHOD ALPHA-NUMERIC CODE
84078	GIARDIA LAMBLIA, 2HSO4 OR SUC GRAD, MICRO, CODE
84079	BACTERIA, CELLUOLYTIC, AEROBIC-ANAEROBIC, RT 5-7, CODE
84080	BACTERIA, HYDROCARBONOCLASTIC, SHAKE INC 32C/WK, CODE
84081	YERSINIA ENTEROCOLITICA, SB BROTH, MAC AGAR,22C, CODE

STORET Code	Description of STORET Parameters Not Suitable for Statistical Analysis
84082	SALMONELLA/SHIGELLA. QUANT OR QUAL. HVF OR SWAB. CODE
84085	ORGANICS. VOLATILE. DETECTED. NUMERIC CODE. CODE
84086	MACROINVERTEBRATE SPECIES NUMERIC CODE
84087	MACROINVERTEBRATE HABITAT CODE
84088	BIOLOGY 1 MACROINVERTEBRATE CODE
84089	BIOLOGY 2 MACROINVERTEBRATE CODE
84094	PHYTOPLANKTON SPECIES CODE. NUMERIC
84095	PHYTOPLANKTON SPECIES CODE. ALPHA
84096	SEVERITY OF NON-PLANKTON ALGAE-MAT COVERAGE CODE
84097	LAGOON MOUTH CONDITION CODE
84098	COLOR OF NON-PLANKTONIC ALGAE CODE
84099	WATER - RELATIVE WATER LEVEL CODE
84100	SEX(1-MALE,2-FEMALE,3-MIXED,4-UNKNOWN) NUM CODE
84101	METAFORM. BENTHIC. ADULT(A). PUPAE(P). LARVAE(L) CODE
84105	OIL-SEPARATOR OBSERVATION ASSESS (0=DID NOT,1=DID)
84106	EVAPORAT/BED OBS ASSESS (0=DID NOT LOOK, 1=DID LOOK)
84107	AREA INSPECTION. VISUAL (0=DID NOT, 1=DID) CODE
84108	DRAIN FIELD INSPECTION ASSESS (0=DID NOT, 1=DID) CODE
84109	SLUDGE BUILD-UP IN WATER (0=DID NOT OBS, 1=OBS) CODE
84110	POND OBSERVATION ASSESS WATER (0=DID NOT, 1=DID) CODE
84111	LITHOLOGIC MODIFIER CODE
84113	WELL INTAKE FINISH CODE
84114	WELL CASING MATERIAL CODE
84115	TYPE OF MATERIAL FROM WHICH OPENING IS MADE CODE
84116	DRILLING FLUID CODE
84117	TYPE OF SURFACE SEAL CODE
84118	METHOD OF DEVELOPMENT CODE
84120	PACKING MATERIAL CODE
84124	METHOD OF EVACUTAION CODE

STORET Code	Description of STORET Parameters Not Suitable for Statistical Analysis
84125	METHOD OF WATER-LEVEL MEASUREMENT CODE
84130	OUTFALL OBSERVATION. VISUAL. Y=YES N=NO CODE
84131	SAMPLING METHOD. CONFIDENCE CODE (A.B.C.D) CODE
84132	STREAMBANK. VEGETATIVE STABILITY RATING CODE
84133	STREAMBANK. STABILITY (BANK EROSION) RATING CODE
84134	PARTICLES. DEGREE SURROUNDED BY FINE SEDIMENT. CODE
84135	STREAMSIDE. (SHORELINE) COVER RATING CODE
84136	CANOPY TYPE CODE
84137	CHANNEL STABILITY RATING CODE (E.G.F.P) CODE
84138	COLIFORM. TOTAL. WATER. WHOLE. MPN. PRES=1. ABSNT=2. CODE
84139	ENTEROBACTER AGGLOMERANS. WTR. MF. PRES=1. ABSNT=2. CODE
84140	KLEBSIELLA PNEUMONIAE. WTR. WH. MF. PRES=1. ABSNT=2. CODE
84143	WELL. PURGING CONDITION CODE
84144	WELL. SELECTION CRITERIA CODE
84145	PROJECT COMPONENT CODE
84146	LAND USE. PREDOMINANT. WITHIN 100 FT OF WELL. CODE
84147	LAND USE. PREDOMINANT. 1/4 MI.RADIUS OF WELL. CODE
84148	LAND USE. PREDMNT.. FRAC.. WITHIN 1/4 MI OF WELL. CODE
84149	LAND USE. CHANGE. LAST 10 YRS. WITHIN 1/4MI WELL. CODE
84150	HABITAT QUALITY INDEX RATING CODE
84151	AQUATIC LIFE. USE CLASSES CODE
84152	STREAM. STAGE CLASS CODE
84153	STREAMBANKS. GRAZING DAMAGE CODE
84154	CHANNEL. MAJOR ALTERATIONS CODE
84155	RIFFLE/RUNS. OCCURRENCE CODE
84156	POOL. DESCRIPTION CODE
84157	SANDBARS. LARGE. OCCURRENCE CODE
84158	LAND USE. NEAR STREAM. PREDOMINANT CODE
84159	STREAM.COVER (INSTREAM SHELTER FOR ADULT FISH). CODE

STORET Code	Description of STORET Parameters Not Suitable for Statistical Analysis
84160	STREAM. DEGRADATION RATING CODE
84161	STREAM. ORDER CODE
84162	LAND RESOURCE AREA CODE
84163	FLOW. STREAM. CLASSIFICATION CODE
84165	DISCHARGE EVENT OBSERVATION. YES=1 NO=0. CODE
84166	STORM HYDROGRAPH. DIRECTION. (RISE,FALL). CODE
84167	MICROSCOPIC EXAMINATION CODE
84168	AVIAN SPECIES ALPHA CODE (BIRDS)
84169	MAMMALIAN ALPHA SPECIES CODE
84170	ALPHA AGE TEXT CODE
84200	LATITUDE/LONGITUDE COORDINATES OF WELL. METHOD CODE
84201	NATIONAL REFERENCE DATUM. ALTITUDE(VERTICAL) CODE
84202	ALTITUDE METHOD CODE
85000	STREAM MILE. ACTUAL MILES
85014	HABITAT. 1970 ACRES THIS TYPE FOR THIS STATION
85015	HAB., ESTIMATED ACRES THIS TYPE THIS STATION
85016	HAB.. ESTIMATED ACRES THIS TYPE THIS STA. BY 1990
85017	HAB.. ESTIMATED ACRES THIS TYPE THIS STA. BY 2000
85018	TYPE CODES: 1=CLEAR CUT/2=SELECT CUT/3=RNGE DEVLP
85019	ACRES. NO. ALTERED FROM 1965-1970 (0-5 YEARS OLD)
85020	ACRES. NO. ALTERED 1960-1965 (5-10 YEARS OLD)
85021	ACRES, NO. ALTERED 1955-1960 (10-15 YEARS OLD)
85022	ACRES. NO. ALTERED 1950-1955 (15-20 YEARS OLD)
85023	ACRES. NO. ALTERED BEFORE 1950 (20+ YEARS OLD)
85024	ACRES.PREDICTED YRLY.AVE.TO BE ALTERED IN FUTURE
85025	LANDOWNERS. CODES FOR ALL IN STATE OF OREGON
85026	ACRES, CURRENT OWNED THIS LANDOWNER THIS STATION
85027	ACRES. ESTIMATED OWNED BY L-O THIS STA. BY 1980
85028	ACRES. ESTIMATED OWNED BY L-O THIS STA. BY 1990

STORET Code	Description of STORET Parameters Not Suitable for Statistical Analysis
85029	ACRES. ESTIMATED OWNED BY L-O THIS STA. BY 2000
85030	LAND USES. CODES FOR ALL IN STATE OF OREGON
85031	ACRES. CURRENT DEDICATED TO THIS USE THIS STATION
85032	ACRES. ESTM. DEDICTD TO THIS USE THIS STA BY 1980
85033	ACRES. ESTM. DEDICTD TO THIS USE THIS STA BY 1990
85034	ACRES. ESTM. DEDICTD TO THIS USE BY YR.2000 --STA.
85035	HAB.. INDICATED ANIMAL USES THIS TYPE IN WINTER
85036	HAB.. INDICATED ANIMAL USES THIS TYPE IN SPRING
85037	HAB.. INDICATED ANIMAL USES THIS TYPE IN SUMMER
85038	HAB.. INDICATED ANIMAL USES THIS TYPE IN FALL
85039	HAB.. INDICATED ANML USES THIS TYPE FOR WINTERING
85040	HAB.. INDICATED ANML USES THIS TYPE FOR FEEDING
85041	HAB.. INDICATED ANML USES TYPE FOR REARING YOUNG
85042	HAB.. INDICATED BIRD USES THIS TYPE FOR NESTING
85043	HAB.. INDICATED ANML USES THIS TYPE FOR SHELTER
85044	HAB.. INDICATED ANML USES THIS TYPE FOR REST AREA
85045	ANML. SHOWS PRESENCE/ABSNC OF COMMENTS ON THIS ANML
85046	HAB..ACRES OCCUPIED BY THIS ANML THIS UNIT & CO.
85050	ANIMALS ARE NOT PRESENT THIS STATION
85051	ANIMALS. ONLY A FEW ARE PRESENT THIS STATION
85052	ANIMALS COMMONLY SEEN: USE MODERATE THIS STATION
85053	ANIMALS FREQUENTLY SEEN: USE HEAVY THIS STATION
85070	OWNERSHIP (.1) AND ACCESS (.2) BY YEAR
85071	PRIVATE OWNERSHIP AND ACCESS MILEAGE
85072	FEDERAL OWNERSHIP AND ACCESS MILEAGE
85073	STATE OWNERSHIP AND ACCESS MILEAGE
85074	COUNTY OWNERSHIP AND ACCESS MILEAGE
85075	CITY OWNERSHIP AND ACCESS MILEAGE
85076	WATER YEAR DATA REFERS TO

STORET Code	Description of STORET Parameters Not Suitable for Statistical Analysis
85077	CALENDAR YEAR DATA REFERS TO
85088	MONTHS POLLUTION IS A PROBLEM JAN THRU JUNE
85089	MONTHS POLLUTION IS A PROBLEM JULY TO DECEMBER
85090	MAN-CAUSED CHANNEL CHANGE IN MILES
85091	STREAM BANK HABITAT DESTROYED IN MILES
85092	STREAMBED SILTED IN MILES
85093	TURBIDITY PROBLEM IN MILES
85094	SEVERITY: 1=ELIMINATES 2=INTERFERES 3=NO PROBLEM
85095	DURATION OF TURBIDITY PROBLEM IN MONTHS
85096	SEASON OF NATURAL DRY CHANNEL 1=SP 2=SU 3=F 4=W
85097	NATURAL DRY CHANNEL IN MILES
85098	MAN-CAUSED DRY CHANNEL SEASON 1=SP 2=SU 3=F 4=W
85099	MAN-CAUSED DRY CHANNEL IN MILES
85100	YEAR BARRIER IS PRESENT
85101	NUMBER OF NATURAL BARRIERS
85102	MILES BLOCKED BY NATURAL BARRIERS
85103	NUMBER OF NATURAL BARRIERS TO BE REMOVED
85104	NUMBER OF DAMS AND MAN CAUSED OBSTRUCTIONS
85105	MILES BLOCKED BY DAMS OR MAN CAUSED OBSTRUCTIONS
85106	NUMBER OF DAMS TO BE ALTERED
85107	MILES OF STREAM OCCUPIED BY IMPOUNDMENT
85108	LOWER END OF SECTION COVERED BY THIS FORM
85109	UPPER END OF SECTION COVERED BY THIS FORM
85110	LOWER LIMIT THIS SPECIES THIS FORM BY RIVER MILE
85111	UPPER LIMIT THIS SPECIES THIS FORM BY RIVER MILE
85112	STREAM SURVEY:1=COMPLETE 2=INCOMPLETE 3=NONE
85113	ABUNDANCE: 1=FSHWY/TAG&R 2=SURVEY 3=EST PLUS 4=EST
85114	ABUNDANCE: N=S&ST 1=ABUNDANT 4=SCARCE RGH FSH 3=SCARCE
85116	SQUARE YARDS OF SPAWNING AREA IN 1970

STORET Code	Description of STORET Parameters Not Suitable for Statistical Analysis
85117	SQUARE YARDS OF SPAWNING AREA IN 1980
85118	SQUARE YARDS OF SPAWNING AREA IN 1990
85119	SQUARE YARDS OF SPAWNING AREA IN 2000
85120	MILES OF REARING AREA IN 1970
85121	MILES OF REARING AREA IN 1980
85122	MILES OF REARING AREA IN 1990
85123	MILES OF REARING AREA IN 2000
85124	CATCH BY SPORT ANGLING IN 1970
85125	RECREATION DAYS SPENT ANGLING IN 1970
85126	RECREATION DAYS SPENT ANGLING IN 1980
85127	RECREATION DAYS SPENT ANGLING IN 1990
85128	RECREATION DAYS SPENT ANGLING IN 2000
85129	CONTRIBUTION TO COMMERCIAL CATCH IN 1970
85130	PERCENT OF TOTAL FISHING DONE FROM BOAT IN 1970
85131	PERCENT OF TOTAL FISHING DONE FROM BANK IN 1970
85132	PERCENT OF TOTAL FISHING DONE WITH LURE IN 1970
85133	PERCENT OF TOTAL FISHING DONE WITH BAIT IN 1970
85134	PERCENT OF TOTAL FISHING DONE WITH A FLY IN 1970
85146	YEAR THIS FACTOR HAS A LIMITING EFFECT
85157	MAN DAYS OF WATER SKIING
85158	SEVERITY: 1=INTERFERES 2=NO INTER. 3=NO ACTIVITY
85159	MAN DAYS OF BOATING OTHER THAN ANGLING
85160	SEVERITY: 1=INTERFERES 2=NO INTER. 3=NO ACTIVITY
85161	MAN DAYS OF SWIMMING
85162	SEVERITY: 1=INTERFERES 2=NO INTER. 3=NO ACTIVITY
85163	SEVERITY: 1=INTERFERES 2=NO INTER. 3=NOT PRESENT
85165	NUMBER OF MONTHS SUSPENDED SOLIDS ARE A PROBLEM
85167	NUMBER OF MONTHS PLANKTON IS A PROBLEM
85168	1=ELIMINATE PROD 2=REDUCE 3=NO INTER. 4=NOT PRES

STORET Code	Description of STORET Parameters Not Suitable for Statistical Analysis
85169	1=ELIMINATE PROD 2=UNDESIRABLE 3=REDUCE 4=NO PROB
85170	1=ELIMINATE PROD 2=UNDESIRABLE 3=REDUCE 4=NO PROB
85171	1=ELIMINATE PROD 2=UNDESIRABLE 3=REDUCE 4=NO PROB
85172	1=ELIMINATE PROD 2=UNDESIRABLE 3=REDUCE 4=NO PROB
85173	1=ELIMINATE PROD 2=UNDESIRABLE 3=REDUCE 4=NO PROB
85174	1=ELIMINATE PROD 2=UNDESIRABLE 3=REDUCE 4=NO PROB
85175	1=ELIMINATE PROD 2=UNDESIRABLE 3=REDUCE 4=NO PROB
85176	1=ELIMINATE PROD 2=UNDESIRABLE 3=REDUCE 4=NO PROB
85177	1=ELIMINATE PROD 2=UNDESIRABLE 3=REDUCE 4=NO PROB
85178	1=ELIMINATE PROD 2=UNDESIRABLE 3=REDUCE 4=NO PROB
85179	YEAR THIS NUMBER OF FACILITIES PRESENT
85180	NUMBER OF BOAT RAMPS
85181	NUMBER OF MOORAGES
85182	NUMBER OF PICNIC AREAS
85183	NUMBER OF CAMP AREAS
85184	NUMBER OF RESORTS
85185	YEAR THIS ZONED AREA PRESENT
85186	ACRES SET ASIDE FOR OTHER BOATING
85187	ACRES SET ASIDE FOR WATER SKIING
85188	MILES OF SHORE LOST TO ACCESS BY HOME SITES
85189	TOTAL MILES OF SHORELINE
85193	WILL RECR BE INC BY RELEASE OF FINGERL 0=NO 1=YES
85195	CATCH AND RECREATION ESTIMATE 1=BEST 4=POOREST
85333	PRECIPITATION-SAMPLE COLLECTION TIME-CODE- NES
85538	GAMMA SCAN DATE (YR.MO.DAY)
85539	DATE OF REPORT (YR.MO.DAY)
85658	TIME NIGHT CO2 HR
85661	TIME. INTERVAL DAY CO2 HR

Appendix F

National EPA Water Quality Criteria Summary [1]

The following table presents the national water quality criteria that were used to assess water quality data on a station-by-station basis and within the entire study area. Criteria are, for the most part, maximum values (except for dissolved oxygen, pH, and as noted). Criteria exist in any of four categories: Fresh Acute, Drinking Water, Marine Acute, and Other. Acute criteria are the highest 1-hour average concentrations which should not result in unacceptable impacts to aquatic organisms in either fresh or marine waters, respectively. The Drinking Water criteria are intended for human consumption; while the Other criteria represents National Park Service or other concerns. Parameters are listed in ascending order by STORET code. It is important to note that similar parameters often have non-consecutive codes. Consequently, scanning the entire list is necessary to obtain the criteria for all parameters of a particular type (eg. lead, copper, etc.). Refer to the Parameter Period of Record Tabulation to obtain the STORET code for any parameter measured in the park.

C.A.S. Number	STORET Code	FRESH ACUTE	DRINKING WATER	MARINE ACUTE	OTHER	PARAMETER DESCRIPTION	UNITS	CATEGORY
	00070				50^1	TURBIDITY, JACKSON CANDLE UNITS	JTU	Physical
	00076				50^1	TURBIDITY, HACH TURBIDIMETER, FORMAZIN TUR. UNITS	FTU	Physical
14808798	00154		250^4			SULFATE (AS S) WHOLE WATER	MG/L	General Inorganic
7782447	00299				$4.0^\#$	OXYGEN, DISSOLVED, ANALYSIS BY PROBE	MG/L	Dissolved Oxygen
7782447	00300				$4.0^\#$	OXYGEN, DISSOLVED	MG/L	Dissolved Oxygen
	00400				$\leq 6.5, \geq 9.0^\#$	PH	SU	Physical
	00403				$\leq 6.5, \geq 9.0^\#$	PH, LAB	SU	Physical
	00406				$\leq 6.5, \geq 9.0^\#$	PH, FIELD	SU	Physical

[1] Sources: (1) U.S. Environmental Protection Agency, Quality Criteria for Water 1995, Final Draft; (2) U.S. Environmental Protection Agency, 40 CFR 141 - National Primary Drinking Water Regulations, and 40 CFR 143 - National Secondary Drinking Water Regulations, July 1, 1994; and (3) Others as Noted in Footnotes.

C.A.S. Number	STORET Code	FRESH ACUTE	DRINKING WATER	MARINE ACUTE	OTHER	PARAMETER DESCRIPTION	UNITS	CATEGORY
471341	00409				<200⁻	ALKALINITY, TOTAL, LOW LEVEL GRAN ANALYSIS	UEQ/L	General Inorganic
17778880	00613		1			NITRITE NITROGEN, DISSOLVED AS N	MG/L	Nitrogen
17778880	00615		1			NITRITE NITROGEN, TOTAL AS N	MG/L	Nitrogen
17778880	00618		10			NITRATE NITROGEN, DISSOLVED AS N	MG/L	Nitrogen
17778880	00620		10			NITRATE NITROGEN, TOTAL AS N	MG/L	Nitrogen
17778880	00628		10			NITRITE + NITRATE, SUSPENDED AS N	MG/L	Nitrogen
17778880	00630		10			NITRITE PLUS NITRATE, TOTAL 1 DET.	MG/L	Nitrogen
17778880	00631		10			NITRITE PLUS NITRATE, DISSOLVED 1 DET.	MG/L	Nitrogen
57125	00718	22	200	1.0		CYANIDE, WEAK ACID, DISSOCIABLE, WATER, WHOLE	UG/L	General Inorganic
57125	00719	22	200	1.0		CYANIDE, FREE,IN WATER&WASTEWATERS, HBG METHOD	UG/L	General Inorganic
57125	00720	0.022	0.2	0.001		CYANIDE, TOTAL	MG/L	General Inorganic
57125	00722	0.022	0.2	0.001		CYANIDE, FREE (AMENABLE TO CHLORINATION)	MG/L	General Inorganic
57125	00723	22	200	1.0		CYANIDE, DISSOLVED STD METHOD	UG/L	General Inorganic
57125	00724	22	200	1.0		CYANIDE COMPLEXED TO A RANGE OF COMPNDS, WATER	UG/L	General Inorganic
16887006	00940	860	250ᵃ			CHLORIDE,TOTAL IN WATER	MG/L	General Inorganic
16887006	00941	860	250ᵃ			CHLORIDE, DISSOLVED IN WATER	MG/L	General Inorganic
14808798	00945		250ᵃ			SULFATE, TOTAL (AS SO4)	MG/L	General Inorganic
14808798	00946		250ᵃ			SULFATE, DISSOLVED (AS SO4)	MG/L	General Inorganic
1332214	00948		7000000			ASBESTOS, WHOLE SAMPLE	CNT/L	General Inorganic
16984488	00950		4.0			FLUORIDE, DISSOLVED AS F	MG/L	General Inorganic

C.A.S. Number	STORET Code	FRESH ACUTE	DRINKING WATER	MARINE ACUTE	OTHER	PARAMETER DESCRIPTION	UNITS	CATEGORY
16984488	00951		4.0			FLUORIDE, TOTAL AS F	MG/L	General Inorganic
7782414	00953		4000			FLUORINE, TOTAL	UG/L	General Inorganic
7440382	00978	360	50	69		ARSENIC, TOTAL RECOVERABLE IN WATER AS AS	UG/L	Metal
7782492	00981	20	50	300		SELENIUM,TOTAL RECOVERABLE IN WATER AS SE	UG/L	Metal
7440280	00982	1400*	2.0	2130*		THALLIUM, TOTAL RECOVERABLE IN WATER AS TL	UG/L	Metal
7782492	00990	20	50	300		SELENITE, TOTAL RECOVERABLE INORGANIC	UG/L	Metal
7440382	00991	360	50	69		ARSENIC , TOTAL RECOVERABLE TRIVALENT INORGANIC	UG/L	Metal
7440382	00995	360	50	69		ARSENIC, INORGANIC DISS	UG/L	Metal
7440382	00996	360	50	69		ARSENIC, INORGANIC SUSP	UG/L	Metal
7440382	00997	360	50	69		ARSENIC, INORGANIC TOT	UG/L	Metal
7440417	00998	130*	4.0			BERYLLIUM,TOTAL RECOVERABLE IN WATER AS BE	UG/L	Metal
7440382	01000	360	50	69		ARSENIC, DISSOLVED	UG/L	Metal
7440382	01001	360	50	69		ARSENIC, SUSPENDED	UG/L	Metal
7440382	01002	360	50	69		ARSENIC, TOTAL	UG/L	Metal
7440393	01005		2000			BARIUM, DISSOLVED	UG/L	Metal
7440393	01006		2000			BARIUM, SUSPENDED	UG/L	Metal
7440393	01007		2000			BARIUM, TOTAL	UG/L	Metal
7440393	01009		2000			BARIUM,TOTAL RECOVERABLE IN WATER AS BA	UG/L	Metal
7440417	01010	130*	4.0			BERYLLIUM, DISSOLVED	UG/L	Metal
7440417	01011	130*	4.0			BERYLLIUM, SUSPENDED	UG/L	Metal

F-3

C.A.S. Number	STORET Code	FRESH ACUTE	DRINKING WATER	MARINE ACUTE	OTHER	PARAMETER DESCRIPTION	UNITS	CATEGORY
7440417	01012	130*	4.0			BERYLLIUM, TOTAL	UG/L	Metal
7440439	01025	3.9+	5.0	43		CADMIUM, DISSOLVED	UG/L	Metal
7440439	01026	3.9+	5.0	43		CADMIUM, SUSPENDED	UG/L	Metal
7440439	01027	3.9+	5.0	43		CADMIUM, TOTAL	UG/L	Metal
7440473	01030		100			CHROMIUM, DISSOLVED	UG/L	Metal
7440473	01031		100			CHROMIUM, SUSPENDED	UG/L	Metal
7440473	01032	16	100	1100		CHROMIUM, HEXAVALENT	UG/L	Metal
16065831	01033	1700+	100	10300*		CHROMIUM, TRI-VAL	UG/L	Metal
7440473	01034		100			CHROMIUM, TOTAL	UG/L	Metal
7440508	01040	18+	1300ª	2.9		COPPER, DISSOLVED	UG/L	Metal
7440508	01041	18+	1300ª	2.9		COPPER, SUSPENDED	UG/L	Metal
7440508	01042	18+	1300ª	2.9		COPPER, TOTAL	UG/L	Metal
7439921	01049	82+	15ª	220		LEAD, DISSOLVED	UG/L	Metal
7439921	01050	82+	15ª	220		LEAD, SUSPENDED	UG/L	Metal
7439921	01051	82+	15ª	220		LEAD, TOTAL	UG/L	Metal
7440280	01057	1400*	2.0	2130*		THALLIUM, DISSOLVED	UG/L	Metal
7440280	01058	1400*	2.0	2130*		THALLIUM, SUSPENDED	UG/L	Metal
7440280	01059	1400*	2.0	2130*		THALLIUM, TOTAL	UG/L	Metal
7440020	01065	1400+	100	75		NICKEL, DISSOLVED	UG/L	Metal
7440020	01066	1400+	100	75		NICKEL, SUSPENDED	UG/L	Metal

F-4

C.A.S. Number	STORET Code	FRESH ACUTE	DRINKING WATER	MARINE ACUTE	OTHER	PARAMETER DESCRIPTION	UNITS	CATEGORY
7440020	01067	1400^t	100	75		NICKEL, TOTAL	UG/L	Metal
7440020	01074	1400^t	100	75		NICKEL, TOTAL RECOVERABLE IN WATER AS NI	UG/L	Metal
7440224	01075	4.1^t	100^s	0.12		SILVER, DISSOLVED	UG/L	Metal
7440224	01076	4.1^t	100^s	0.12		SILVER, SUSPENDED	UG/L	Metal
7440224	01077	4.1^t	100^s	0.12		SILVER, TOTAL	UG/L	Metal
7440224	01079	4.1^t	100^s	0.12		SILVER, TOTAL RECOVERABLE IN WATER AS AG	UG/L	Metal
7440508	01089	0.018^t	1.3^a	0.0029		COPPER AS SUSPENDED BLACK OXIDE IN WATER	MG/L	General Inorganic
7440666	01090	120^t	5000^p	95		ZINC, DISSOLVED	UG/L	Metal
7440666	01091	120^t	5000^s	95		ZINC, SUSPENDED	UG/L	Metal
7440666	01092	120^t	5000^s	95		ZINC, TOTAL	UG/L	Metal
7440666	01094	120^t	5000^s	95		ZINC, TOTAL RECOVERABLE IN WATER AS ZN	UG/L	Metal
7440360	01095	88^p	6.0	1500^p		ANTIMONY, DISSOLVED	UG/L	Metal
7440360	01096	88^p	6.0	1500^p		ANTIMONY, SUSPENDED	UG/L	Metal
7440360	01097	88^p	6.0	1500^p		ANTIMONY, TOTAL	UG/L	Metal
7440439	01113	3.9^t	5.0	43		CADMIUM,TOTAL RECOVERABLE IN WATER AS CD	UG/L	Metal
7439921	01114	82^t	15^a	220		LEAD, TOTAL RECOVERABLE IN WATER AS PB	UG/L	Metal
7440473	01118		100			CHROMIUM TOTAL RECOVERABLE IN WATER AS CR	UG/L	Metal
7440508	01119	18^t	1300^a	2.9		COPPER, TOTAL RECOVERABLE IN WATER AS CU	UG/L	Metal
7440280	01124	1400*	2.0	2130*		THALLIUM, ACID SOLUBLE, WATER, WHOLE	UG/L	Metal
7440280	01128	1400*	2.0	2130*		THALLIUM, TOTAL RECOVERABLE <95%	UG/L	Metal

F-5

C.A.S. Number	STORET Code	FRESH ACUTE	DRINKING WATER	MARINE ACUTE	OTHER	PARAMETER DESCRIPTION	UNITS	CATEGORY
7782492	01145	20	50	300		SELENIUM, DISSOLVED	UG/L	Metal
7782492	01146	20	50	300		SELENIUM, SUSPENDED	UG/L	Metal
7782492	01147	20	50	300		SELENIUM, TOTAL	UG/L	Metal
7782492	01167	20	50	300		SELENIUM, ACID SOLUBLE, WATER, WHOLE	UG/L	Metal
18540299	01220	16	100	1100		CHROMIUM, HEXAVALENT, DISSOLVED	UG/L	Metal
7440360	01268	88^P	6.0	1500^P		ANTIMONY (SB), WATER, TOTAL RECOVERABLE	UG/L	Metal
57125	01291	22	200	1.0		CYANIDE, FILTERABLE, TOTAL IN WATER	UG/L	General Inorganic
7440666	01303	0.120$^+$	5.0^a	0.095		ZINC, POTENTIALLY DISSOLVED WATER	MG/L	Metal
7440224	01304	0.0041$^+$	0.1^a	0.00012		SILVER, POTENTIALLY DISSOLVED WATER	MG/L	Metal
7440508	01306	0.018$^+$	1.3^a	0.0029		COPPER, POTENTIALLY DISSOLVED WATER	MG/L	Metal
18540299	01307	0.016	0.1	1.1		CHROMIUM, HEXAVALENT, POTENTIALLY DISSOLVED	MG/L	Metal
7440382	01309	0.36	0.05	0.069		ARSENIC, POTENTIALLY, DISSOLVED, WATER	MG/L	Metal
7440393	01311		2.0			BARIUM, POTENTIALLY, DISSOLVED, WATER	MG/L	Metal
7440417	01312	0.13*	0.004			BERYLLIUM, POTENTIALLY, DISSOLVED, WATER	MG/L	Metal
7440439	01313	0.0039$^+$	0.005	0.043		CADMIUM, POTENTIALLY, DISSOLVED, WATER	MG/L	Metal
16065831	01314	1.7$^+$	0.1	10.3*		CHROMIUM, TRIVALENT, POTENTIALLY DISSOLVED	MG/L	Metal
7439921	01318	0.082$^+$	0.015^a	0.220		LEAD, POTENTIALLY, DISSOLVED, WATER	MG/L	Metal
7439976	01321	0.0024	0.002	0.0021		MERCURY, POTENTIALLY, DISSOLVED, WATER	MG/L	Metal
7440020	01322	1.4$^+$	0.1	0.075		NICKEL, POTENTIALLY, DISSOLVED, WATER	MG/L	Metal
7782492	01323	0.020	0.050	0.300		SELENIUM, POTENTIALLY, DISSOLVED, WATER	MG/L	Metal

C.A.S. Number	STORET Code	FRESH ACUTE	DRINKING WATER	MARINE ACUTE	OTHER	PARAMETER DESCRIPTION	UNITS	CATEGORY
7440280	01324	1.4*	0.002	2.13*		THALLIUM, POTENTIALLY, DISSOLVED, WATER	MG/L	Metal
7440611	01326		0.020^c			URANIUM, POTENTIALLY DISSOLVED, WATER	MG/L	Metal
7440224	01523	4.1^+	100^s	0.12		SILVER, IONIC	UG/L	Metal
50328	03648		0.2			BENZO (A) PYRENE, LIQUID FRACTION, ELUTRIATE	UG/L	General Organic
122349	04035		4.0			SIMAZINE, DISSOLVED, WATER, TOTAL RECOVERABLE	UG/L	Pesticide
10028178	04124		20^c			TRITIUM, TOTAL, WATER	PC/ML	Radiological
10028178	07000		20000^c			TRITIUM, TOTAL	PC/L	Radiological
10028178	07005		20000^c			TRITIUM, DISSOLVED	PC/L	Radiological
10028178	07010		20000^c			TRITIUM, SUSPENDED	PC/L	Radiological
	09501		5.0			RADIUM 226, TOTAL	PC/L	Radiological
	09503		5.0			RADIUM 226, DISSOLVED	PC/L	Radiological
	09505		5.0			RADIUM 226, SUSPENDED	PC/L	Radiological
	11500		5.0			RADIUM 226 + RADIUM 228, DISSOLVED	PC/L	Radiological
	11501		5.0			RADIUM 228, TOTAL	PC/L	Radiological
	11503		5.0			RADIUM 226 + RADIUM 228, TOTAL	PC/L	Radiological
10098972	13501		8.0^c			STRONTIUM 90, TOTAL	PC/L	Radiological
10098972	13503		8.0^c			STRONTIUM 90, DISSOLVED	PC/L	Radiological
10098972	13505		8.0^c			STRONTIUM 90, SUSPENDED	PC/L	Radiological
7782492	22675	20	50	300		SELENIUM, DISSOLVED ORGANIC	UG/L	Metal
7782492	22676	20	50	300		SELENIUM, HEXAVALENT, DISSOLVED	UG/L	Metal

C.A.S. Number	STORET Code	FRESH ACUTE	DRINKING WATER	MARINE ACUTE	OTHER	PARAMETER DESCRIPTION	UNITS	CATEGORY
7782492	22677	20	50	300		SELENIUM, TETRAVALENT, DISSOLVED	UG/L	Metal
7440382	22678	360	50	69		ARSENIC, DISSOLVED ORGANIC	UG/L	Metal
7440382	22679	850*	50	2319*		ARSENIC, PENTAVALENT,DISSOLVED	UG/L	Metal
7440382	22680	360	50	69		ARSENIC, TRIVALENT, DISSOLVED	UG/L	Metal
7440611	22703		20[c]			URANIUM, NATURAL DISSOLVED	UG/L	Metal
7440611	22705		20[c]			URANIUM, NATURAL SUSPENDED	UG/L	Metal
7440611	22706		20[c]			URANIUM, TOTAL AS U308	UG/L	Metal
7440611	22708		0.020[c]			URANIUM, NATURAL, TOTAL	MG/L	Radiological
7440611	28011		20[c]			URANIUM, NATURAL, TOTAL	UG/L	Radiological
88857	30191		7.0			DINOSEB, WATER, WHOLE RECOVERABLE	UG/L	Pesticide
75990	30200		200			DALAPON, WATER, WHOLE RECOVERABLE	UG/L	Pesticide
106934	30203		0.05			ETHANE, 1,2-DIBROMO-, WATER, WHOLE, RECOVERABLE	UG/L	Pesticide
	31501		1.0[a]		1000[b]	COLIFORM, TOTAL, MEMBRANE FILTER, IMMED.	CFU/100ML	Bacteriological
	31503		1.0[a]		1000[b]	COLIFORM, TOTAL, MEMBRANE FILTER, DELAY. M-ENDO	CFU/100ML	Bacteriological
	31504		1.0[a]		1000[b]	COLIFORM, TOTAL, MEMBRANE FILTER, IMMED. LES-ENDO	CFU/100ML	Bacteriological
	31505		1.0[a]		1000[b]	COLIFORM, TOTAL, MPN, CONF. TEST 35C (TUBE 31506)	MPN/100ML	Bacteriological
	31506		1.0[a]		1000[b]	COLIFORM, TOTAL, MPN, CONF. TEST, TUBE CONFIG	MPN/100ML	Bacteriological
	31507		1.0[a]		1000[b]	COLIFORM, TOTAL, MPN, COMP. TEST 35C (TUBE 31508)	MPN/100ML	Bacteriological
	31508		1.0[a]		1000[b]	COLIFORM, TOTAL, MPN, COMP. TEST, TUBE CONFIG	MPN/100ML	Bacteriological
	31613				200	FECAL COLIFORM, MEMBRANE FILTER, AGAR	CFU/100ML	Bacteriological

C.A.S. Number	STORET Code	FRESH ACUTE	DRINKING WATER	MARINE ACUTE	OTHER	PARAMETER DESCRIPTION	UNITS	CATEGORY
	31614				200	FECAL COLIFORM, MPN, TUBE CONFIGURATION	MPN/100ML	Bacteriological
	31615				200	FECAL COLIFORM, MPN, EC MED. 44.5C (TUBE 31614)	MPN/100ML	Bacteriological
	31616				200	FECAL COLIFORM, MEMBRANE FILTER, 44.5C	CFU/100ML	Bacteriological
	31617				200	FECAL COLIFORM, MPN, EIJKMAN, 44.5C (TUBE 31618)	MPN/100ML	Bacteriological
	31625				200	FECAL COLIFORM, MF, M-FC, 0.7 UM	CFU/100ML	Bacteriological
	31648				126	E. COLI, MTEC, MF	CFU/100ML	Bacteriological
	31649				33	ENTEROCOCCI, ME, MF	CFU/100ML	Bacteriological
67663	32003	28900*	100^l			CARBON CHLOROFORM AND CARBON ALCOHOL EXTRS.,TOTAL	UG/L	General Organic
67663	32005	28900*	100^l			CARBON CHLOROFORM EXTRACTABLES	UG/L	General Organic
67663	32021	28900*	100^l			CARBON CHLOROFORM EXTRACTS, ETHER INSOLUBLES OF	UG/L	General Organic
67663	32022	28900*	100^l			CARBON CHLOROFORM EXTRACTS, WATER SOLUBLES OF	UG/L	General Organic
75274	32101		100^l			BROMODICHLOROMETHANE, WHOLE WATER	UG/L	General Organic
56235	32102	35200*	5.0	50000*		CARBON TETRACHLORIDE, WHOLE WATER	UG/L	General Organic
107062	32103	118000*	5.0	113000*		1,2-DICHLOROETHANE, WHOLE WATER	UG/L	General Organic
75252	32104		100^l			BROMOFORM, WHOLE WATER	UG/L	General Organic
124481	32105		100^l			DIBROMOCHLOROMETHANE, WHOLE WATER	UG/L	General Organic
67663	32106	28900*	100^l			CHLOROFORM, WHOLE WATER	UG/L	General Organic
56235	32260	35.2*	0.005	50*		CARBON TETRACHLORIDE EXTRACTABLES	MG/L	General Organic
67663	32270	28.9*	0.1^l			CHLOROFORM EXTRACTABLES TOTAL	MG/L	General Organic
108883	34010	17500*	1000	6300*		TOLUENE IN WTR SMPLE GC-MS, HEXADECONE EXTR.	UG/L	General Organic

F-9

C.A.S. Number	STORET Code	FRESH ACUTE	DRINKING WATER	MARINE ACUTE	OTHER	PARAMETER DESCRIPTION	UNITS	CATEGORY
1330207	34020		10000			XYLENES IN WTR SMPLE GC-MS, HEXADECONE EXTR.	UG/L	General Organic
83329	34205	1700*		970*		ACENAPHTHENE, TOTAL	UG/L	General Organic
83329	34206	1700*		970*		ACENAPHTHENE, DISSOLVED	UG/L	General Organic
83329	34207	1700*		970*		ACENAPHTHENE, SUSPENDED	UG/L	General Organic
107028	34210	68*		55*		ACROLEIN, TOTAL	UG/L	Pesticide
107028	34211	68*		55*		ACROLEIN, DISSOLVED	UG/L	Pesticide
107028	34212	68*		55*		ACROLEIN, SUSPENDED	UG/L	Pesticide
107131	34215	7550*				ACRYLONITRILE, TOTAL	UG/L	General Organic
107131	34216	7550*				ACRYLONITRILE, DISSOLVED	UG/L	General Organic
107131	34217	7550*				ACRYLONITRILE, SUSPENDED	UG/L	General Organic
71432	34235	5300*	5.0	5100*		BENZENE, DISSOLVED	UG/L	General Organic
71432	34236	5300*	5.0	5100*		BENZENE, SUSPENDED	UG/L	General Organic
92875	34239	2500*				BENZIDINE, DISSOLVED	UG/L	General Organic
92875	34240	2500*				BENZIDINE, SUSPENDED	UG/L	General Organic
58899	34265	2.0	0.2	0.16		R-BHC (LINDANE) GAMMA, DISSOLVED	UG/L	Pesticide
58899	34266	2.0	0.2	0.16		R-BHC (LINDANE) GAMMA, SUSPENDED	UG/L	Pesticide
75252	34288		100'			BROMOFORM, DISSOLVED	UG/L	General Organic
75252	34289		100'			BROMOFORM, SUSPENDED	UG/L	General Organic
56235	34297	35200*	5.0	50000*		CARBON TETRACHLORIDE, DISSOLVED	UG/L	General Organic
56235	34298	35200*	5.0	50000*		CARBON TETRACHLORIDE, SUSPENDED	UG/L	General Organic

C.A.S. Number	STORET Code	FRESH ACUTE	DRINKING WATER	MARINE ACUTE	OTHER	PARAMETER DESCRIPTION	UNITS	CATEGORY
108907	34301		100			CHLOROBENZENE, TOTAL	UG/L	General Organic
108907	34302		100			CHLOROBENZENE, DISSOLVED	UG/L	General Organic
108907	34303		100			CHLOROBENZENE, SUSPENDED	UG/L	General Organic
124481	34306		100^t			CHLORODIBROMOMETHANE, TOTAL	UG/L	General Organic
124481	34307		100^t			CHLORODIBROMOMETHANE, DISSOLVED	UG/L	General Organic
124481	34308		100^t			CHLORODIBROMOMETHANE, SUSPENDED	UG/L	General Organic
67663	34316	28900*	100^t			CHLOROFORM, DISSOLVED	UG/L	General Organic
67663	34317	28900*	100^t			CHLOROFORM, SUSPENDED	UG/L	General Organic
57125	34325	0.022	0.2	0.001		CYANIDE, SUSPENDED	MG/L	General Inorganic
75274	34328		100^t			DICHLOROBROMOMETHANE, DISSOLVED	UG/L	General Organic
75274	34329		100^t			DICHLOROBROMOMETHANE, SUSPENDED	UG/L	General Organic
122667	34346	270*				1,2-DIPHENYLHYDRAZINE, TOTAL	UG/L	General Organic
122667	34347	270*				1,2-DIPHENYLHYDRAZINE, DISSOLVED	UG/L	General Organic
122667	34348	270*				1,2-DIPHENYLHYDRAZINE, SUSPENDED	UG/L	General Organic
33213659	34356	0.22		0.034		ENDOSULFAN, BETA, TOTAL	UG/L	Pesticide
33213659	34357	0.22		0.034		ENDOSULFAN, BETA, DISSOLVED	UG/L	Pesticide
33213659	34358	0.22		0.034		ENDOSULFAN, BETA, SUSPENDED	UG/L	Pesticide
959988	34361	0.22		0.034		ENDOSULFAN, ALPHA, TOTAL	UG/L	Pesticide
959988	34362	0.22		0.034		ENDOSULFAN, ALPHA, DISSOLVED	UG/L	Pesticide
959988	34363	0.22		0.034		ENDOSULFAN, ALPHA, SUSPENDED	UG/L	Pesticide

C.A.S. Number	STORET Code	FRESH ACUTE	DRINKING WATER	MARINE ACUTE	OTHER	PARAMETER DESCRIPTION	UNITS	CATEGORY
100414	34371	32000*	700	430*		ETHYLBENZENE, TOTAL	UG/L	General Organic
100414	34372	32000*	700	430*		ETHYLBENZENE, DISSOLVED	UG/L	General Organic
100414	34373	32000*	700	430*		ETHYLBENZENE, SUSPENDED	UG/L	General Organic
206440	34376	3980*		40*		FLUORANTHENE, TOTAL	UG/L	General Organic
206440	34377	3980*		40*		FLUORANTHENE, DISSOLVED	UG/L	General Organic
206440	34378	3980*		40*		FLUORANTHENE, SUSPENDED	UG/L	General Organic
77474	34386	7.0*	50	7.0*		HEXACHLOROCYCLOPENTADIENE, TOTAL	UG/L	General Organic
77474	34387	7.0*	50	7.0*		HEXACHLOROCYCLOPENTADIENE, DISSOLVED	UG/L	General Organic
77474	34388	7.0*	50	7.0*		HEXACHLOROCYCLOPENTADIENE, SUSPENDED	UG/L	General Organic
87683	34391	90*		32*		HEXACHLOROBUTADIENE, TOTAL	UG/L	General Organic
87683	34392	90*		32*		HEXACHLOROBUTADIENE, DISSOLVED	UG/L	General Organic
87683	34393	90*		32*		HEXACHLOROBUTADIENE, SUSPENDED	UG/L	General Organic
67721	34396	980*		940*		HEXACHLOROETHANE, TOTAL	UG/L	General Organic
67721	34397	980*		940*		HEXACHLOROETHANE, DISSOLVED	UG/L	General Organic
67721	34398	980*		940*		HEXACHLOROETHANE, SUSPENDED	UG/L	General Organic
118741	34401	6.0[D]	1.0			HEXACHLOROBENZENE, DISSOLVED	UG/L.	General Organic
118741	34402	6.0[D]	1.0			HEXACHLOROBENZENE, SUSPENDED	UG/L	General Organic
193395	34403		0.40[c]			INDENO (1,2,3-CD) PYRENE, TOTAL	UG/L	General Organic
193395	34404		0.40[c]			INDENO (1,2,3-CD) PYRENE, DISSOLVED	UG/L	General Organic
193395	34405		0.40[c]			INDENO (1,2,3-CD) PYRENE, SUSPENDED	UG/L	General Organic

F-12

C.A.S. Number	STORET Code	FRESH ACUTE	DRINKING WATER	MARINE ACUTE	OTHER	PARAMETER DESCRIPTION	UNITS	CATEGORY
78591	34408	117000*		12900*		ISOPHORONE, TOTAL	UG/L	Pesticide
78591	34409	117000*		12900*		ISOPHORONE, DISSOLVED	UG/L	Pesticide
78591	34410	117000*		12900*		ISOPHORONE, SUSPENDED	UG/L	Pesticide
75092	34423		5.0			METHYLENE CHLORIDE, TOTAL	UG/L	General Organic
75092	34424		5.0			METHYLENE CHLORIDE, DISSOLVED	UG/L	General Organic
75092	34425		5.0			METHYLENE CHLORIDE, SUSPENDED	UG/L	General Organic
91203	34443	2300*		2350*		NAPHTHALENE, DISSOLVED	UG/L	General Organic
91203	34444	2300*		2350*		NAPHTHALENE, SUSPENDED	UG/L	General Organic
98953	34447	27000*		6680*		NITROBENZENE, TOTAL	UG/L	General Organic
98953	34448	27000*		6680*		NITROBENZENE, DISSOLVED	UG/L	General Organic
98953	34449	27000*		6680*		NITROBENZENE, SUSPENDED	UG/L	General Organic
59507	34452	30*				PARACHLOROMETA CRESOL, TOTAL	UG/L	General Organic
59507	34453	30*				PARACHLOROMETA CRESOL, DISSOLVED	UG/L	General Organic
59507	34454	30*				PARACHLOROMETA CRESOL, SUSPENDED	UG/L	General Organic
87865	34459	20***	1.0	13		PCP (PENTACHLOROPHENOL), DISSOLVED	UG/L	Pesticide
87865	34460	20***	1.0	13		PCP (PENTACHLOROPHENOL), SUSPENDEID	UG/L	Pesticide
85018	34461	30^P		7.7^b		PHENANTHRENE, TOTAL	UG/L	General Organic
85018	34462	30^P		7.7^b		PHENANTHRENE, DISSOLVED	UG/L	General Organic
85018	34463	30^P		7.7^b		PHENANTHRENE, SUSPENDED	UG/L	General Organic
108952	34466	10200*		5800*		PHENOL, DISSOLVED	UG/L	General Organic

C.A.S. Number	STORET Code	FRESH ACUTE	DRINKING WATER	MARINE ACUTE	OTHER	PARAMETER DESCRIPTION	UNITS	CATEGORY
108952	34467	10200*		5800*		PHENOL, SUSPENDED	UG/L	General Organic
127184	34475	5280*	5.0	10200*		TETRACHLOROETHYLENE, TOTAL	UG/L	General Organic
127184	34476	5280*	5.0	10200*		TETRACHLOROETHYLENE, DISSOLVED	UG/L	General Organic
127184	34477	5280*	5.0	10200*		TETRACHLOROETHYLENE, SUSPENDED	UG/L	General Organic
108883	34481	17500*	1000	6300*		TOLUENE, DISSOLVED	UG/L	General Organic
108883	34482	17500*	1000	6300*		TOLUENE, SUSPENDED	UG/L	General Organic
79016	34485	45000*	5.0	2000*		TRICHLOROETHYLENE, DISSOLVED	UG/L	General Organic
79016	34486	45000*	5.0	2000*		TRICHLOROETHYLENE, SUSPENDED	UG/L	General Organic
75014	34493		2.0			VINYL CHLORIDE, DISSOLVED	UG/L	General Organic
75014	34494		2.0			VINYL CHLORIDE, SUSPENDED	UG/L	General Organic
75354	34501		7.0			1,1-DICHLOROETHYLENE, TOTAL	UG/L	General Organic
75354	34502		7.0			1,1-DICHLOROETHYLENE, DISSOLVED	UG/L	General Organic
75354	34503		7.0			1,1-DICHLOROETHYLENE, SUSPENDED	UG/L	General Organic
71556	34506		200	31200*		1,1,1-TRICHLOROETHANE, TOTAL	UG/L	General Organic
71556	34507		200	31200*		1,1,1-TRICHLOROETHANE, DISSOLVED	UG/L	General Organic
71556	34508		200	31200*		1,1,1-TRICHLOROETHANE, SUSPENDED	UG/L	General Organic
79005	34511		5.0			1,1,2-TRICHLOROETHANE, TOTAL	UG/L	General Organic
79005	34512		5.0			1,1,2-TRICHLOROETHANE, DISSOLVED	UG/L	General Organic
79005	34513		5.0			1,1,2-TRICHLOROETHANE, SUSPENDED	UG/L	General Organic
79345	34516			9020*		1,1,2,2-TETRACHLOROETHANE, TOTAL	UG/L	General Organic

C.A.S. Number	STORET Code	FRESH ACUTE	DRINKING WATER	MARINE ACUTE	OTHER	PARAMETER DESCRIPTION	UNITS	CATEGORY
79345	34517			9020*		1,1,2,2-TETRACHLOROETHANE, DISSOLVED	UG/L	General Organic
79345	34518			9020*		1,1,2,2-TETRACHLOROETHANE, SUSPENDED	UG/L	General Organic
107062	34531	118000*	5.0	113000*		1,2-DICHLOROETHANE, TOTAL	UG/L	General Organic
107062	34532	118000*	5.0	113000*		1,2-DICHLOROETHANE, DISSOLVED	UG/L	General Organic
107062	34533	118000*	5.0	113000*		1,2-DICHLOROETHANE, SUSPENDED	UG/L	General Organic
95501	34536		600			1,2-DICHLOROBENZENE, TOTAL	UG/L	General Organic
95501	34537		600			1,2-DICHLOROBENZENE, DISSOLVED	UG/L	General Organic
95501	34538		600			1,2-DICHLOROBENZENE, SUSPENDED	UG/L	General Organic
78875	34541		5.0			1,2-DICHLOROPROPANE, TOTAL	UG/L	General Organic
78875	34542		5.0			1,2-DICHLOROPROPANE, DISSOLVED	UG/L	General Organic
78875	34543		5.0			1,2-DICHLOROPROPANE, SUSPENDED	UG/L	General Organic
156605	34546		100			TRANS-1,2-DICHLOROETHENE, TOTAL, IN WATER	UG/L	General Organic
156605	34547		100			TRANS-1,2-DICHLOROETHENE, DISSOLVED	UG/L	General Organic
156605	34548		100			TRANS-1,2-DICHLOROETHENE, SUSPENDED	UG/L	General Organic
120821	34551		70			1,2,4-TRICHLOROBENZENE, TOTAL	UG/L	General Organic
120821	34552		70			1,2,4-TRICHLOROBENZENE, DISSOLVED	UG/L	General Organic
120821	34553		70			1,2,4-TRICHLOROBENZENE, SUSPENDED	UG/L	General Organic
541731	34566		600			1,3-DICHLOROBENZENE, TOTAL	UG/L	General Organic
541731	34567		600			1,3-DICHLOROBENZENE, DISSOLVED	UG/L	General Organic
541731	34568		600			1,3-DICHLOROBENZENE, SUSPENDED	UG/L	General Organic

F-15

C.A.S. Number	STORET Code	FRESH ACUTE	DRINKING WATER	MARINE ACUTE	OTHER	PARAMETER DESCRIPTION	UNITS	CATEGORY
106467	34571		75			1,4-DICHLOROBENZENE, TOTAL	UG/L	General Organic
106467	34572		75			1,4-DICHLOROBENZENE, DISSOLVED	UG/L	General Organic
106467	34573		75			1,4-DICHLOROBENZENE, SUSPENDED	UG/L	General Organic
95578	34586	4380*				2-CHLOROPHENOL, TOTAL	UG/L	General Organic
95578	34587	4380*				2-CHLOROPHENOL, DISSOLVED	UG/L	General Organic
95578	34588	4380*				2-CHLOROPHENOL, SUSPENDED	UG/L	General Organic
120832	34601	2020*				2,4-DICHLOROPHENOL, TOTAL	UG/L	General Organic
120832	34602	2020*				2,4-DICHLOROPHENOL, DISSOLVED	UG/L	General Organic
120832	34603	2020*				2,4-DICHLOROPHENOL, SUSPENDED	UG/L	General Organic
105679	34606	2120*				2,4-DIMETHYLPHENOL, TOTAL	UG/L	General Organic
105679	34607	2120*				2,4-DIMETHYLPHENOL, DISSOLVED	UG/L	General Organic
105679	34608	2120*				2,4-DIMETHYLPHENOL, SUSPENDED	UG/L	General Organic
121142	34611	330*		590*		2,4-DINITROTOLUENE, TOTAL	UG/L	General Organic
121142	34612	330*		590*		2,4-DINITROTOLUENE, DISSOLVED	UG/L	General Organic
121142	34613	330*		590*		2,4-DINITROTOLUENE, SUSPENDED	UG/L	General Organic
72548	34651	0.6*		3.6*		P,P'-DDD, DISSOLVED	UG/L	Pesticide
72548	34652	0.6*		3.6*		P,P'-DDD, SUSPENDED	UG/L	Pesticide
72559	34653	1050*		14*		P,P'-DDE, DISSOLVED	UG/L	Pesticide
72559	34654	1050*		14*		P,P'-DDE, SUSPENDED	UG/L	Pesticide
50293	34655	1.1		0.13		P,P'-DDT, DISSOLVED	UG/L	Pesticide

C.A.S. Number	STORET Code	FRESH ACUTE	DRINKING WATER	MARINE ACUTE	OTHER	PARAMETER DESCRIPTION	UNITS	CATEGORY
50293	34656	1.1		0.13		P,P'-DDT, SUSPENDED	UG/L	Pesticide
1746016	34675	0.01*	0.00003			2,3,7,8-TETRACHLORODIBENZO-P-DIOXIN(TCDD), TOT	UG/L	General Organic
1746016	34676	0.01*	0.00003			2,3,7,8-TETRACHLORODIBENZO-P-DIOXIN(TCDD), DISS	UG/L	General Organic
1746016	34677	0.01*	0.00003			2,3,7,8-TETRACHLORODIBENZO-P-DIOXIN(TCDD), SUSP	UG/L	General Organic
108952	34694	10200*		5800*		PHENOL (C6H5OH) - SINGLE COMPOUND, TOTAL	UG/L	General Organic
91203	34696	2300*		2350*		NAPHTHALENE, TOTAL	UG/L	General Organic
75990	38432		200			DALAPON, WATER, TOTAL	UG/L	Pesticide
75990	38433		200			DALAPON, WATER, DISSOLVED	UG/L	Pesticide
75990	38434		200			DALAPON, WATER, SUSPENDED	UG/L	Pesticide
96128	38437		0.2			DIBROMOCHLOROPROPANE, WATER, TOTAL	UG/L	Pesticide
96128	38438		0.2			DIBROMOCHLOROPROPANE, WATER, DISSOLVED	UG/L	Pesticide
96128	38439		0.2			DIBROMOCHLOROPROPANE WATER, SUSPENDED	UG/L	Pesticide
96128	38760		0.2			DBCP, WATER, TOTAL	UG/L	Pesticide
96128	38761		0.2			DBCP, WATER, DISSOLVED	UG/L	Pesticide
96128	38762		0.2			DBCP, WATER, SUSPENDED	UG/L	Pesticide
88857	38779		7.0			DINOSEB, DISSOLVED	UG/L	Pesticide
88857	38780		7.0			DINOSEB, SUSPENDED	UG/L	Pesticide
23135220	38865		200			OXAMYL, TOTAL	UG/L	Pesticide
23135220	38866		200			OXAMYL, DISSOLVED	UG/L	Pesticide
23135220	38867		200			OXAMYL, SUSPENDED	UG/L	Pesticide

C.A.S. Number	STORET Code	FRESH ACUTE	DRINKING WATER	MARINE ACUTE	OTHER	PARAMETER DESCRIPTION	UNITS	CATEGORY
145733	38926		100			ENDOTHALL, WHOLE WATER SAMPLE	UG/L	Pesticide
2921882	38932	0.083*		0.011		CHLORPYRIFOS, TOTAL RECOVERABLE	UG/L	Pesticide
2921882	38933	0.083*		0.011		CHLORPYRIFOS, DISSOLVED	UG/L	Pesticide
2163806	38935		50			MONOSODIUM METHANEARSONATE (MSMA)	UG/L	Pesticide
2921882	39012	0.083*		0.011		DURSBAN, FLAME PHOTOMETRIC, WATER SAMPLE	UG/L	Pesticide
56382	39015	0.065				ETHYLPARATHION, FLAME IONIZATION, WATER SAMPLE	UG/L	Pesticide
122349	39025		4.0			SIMAZINE, COULSON CONDUCTIVITY WATER SAMPLE	UG/L	Pesticide
87865	39032	20***	1.0	13		PCP (PENTACHLOROPHENOL) WHOLE WATER SAMPLE	UG/L	Pesticide
1912249	39033		3.0			ATRAZINE IN WHOLE WATER SAMPLE	UG/L	Pesticide
118741	39039	6.0^p	1.0			HEXACHLOROBENZENE WATER SAMPLE, ELECTRON CPT	UG/L	Pesticide
93721	39045		50			2,4,5-TP INCLUDES ACIDS & SALTS WATER SAMPLE	UG/L	Pesticide
116063	39053		3.0			ALDICARB IN WHOLE WATER	UG/L	Pesticide
122349	39055		4.0			SIMAZINE IN WHOLE WATER	UG/L	Pesticide
117817	39100	2000*	6.0			BIS(2-ETHYLHEXYL) PHTHALATE, WHOLE WATER	UG/L	General Organic
117817	39103	2000*	6.0			BIS(2-ETHYLHEXYL) PHTHALATE, DISSOLVED	UG/L	General Organic
117817	39104	2000*	6.0			BIS(2-ETHYLHEXYL) PHTHALATE, SUSPENDED	UG/L	General Organic
	39117	0.94*		2.994*		PHTHLATE ESTERS IN WATER	MG/L	General Organic
75014	39175		2.0			VINYL CHLORIDE-WHOLE WATER SAMPLE	UG/L	General Organic
79016	39180	45000*	5.0	2000*		TRICHLOROETHYLENE-WHOLE WATER SAMPLE	UG/L	General Organic
50293	39300	1.1		0.13		P,P' DDT IN WHOLE WATER SAMPLE	UG/L	Pesticide

C.A.S. Number	STORET Code	FRESH ACUTE	DRINKING WATER	MARINE ACUTE	OTHER	PARAMETER DESCRIPTION	UNITS	CATEGORY
72548	39310	0.6*		3.6*		P,P' DDD IN WHOLE WATER SAMPLE	UG/L	Pesticide
72559	39320	1050*		14*		P,P' DDE IN WHOLE WATER SAMPLE	UG/L	Pesticide
309002	39330	3.0		1.3		ALDRIN IN WHOLE WATER SAMPLE	UG/L	Pesticide
309002	39331	3.0		1.3		ALDRIN IN FILT. FRAC. OF WAT. SAMP.	UG/L	Pesticide
309002	39332	3.0		1.3		ALDRIN IN SUSP. FRAC. OF WAT. SAMP.	UG/L	Pesticide
58899	39340	2.0		0.16		GAMMA-BHC(LINDANE), WHOLE WATER	UG/L	Pesticide
58899	39341	2.0	0.2	0.16		GAMMA-BHC(LINDANE), DISSOLVED	UG/L	Pesticide
58899	39342	2.0	0.2	0.16		GAMMA-BHC(LINDANE), SUSPENDED	UG/L	Pesticide
57749	39350	2.4	0.2	0.09		CHLORDANE(TECH MIX & METABS), WHOLE WATER	UG/L	Pesticide
57749	39352	2.4	2.0	0.09		CHLORDANE(TECH MIX & METABS), DISSOLVED	UG/L	Pesticide
57749	39353	2.4	2.0	0.09		CHLORDANE(TECH MIX & METABS), SUSPENDED	UG/L	Pesticide
72548	39360	0.6*	2.0	3.6*		DDD IN WHOLE WATER SAMPLE	UG/L	Pesticide
72548	39361	0.6*		3.6*		DDD IN FILT. FRAC. OF WATER SMAPLE	UG/L	Pesticide
72548	39362	0.6*		3.6*		DDD IN SUSP. FRAC. OF WATER SAMPLE	UG/L	Pesticide
72559	39365	1050*		14*		DDE IN WHOLE WATER SAMPLE	UG/L	Pesticide
72559	39366	1050*		14*		DDE IN FILT. FRAC. OF WATER SAMPLE	UG/L	Pesticide
72559	39367	1050*		14*		DDE IN SUSP. FRAC. OF WATER SAMPLE	UG/L	Pesticide
50293	39370	1.1		0.13		DDT IN WHOLE WATER SAMPLE	UG/L	Pesticide
50293	39371	1.1		0.13		DDT IN FILT. FRAC. OF WATER SAMPLE	UG/L	Pesticide
50293	39372	1.1		0.13		DDT IN SUSP. FRAC. OF WATER SAMPLE	UG/L	Pesticide

C.A.S. Number	STORET Code	FRESH ACUTE	DRINKING WATER	MARINE ACUTE	OTHER	PARAMETER DESCRIPTION	UNITS	CATEGORY
60571	39380	2.5		0.71		DIELDRIN IN WHOLE WATER SAMPLE	UG/L	Pesticide
60571	39381	2.5		0.71		DIELDRIN IN FILT. FRAC. OF WATER SAMPLE	UG/L	Pesticide
60571	39382	2.5		0.71		DIELDRIN IN SUSP. FRAC. OF WATER SAMPLE	UG/L	Pesticide
115297	39388	0.22		0.034		ENDOSULFAN IN WHOLE WATER SAMPLE	UG/L	Pesticide
72208	39390	0.18	2.0	0.037		ENDRIN IN WHOLE WATER SAMPLE	UG/L	Pesticide
72208	39391	0.18	2.0	0.037		ENDRIN IN FILT. FRAC. OF WATER SAMPLE	UG/L	Pesticide
72208	39392	0.18	2.0	0.037		ENDRIN IN SUSP. FRAC. OF WATER SAMPLE	UG/L	Pesticide
8001352	39400	0.73	3.0	0.21		TOXAPHENE IN WHOLE WATER SAMPLE	UG/L	Pesticide
8001352	39401	0.73	3.0	0.21		TOXAPHENE IN FILT. FRAC. OF WATER SAMPLE	UG/L	Pesticide
8001352	39402	0.73	3.0	0.21		TOXAPHENE IN SUSP. FRAC. OF WATER SAMPLE	UG/L	Pesticide
76448	39410	0.52	0.4	0.053		HEPTACHLOR IN WHOLE WATER SAMPLE	UG/L	Pesticide
76448	39411	0.52	0.4	0.053		HEPTACHLOR IN FILT. FRAC. OF WATER SAMPLE	UG/L	Pesticide
76448	39412	0.52	0.4	0.053		HEPTACHLOR IN SUSP. FRAC. OF WATER SAMPLE	UG/L	Pesticide
1024573	39420	0.52	0.2	0.053		HEPTACHLOR EPOXIDE IN WHOLE WATER SAMPLE	UG/L	Pesticide
1024573	39421	0.52	0.2	0.053		HEPTACHLOR EPOXIDE IN FILT. FRAC. WATER SAMPLE	UG/L	Pesticide
1024573	39422	0.52	0.2	0.053		HEPTACHLOR EPOXIDE IN SUSP. FRAC. WATER SAMPLE	UG/L	Pesticide
72435	39478		40			METHOXYCHLOR IN WHOLE WATER DISSOLVED	UG/L	Pesticide
72435	39479		40			METHOXYCHLOR IN WHOLE WATER SUSPENDED	UG/L	Pesticide
72435	39480		40			METHOXYCHLOR IN WHOLE WATER SAMPLE	UG/L	Pesticide
56382	39540	0.065				PARATHION IN WHOLE WATER SAMPLE	UG/L	Pesticide

C.A.S. Number	STORET Code	FRESH ACUTE	DRINKING WATER	MARINE ACUTE	OTHER	PARAMETER DESCRIPTION	UNITS	CATEGORY
56382	39542	0.065				PARATHION IN FILT. FRAC. OF WATER SAMPLE	UG/L	Pesticide
56382	39543	0.065				PARATHION IN SUSP. FRAC. OF WATER SAMPLE	UG/L	Pesticide
1912249	39630		3.0			ATRAZINE(AATREX) IN WHOLE WATER SAMPLE	UG/L	Pesticide
1912249	39632		3.0			ATRAZINE DISSOLVED IN WATER	PPB	Pesticide
118741	39700	6.0^D	1.0			HEXACHLOROBENZENE IN WHOLE WATER SAMPLE	UG/L	General Organic
87683	39702	90*		32*		HEXACHLOROBUTADIENE IN WHOLE WATER SAMPLE	UG/L	General Organic
1918021	39720		500			PICLORAM IN WHOLE WATER SAMPLE	UG/L	Pesticide
94757	39730		70			2,4-D IN WHOLE WATER SAMPLE	UG/L	Pesticide
94757	39732		70			2,4-D IN FILT. FRAC. OF WATER SAMPLE	UG/L	Pesticide
94757	39733		70			2,4-D IN SUSP. FRAC. OF WATER SAMPLE	UG/L	Pesticide
93721	39760		50			SILVEX IN WHOLE WATER SAMPLE	UG/L	Pesticide
93721	39762		50			SILVEX IN FILT. FRAC. OF WATER SAMPLE	UG/L	Pesticide
93721	39763		50			SILVEX IN SUSP. FRAC. OF WATER SAMPLE	UG/L	Pesticide
58899	39782	2.0	0.2	0.16		LINDANE IN WHOLE WATER SAMPLE	UG/L	Pesticide
1071836	39941	0.019	700			ROUNDUP IN WHOLE WATER SAMPLE (GLYPHOSATE)	UG/L	Pesticide
7782505	45650			0.013		CHLORINE, IN ORGANIC COMPOUNDS, WATER, WHOLE	MG/L	General Inorganic
56382	46315	0.065				ETHYL PARATHION IN WHOLE WATER SAMPLE	UG/L	Pesticide
58899	46322	2.0	0.2	0.16		LINDANE PLUS ISOMERS IN WHOLE WATER SAMPLE	UG/L	Pesticide
76448	46326	0.52	0.4	0.053		HEPTACHLOR AND METABOLITES IN WHOLE H2O SAMPLE	UG/L	Pesticide
15972608	46342		2.0			ALACHLOR (LASSO), WATER, DISSOLVED	UG/L	Pesticide

C.A.S. Number	STORET Code	FRESH ACUTE	DRINKING WATER	MARINE ACUTE	OTHER	PARAMETER DESCRIPTION	UNITS	CATEGORY
7782505	46472	0.019		0.013		CHLORINE, TOTAL RESIDUAL, AVERAGE VALUE, WATER	MG/L	General Inorganic
7782505	46473	0.019		0.013		CHLORINE, FREE AVAILABLE, AVERAGE VALUE, WATER	MG/L	General Inorganic
57125	46479	22	200	1.0		CYANIDE, DISSOLVED, WATER	UG/L	General Inorganic
7440382	46551	360	50	69		ARSENIC, FIELD ACIDIFIED W/HNO3. LAB FILTERED	UG/L	Metal
7440393	46558		2000			BARIUM, FIELD ACIDIFIED W/HNO3-LAB FILT	UG/L	Metal
7440439	46559	3.9*	5.0	43		CADMIUM,FIELD ACIDIFIED-HNO3-LAB FILTER	UG/L	Metal
7440473	46560		100			CHROMIUM, FIELD ACIDIFIED-HNO3-LAB FILT.	UG/L	Metal
7440508	46562	18*	1300*	2.9		COPPER, FIELD ACIDIFIED-HNO3- LAB FILTER.	UG/L	Metal
7439921	46564	82*	15*	220		LEAD, FIELD ACIDIFIED-HNO3-LAB FILTERED	UG/L	Metal
7440224	46566	4.1*	100*	0.12		SILVER, FIELD ACIDIFIED-HNO3-LAB FILTER.	UG/L	Metal
7440666	46567	120*	5000*	95		ZINC, EXTRACTABLE, FIELD ACID W/HNO3.LAB FILTR	UG/L	Metal
56382	49011	0.065				UNKNOWNS AS PARATHION IN WHOLE WATER SAMPLE	UG/L	Pesticide
7782505	50058	0.019		0.013		CHLORINE DOSE	MG/L	General Inorganic
7782505	50060	0.019		0.013		CHLORINE, TOTAL RESIDUAL	MG/L	General Inorganic
7782505	50064	0.019		0.013		CHLORINE, FREE AVAILABLE	MG/L	General Inorganic
7782505	50066	0.019		0.013		CHLORINE, COMBINED AVAILABLE	MG/L	General Inorganic
7782505	50074	0.019		0.013		CHLORITE, WHOLE WATER	MG/L	General Inorganic
	61215				200	FECAL COLIFORM, GENERAL #/100ML	#/100ML	Bacteriological
16887006	70352	860	250*			CHLORIDE, ORGANIC	MG/L	General Organic
14797558	71850		44			NITRATE NITROGEN, TOTAL (AS NO3)	MG/L	Nitrogen

C.A.S. Number	STORET Code	FRESH ACUTE	DRINKING WATER	MARINE ACUTE	OTHER	PARAMETER DESCRIPTION	UNITS	CATEGORY
14797558	71851		44			NITRATE NITROGEN, DISSOLVED (AS NO3)	MG/L	Nitrogen
14797650	71855		3.3			NITRITE NITROGEN, TOTAL (AS NO2)	MG/L	Nitrogen
14797650	71856		3.3			NITRITE NITROGEN, DISSOLVED (AS NO2)	MG/L	Nitrogen
7439976	71890	2.4	2.0	2.1		MERCURY, DISSOLVED	UG/L	Metal
7439976	71895	2.4	2.0	2.1		MERCURY, SUSPENDED	UG/L	Metal
7439976	71900	2.4	2.0	2.1		MERCURY, TOTAL	UG/L	Metal
7439976	71901	2.4	2.0	2.1		MERCURY,TOTAL RECOVERABLE IN WATER AS HG	UG/L	Metal
7440439	71946	3.9*	5.0	43		CADMIUM, EXTRACTABLE	UG/L	Metal
7440473	71947		100			CHROMIUM, EXTRACTABLE	UG/L	Metal
7439921	71949	82*	15ᵃ	220		LEAD, EXTRACTABLE	UG/L	Metal
7440666	71950	120'	5000ᶜ	95		ZINC, EXTRACTABLE	UG/L	Metal
7440508	71951	18'	1300ᵇ	2.9		COPPER, EXTRACTABLE	UG/L	Metal
1336363	76011	2000	500	10000		PCBS, SUSPENDED, WATER	NG/L	General Organic
1336363	76012	2000	500	10000		PCBS, TOTAL RECOVERABLE, WATER	NG/L	General Organic
156592	77093		70			CIS-1,2-DICHLOROETHYLENE, WHOLE WATER	UG/L	General Organic
100425	77128		100			STYRENE, WHOLE WATER	UG/L	General Organic
106489	77296			29700*		P-CHLOROPHENOL, WHOLE WATER	UG/L	General Organic
106934	77651		0.05			1,2-DIBROMOETHANE, WHOLE WATER	UG/L	General Organic
95954	77687	100ᵉ		240ᵉ		2,4,5-TRICHLOROPHENOL, WHOLE WATER	UG/L	General Organic
935955	77769			440*		2,3,5,6-TETRACHLOROPHENOL, WHOLE WATER	UG/L	General Organic

C.A.S. Number	STORET Code	FRESH ACUTE	DRINKING WATER	MARINE ACUTE	OTHER	PARAMETER DESCRIPTION	UNITS	CATEGORY
103231	77903		400			BIS (2-ETHYLHEXYL) ADIPATE, WHOLE WATER	UG/L	General Organic
18540299	78247	16	100	1100		CHROMIUM, HEXAVALENT, TOTAL RECOVERABLE	UG/L	Metal
57125	78248	22	200	1.0		CYANIDE, TOTAL RECOVERABLE, WATER, WHOLE	UG/L	Metal
	78456	11*		12*		HALOMETHANES, SUMMATION, WHOLE WATER	MG/L	General Organic
14808798	78462		250*			SULFATE, WATER, DISSOLVED AS S	MG/L	Metal
85007	78885		20			DIQUAT DIBROMIDE (REGLONE) WHOLE WATER SAMPLE	UG/L	Pesticide
7440611	80020		20^c			URANIUM, DISS. BY EXTRACTION FLUOROMETRIC	UG/L	Radiological
16065831	80357	1700	100	10300*		CHROMIUM, TRIVALENT, DISSOLVED	UG/L	Metal
57125	81208	0.022	0.2	0.001		CYANIDE,FREE (NOT AMENABLE TO CHLORINATION)	MG/L	General Inorganic
608731	81283	100*		0.34*		BENZENEHEXACHLORIDE, WHOLE WATER	UG/L	Pesticide
88857	81287		7.0			DNBP(C10H12N2O5), WHOLE WATER SAMPLE	UG/L	Pesticide
26638197	81327	23000*	5.0	10300*		DICHLOROPROPANE, WHOLE WATER SAMPLE	UG/L	General Organic
25321226	81333	1120*		1970*		DICHLOROBENZENE ISOMER, WHOLE WATER SAMPLE	UG/L	General Organic
2921882	81403	0.083		0.011		DURSBAN (CHLOROPYRIFOS) WHOLE WATER SAMPLE	UG/L	Pesticide
1563662	81405		40			CARBOFURAN (EURADAN) WHOLE WATER SAMPLE	UG/L	Pesticide
76017	81501	7.240*		390*		PENTACHLOROETHANE, WHOLE WATER SAMPLE	UG/L	General Organic
25321226	81524	1120*		1970*		DICHLOROBENZENE, WHOLE WATER SAMPLE	UG/L	General Organic
25322207	81549	93.20*				TETRACHLOROETHANE, WHOLE WATER SAMPLE	UG/L	General Organic
26638197	81703	23*	0.005*	10.3*		DICHLOROPROPANE, WHOLE WATER SAMPLE	MG/L	General Organic
7440508	81750	18*	1300#	2.9		COPPER, INTERSTITIAL WATERFROM SEDIMENTS	UG/L	Metal

C.A.S. Number	STORET Code	FRESH ACUTE	DRINKING WATER	MARINE ACUTE	OTHER	PARAMETER DESCRIPTION	UNITS	CATEGORY
7440020	81752	1400¹	100	75		NICKEL, INTERSTITIAL WATER FROM SEDIMENTS	UG/L	Metal
7440666	81754	120¹	5000¹	95		ZINC, INTERSTITIAL WATER FROM SEDIMENTS	UG/L	Metal
25323891	81853	18000¹				TRICHLOROETHANE, WHOLE WATER SAMPLE	UG/L	General Organic
7439976	81931	2.4	2.0	2.1		MERCURY (HG) SUSPENDED FRACTION OF WATER	UG/G	Metal
7440666	81933	120¹	5000⁵	95		ZINC (ZN) SUSPENDED FRACTION OF WATER	UG/G	Metal
7439921	81936	82¹	15ᵃ	220		LEAD (PB) DISSOLVED CATIONIC SPECIES	UG/L	Metal
7440439	81937	3.9¹	5.0	43		CADMIUM (CD) DISSOLVED CATIONIC SPECIES	UG/L	Metal
7440473	81938		100			CHROMIUM (CR) DISSOLVED CATIONIC SPECIES	UG/L	Metal
7440508	81939	18¹	1300ᵃ	2.9		COPPER (CU) DISSOLVED CATIONIC SPECIES	UG/L	Metal
7440666	81940	120¹	5000⁵	95		ZINC (ZN) DISSOLVED CATIONIC SPECIES	UG/L	Metal
7440473	81941		100			CHROMIUM (CR) DISSOLVED ANIONIC SPECIES	UG/L	Metal
7440508	81942	18¹	1300ᵃ	2.9		COPPER (CU) DISSOLVED ANIONIC SPECIES	UG/L	Metal
7440666	81943	120¹	5000⁵	95		ZINC (ZN) DISSOLVED ANIONIC SPECIES	UG/L	Metal
	82078				50¹	TURBIDITY, FIELD	NTU	Physical
	82079				50¹	TURBIDITY, LAB	NTU	Physical
88857	82226		7.0			2 SECONDARY BUTYL 4,6-DINITROPHENOL	UG/L	Pesticide
16887006	82295	860000	250000¹			CHLORIDE DISSOLVED AS CL IN WATER	UG/L	General Inorganic
72435	82350		40			METHOXYCHLOR, DISSOLVED IN WATER	UG/L	Pesticide
72435	82351		40			METHOXYCHLOR, SUSPENDED IN WATER	UG/L	Pesticide
115297	82354	0.22		0.034		ENDOSULFAN, DISSOLVED IN WATER	UG/L	Pesticide

C.A.S. Number	STORET Code	FRESH ACUTE	DRINKING WATER	MARINE ACUTE	OTHER	PARAMETER DESCRIPTION	UNITS	CATEGORY
115297	82355	0.22		0.034		ENDOSULFAN, SUSPENDED IN WATER	UG/L	Pesticide
57125	82573	0.022	0.2	0.001		CYANIDE/CHLORINATION IN WATER	MG/L	General Inorganic
1646873	82586		4.0			ALDICARB SULFOXIDE, WATER, TOTAL RECOVERABLE	UG/L	General Organic
1646884	82587		2.0			ALDICARB SULFONE, WHOLE WATER, TOTAL RECOVERABLE	UG/L	General Organic
23135220	82613		200			OXAMYL, WHOLE WATER, TOTAL RECOVERABLE	UG/L	Pesticide
1563662	82615		40			CARBOFURAN, WHOLE WATER, TOTAL RECOVERABLE	UG/L	Pesticide
116063	82619		3.0			ALDICARB, WHOLE WATER, TOTAL RECOVERABLE	UG/L	Pesticide
33213659	82624	0.22		0.034		ENDOSULFAN, BETA, WH WATER, TOTAL RECOVERABLE	UG/L	Pesticide
96128	82625		0.2			DIBROMOCHLOROPROPANE, WATER, TOTAL RECOVERABLE	UG/L	Pesticide
7440382	82702	360	50	69		ARSENIC, FIELD ACIDIFIED, DECANTED, WATER	UG/L	Metal
7440393	82703		2			BARIUM, FIELD ACIDIFIED, DECANTED, WATER	MG/L	Metal
7440417	82704	130*	4.0			BERYLLIUM, FIELD ACIDIFIED, DECANTED, WATER	UG/L	Metal
7440439	82705	3.9+	5.0	43		CADMIUM, FIELD ACIDIFIED, DECANTED, WATER	UG/L	Metal
7440473	82706		100			CHROMIUM, FIELD ACIDIFIED, DECANTED, WATER	UG/L	Metal
7440508	82708	18+	1300a	2.9		COPPER, FIELD ACIDIFIED, DECANTED, WATER	UG/L	Metal
7439921	82711	82+	15a	220		LEAD, FIELD ACIDIFIED, DECANTED, WATER	UG/L	Metal
7439976	82713	2.4	2.0	2.1		MERCURY, FIELD ACIDIFIED, DECANTED, WATER	UG/L	Metal
7440020	82715	1400+	100	75		NICKEL, FIELD ACIDIFIED, DECANTED, WATER	UG/L	Metal
7440224	82716	4.1+	100a	0.12		SILVER, FIELD ACIDIFIED, DECANTED, WATER	UG/L	Metal
7440666	82719	120+	5000a	95		ZINC, FIELD ACIDIFIED, DECANTED, WATER	UG/L	Metal

Footnote Key:

* Insufficient Data to Develop Criteria. Value Presented is the L.O.E.L. - Lowest Observed Effect Level.

+ Hardness Dependent Criteria (100 mg/L CaCO₃ Used).

*** pH Dependent Criteria (7.8 pH Used).

= Rule of thumb criterion used by the NPS Air Quality Division for determining sensitivity to acid deposition.

^ Freshwater bathing criterion, EPA geometric mean based on at least 5 samples equally spaced over a 30-day period; Enterococci marine water bathing criterion 35 CFU/100 ml.

EPA freshwater aquatic life chronic criterion; marine criterion is ≤6.5, ≥8.5.

' Arizona state standard.

a EPA action level, 40 CFR 141.80.

b California and Florida state bathing water standards.

c A Compilation of Water Quality Goals, California Regional Water Quality Control Board Central Valley Region, Sacramento, California, September, 1991.

m Total coliform drinking water maximum contaminant level (1 cfu/100ml or 1 mpn/100ml) was not used in water quality criteria comparisons.

p Proposed Criterion.

r Average annual concentration assumed to produce a total body or organ dose of 4 mrem/year, 40 CFR 141.16.

s EPA National Secondary Drinking Water Regulation, 40 CFR 143.

t The maximum contaminant level for the sum of the concentrations of trihalomethanes is 100 μg/L, 40 CFR 141.12.

u Coldwater criterion one day minimum; warmwater criterion seven day mean minimum.

F-27

Appendix G

Inventory Data Evaluation and Analysis (IDEA)
Servicewide Inventory and Monitoring Program "Level I"
Parameter Groups

The following table provides the Servicewide Inventory and Monitoring Program's "Level I" water quality inventory parameter groups (National Park Service 1993). In order to determine the presence and/or absence of data for each of these parameter groups in the park. the parameter groups had to be defined by STORET parameter codes. This table provides the STORET codes and parameter descriptions for each parameter comprising one of the Servicewide Inventory and Monitoring Program's "Level I" water quality parameter groups. Additional parameters could have been incorporated into each group. but an effort was made to represent each group with the parameters deemed to most likely occur in STORET and parks. The Toxic Elements Parameter Group was defined as the EPA's Clean Water Act Section 304(a) Priority Toxic Pollutants (40 CFR 131.36). Parameters are listed in ascending order of STORET code within each parameter group. It is important to note that similar parameters often have non-consecutive codes. Consequently. scanning the entire list is necessary to find all the parameters of a particular type (eg. lead. copper. etc.). Refer to the Parameter Period of Record Tabulation to obtain the STORET code for any parameter measured in the park.

STORET Code	Water Temperature Parameter Group	C.A.S. Number
00010	TEMPERATURE. WATER (DEGREES CENTIGRADE)	-
00011	TEMPERATURE. WATER (DEGREES FAHRENHEIT)	-

STORET Code	Flow Parameter Group[1]	C.A.S. Number
00056	FLOW RATE. GALLONS/DAY	-
00058	FLOW RATE. GALLONS/MIN.	-
00059	FLOW RATE. INSTANTANEOUS. GALLONS/MINUTE	-
00060	FLOW. STREAM. MEAN DAILY CFS	-
00061	FLOW. STREAM. INSTANTANEOUS CFS	-
00065	STAGE. STREAM (FEET)	-
00067	TIDE STAGE CODE	-
00072	STAGE. STREAM (METERS)	-

---

[1]Tide stage is included in the Flow Parameter Group for coastal parks.

STORET Code	Clarity/Turbidity Parameter Group	C.A.S. Number
00070	TURBIDITY. (JACKSON CANDLE UNITS)	-
00075	TURBIDITY. HELLIGE (PPM AS SILICON DIOXIDE)	-
00076	TURBIDITY. HACH TURBIDIMETER (FORMAZIN TURB UNIT)	-
00077	TRANSPARENCY. SECCHI DISC (INCHES)	-
00078	TRANSPARENCY. SECCHI DISC (METERS)	-
00530	RESIDUE. TOTAL NONFILTRABLE (MG/L)	-
82078	TURBIDITY. FIELD NEPHELOMETRIC TURBIDITY UNITS NTU	-
82079	TURBIDITY. LAB NEPHELOMETRIC TURBIDITY UNITS. NTU	-

STORET Code	Conductivity Parameter Group	C.A.S. Number
00094	SPECIFIC CONDUCTANCE. FIELD (UMHOS/CM @ 25C)	-
00095	SPECIFIC CONDUCTANCE (UMHOS/CM @ 25C)	-
00096	SALINITY AT 25 DEGREES C (MG/ML)	-
00480	SALINITY - PARTS PER THOUSAND	-

STORET Code	Dissolved Oxygen Parameter Group	C.A.S. Number
00299	OXYGEN. DISSOLVED. ANALYSIS BY PROBE (MG/L)	7782447
00300	OXYGEN. DISSOLVED (MG/L)	7782447
00301	OXYGEN. DISSOLVED. PERCENT OF SATURATION	7782447
00389	OXYGEN. DISSOLVED. LAB ANAL. BY PROBE OF FIELD SAMPLE (MG/L)	7782447

STORET Code	pH Parameter Group	C.A.S. Number
00400	PH (STANDARD UNITS)	-
00403	PH. LAB (STANDARD UNITS)	-
00406	PH. FIELD (STANDARD UNITS)	-

STORET Code	Alkalinity Parameter Group	C.A.S. Number
00409	ALKALINITY. TOTAL. LOW LEVEL GRAN ANALYSIS (µEQ/L)	471341
00410	ALKALINITY. TOTAL (MG/L AS CACO3)	471341
00415	ALKALINITY. PHENOLPHTHALEIN (MG/L)	77098
00430	ALKALINITY. CARBONATE (MG/L AS CACO3)	471341
00435	ACIDITY. TOTAL (MG/L AS CACO3)	471341
00440	BICARBONATE ION (MG/L AS HCO3)	71523
00445	CARBONATE ION (MG/L AS CO3)	3812326

STORET Code	Nitrate/Nitrogen Parameter Group	C.A.S. Number
00600	NITROGEN. TOTAL (MG/L AS N)	17778880
00602	NITROGEN. DISSOLVED (MG/L AS N)	17778880
00605	NITROGEN. ORGANIC. TOTAL (MG/L AS N)	17778880
00607	NITROGEN. ORGANIC. DISSOLVED (MG/L AS N)	17778880
00608	NITROGEN. AMMONIA. DISSOLVED (MG/L AS N)	17778880
00610	NITROGEN. AMMONIA. TOTAL (MG/L AS N)	17778880
00612	AMMONIA. UNIONZED (MG/L AS N)	7664417
00618	NITRATE NITROGEN. DISSOLVED (MG/L AS N)	17778880
00620	NITRATE NITROGEN. TOTAL (MG/L AS N)	17778880
00623	NITROGEN. KJELDAHL. DISSOLVED (MG/L AS N)	17778880
00625	NITROGEN. KJELDAHL. TOTAL (MG/L AS N)	17778880
00630	NITRITE PLUS NITRATE. TOTAL 1 DET. (MG/L AS N)	17778880
00631	NITRITE PLUS NITRATE. DISSOLVED 1 DET. (MG/L AS N)	17778880
71845	NITROGEN. AMMONIA. TOTAL (MG/L AS NH4)	14798039
71846	NITROGEN. AMMONIA. DISSOLVED (MG/L AS NH4)	14798039
71850	NITRATE NITROGEN. TOTAL (MG/L AS NO3)	14797558
71851	NITRATE NITROGEN. DISSOLVED (MG/L AS NO3)	14797558
71855	NITRITE NITROGEN. TOTAL (MG/L AS NO2)	14797650
71856	NITRITE NITROGEN. DISSOLVED (MG/L AS NO2)	14797650

STORET Code	Phosphate/Phosphorus Parameter Group	C.A.S. Number
00650	PHOSPHATE. TOTAL (MG/L AS PO4)	14265442
00655	PHOSPHATE. POLY (MG/L AS PO4)	14265442
00660	PHOSPHATE. ORTHO (MG/L AS PO4)	14265442
00665	PHOSPHORUS. TOTAL (MG/L AS P)	7723140
00666	PHOSPHORUS. DISSOLVED (MG/L AS P)	7723140
00670	PHOSPHORUS. TOTAL ORGANIC (MG/L AS P)	7723140
00671	PHOSPHORUS. DISSOLVED ORTHOPHOSPHATE (MG/L AS P)	7723140
70505	PHOSPHORUS. TOTAL. COLORIMETRIC METHOD (MG/L AS P)	7723140
70507	PHOSPHORUS. IN TOTAL ORTHOPHOSPHATE (MG/L AS P)	7723140

STORET Code	Sulfates/Total Dissolved Solids/Hardness Parameter Group	C.A.S. Number
00900	HARDNESS. TOTAL (MG/L AS CACO3)	471341
00945	SULFATE. TOTAL (MG/L AS SO4)	14808798
00946	SULFATE. DISSOLVED (MG/L AS SO4)	14808798
70300	RESIDUE. TOTAL FILTRABLE (DRIED AT 180C). (MG/L)	-

STORET Code	Chlorophyll Parameter Group	C.A.S. Number
32209	CHLOROPHYLL A (UG/L) FLUOROMETRIC CORRECTED	479618
32210	CHLOROPHYLL A (UG/L) TRICHROMATIC UNCORRECTED	479618
32211	CHLOROPHYLL A (UG/L) SPECTROPHOTOMETRIC ACID METH.	479618
32217	CHLOROPHYLL A (UG/L) FLUOROMETRIC UNCORRECTED	479618
32223	CHLOROPHYLL A (MG/M2) SPECTROPHOTOMETRIC CORRECTED	479618
32228	CHLOROPHYLL A (MG/M2) PERIPHYTON SPECTRO.	479618
32229	CHLOROPHYLL A (MG/M2) FLUOR. CORRECTED. SUBSTRATER	479618
32230	CHLOROPHYLL A (MG/L)	479618

STORET Code	Bacteria Parameter Group	C.A.S. Number
00111	RATIO OF FECAL COLIFORM TO FECAL STREPTOCOCCI	-
31501	COLIFORM. TOT. MEMBRANE FILTER. IMMED.. M-ENDO MED.35C	-
31503	COLIFORM. TOT. MEMBRANE FILTER. DELAY. M-ENDO MED. 35C	-
31504	COLIFORM. TOT. MEMBRANE FILTER. IMMED.. LES-ENDO AGAR. 35C	-
31505	COLIFORM. TOT. MPN. CONFIRMED TEST.35C(TUBE 31506)	-
31506	COLIFORM. TOT. MPN. CONFIRMED TEST. TUBE CONFIG.	-
31507	COLIFORM. TOT. MPN. COMPLETED TEST.35C(TUBE 31508)	-
31508	COLIFORM. TOT. MPN. COMPLETED TEST. TUBE CONFIG.	-
31613	FECAL COLIFORM. MEMBR. FILTER.M-FC AGAR.44.5C.24HR	-
31614	FECAL COLIFORM. MPN. TUBE CONFIGURATION	-
31615	FECAL COLIFORM. MPN. EC MED. 44.5C (TUBE 31614)	-
31616	FECAL COLIFORM. MEMBR FILTER. M-FC BROTH. 44.5C	-
31617	FECAL COLIFORM. MPN.EIJKMAN TEST.44.5C(TUBE 31618)	-
31625	FECAL COLIFORM. MF. M-FC. 0.7 UM	-
31648	E. COLI - MTEC-MF	-
31649	ENTEROCOCCI- ME-MF	-
31673	FECAL STREPTOCOCCI. MBR FILT. KF AGAR. 35C. 48HR	-
31676	FECAL STREPTOCOCCI. MPN. KF BROTH. TUBE CONFIG.	-
31677	FECAL STREPTOCOCCI. MPN. AD-EVA. 35C (TUBE 31678)	-
31751	PLATE COUNT. TOTAL. TPC AGAR. 35C. 24 HRS	-
61214	FECAL STREPTOCOCCI. GENERAL #/100ML	-
61215	FECAL COLIFORM. GENERAL #/100ML	-

STORET Code	Toxic Elements (EPA Section 304(a) Priority Toxic Pollutants)	C.A.S. Number
00718	CYANIDE. WEAK ACID. DISSOC. WATER. WHOLE (UG/L)	57125
00719	CYANIDE. FREE. IN WATER & WASTEWATERS. HBG (UG/L)	57125
00720	CYANIDE. TOTAL (MG/L AS CN)	57125
00722	CYANIDE. FREE (AMENABLE TO CHLORINATION) (MG/L)	57125

STORET Code	Toxic Elements (EPA Section 304(a) Priority Toxic Pollutants) cont.-	C.A.S. Number
00723	CYANIDE. DISSOLVED STD METHOD (UG/L)	57125
00724	CYANIDE COMPLEXED TO A RANGE OF COMPNDS (UG/L)	57125
00969	CHRYSOTILE ASBESTOS FIBERS/LITER	1332214
00973	AMPHIBOLE ASBESTOS FIBERS/LITER	1332214
00976	AMBIGUOUS ASBESTOS FIBERS/LITER	1332214
00977	NON-AMPHIBOLE NON-CHRYSOTILE ASBESTOS FIBERS/LITER	1332214
00978	ARSENIC. TOTAL RECOVERABLE IN WATER AS AS	7440382
00981	SELENIUM. TOTAL RECOVERABLE IN WATER AS SE (UG/L)	7782492
00982	THALLIUM. TOTAL RECOVERABLE IN WATER AS (UG/L)	7440280
00990	SELENITE. TOTAL RECOVERABLE INORGANIC (UG/L)	7782492
00991	ARSENIC. TOTAL RECOVER. TRIVALENT INORGANIC (UG/L)	7440382
00995	ARSENIC. INORGANIC DISSOLVED (UG/L AS AS)	7440382
00996	ARSENIC. INORGANIC SUSPENDED (UG/L AS AS)	7440382
00997	ARSENIC. INORGANIC TOTAL (UG/L AS AS)	7440382
00998	BERYLLIUM. TOTAL RECOVERABLE IN WATER AS BE (UG/L)	7440417
01000	ARSENIC. DISSOLVED (UG/L AS AS)	7440382
01001	ARSENIC. SUSPENDED (UG/L AS AS)	7440382
01002	ARSENIC. TOTAL (UG/L AS AS)	7440382
01010	BERYLLIUM. DISSOLVED (UG/L AS BE)	7440417
01011	BERYLLIUM. SUSPENDED (UG/L AS BE)	7440417
01012	BERYLLIUM. TOTAL (UG/L AS BE)	7440417
01025	CADMIUM. DISSOLVED (UG/L AS CD)	7440439
01026	CADMIUM. SUSPENDED (UG/L AS CD)	7440439
01027	CADMIUM. TOTAL (UG/L AS CD)	7440439
01030	CHROMIUM. DISSOLVED (UG/L AS CR)	7440473
01031	CHROMIUM. SUSPENDED (UG/L AS CR)	7440473
01032	CHROMIUM. HEXAVALENT (UG/L AS CR)	7440473
01033	CHROMIUM. TRI-VAL (UG/L AS CR)	16065831
01034	CHROMIUM. TOTAL (UG/L AS CR)	7440473

STORET Code	Toxic Elements (EPA Section 304(a) Priority Toxic Pollutants) cont.-	C.A.S. Number
01040	COPPER. DISSOLVED (UG/L AS CU)	7440508
01041	COPPER. SUSPENDED (UG/L AS CU)	7440508
01042	COPPER. TOTAL (UG/L AS CU)	7440508
01049	LEAD. DISSOLVED (UG/L AS PB)	7439921
01050	LEAD. SUSPENDED (UG/L AS PB)	7439921
01051	LEAD. TOTAL (UG/L AS PB)	7439921
01057	THALLIUM. DISSOLVED (UG/L AS TL)	7440280
01058	THALLIUM. SUSPENDED (UG/L AS TL)	7440280
01059	THALLIUM. TOTAL (UG/L AS TL)	7440280
01065	NICKEL. DISSOLVED (UG/L AS NI)	7440020
01066	NICKEL. SUSPENDED (UG/L AS NI)	7440020
01067	NICKEL. TOTAL (UG/L AS NI)	7440020
01074	NICKEL. TOTAL RECOVERABLE IN WATER AS NI (UG/L)	7440020
01075	SILVER. DISSOLVED (UG/L AS AG)	7440224
01076	SILVER. SUSPENDED (UG/L AS AG)	7440224
01077	SILVER. TOTAL (UG/L AS AG)	7440224
01079	SILVER. TOTAL RECOVERABLE IN WATER AS AG (UG/L)	7440224
01089	COPPER AS SUSPENDED BLACK OXIDE IN WATER (MG/L)	7440508
01090	ZINC. DISSOLVED (UG/L AS ZN)	7440666
01091	ZINC. SUSPENDED (UG/L ZN)	7440666
01092	ZINC. TOTAL (UG/L AS ZN)	7440666
01094	ZINC. TOTAL RECOVERABLE IN WATER AS ZN (UG/L)	7440666
01095	ANTIMONY. DISSOLVED (UG/L AS SB)	7440360
01096	ANTIMONY. SUSPENDED (UG/L AS SB)	7440360
01097	ANTIMONY. TOTAL (UG/L AS SB)	7440360
01113	CADMIUM. TOTAL RECOVERABLE IN WATER AS CD (UG/L)	7440439
01114	LEAD. TOTAL RECOVERABLE IN WATER AS PB (UG/L)	7439921
01118	CHROMIUM. TOTAL RECOVERABLE IN WATER AS CR (UG/L)	7440473
01119	COPPER.TOTAL RECOVERABLE IN WATER AS CU (UG/L)	7440508

STORET Code	Toxic Elements (EPA Section 304(a) Priority Toxic Pollutants) cont.-	C.A.S. Number
01124	THALLIUM. ACID SOLUBLE. WATER. WHOLE (UG/L)	7440280
01128	THALLIUM.TOTAL RECOVERABLE <95%, UG/L AS TL	7440280
01138	SELENIUM. IN WATER. LBS/DAY	7782492
01145	SELENIUM. DISSOLVED (UG/L AS SE)	7782492
01146	SELENIUM. SUSPENDED (UG/L AS SE)	7782492
01147	SELENIUM. TOTAL (UG/L AS SE)	7782492
01167	SELENIUM. ACID SOLUBLE. WATER, WHOLE (UG/L)	7782492
01220	CHROMIUM. HEXAVALENT. DISSOLVED IN (UG/L AS CR)	18540299
01252	ARSENIC, LB/DAY/CFS STREAM FLOW	7440382
01253	CADMIUM. LB/DAY/CFS STREAM FLOW	7440439
01254	CHROMIUM. TOTAL (LBS/DAY/CFS STREAM FLOW)	7740473
01255	CHROMIUM. HEXAVALENT. LB/DAY/CFS STREAM FLOW	18540299
01256	COPPER. LB/DAY/CFS STREAM FLOW	7440508
01257	CYANIDE LB/DAY/CFS STREAM FLOW	57125
01259	LEAD. LB/DAY/CFS STREAM FLOW	7439921
01260	MERCURY. LB/DAY/CFS STREAM FLOW	7439976
01261	NICKEL, LB/DAY/CFS STREAM FLOW	7440020
01263	SILVER. LB/DAY/CFS STREAM FLOW	7440224
01264	ZINC LB/DAY/CFS STREAM FLOW	7440666
01268	ANTIMONY, (SB), WATER. TOTAL RECOVERABLE (UG/L)	7440360
01291	CYANIDE, FILTERABLE. TOTAL IN WATER (UG/L)	57125
01303	ZINC. POTENTIALLY DISSOLVED WATER (MG/L)	7440666
01304	SILVER. POTENTIALLY DISSOLVED WATER (MG/L)	7440224
01306	COPPER. POTENTIALLY DISSOLVED WATER (MG/L)	7440508
01307	CHROMIUM. HEXAVALENT, POTENT. DISS. WATER (MG/L)	18540299
01309	ARSENIC, POTENTIALLY, DISSOLVED. WATER (MG/L)	7440382
01312	BERYLLIUM. POTENTIALLY. DISSOLVED, WATER (MG/L)	7440417
01313	CADMIUM. POTENTIALLY. DISSOLVED. WATER (MG/L)	7440439

STORET Code	Toxic Elements (EPA Section 304(a) Priority Toxic Pollutants) cont.-	C.A.S. Number
01314	CHROMIUM. TRIVALENT. POTENT.. DISS.. WATER (MG/L)	16065831
01318	LEAD. POTENTIALLY. DISSOLVED. WATER (MG/L)	7439921
01321	MERCURY. POTENTIALLY. DISSOLVED. WATER (MG/L)	7439976
01322	NICKEL. POTENTIALLY. DISSOLVED. WATER (MG/L)	7440020
01323	SELENIUM. POTENTIALLY. DISSOLVED. WATER (MG/L)	7782492
01324	THALLIUM. POTENTIALLY. DISSOLVED. WATER (MG/L)	7440280
01523	SILVER. IONIC (UG/L)	7440224
22675	SELENIUM. DISSOLVED ORGANIC (UG/L)	7782492
22676	SELENIUM. HEXAVALENT. DISSOLVED (UG/L)	7782492
22677	SELENIUM. TETRAVALENT. DISSOLVED	7782492
22678	ARSENIC. DISSOLVED ORGANIC (UG/L)	7440382
22679	ARSENIC. PENTAVALENT. DISSOLVED (UG/L)	7440382
22680	ARSENIC. TRIVALENT. DISSOLVED (UG/L)	7440382
30197	2-CHLOROETHYLVINYL ETHER.WATER.WHL.RECOVER (UG/L)	110758
30201	CHLOROMETHANE. WATER. WHOLE. RECOVERABLE (UG/L)	74873
30202	BROMOMETHANE. WATER. WHOLE. RECOVERABLE (UG/L)	74839
32003	CARBON CHLOROFORM AND CARBON ALCOHOL EXT. (UG/L)	67663
32005	CARBON CHLOROFORM EXTRACTABLES (UG/L)	67663
32021	CARBON CHLOROFORM EXTRACTS. ETHER INSOLUBLE (UG/L)	67663
32022	CARBON CHLOROFORM EXTRACTS. WATER SOLUBLES (UG/L)	67663
32101	BROMODICHLOROMETHANE. WHOLE WATER (UG/L)	75274
32102	CARBON TETRACHLORIDE. WHOLE WATER. (UG/L)	56235
32103	1.2-DICHLOROETHANE. WHOLE WATER (UG/L)	107062
32104	BROMOFORM. WHOLE WATER. (UG/L)	75252
32105	DIBROMOCHLOROMETHANE. WHOLE WATER. (UG/L)	124481
32106	CHLOROFORM. WHOLE WATER (UG/L)	67663
32260	CARBON TETRACHLORIDE EXTRACTABLES (MG/L)	56235
32270	CHLOROFORM EXTRACTABLES TOTAL IN MG PER LITER	67663

STORET Code	Toxic Elements (EPA Section 304(a) Priority Toxic Pollutants) cont.-	C.A.S. Number
34010	TOLUENE IN WTR SMPLE GC-MS. HEXADECONE EXT. (UG/L)	108883
34030	BENZENE IN WTR SMPLE GC-MS. HEXADECONE EXT. (UG/L)	71432
34198	BHC-DELTA. WATER. WHOLE (LBS/DAY)	319868
34200	ACENAPHTHYLENE. TOTAL (UG/L)	208968
34201	ACENAPHTHYLENE. DISSOLVED (UG/L)	208968
34202	ACENAPHTHYLENE. SUSPENDED (UG/L)	208968
34205	ACENAPHTHENE. TOTAL (UG/L)	83329
34206	ACENAPHTHENE. DISSOLVED (UG/L)	83329
34207	ACENAPHTHENE. SUSPENDED (UG/L)	83329
34210	ACROLEIN. TOTAL (UG/L)	107028
34211	ACROLEIN. DISSOLVED (UG/L)	107028
34212	ACROLEIN. SUSPENDED (UG/L)	107028
34215	ACRYLONITRILE. TOTAL (UG/L)	107131
34216	ACRYLONITRILE. DISSOLVED (UG/L)	107131
34217	ACRYLONITRILE. SUSPENDED (UG/L)	107131
34220	ANTHRACENE. TOTAL (UG/L)	120127
34221	ANTHRACENE. DISSOLVED (UG/L)	120127
34222	ANTHRACENE. SUSPENDED (UG/L)	120127
34225	ASBESTOS (FIBROUS) TOTAL (UG/L)	1332214
34226	ASBESTOS (FIBROUS) DISSOLVED (UG/L)	1332214
34227	ASBESTOS (FIBROUS) SUSPENDED (UG/L)	1332214
34230	BENZO(B)FLUORANTHENE. WHOLE WATER (UG/L)	205992
34231	BENZO(B)FLUORANTHENE. DISSOLVED (UG/L)	205992
34232	BENZO(B)FLUORANTHENE. SUSPENDED (UG/L)	205992
34235	BENZENE. DISSOLVED (UG/L)	71432
34236	BENZENE. SUSPENDED (UG/L)	71432
34239	BENZIDINE. DISSOLVED (UG/L)	92875
34240	BENZIDINE. SUSPENDED (UG/L)	92875

STORET Code	Toxic Elements (EPA Section 304(a) Priority Toxic Pollutants) cont.-	C.A.S. Number
34242	BENZO(K)FLUORANTHENE. TOTAL (UG/L)	207089
34243	BENZO(K)FLUORANTHENE. DISSOLVED (UG/L)	207089
34244	BENZO(K)FLUORANTHENE. SUSPENDED (UG/L)	207089
34247	BENZO-A-PYRENE. TOTAL (UG/L)	50328
34248	BENZO-A-PYRENE. DISSOLVED (UG/L)	50328
34249	BENZO-A-PYRENE. SUSPENDED (UG/L)	50328
34253	A-BHC-ALPHA. DISSOLVED (UG/L)	319846
34254	A-BHC-ALPHA. SUSPENDED (UG/L)	319846
34255	B-BHC-BETA. DISSOLVED (UG/L)	319857
34256	B-BHC-BETA. SUSPENDED (UG/L)	319857
34259	DELTA BENZENE HEXACHLORIDE. TOTAL (UG/L)	319868
34260	DELTA BENZENE HEXACHLORIDE. DISSOLVED (UG/L)	319868
34261	DELTA BENZENE HEXACHLORIDE. SUSPENDED (UG/L)	319868
34265	R-BHC (LINDANE) GAMMA. DISSOLVED (UG/L)	58899
34266	R-BHC (LINDANE) GAMMA. SUSPENDED (UG/L)	58899
34273	BIS (2-CHLOROETHYL) ETHER. TOTAL (UG/L)	111444
34274	BIS (2-CHLOROETHYL) ETHER. DISSOLVED (UG/L)	111444
34275	BIS (2-CHLOROETHYL) ETHER. SUSPENDED (UG/L)	111444
34278	BIS (2-CHLOROETHOXY) METHANE. TOTAL (UG/L)	111911
34279	BIS (2-CHLOROETHOXY) METHANE. DISSOLVED (UG/L)	111911
34280	BIS (2-CHLOROETHOXY) METHANE. SUSPENDED (UG/L)	111911
34288	BROMOFORM. DISSOLVED (UG/L)	75252
34289	BROMOFORM. SUSPENDED (UG/L)	75252
34292	N-BUTYL BENZYL PHTHALATE. WHOLE WATER (UG/L)	85687
34293	N-BUTYL BENZYL PHTHALATE. DISSOLVED (UG/L)	85687
34294	N-BUTYL BENZYL PHTHALATE. SUSPENDED (UG/L)	85687
34297	CARBON TETRACHLORIDE. DISSOLVED (UG/L)	56235
34298	CARBON TETRACHLORIDE. SUSPENDED (UG/L)	56235

STORET Code	Toxic Elements (EPA Section 304(a) Priority Toxic Pollutants) cont.-	C.A.S. Number
34301	CHLOROBENZENE. TOTAL (UG/L)	108907
34302	CHLOROBENZENE. DISSOLVED (UG/L)	108907
34303	CHLOROBENZENE. SUSPENDED (UG/L)	108907
34306	CHLORODIBROMOMETHANE. TOTAL (UG/L)	124481
34307	CHLORODIBROMOMETHANE. DISSOLVED (UG/L)	124481
34308	CHLORODIBROMOMETHANE. SUSPENDED (UG/L)	124481
34311	CHLOROETHANE. TOTAL (UG/L)	75003
34312	CHLOROETHANE. DISSOLVED (UG/L)	75003
34313	CHLOROETHANE. SUSPENDED (UG/L)	75003
34316	CHLOROFORM. DISSOLVED (UG/L)	67663
34317	CHLOROFORM. SUSPENDED (UG/L)	67663
34320	CHRYSENE. TOTAL (UG/L)	218019
34321	CHRYSENE. DISSOLVED (UG/L)	218019
34322	CHRYSENE. SUSPENDED (UG/L)	218019
34325	CYANIDE. SUSPENDED (MG/L)	57125
34327	DI-N-BUTYL PHTHALATE. DISSOLVED (UG/L)	84742
34328	DICHLOROBROMOMETHANE. DISSOLVED (UG/L)	75274
34329	DICHLOROBROMOMETHANE. SUSPENDED (UG/L)	75274
34336	DIETHYL PHTHALATE. TOTAL (UG/L)	84662
34337	DIETHYL PHTHALATE. DISSOLVED (UG/L)	84662
34338	DIETHYL PHTHALATE. SUSPENDED (UG/L)	84662
34341	DIMETHYL PHTHALATE. TOTAL (UG/L)	131113
34342	DIMETHYL PHTHALATE. DISSOLVED (UG/L)	131113
34343	DIMETHYL PHTHALATE. SUSPENDED (UG/L)	131113
34346	1.2-DIPHENYLHYDRAZINE. TOTAL (UG/L)	122667
34347	1.2-DIPHENYLHYDRAZINE. DISSOLVED (UG/L)	122667
34348	1.2-DIPHENYLHYDRAZINE. SUSPENDED (UG/L)	122667
34351	ENDOSULFAN SULFATE. TOTAL (UG/L)	1031078

STORET Code	Toxic Elements (EPA Section 304(a) Priority Toxic Pollutants) cont.-	C.A.S. Number
34352	ENDOSULFAN SULFATE. DISSOLVED (UG/L)	1031078
34353	ENDOSULFAN SULFATE. SUSPENDED (UG/L)	1031078
34356	ENDOSULFAN. BETA. TOTAL (UG/L)	33213659
34357	ENDOSULFAN. BETA. DISSOLVED (UG/L)	33213659
34358	ENDOSULFAN. BETA. SUSPENDED (UG/L)	33213659
34361	ENDOSULFAN. ALPHA. TOTAL (UG/L)	959988
34362	ENDOSULFAN. ALPHA. DISSOLVED (UG/L)	959988
34363	ENDOSULFAN. ALPHA. SUSPENDED (UG/L)	959988
34371	ETHYLBENZENE. TOTAL (UG/L)	100414
34372	ETHYLBENZENE. DISSOLVED (UG/L)	100414
34373	ETHYLBENZENE. SUSPENDED (UG/L)	100414
34376	FLUORANTHENE. TOTAL (UG/L)	206440
34377	FLUORANTHENE. DISSOLVED (UG/L)	206440
34378	FLUORANTHENE. SUSPENDED (UG/L)	206440
34381	FLUORENE. TOTAL (UG/L)	86737
34382	FLUORENE. DISSOLVED (UG/L)	86737
34383	FLUORENE. SUSPENDED (UG/L)	86737
34386	HEXACHLOROCYCLOPENTADIENE. TOTAL (UG/L)	77474
34387	HEXACHLOROCYCLOPENTADIENE. DISSOLVED (UG/L)	77474
34388	HEXACHLOROCYCLOPENTADIENE. SUSPENDED (UG/L)	77474
34391	HEXACHLOROBUTADIENE. TOTAL (UG/L)	87683
34392	HEXACHLOROBUTADIENE. DISSOLVED (UG/L)	87683
34393	HEXACHLOROBUTADIENE. SUSPENDED (UG/L)	87683
34396	HEXACHLOROETHANE. TOTAL (UG/L)	67721
34397	HEXACHLOROETHANE. DISSOLVED (UG/L)	67721
34398	HEXACHLOROETHANE. SUSPENDED (UG/L)	67721
34401	HEXACHLOROBENZENE. DISSOLVED (UG/L)	118741
34402	HEXACHLOROBENZENE. SUSPENDED (UG/L)	118741

STORET Code	Toxic Elements (EPA Section 304(a) Priority Toxic Pollutants) cont.-	C.A.S. Number
34403	INDENO (1.2.3-CD) PYRENE. TOTAL (UG/L)	193395
34404	INDENO (1.2.3-CD) PYRENE. DISSOLVED (UG/L)	193395
34405	INDENO (1.2.3-CD) PYRENE. SUSPENDED (UG/L)	193395
34408	ISOPHORONE. TOTAL (UG/L)	78591
34409	ISOPHORONE. DISSOLVED (UG/L)	78591
34410	ISOPHORONE. SUSPENDED (UG/L)	78591
34413	METHYL BROMIDE. TOTAL (UG/L)	74839
34414	METHYL BROMIDE. DISSOLVED (UG/L)	74839
34415	METHYL BROMIDE. SUSPENDED (UG/L)	74839
34418	METHYL CHLORIDE. TOTAL (UG/L)	74873
34419	METHYL CHLORIDE. DISSOLVED (UG/L)	74873
34420	METHYL CHLORIDE. SUSPENDED (UG/L)	74873
34423	METHYLENE CHLORIDE. TOTAL (UG/L)	75092
34424	METHYLENE CHLORIDE. DISSOLVED (UG/L)	75092
34425	METHYLENE CHLORIDE. SUSPENDED (UG/L)	75092
34428	N-NITROSODI-N-PROPYLAMINE. TOTAL (UG/L)	621647
34429	N-NITROSODI-N-PROPYLAMINE. DISSOLVED (UG/L)	621647
34430	N-NITROSODI-N-PROPYLAMINE. SUSPENDED (UG/L)	621647
34433	N-NITROSODIPHENYLAMINE. TOTAL (UG/L)	86306
34434	N-NITROSODIPHENYLAMINE. DISSOLVED (UG/L)	86306
34435	N-NITROSODIPHENYLAMINE. SUSPENDED (UG/L)	86306
34438	N-NITROSODIMETHYLAMINE. TOTAL (UG/L)	62759
34439	N-NITROSODIMETHYLAMINE. DISSOLVED (UG/L)	62759
34440	N-NITROSODIMETHYLAMINE. SUSPENDED (UG/L)	62759
34443	NAPHTHALENE. DISSOLVED (UG/L)	91203
34444	NAPHTHALENE. SUSPENDED (UG/L)	91203
34447	NITROBENZENE. TOTAL (UG/L)	98953
34448	NITROBENZENE. DISSOLVED (UG/L)	98953

STORET Code	Toxic Elements (EPA Section 304(a) Priority Toxic Pollutants) cont.-	C.A.S. Number
34449	NITROBENZENE. SUSPENDED (UG/L)	98953
34452	PARACHLOROMETA CRESOL. TOTAL (UG/L)	59507
34453	PARACHLOROMETA CRESOL. DISSOLVED (UG/L)	59507
34454	PARACHLOROMETA CRESOL. SUSPENDED (UG/L)	59507
34457	PCB - 1242. DISSOLVED (UG/L)	53469219
34458	PCB - 1242. SUSPENDED (UG/L)	53469219
34459	PCP (PENTACHLOROPHENOL). DISSOLVED (UG/L)	87865
34460	PCP (PENTACHLOROPHENOL). SUSPENDED (UG/L)	87865
34461	PHENANTHRENE. TOTAL (UG/L)	85018
34462	PHENANTHRENE. DISSOLVED (UG/L)	85018
34463	PHENANTHRENE. SUSPENDED (UG/L)	85018
34466	PHENOL. DISSOLVED (UG/L)	108952
34467	PHENOL. SUSPENDED (UG/L)	108952
34469	PYRENE. TOTAL (UG/L)	129000
34470	PYRENE. DISSOLVED (UG/L)	129000
34471	PYRENE. SUSPENDED (UG/L)	129000
34475	TETRACHLOROETHYLENE. TOTAL (UG/L)	127184
34476	TETRACHLOROETHYLENE. DISSOLVED (UG/L)	127184
34477	TETRACHLOROETHYLENE. SUSPENDED (UG/L)	127184
34481	TOLUENE. DISSOLVED (UG/L)	108883
34482	TOLUENE. SUSPENDED (UG/L)	108883
34485	TRICHLOROETHYLENE. DISSOLVED (UG/L)	79016
34486	TRICHLOROETHYLENE. SUSPENDED (UG/L)	79016
34493	VINYL CHLORIDE. DISSOLVED (UG/L)	75014
34494	VINYL CHLORIDE. SUSPENDED (UG/L)	75014
34496	1.1-DICHLOROETHANE. TOTAL (UG/L)	75343
34497	1.1-DICHLOROETHANE. DISSOLVED (UG/L)	75343
34498	1.1-DICHLOROETHANE. SUSPENDED (UG/L)	75343

STORET Code	Toxic Elements (EPA Section 304(a) Priority Toxic Pollutants) cont.-	C.A.S. Number
34501	1.1-DICHLOROETHYLENE. TOTAL (UG/L)	75354
34502	1.1-DICHLOROETHYLENE. DISSOLVED (UG/L)	75354
34503	1.1-DICHLOROETHYLENE. SUSPENDED (UG/L)	75354
34506	1.1.1-TRICHLOROETHANE. TOTAL (UG/L)	71556
34507	1.1.1-TRICHLOROETHANE. DISSOLVED (UG/L)	71556
34508	1.1.1-TRICHLOROETHANE. SUSPENDED (UG/L)	71556
34511	1.1.2-TRICHLOROETHANE. TOTAL (UG/L)	79005
34512	1.1.2-TRICHLOROETHANE. DISSOLVED (UG/L)	79005
34513	1.1.2-TRICHLOROETHANE. SUSPENDED (UG/L)	79005
34516	1.1.2.2-TETRACHLOROETHANE. TOTAL (UG/L)	79345
34517	1.1.2.2-TETRACHLOROETHANE, DISSOLVED (UG/L)	79345
34518	1.1.2.2-TETRACHLOROETHANE. SUSPENDED (UG/L)	79345
34521	BENZO(GHI)PERYLENE1.12-BENZOPERYLENE. TOTAL (UG/L)	191242
34522	BENZO(GHI)PERYLENE1.12-BENZOPERYLENE. DISS. (UG/L)	191242
34523	BENZO(GHI)PERYLENE1.12-BENZOPERYLENE. SUSP. (UG/L)	191242
34526	BENZO(A)ANTHRACENE1.2-BENZANTHRACENE. TOTAL (UG/L)	56553
34527	BENZO(A)ANTHRACENE1.2-BENZANTHRACENE. DISS. (UG/L)	56553
34528	BENZO(A)ANTHRACENE1.2-BENZANTHRACENE. SUSP. (UG/L)	56553
34531	1.2-DICHLOROETHANE. TOTAL (UG/L)	107062
34532	1.2-DICHLOROETHANE. DISSOLVED (UG/L)	107062
34533	1.2-DICHLOROETHANE. SUSPENDED (UG/L)	107062
34536	1.2-DICHLOROBENZENE. TOTAL (UG/L)	95501
34537	1.2-DICHLOROBENZENE. DISSOLVED (UG/L)	95501
34538	1.2-DICHLOROBENZENE. SUSPENDED (UG/L)	95501
34541	1.2-DICHLOROPROPANE. TOTAL (UG/L)	78875
34542	1.2-DICHLOROPROPANE. DISSOLVED (UG/L)	78875
34543	1.2-DICHLOROPROPANE. SUSPENDED (UG/L)	78875
34546	TRANS-1.2-DICHLOROETHENE. TOTAL, IN WATER (UG/L)	156605

STORET Code	Toxic Elements (EPA Section 304(a) Priority Toxic Pollutants) cont.-	C.A.S. Number
34547	TRANS-1.2-DICHLOROETHENE. DISSOLVED (UG/L)	156605
34548	TRANS-1.2-DICHLOROETHENE. SUSPENDED (UG/L)	156605
34551	1.2.4-TRICHLOROBENZENE. TOTAL (UG/L)	120821
34552	1.2.4-TRICHLOROBENZENE. DISSOLVED (UG/L)	120821
34553	1.2.4-TRICHLOROBENZENE. SUSPENDED (UG/L)	120821
34556	1.2.5.6-DIBENZANTHRACENE. TOTAL (UG/L)	53703
34557	1.2.5.6-DIBENZANTHRACENE. DISSOLVED (UG/L)	53703
34558	1.2.5.6-DIBENZANTHRACENE. SUSPENDED (UG/L)	53703
34561	1.3-DICHLOROPROPENE. TOTAL (UG/L)	542756
34562	1.3-DICHLOROPROPENE. DISSOLVED (UG/L)	542756
34563	1.3-DICHLOROPROPENE. SUSPENDED (UG/L)	542756
34566	1.3-DICHLOROBENZENE. TOTAL (UG/L)	541731
34567	1.3-DICHLOROBENZENE. DISSOLVED (UG/L)	541731
34568	1.3-DICHLOROBENZENE. SUSPENDED (UG/L)	541731
34571	1.4-DICHLOROBENZENE. TOTAL (UG/L)	106467
34572	1.4-DICHLOROBENZENE. DISSOLVED (UG/L)	106467
34573	1.4-DICHLOROBENZENE. SUSPENDED (UG/L)	106467
34576	2-CHLOROETHYL VINYL ETHER. TOTAL (UG/L)	110758
34577	2-CHLOROETHYL VINYL ETHER. DISSOLVED (UG/L)	110758
34578	2-CHLOROETHYL VINYL ETHER. SUSPENDED (UG/L)	110758
34581	2-CHLORONAPHTHALENE. TOTAL (UG/L)	91587
34582	2-CHLORONAPHTHALENE. DISSOLVED (UG/L)	91587
34583	2-CHLORONAPHTHALENE. SUSPENDED (UG/L)	91587
34586	2-CHLOROPHENOL. TOTAL (UG/L)	95578
34587	2-CHLOROPHENOL. DISSOLVED (UG/L)	95578
34588	2-CHLOROPHENOL. SUSPENDED (UG/L)	95578
34591	2-NITROPHENOL. TOTAL (UG/L)	88755
34592	2-NITROPHENOL. DISSOLVED (UG/L)	88755

STORET Code	Toxic Elements (EPA Section 304(a) Priority Toxic Pollutants) cont.-	C.A.S. Number
34593	2-NITROPHENOL. SUSPENDED (UG/L)	88755
34596	DI-N-OCTYL PHTHALATE. TOTAL (UG/L)	117840
34597	DI-N-OCTYL PHTHALATE. DISSOLVED (UG/L)	117840
34598	DI-N-OCTYL PHTHALATE. SUSPENDED (UG/L)	117840
34601	2.4-DICHLOROPHENOL. TOTAL (UG/L)	120832
34602	2.4-DICHLOROPHENOL. DISSOLVED (UG/L)	120832
34603	2.4-DICHLOROPHENOL. SUSPENDED (UG/L)	120832
34606	2.4-DIMETHYLPHENOL. TOTAL (UG/L)	105679
34607	2.4-DIMETHYLPHENOL. DISSOLVED (UG/L)	105679
34608	2.4-DIMETHYLPHENOL. SUSPENDED (UG/L)	105679
34611	2.4-DINITROTOLUENE. TOTAL (UG/L)	121142
34612	2.4-DINITROTOLUENE. DISSOLVED (UG/L)	121142
34613	2.4-DINITROTOLUENE. SUSPENDED (UG/L)	121142
34616	2.4-DINITROPHENOL. TOTAL (UG/L)	51285
34617	2.4-DINITROPHENOL. DISSOLVED (UG/L)	51285
34618	2.4-DINITROPHENOL. SUSPENDED (UG/L)	51285
34621	2.4.6-TRICHLOROPHENOL. TOTAL (UG/L)	88062
34622	2.4.6-TRICHLOROPHENOL. DISSOLVED (UG/L)	88062
34623	2.4.6-TRICHLOROPHENOL. SUSPENDED (UG/L)	88062
34626	2.6-DINITROTOLUENE. TOTAL (UG/L)	606202
34627	2.6-DINITROTOLUENE. DISSOLVED (UG/L)	606202
34628	2.6-DINITROTOLUENE. SUSPENDED (UG/L)	606202
34631	3.3'-DICHLOROBENZIDINE. TOTAL (UG/L)	91941
34632	3.3'-DICHLOROBENZIDINE. DISSOLVED (UG/L)	91941
34633	3.3'-DICHLOROBENZIDINE. SUSPENDED (UG/L)	91941
34636	4-BROMOPHENYL PHENYL ETHER. TOTAL (UG/L)	101553
34637	4-BROMOPHENYL PHENYL ETHER. DISSOLVED (UG/L)	101553
34638	4-BROMOPHENYL PHENYL ETHER. SUSPENDED (UG/L)	101553

STORET Code	Toxic Elements (EPA Section 304(a) Priority Toxic Pollutants) cont.-	C.A.S. Number
34641	4-CHLOROPHENYL PHENYL ETHER. TOTAL (UG/L)	7005723
34642	4-CHLOROPHENYL PHENYL ETHER. DISSOLVED (UG/L)	7005723
34643	4-CHLOROPHENYL PHENYL ETHER. SUSPENDED (UG/L)	7005723
34646	4-NITROPHENOL. TOTAL (UG/L)	100027
34647	4-NITROPHENOL. DISSOLVED (UG/L)	100027
34648	4-NITROPHENOL. SUSPENDED (UG/L)	100027
34651	P.P'-DDD. DISSOLVED (UG/L)	72548
34652	P.P'-DDD. SUSPENDED (UG/L)	72548
34653	P.P'-DDE. DISSOLVED (UG/L)	72559
34654	P.P'-DDE. SUSPENDED (UG/L)	72559
34655	P.P'-DDT. DISSOLVED (UG/L)	50293
34656	P.P'-DDT. SUSPENDED (UG/L)	50293
34657	DNOC (4.6-DINITRO-ORTHO-CRESOL). TOTAL (UG/L)	534521
34658	DNOC (4.6-DINITRO-ORTHO-CRESOL). DISSOLVED (UG/L)	534521
34659	DNOC (4.6-DINITRO-ORTHO-CRESOL). SUSPENDED (UG/L)	534521
34662	PCB - 1221. DISSOLVED (UG/L)	11104282
34663	PCB - 1221. SUSPENDED (UG/L)	11104282
34665	PCB - 1232. DISSOLVED (UG/L)	11141165
34666	PCB - 1232. SUSPENDED (UG/L)	11141165
34671	PCB - 1016. TOTAL (UG/L)	12674112
34672	PCB - 1016. DISSOLVED (UG/L)	12674112
34673	PCB - 1016. SUSPENDED (UG/L)	12674112
34675	2.3.7.8-TETRACHLORODIBENZO-PDIOXIN(TCDD).TOT(UG/L)	1746016
34676	2.3.7.8-TETRACHLORODIBENZO-PDIOXIN(TCDD)DISS(UG/L)	1746016
34677	2.3.7.8-TETRACHLORODIBENZO-PDIOXIN(TCDD)SUSP(UG/L)	1746016
34694	PHENOL(C6H5OH)-SINGLE COMPOUND TOTAL (UG/L)	108952
34696	NAPHTHALENE. TOTAL (UG/L)	91203
34750	2.3.7.8-TETRACHLORODIBENZO-PDIOXIN(TCDD)TOT(PG/L)	1746016

STORET Code	Toxic Elements (EPA Section 304(a) Priority Toxic Pollutants) cont.-	C.A.S. Number
34751	2,3,7,8-TETRACHLORODIBENZO-PDIOXIN(TCDD)DISS(PG/L)	1746016
34752	2,3,7,8-TETRACHLORODIBENZO-PDIOXIN(TCDD)SUSP(PG/L)	1746016
39032	PCP (PENTACHLOROPHENOL) WHOLE WATER SAMPLE (UG/L)	87865
39039	HEXACHLOROBENZENE WATER SAMPLE,ELECTRON CPT (UG/L)	118741
39100	BIS(2-ETHYLHEXYL) PHTHALATE, WHOLE WATER (UG/L)	117817
39103	BIS(2-ETHYLHEXYL) PHTHALATE, DISSOLVED, (UG/L)	117817
39104	BIS(2-ETHYLHEXYL) PHTHALATE, SUSPENDED, (UG/L)	117817
39107	PHTHALATES,DIETHYLHEXYL SUS.FRAC.WTR DWT (MG/KG)	117817
39110	DI-N-BUTYL PHTHALATE, WHOLE WATER (UG/L)	84742
39114	DI-N-BUTYL PHTHALATE, SUSPENDED (UG/L)	84742
39115	PHTHALATES,DIBUTYL SUS.FRAC.WATER DWT (UG/KG)	84742
39120	BENZIDINE IN WHOLE WATER SAMPLE (UG/L)	92875
39175	VINYL CHLORIDE-WHOLE WATER SAMPLE (UG/L)	75014
39180	TRICHLOROETHYLENE-WHOLE WATER SAMPLE (UG/L)	79016
39300	P,P' DDT IN WHOLE WATER SAMPLE (UG/L)	50293
39310	P,P' DDD IN WHOLE WATER SAMPLE (UG/L)	72548
39320	P,P' DDE IN WHOLE WATER SAMPLE (UG/L)	72559
39330	ALDRIN IN WHOLE WATER SAMPLE (UG/L)	309002
39331	ALDRIN IN FILT. FRAC. OF WAT. SAMP. (UG/L)	309002
39332	ALDRIN IN SUSP. FRAC. OF WAT. SAMP. (UG/L)	309002
39336	BHC-ALPHA, WATER, WHOLE (LBS/DAY)	319846
39337	ALPHA BENZENE HEXACHLORIDE IN WHOLE WATER (UG/L)	319846
39338	BETA BENZENE HEXACHLORIDE IN WHOLE WATER (UG/L)	319857
39340	GAMMA-BHC(LINDANE), WHOLE WATER (UG/L)	58899
39341	GAMMA-BHC(LINDANE), DISSOLVED (UG/L)	58899
39342	GAMMA-BHC(LINDANE), SUSPENDED (UG/L)	58899
39344	BHC-GAMMA, WATER, WHOLE (LBS/DAY)	58899
39350	CHLORDANE(TECH MIX & METABS), WHOLE WATER (UG/L)	57749

STORET Code	Toxic Elements (EPA Section 304(a) Priority Toxic Pollutants) cont.-	C.A.S. Number
39352	CHLORDANE(TECH MIX & METABS). DISSOLVED (UG/L)	57749
39353	CHLORDANE(TECH MIX & METABS). SUSPENDED (UG/L)	57749
39360	DDD IN WHOLE WATER SAMPLE (UG/L)	72548
39361	DDD IN FILT. FRAC. OF WATER SMAPLE (UG/L)	72548
39362	DDD IN SUSP. FRAC. OF WATER SAMPLE (UG/L)	72548
39365	DDE IN WHOLE WATER SAMPLE (UG/L)	72559
39366	DDE IN FILT. FRAC. OF WATER SAMPLE (UG/L)	72559
39367	DDE IN SUSP. FRAC. OF WATER SAMPLE (UG/L)	72559
39370	DDT IN WHOLE WATER SAMPLE (UG/L)	50293
39371	DDT IN FILT. FRAC. OF WATER SAMPLE (UG/L)	50293
39372	DDT IN SUSP. FRAC. OF WATER SAMPLE (UG/L)	50293
39380	DIELDRIN IN WHOLE WATER SAMPLE (UG/L)	60571
39381	DIELDRIN IN FILT. FRAC. OF WATER SAMPLE (UG/L)	60571
39382	DIELDRIN IN SUSP. FRAC. OF WATER SAMPLE (UG/L)	60571
39390	ENDRIN IN WHOLE WATER SAMPLE (UG/L)	72208
39391	ENDRIN IN FILT. FRAC. OF WATER SAMPLE (UG/L)	72208
39392	ENDRIN IN SUSP. FRAC. OF WATER SAMPLE (UG/L)	72208
39400	TOXAPHENE IN WHOLE WATER SAMPLE (UG/L)	8001352
39401	TOXAPHENE IN FILT. FRAC. OF WATER SAMPLE (UG/L)	8001352
39402	TOXAPHENE IN SUSP. FRAC. OF WATER SAMPLE (UG/L)	8001352
39410	HEPTACHLOR IN WHOLE WATER SAMPLE (UG/L)	76448
39411	HEPTACHLOR IN FILT. FRAC. OF WATER SAMPLE (UG/L)	76448
39412	HEPTACHLOR IN SUSP. FRAC. OF WATER SAMPLE (UG/L)	76448
39420	HEPTACHLOR EPOXIDE IN WHOLE WATER SAMPLE (UG/L)	1024573
39421	HEPTACHLOR EPOXIDE IN FILT. FRAC. WAT. SAM. (UG/L)	1024573
39422	HEPTACHLOR EPOXIDE IN SUSP. FRAC. WAT. SAM. (UG/L)	1024573
39488	PCB - 1221 IN THE WHOLE WATER SAMPLE (UG/L)	11104282
39492	PCB - 1232 PCB SERIES WHOLE WATER SAMPLE (UG/L)	11141165

STORET Code	Toxic Elements (EPA Section 304(a) Priority Toxic Pollutants) cont.-	C.A.S. Number
39496	PCB - 1242 PCB SERIES WHOLE WATER SAMPLE (UG/L)	53469219
39500	PCB - 1248 PCB SERIES WHOLE WATER SAMPLE (UG/L)	12672296
39501	PCB - 1248 IN FILT. FRAC. OF WATER SAMPLE (UG/L)	12672296
39502	PCB - 1248 IN SUSP. FRAC. OF WATER SAMPLE (UG/L)	12672296
39504	PCB - 1254 PCB SERIES WHOLE WATER SAMPLE (UG/L)	11097691
39505	PCB - 1254 IN FILT. FRAC. OF WATER SAMPLE (UG/L)	11097691
39506	PCB - 1254 IN SUSP. FRAC. OF WATER SAMPLE (UG/L)	11097691
39508	PCB - 1260 PCB SERIES WHOLE WATER SAMPLE (UG/L)	11096825
39509	PCB - 1260 IN FILT. FRAC. OF WATER SAMPLE (UG/L)	11096825
39510	PCB - 1260 IN SUSP. FRAC. OF WATER SAMPLE (UG/L)	11096825
39700	HEXACHLOROBENZENE IN WHOLE WATER SAMPLE (UG/L)	118741
39702	HEXACHLOROBUTADIENE IN WHOLE WATER SAMPLE (UG/L)	87683
39782	LINDANE IN WHOLE WATER SAMPLE (UG/L)	58899
39920	DNOC IN WHOLE WATER SAMPLE (UG/L)	534521
46322	LINDANE PLUS ISOMERS IN WHOLE WATER SAMPLE (UG/L)	58899
46323	DELTA-BHC IN WHOLE WATER SAMPLE (UG/L)	319868
46326	HEPTACHLOR AND METABOLITES IN WH. H2O SAMP. (UG/L)	76448
46479	CYANIDE. DISSOLVED. WATER (UG/L)	57125
46551	ARSENIC, FIELD ACIDIFIED W/HNO3. LAB FILT. (UG/L)	7440382
46559	CADMIUM. FIELD ACIDIFIED-HNO3-LAB FILTER (UG/L-CD)	7440439
46560	CHROMIUM. FIELD ACIDIFIED-HN03-LAB FILT. (UG/L-CR)	7440473
46562	COPPER. FIELD ACIDIFIED-HNO3-LAB FILTER. (UG/L-CU)	7440508
46564	LEAD. FIELD ACIDIFIED-HNO3-LAB FILTERED (UG/L-PB)	7439921
46566	SILVER. FIELD ACIDIFIED-HNO3-LAB FILTER.(UG/L-AG)	7440224
46567	ZINC, EXTRACT. FIELD ACID W/HNO3. LAB FILT. (UG/L)	7440666
70012	PARACHLOROMETA CRESOL. WATER. WHOLE (LBS/DAY)	59507
70017	HEXACHLOROCYCLOPENTADIENE. WATER. WHOLE (LBS/DAY)	77474
70021	LEAD. (TCLP). WATER. TOTAL (MG/L)	7439921

STORET Code	Toxic Elements (EPA Section 304(a) Priority Toxic Pollutants) cont.-	C.A.S. Number
71890	MERCURY. DISSOLVED (UG/L AS HG)	7439976
71895	MERCURY. SUSPENDED (UG/L AS HG)	7439976
71900	MERCURY. TOTAL (UG/L AS HG)	7439976
71901	MERCURY. TOTAL RECOVERABLE IN WATER AS HG (UG/L)	7439976
71946	CADMIUM. EXTRACTABLE (UG/L AS CD)	7440439
71947	CHROMIUM. EXTRACTABLE (UG/L AS CR)	7440473
71949	LEAD. EXTRACTABLE (UG/L AS PB)	7439921
71950	ZINC. EXTRACTABLE (UG/L AS ZN)	7440666
71951	COPPER. EXTRACTABLE (UG/L AS CU)	7440508
73063	CHLOROGUAIACOL.4-. TOTAL. WATER (UG/L)	16766306
73522	PROPANE. 2.2'-OXYBIS(1-CHLORO)- TOTAL (UG/L)	108601
77163	1.3-DICHLOROPROPENE-1. WHOLE WATER (UG/L)	542756
77354	1.1-DICHLORO-2.2-DIFLUOROETHANE WHOLE WATER (UG/L)	471432
77771	3-CHLORO-4-HYDROXYBENZOPHENONE. WHOLE WATER (UG/L)	55191203
78113	ETHYL BENZENE WHOLE WATER SAMPLE (UG/L)	100414
78124	BENZENE IN WATER (VOLATILE ANALYSIS) (UG/L)	71432
78131	TOLUENE IN WHOLE WATER (VOLATILE ANALYSIS) (UG/L)	108883
78208	2.4-DINITRO-O-CRESOL IN WHOLE WATER SAMPLE (UG/L)	534521
78247	CHROMIUM. HEXAVALENT. TOTAL RECOVERABLE. WT (UG/L)	18540299
78248	CYANIDE. TOTAL RECOVERABLE. WATER. WHOLE (UG/L)	57125
80357	CHROMIUM. TRIVALENT. DISSOLVED. AS CR	16065831
81208	CYANIDE. FREE (NOT AMEN. TO CHLORINATION) (MG/L)	57125
81210	CYANIDE - STATE OF ILLINOIS (MG/L)	57125
81214	CADMIUM - STATE OF ILLINOIS (MG/L)-COLD	7440439
81215	CHROMIUM - STATE OF ILLINOIS (MG/L). COLD DIGEST	18540299
81216	CHROMIUM(TRI)-STATE OF ILLINOIS (MG/L)-COLD DIGEST	16065831
81217	CHROMIUM. TOTAL - STATE OF ILLINOIS (MG/L) COLD DIGEST	7440473
81218	COPPER. STATE OF ILLINOIS. MG/L. COLD DIGEST	7440508

STORET Code	Toxic Elements (EPA Section 304(a) Priority Toxic Pollutants) cont.-	C.A.S. Number
81220	LEAD, STATE OF ILLINOIS, MG/L, COLD DIGEST	7439921
81222	NICKEL - STATE OF ILLINOIS, MG/L, COLD DIGEST	7440020
81223	SILVER, STATE OF ILLINOIS, MG/L, COLD DIGEST	7440224
81224	ZINC - STATE OF ILLINOIS, MG/L, COLD DIGEST	7440666
81642	SILVER (AG) IN WATER POUNDS PER DAY (LBS/DAY)	7440224
81750	COPPER, INTERSTITIAL WATER FROM SEDIMENTS (UG/L)	7440508
81751	LEAD, INTERSTITIAL WATER FROM SEDIMENTS (UG/L)	7439921
81752	NICKEL, INTERSTITIAL WATER FROM SEDIMENTS (UG/L)	7440020
81753	CADMIUM, INTERSTITIAL WATER FROM SEDIMENT	7440439
81754	ZINC, INTERSTITIAL WATER FROM SEDIMENTS (UG/L)	7440666
81766	HEPTACHLOR EPOXIDE IN EPILITHIC ALGAE SED. (UG/KG)	1024573
81931	MERCURY (HG) SUSPENDED FRACTION OF WATER (UG/G)	7439976
81932	CADMIUM (CD) SUSPENDED FRACTION OF WATER (UG/G)	7440439
81933	ZINC (ZN) SUSPENDED FRACTION OF WATER (UG/G)	7440666
81934	LEAD (PB) SUSPENDED FRACTION OF WATER (UG/G)	7439921
81936	LEAD (PB) DISSOLVED CATIONIC SPECIES (UG/L)	7439921
81937	CADMIUM (CD) DISSOLVED CATIONIC SPECIES (UG/L)	7440439
81938	CHROMIUM, DISSOLVED CATIONIC SPECIES (UG/L)	7440473
81939	COPPER (CU) DISSOLVED CATIONIC SPECIES (UG/L)	7440508
81940	ZINC (ZN) DISSOLVED CATIONIC SPECIES (UG/L)	7440666
81941	CHROMIUM, DISSOLVED ANIONIC SPECIES (UG/L)	7440473
81942	COPPER (CU) DISSOLVED ANIONIC SPECIES (UG/L)	7440508
81943	ZINC (ZN) DISSOLVED ANIONIC SPECIES (UG/L)	7440666
82058	CHROMIUM, TOTAL, PERCENT REMOVAL	7440473
82399	CHROMIUM, HEXAVALENT (KG/BATCH)	18540299
82512	M,P-DICHLOROBENZENE (MEASURES 1,3&1,4) TOT. (UG/L)	541731
82573	CYANIDE/CHLORINATION IN WATER (MG/L)	57125
82621	HEXACHLOROBENZENE, WATER, TOTAL RECOVER. (UG/L)	118741

STORET Code	Toxic Elements (EPA Section 304(a) Priority Toxic Pollutants) cont.-	C.A.S. Number
82622	ENDRIN ALDEHYDE. WH. WATER. TOTAL RECOVER. (UG/L)	7421934
82623	ENDOSULFAN SULFATE. WATER. TOTAL RECOVER. (UG/L)	1031078
82624	ENDOSULFAN. BETA. WH. WATER. TOTAL RECOVER. (UG/L)	33213659
82626	1,2-DIPHENYLHYDRAZINE. WATER. TOTAL RECOVER. (UG/L)	122667
82627	PARACHLOROMETA CRESOL. WATER. TOTAL RECOVER. (UG/L)	59507
82702	ARSENIC. FIELD ACIDIFIED. DECANTED. WATER (UG/L)	7440382
82704	BERYLLIUM. FIELD ACIDIFIED. DECANTED. WATER (UG/L)	7440417
82705	CADMIUM. FIELD ACIDIFIED. DECANTED. WATER (UG/L)	7440439
82706	CHROMIUM. FIELD ACIDIFIED. DECANTED. WATER (UG/L)	7440473
82708	COPPER. FIELD ACIDIFIED. DECANTED. WATER (UG/L)	7440508
82711	LEAD. FIELD ACIDIFIED. DECANTED. WATER (UG/L)	7439921
82713	MERCURY. FIELD ACIDIFIED. DECANTED. WATER (UG/L)	7439976
82715	NICKEL. FIELD ACIDIFIED. DECANTED. WATER (UG/L)	7440020
82716	SILVER. FIELD ACIDIFIED. DECANTED. WATER (UG/L)	7440224
82719	ZINC. FIELD ACIDIFIED. DECANTED. WATER (UG/L)	7440666
85006	ZINC. TOTAL - (#/DAY)	7440666
85007	CHROMIUM. TOTAL (#/DAY)	7440473
85010	NICKEL. TOTAL - (#/DAY)	7440020
85013	MERCURY. TOTAL - (#/DAY)	7439976

Appendix H

Literature Cited

Code of Federal Regulations. 1994. Protection of Environment. 40 CFR Parts 100 to 149. Revised as of July 1, 1994. Published by the Office of the Federal Register, National Archives and Records Administration. U.S. Government Printing Office, Washington, D.C. 20402.

Gilbert, R. O. 1987. Statistical Methods for Environmental Pollution Monitoring. Van Nostrand Reinhold Co., New York, NY. 320p.

GKY and Associates. 1990. Dam Inventory Database and Retrieval Software: Final Report. U.S. Environmental Protection Agency, Water Quality Analysis Branch. Under Contract #68-03-3339.

Kunkle, S. and J. Wilson. 1984. Specific Conductance and pH Measurements in Surface Waters: An Introduction for Park Natural Resource Specialists. Water Resources Field Support Laboratory Report No. 84-3. National Park Service, Water Resources Division, Fort Collins, Colorado 80525. 51p.

National Park Service. 1993. Strategic Plan for Conducting Baseline Natural Resource Inventories in the National Park Service. National Park Service, Washington Office, Servicewide Inventory and Monitoring Program, Washington, D.C. Unpublished. 17p.

U.S. Environmental Protection Agency. 1995. Quality Criteria for Water 1995. Final Draft. Office of Water Regulations and Standards, Washington, D.C.

U.S. Environmental Protection Agency. 1989. STORET User Handbook. U.S. Environmental Proteciton Agency, Office of Water, Washington, D.C. 20460.

U.S. Environmental Protection Agency. 1992. Office of Water Environmental and Program Information Systems Compendium. U.S. Environmental Protection Agency, Office of Water, Washington, D.C. 20460. 152p.

U.S. Environmental Protection Agency. 1993. Technical Description of the Reach File. U.S. Environmental Protection Agency, Office of Water, Washington, D.C. 20460. 23p.

U.S. Geological Survey. 1982. A U.S. Geological Survey Data Standard: Codes for the Identification of Hydrologic Units in the United States and Caribbean Outlying Areas. Geological Survey Circular 878-A. U.S. Geological Survey, Water Resources Division, Reston, VA. 22092. 115p.

U.S. Geological Survey 1992. Hydro-Climatic Data Network: A U.S. Geological Survey Streamflow Data Set for the United States for the Study of Climate Variations 1874-1988. Open File Report 92-129/USGS Water Supply Paper No. 2406. U.S. Geological Survey, Water Resources Division, Reston, VA. 22092. 193p.

Ward, R. C., J. C. Loftis, and G. B. McBride. 1990. Design of Water Quality Monitoring Systems. Van Nostrand Reinhold Co., New York, NY. 231p.

Appendix I

Selected General Water Quality References

American Public Health Association. 1989. Standard Methods for the Examination of Water and Wastewater (17th ed.). Washington, D.C. 1476p.

Drever, J. I. 1982. The Geochemistry of Natural Waters. Prentice-Hall, Inc., Englewood Cliffs, NJ. 388p.

Dunne, T. and L. B. Leopold. 1978. Water in Environmental Planning. W.H. Freeman and Company, San Francisco, CA. 818p.

Everett, L. G. 1980. Groundwater Monitoring. General Electric Co., Schenectady, NY. 440p.

Fetter, C. W. 1988. Applied Hydrogeology (2nd ed.). MacMillan Publishing Co., New York, NY. 592p.

Flora, M. D., T. E. Ricketts, J. Wilson, and S. Kunkle. 1984. Water Quality Criteria: An Overview for Park Natural Resource Specialists. WRFSL Report No. 84-4. National Park Service, Water Resources Field Support Laboratory, Fort Collins, CO. 46p.

Gilbert, R. O. 1987. Statistical Methods for Environmental Pollution Monitoring. Van Nostrand Reinhold Co., New York, NY. 320p.

Hem, J. D. 1985. Study and Interpretation of the Chemical Characteristics of Natural Water (3rd ed.). U.S. Geological Survey Water-Supply Paper 2254. U.S. Government Printing Office, Washington, D.C. 263p.

Kunkle, S., W. S. Johnson, and M. Flora. 1987. Monitoring Stream Water Quality for Land-Use Impacts: A Training Manual for Natural Resource Management Specialists. Water Resources Division, National Park Service, Fort Collins, CO. 102p.

Kunkle, S. and J. Wilson. 1984. Specific Conductance and pH Measurements in Surface Waters: An Introduction for Park Natural Resource Specialists. Water Resources Field Support Laboratory Report No. 84-3. National Park Service, Water Resources Division, Fort Collins, Colorado 80525. 51p.

Merritt, R. W., and K. W. Cummins (eds.). 1984. An Introduction to the Aquatic Insects of North America (2nd ed.). Kendall/Hunt Publishing Co., Dubuque, IA. 44p.

Morel, F. M. 1983. Principles of Aquatic Chemistry. John Wiley & Sons, Inc., New York, NY. 446p.

Nielsen, D. M. (ed.). 1991. Practical Handbook of Ground-Water Monitoring. Lewis Publishers, Inc. Chelsea, MI. 717p.

Ponce, S. L. 1980a. Statistical Methods Commonly Used in Water Quality Data Analysis. WSDG Technical Paper WSDG-TP-00001. U.S. Department of Agriculture, Forest Service, Watershed Systems Development Group, Fort Collins, CO. 136p.

Ponce, S. L. 1980b. Water Quality Monitoring Programs. WSDG Technical Paper WSDG-TP-00002. U.S. Department of Agriculture, Forest Service, Watershed Systems Development Group, Fort Collins, CO. 68p.

Rand, G. M. and S. R. Petrocelli (eds.). 1985. Fundamentals of Aquatic Toxicology. Hemisphere Publishing Co., New York, NY. 666p.

Rantz, S. E. and others. 1982. Measurement and Computation of Streamflow: Volume 1. Measurement of Stage and Discharge. Volume 2. Computation of Discharge. U.S. Department of the Interior, Geological Survey Water Supply Paper 2175. 631p.

Stednick, J.D. and D. M. Gilbert. 1998. Water Quality Inventory Protocol: Riverine Environments. National Park Service, Water Resources Division Technical Report NPS/NRWRD/NRTR-98/177. Fort Collins, CO. 103p.

Stednick, J. D. 1991. Wildland Water Quality Sampling and Analysis. Academic Press, Inc., San Diego, CA. 217p.

United Nations Educational, Scientific and Cultural Organization (UNESCO). 1978. Water Quality Surveys: A Guide for the Collection and Interpretation of Water Quality Data. IHD-WHO Working Group on the Quality of Water, Paris, France. 350p.

U.S. Department of the Interior. 1977. National Handbook of Recommended Methods for Water-Data Acquisition. U.S. Geological Survey, Office of Water-Data Coordination, Reston, VA. 990p.

U.S. Environmental Protection Agency. 1978. Microbiological Methods for Monitoring the Environment: Water and Wastes. R. H. Border, J. A. Winter, and P. W. Scarpino. EPA-600/8-78-017. Office of Research and Development, Environmental Monitoring Systems Laboratory, Cincinnati, OH. 338p.

U.S. Environmental Protection Agency. 1979b. Methods for Chemical Analysis of Water and Wastes. EPA-600/4-79-020. (Revised March 1983). Office of Research and Development, Environmental Monitoring Systems Laboratory, Cincinnati, OH. 460p.

U.S. Environmental Protection Agency. 1983. Water Quality Standards Handbook. Office of Water Regulations and Standards, Washington, D.C. 218p.

U.S. Environmental Protection Agency. 1995. Quality Criteria for Water 1995. Final Draft. Office of Water Regulations and Standards, Washington, D.C.

U.S. Environmental Protection Agency. 1989. Rapid Bioassessment Protocols for Use in Streams and Rivers: Benthic Macroinvertebrates and Fish. J. L. Plafkin, M. T. Barbour, K. D. Porter, S. K. Gross, and R. M. Hughes. EPA-444/4-89-001. Office of Water Regulations and Standards, Assessment and Watershed Protection Division, Washington, D.C. 162p.

U.S. Environmental Protection Agency. 1990. Macroinvertebrate Field and Laboratory Methods for Evaluating the Biological Integrity of Surface Waters. D. J. Klemm, P. A. Lewis, F. Fulk, and J. M. Lazorchak. EPA-600/4-90-030. Office of Research and Development, Environmental Monitoring Systems Laboratory, Cincinnati, OH. 256p.

U.S. Environmental Protection Agency. 1991a. Methods for Measuring the Acute Toxicity of Effluents and Receiving Waters to Freshwater and Marine Organisms (4th ed.). C. I. Weber, ed. EPA-600/4-90-027. Office of Research and Development, Environmental Monitoring Systems Laboratory, Cincinnati, OH. 293p.

U.S. Environmental Protection Agency. 1991b. Monitoring Guidelines to Evaluate Effects of Forestry Activities on Streams in the Pacific Northwest and Alaska. L. H. MacDonald, A. W.Smart, and R. C. Wissmar. EPA-910/9-91-001. Region 10, Seattle, WA. 162p.

U.S. Environmental Protection Agency. 1993. Guide to Federal Water Quality Programs and Information. T. Stuart and N. P. Ross. EPA-230-B-93-001. Office of Strategic Planning and Environmental Data, Environmental Statistics and Information Division. Washington, D.C. 194p.

Verschueren, K. 1983. Handbook of Environmental Data on Organic Chemicals (2nd ed.). Van Nostrand Reinhold Co., New York, NY. 1310p.

Viessman W. and M. J. Hammer. 1985. Water Supply and Pollution Control (4th ed.). Harper and Row, Publishers, Inc. New York, NY. 797p.

Ward, R. C., J. C. Loftis, and G. B. McBride. 1990. Design of Water Quality Monitoring Systems. Van Nostrand Reinhold Co., New York, NY. 231p.

Wetzel, R. G. 1983. Limnology (2nd ed.). Sanders College Publishing, Philadelphia, PA. 767p.

As the nation's principal conservation agency, the Department of the Interior has the responsibility for most of our nationally owned public lands and natural and cultural resources. This includes fostering wise use of our land and water resources, protecting our fish and wildlife, preserving the environmental and cultural values of our national parks and historical places, and providing for enjoyment of life through outdoor recreation. The Department assesses our energy and mineral resources and works to ensure that their development is in the best interests of all our people. The Department also promotes the goals of the Take Pride in America campaign by encouraging stewardship and citizen responsibility for the public lands and promoting citizen participation in their care. The Department also has a major responsibility for American Indian reservation communities and for people who live in island territories under U.S. administration.

NPS D-79

October 1998